REACTIVE SPECIES DETECTION IN BIOLOGY

REACTIVE SPECIES DETECTION IN BIOLOGY

From Fluorescence to Electron Paramagnetic Resonance Spectroscopy

FREDERICK A. VILLAMENA
Department of Biological Chemistry and Pharmacology
The Ohio State University, OH, United States

ELSEVIER

AMSTERDAM • BOSTON • HEIDELBERG • LONDON • NEW YORK • OXFORD
PARIS • SAN DIEGO • SAN FRANCISCO • SINGAPORE • SYDNEY • TOKYO

Elsevier
Radarweg 29, PO Box 211, 1000 AE Amsterdam, Netherlands
The Boulevard, Langford Lane, Kidlington, Oxford OX5 1GB, United Kingdom
50 Hampshire Street, 5th Floor, Cambridge, MA 02139, United States

Notices

Knowledge and best practice in this field are constantly changing. As new research and experience broaden our understanding, changes in research methods, professional practices, or medical treatment may become necessary.

Practitioners and researchers must always rely on their own experience and knowledge in evaluating and using any information, methods, compounds, or experiments described herein. In using such information or methods they should be mindful of their own safety and the safety of others, including parties for whom they have a professional responsibility.

To the fullest extent of the law, neither the Publisher nor the authors, contributors, or editors, assume any liability for any injury and/or damage to persons or property as a matter of products liability, negligence or otherwise, or from any use or operation of any methods, products, instructions, or ideas contained in the material herein.

British Library Cataloguing-in-Publication Data
A catalogue record for this book is available from the British Library

Library of Congress Cataloging-in-Publication Data
A catalog record for this book is available from the Library of Congress

ISBN: 978-0-12-420017-3

Publisher: Cathleen Sether
Acquisition Editor: Kathryn Morrissey
Editorial Project Manager: Anneka Hess
Production Project Manager: Anitha Sivaraj
Designer: Maria Ines Cruz

Typeset by MPS Limited, Chennai, India

DEDICATION

To Prof. DeLanson R. Crist, my mentor, father, and friend

And to my parents, Prisco and Tess,
for all your love and sacrifices.

CONTENTS

PREFACE

Life is a sea of electrons that when in the state of equilibrium thrives but causes havoc when perturbed. But this oversimplistic analogy leaves a multitude of unanswered questions and is ridden with complexities. It is encouraging that current advancements in biochemistry and biological chemistry have allowed modern-day biomedical investigators to understand some of these complexities and get ever closer to unraveling some of the most fundamental questions in disease development. Most often overlooked are the analytical tools employed to probe these mysteries. A research area by itself, it owes its modern progress to chemists, biologists, engineers, and biomedical researchers for the development of innovative molecular probes and imaging agents, efficient methodologies, and state-of-the-art instrumentation. This book is therefore a tribute to them: without their tireless dedication, our understanding of disease pathogenesis would not be in the state it is today. At the heart of this perturbation of a sea of electrons is the reduction and oxidation chemistry that involves a variety of players, the most important of which is molecular oxygen; when not properly metabolized, it results in the production of reactive species. These reactive species are like an army of destructive forces that destroy, inactivate, or cause only partial functioning of key biomolecular systems essential for normal functions of living systems. Knowledge of reactive species' location, their origin, and their identity as well as their target molecules and molecular trails they leave as biomarkers are critical and can only be accomplished through the use of both earlier and modern sophisticated analytical tools. Reactive species detection is therefore important for identifying specific molecular and cellular pathways leading to oxidative stress that could lead the way to the development of genetic, molecular, and pharmacological approaches, as well as diagnostic tools to prevent or ameliorate free radical–mediated diseases.

This book is basically divided into four major parts: introduction, chemistry of reactive species, their biology, and their modes of detection. Chapter 1, Introduction, provides a historical account of how oxygen and its reactive metabolites have been implicated in the pathogenesis of diseases. It also describes how important discoveries of the last century were aided by the use of analytical tools to characterize reactive species as important biological mediators and how bioassays can complement such findings. Chapter 2, Chemistry of Reactive Species, will only present biologically relevant reactions of reactive species in an effort to give readers a sense of their relative reactivity, specificity, and selectivity to various biomolecules. In Chapter 3, Reactive Species in Biological Systems, shows various sources of reactive species in biological settings, either from exogenous sources or as generated from enzymes, and it will discuss various sources of reactive species production from cellular compartments. Lastly, Chapters 4, Fluorescence Technique; Chapter 5, EPR Spin Trapping; Chapter 6, UV–Vis Absorption and Chemiluminescence Techniques; Chapter 7, Electrochemical, Mass Spectroscopic, Immunochemical, and Nuclear Magnetic Resonance Techniques; showcase conventional and modern analytical techniques that are often if not widely employed in the detection of biological reactive species such as fluorescence, electron paramagnetic resonance (EPR) spin trapping, UV–Vis light and chemiluminescence, and electrochemical, mass spectrometric, immunochemical, and other magnetic resonance techniques with the exception of some techniques that are too specialized (e.g., EPR and positron emission tomography imaging) or redundant in their principles and chemistry to those already discussed here.

I hope readers will find this first edition a one-stop kind of reference material for fundamental principles, limitations, and applications of the various analytical technique used in reactive species detection. Since this book will only cover so much information, readers are encouraged to consult the references cited herein for more detailed discussions and descriptions of the topics discussed.

This book would not have been possible without the dedication and passion of investigators working in the field of free radical research, whether in the synthesis of new molecules or the development of methodologies and instrumentations, as well as biomedical researchers who continuously validate these tools for their effective application in biological systems.

Frederick A. Villamena

Columbus, OH, United States
October, 2016

CHAPTER 1

Introduction

Respiration has been commonly known for centuries as an essential process for survival because it provides the "fuel" for the normal functioning of animal organs. This fuel found in air was first known as *dephlogisticated air* on its discovery in the 1770s by scientists Joseph Priestly and Carl Wilhelm Scheele.[1,2] Antoine Lavoisier later coined the term *oxygen* and extended the theory of combustion to introduce the idea of respiration as a biological process in which inhaled oxygen is used in the oxidation of carbon and hydrogen from food to give carbonic acid, thus relating respiration as a combustion process.[3,4] Studies from 1940s and 1950s on the enzymatic metabolism of oxygen provided molecular bases for oxygen's diverse biological functions through enzyme-catalyzed transfer of oxygen atom to a substrate, or electron-transfer reactions to oxygen leading to the formation of reactive oxygen species (ROS) or water.[5] This enzyme-catalyzed reduction of oxygen was later found to have beneficial as well as detrimental effects on cellular function. Without a doubt, the progress made in understanding oxygen metabolism owes a debt to the development of electrochemical techniques for analyzing oxygen in biological fluids and tissues as well as whole animals. Oxygen sensor development was described[6] dating as far back as 1938 when the first biological application of platinum electrodes was demonstrated for the purpose of monitoring oxygen to study photosynthesis, and this was followed by the use of a Clark-type electrode that allowed for the measurement of oxygen tension (pO_2) in an in vivo system, air, blood, and cell cultures, which then led to further innovation that exhibited high accuracy.

The absorption of oxygen and its transformation to carbonic anhydride (carbon dioxide) during respiration has thus been the paradigm of oxygen metabolism.[7] In the early 1900s, the relationship between oxygen and disease was suggested by Todd,[8] whereby the human body is in a state of chemical equilibrium between the processes of oxidation and reduction; when this equilibrium shifts toward the formation of more reduced species than oxidized ones, the body could lose resistance to diseases, and hence resupplying the body with oxygen in the form of ozonized air or oxidized oils is used as a therapeutic means of counteracting diseases such as tuberculosis or Bright's disease. Oxygen therapy was employed for a variety of diseases; e.g., patients with respiratory disease such as pneumonia exhibited excellent therapeutic effect when such therapy was introduced soon after diagnosis.[9] It also became apparent that

Reactive Species Detection in Biology
DOI: http://dx.doi.org/10.1016/B978-0-12-420017-3.00001-3

oxygen exhibits toxicity against bacterial pneumococcus type I[10] as well as in protozoans,[11] brain respiration,[12,13] or in whole animals causing pulmonary damage.[14] In humans, oxygen results in the reduction of blood-flow rate to the brain when inhaled at high atmospheric pressure,[15] causing cerebral complications as well as diminished overall cardiac output and changes in alveoli that result in edema, transudation, and fibrinous deposits.[16]

The formation of hydroxyl radicals from water under ionizing radiation had long been implicated for the radicals' biological actions and toxicity.[17] Soon thereafter, chemical agents that have radiation-like properties were implicated in the initiation[18] of cancer or tumor inhibition via chromosome alteration through formation of free radicals.[19] Evidence supported the idea that free radicals formed radiolytically were toxic because they were found to diffuse inside the cells when generated extracellularly.[20] Moreover, it was also demonstrated that x-ray radiation inhibited glutathione metabolism inside the cells, and this inhibition was decreased at low oxygen concentration and on addition of catalase, which suggested the involvement of oxygen-derived reactive species such as H_2O_2.[21] The link between radiation and oxygen levels on their cellular toxicity had become more apparent by their inactivation of T_2 bacteriophage and by the observation that thiol compounds such as thiourea could compete with oxygen-derived radicals, thereby protecting the phages from radiation.[22] Using electron paramagnetic resonance (EPR), radiation damage to DNA or RNA was reported to produce paramagnetic nucleic acids at 77K.[23] This finding was consistent with the increased effect of radiation on DNA inactivation in the presence of oxygen and protection in the presence of the thiol cysteamine[24,25] with mutations successfully induced in the cell nucleus of onion rootlets, e.g., by hydroxyl radical and x-irradiation.[26]

While the effect of irradiation-mediated free radical formation on nucleotides was not as pronounced as in proteins and peptides, it became clear that free radicals formed enzymatically could have profound consequences on protein function. Metabolic hydroxylation of aromatic amino acids has long been suspected as a biosynthetic process for the conversation of phenylalanine to tyrosine, tyrosine to 3,4-dihydroxyphenylalanine, kynurenine to 3-hydroxykynurenine, and tryptophan to 5-hydroxytryptophan.[27] In the 1950s, metabolic hydroxylation of aromatic compounds such as *N*-2-fluorenylacetamide in guinea pigs and rats was demonstrated and believed to be a detoxification mechanism.[28] This conversion was duplicated in cell-free in vitro studies involving ferrous ion, ascorbic acid, oxygen, and a chelating agent (ethylenediaminetetraacetic acid) under physiological conditions, showing that hydroxylation of aromatic compounds could indeed be mediated by free radial reaction, specifically that of hydroxyl radicals.[29] Not long after, it was proposed that biological hydroxylation occurs via activation of oxygen by peroxidase[30,31] and by other hydroxylating systems found in liver microsomes that require triphosphopyridine nucleotides and oxygen for their activity.[32] Altogether, Harman[33] proposed the role of oxygen, metals ions, oxidative enzymes, and radiation on the development of age-related and degenerative diseases through generation of reactive oxygen species, and these propositions became the foundation of today's widely accepted free radical theory of aging. Subsequently, the link between free radicals and the development of atherosclerosis,[34,35] cancer,[36] and neurodegeneration[37] was proposed, and the role of proper nutrition and lower metabolic demand were seen as essential for the slower progression of free radical–mediated reactions in the body.[38,39] While free radicals such as semiquinones were identified using EPR as integral components of the mitochondrial respiratory chain,[40,41] evidence for the production of ROS such as superoxide and hydrogen peroxide by mitochondria through electron transfer had become more compelling.[42–44]

Although the existence of superoxide as an inorganic species is known dating as far back as the late 1890s, its paramagnetic character was not established until the 1930s.[45] For the next four decades, studies on superoxide were mostly focused on their chemistry with metals and nonmetals and, for the first time, its characterization through EPR spectroscopy.[46] Only in the late 1960s did the idea become acceptable that superoxide could also be generated in biological system, an idea helped by the discovery of superoxide generation from enzymatic systems. The idea was first introduced by McCord and Fridovich,[47] whose seminal study demonstrated the production of superoxide from xanthine oxidase and xanthine; they found this formed species capable of reducing cytochrome c and initiating the sulfite oxidation reaction. In further support of this evidence, superoxide formation from xanthine oxidase was confirmed using EPR spectroscopy at pH 10; signal intensity was shown to be dependent on oxygen concentration and not the enzyme itself.[48] The oxygen origin of superoxide from xanthine oxidase was unequivocally confirmed using ^{17}O-labeled O_2 and EPR spectroscopy giving the 11-line or 6-line EPR spectra for the formed $^{17}O_2^{\bullet -}$ or $^{17}O^{16}O^{\bullet -}$, respectively.[49] Generation of superoxide from oxygen was achieved through electrolytic reduction of oxygen in aqueous solution.[50] In addition, electrochemically generated chemiluminescence of lucigenin showed evidence of superoxide-mediated light emission, paving the way for the development of chemiluminescence probes for superoxide.[51] Excited-state oxygen resulting from the oxidation of xanthine by xanthine oxidase can also induce chemiluminescence via recombination of ROS, probably that of superoxide,[52] and this finding was further supported by evidence showing that singlet oxygen–sensitized fluorescence from organic compounds is mediated by superoxide, which suggests the possible enzyme-mediated formation of singlet oxygen.[53]

The chemistry of superoxide enzymatic formation and decomposition then became of interest to investigators who wanted to know whether the mechanism is ligand mediated or metal mediated. It was demonstrated that superoxide is decomposed by erythrocuprein and ferricytochrome c and is formed during the oxidation of reduced flavin[54] rather the iron heme of flavoproteins.[55] This was later supported by studies showing that the one-electron reduction of oxygen by reduced flavins and quinones results in the formation superoxide.[56] However, the formation of superoxide from the reaction of oxygen with reduced iron-sulfur proteins from plant ferredoxins that are flavin free was also reported,[57] indicating that oxygen reduction can occur via electron transfer not only from organic radicals but also from low-valent metal ions.[56] The enzyme superoxide dismutase (SOD) was then proposed to catalyze the dismutation of superoxide to oxygen and hydrogen peroxide and had been a gold standard as a competitive inhibitor for the investigation of superoxide-mediated reactions such as the oxidation of epinephrine to adrenochrome by xanthine oxidase and the reduction of ferricytochrome c or tetranitromethane.[58] The reduction of ferricytochrome c was found to be augmented by electron carriers such as flavin adenine dinucleotide, menadione, or flavin mononucleotide.[59] The characterization of SOD in bovine heart and its ubiquity in mammalian tissues suggest the important role SOD plays in the regulation of ROS in biological systems.[60] It became clearer that the cause of oxygen toxicity is not oxygen itself but the oxygen-derived reactive species such as superoxide, hydrogen peroxide, and hydroxyl radicals.

Detection of reactive species in biological systems could date as far back as the late 1800s and early 1900s through visual observations of color changes in test tubes or spot plates. For example, the detection of hydrogen peroxide in plants and milk employed the use of various reagents that impart color changes with oxidation.[61–63] The introduction of commercial ultraviolet visible spectrum (UV vis) spectrophotometers in the early 1940s by Arnold O. Beckman[64] allowed for the detection of oxygen-derived radicals (e.g., superoxide and hydroxyl) in biological systems. Since then, ROS detection focused on the use of spectrophotometric techniques that allow detection at the UV region and providing a more accurate, reproducible, time-saving, and reduced sample size for ROS analysis. Detection of superoxide became possible through superoxide dismutation by superoxide dismutase or reduction by ferricytochrome c.[56,59,65] Several other techniques has been employed for the analysis of H_2O_2 and superoxide radical such as electrochemistry,[66] fluorescence,[67] chemiluminescence,[68] and EPR spin trapping.[69] Hydroxyl radical detection involved the analysis of gaseous substances such as methane on hydroxyl radical reaction with dimethyl sulfoxide from human phagocytes as analyzed by mass spectroscopy[70] or ^{14}C-carbon dioxide from the decarboxylation of [^{14}C]benzoic acid on oxidation by hydroxyl radicals, also from human ganulocytes, using an ionization chamber-electrometer.[71] Concurrently, the hydroxyl radical was identified through hydroxylation of 5,5-dimethyl-pyrroline N-oxide (DMPO) by EPR spin trapping in respiring rat heat mitochondria[44] or through hydroxylation of salicylate as measured by colorimetric or gas chromatographic (GC) assays.[72]

Analytical techniques have a profound utility in resolving some of the most confounding questions in science but sometimes do not provide unequivocal evidence of their formation and therefore need to be complemented by biological assays. For example, the question whether the endothelium-derived relaxing factor (EDRF) is nitric oxide (NO). Their pharmacological properties were found to be similar[73,74] as well as their in vivo half-life of 3–5 s[73] or 41 s in Krebs solution.[75] However, subsequent findings show that the amount of nitric oxide detected by chemiluminescence was significantly less than that required to account for the detector vessel relaxation,[76] suggesting that NO is not the sole component of EDRF. Through the use of the same chemiluminescence technique, it was proposed that EDRF resembles that of *S*-nitrosocysteine (or *S*-nitrosothiols in general) in terms of its vasodilator potencies,[77] which is further supported by previous findings that show that exogenously introduced nitrosocysteine appears to be more similar to EDRF than free NO in terms of vasodilation per amount of contained NO.[78] Other perplexing data revealed through continuous-flow spectrophotometric detection using diazotization reaction and did not detect nitrogen oxides and lacked the specificity to distinguish NO from other nitrogen oxides released from cultured endothelial cells by bradykinin, adenosine triphosphate, or calcium ionophore, whereas the bioassay readily detected EDRF release.[79] Using EPR or spin trapping, the adduct formed from activation of muscarinic receptors by carbamylcholine did not resemble the free radicals originating from NO or hydroxylamine but is similar to that of L-arginine-NADPH-Ca^{2+}–derived NO using 3,5-dibromo-4-nitrosobenzene sulfonate as a spin trap,[80] which further supports that L-arginine is the precursor of EDRF.[81] However, the triplet signal observed for hemoglobin-NO (Hb-NO) complex cannot be duplicated from effluents collected from acetylcholine-stimulated release of EDRF from intact femoral arteries.[82] EPR spin-trapping studies of NO using Hb corroborated the previous findings that EDRF resembles *S*-nitrosothiols and not NO at concentrations required to cause vasorelaxation similar to EDRF.[83] In spite of these equivocal data, bioassay methods provided evidence that supports initial findings that NO is EDRF such that it causes increased intracellular cGMP response.[84] For example, evidences derived from chemical and biological assays showed that EDRF inhibition by L-NMMA and L-NAME[85]; potentiation of vasorelaxation by PDE isoenzymes[86]; augmentation of EDRF activity through increased O_2[87]; inactivation of EDRF by pyrogallol or superoxide, stabilization by SOD and inhibition by oxyHb or K^+[74]; dependence of EDRF production from Ca^{2+} concentration and CaM[88]; EDRF inhibition of platelet activity[89]; and dependence of nitrite levels on EDRF production in biological fluids,[90,91] therefore, further supporting the fact that EDRF components is NO-derived. Other evidence also links EDRF to other substances such as the nonprostanoid endothelin-1[92] and most recently H_2S.[93]

The interplay between superoxide and nitric oxide in biological system has a profound implication on the normal function of the cell. Superoxide radical dismutation to hydrogen peroxide ultimately leads to the formation of hydroxyl radicals via

Fe-catalyzed or uncatalyzed Haber-Weiss processes.[94] The formation of these reactive intermediates, although critical in the modulation of cellular function, signaling, and immune response, in unregulated levels are detrimental to the integrity of key biomolecular systems and lead to cellular injury. Similarly, normal levels of nitric oxide are important mediators of physiological functions but when below or above normal levels could compromise cellular functions due to low NO bioavailabity or could cause toxicity, respectively. The paradoxical effect of NO at high levels confounds investigators since one would expect a more beneficial effect, but evidence showed that mice with long-term exposure to nitric oxide suffered degenerative and necrotic changes followed by bronchiolar epitheliums of the alveolar,[95] lysis of tumor cells on activation of endothelial cells by cytokines,[96] or chemically generated NO from NO-donating drugs causes pancreatic islet cells lysis in a time- and concentration-dependent manner.[97] The link between oxygen, nitric oxide, and superoxide toxicity in vivo was hypothesized in the seminal work by Oury et al. through the use of transgenic mice models of overexpressing human extracellular superoxide dismutase, endogenous nitric oxide synthase (NOS) inhibition and hyperbaric oxygen conditions, and showed that nitric oxide plays an important role in decreasing oxygen toxicity in the central nervous system by oxygen-mediated NO inactivation perhaps via superoxide formation.[98] Although the reaction of superoxide and NO could be considered protective due to NO's sequestration of superoxide, the product formed from the reaction of these two paramagnetic species (i.e., peroxynitrite)[99] was proposed to be oxidative in nature by decomposing two potent oxidants (i.e., hydroxyl and nitric oxide), and this could have significant implications in the initiation of oxidative damage as mediated by activated leukocytes and myocardial or neural ischemia or reperfusion injury,[100–105] as well as in the formation of atherosclerotic lesions.[106–108] Although it was proposed a decade earlier that peroxynitrite is produced from direct oxidation of NO by oxymyoglobin,[109] this opened up the possibility that peroxynitrite may be produced in biological systems. Beckman et al.[110] proposed the implication of peroxynitrite formation for endothelial injury through production of hydroxyl radical. Following their discoveries, mechanistic insights into the peroxynitrite cytotoxicity had become clearer by demonstrating its reactivity with protein thiols,[111] its formation of protein nitration,[112] its role in the initiation of lipid peroxidation[113] and DNA oxidation.[114] As with other reactive species, increasing evidence of peroxynitrite importance in the initiation of oxidative damage has become increasingly apparent and necessitating the development of analytical tools for its detection in biological systems. Early evidence of peroxynitrite formation in chemical systems mainly involved high-performance liquid chromatography (HPLC) analysis of hydroxylation of phenylalanine to yield isomeric mixtures of tyrosine products along with products of the nitration of phenylalanine and tyrosine, suggesting the formation of hydroxyl and nitrogen dioxide radical intermediates.[115,116] EPR spin-trapping studies also confirmed the formation of these decomposition species using DMPO as a spin trap [117] or using oxyhemoglobin photometrically.[118] However, in vivo detection of peroxynitrite proved to be a challenge due to its high reactivity, hence short in vivo half-life,[119] but with the development of antinitrotyrosine antibodies, the in vivo evidence of peroxynitrite formation can be correlated with levels of nitrotyrosine in tissues as analyzed by immunohistochemical techniques.[120,121] Other techniques had since been applied for the analysis of peroxynitrite in vivo using fluorescent probes[122] or HPLC,[123,124] GC, or mass spectroscopic (MS) assay.[125]

This introduction gives us just a short glimpse of the history of oxygen, reactive species derived from it, and the importance of detection techniques, whether crude or state of the art, in the understanding of their role in biology. Through the design of innovative analytical tools such as instrumentation and probes for reactive species detection, we are getting closer to unraveling the mysteries underlying the role of oxygen and its reactive derivatives in the pathogenesis of diseases that remain to be fully understood and in developing innovative therapeutic strategies to minimize if not prevent oxygen- or reactive species–mediated cellular injuries.

REFERENCES

1. Cassebaum H, Schufle JA. Scheele's priority for the discovery of oxygen. *J Chem Educ* 1975;**52**:442.
2. Neville RG. Steps leading to the discovery of oxygen, 1774. Bicentennial tribute to Joseph Priestley. *J Chem Educ* 1974;**51**:428.
3. Karamanou M, Tsoucalas G, Androutsos G. Hallmarks in the study of respiratory physiology and the crucial role of Antoine-Laurent de Lavoisier (1743–1794). *Am J Physiol* 2013;**305**:L591.
4. West JB. The collaboration of Antoine and Marie-Anne Lavoisier and the first measurements of human oxygen consumption. *Am J Physiol* 2013;**305**:L775.
5. Mason HS. Mechanisms of oxygen metabolism. *Science* 1957;**125**:1185.
6. Severinghaus JW, Astrup PB. History of blood gas analysis. V. Oxygen measurement. *J Clin Monit* 1986;**2**:174.
7. Marcet W. Researches on the Absorption of Oxygen and Formation of Carbonic Acid in Ordinary Human Respiration, and in the Respiration of Air Containing an Excess of Carbonic Acid. *Proc R Soc Lond* 1891;**50**:58.
8. Todd J, et al. Experiments with oxygen on disease. Tuberculosis, Bright's disease. *Monograph* 1916. viii + 225 pages.
9. Evans JH, Durshordwe CJ. Indications for oxygen therapy in respiratory diseases. *Anesth Analg* 1935;**14**:162.
10. Bean JW. Oxygen poisoning in microörganisms and its relation to the toxicity of oxygen at high pressure on mammalian tissue. *J Cell Comp Physiol* 1941;**17**:277.
11. Cleveland LR. Toxicity of oxygen for protozoa in vivo and in vitro: animals defaunated without injury. *Biol Bull* 1925;**48**:455.
12. Dickens F. Toxic effects of oxygen on brain metabolism and on tissue enzymes. 1. Brain metabolism. *Biochem J* 1946;**40**:145.
13. Mann PJG, Quastel JH. Toxic effects of oxygen and of hydrogen peroxide on brain metabolism. *Biochem J* 1946;**40**:139.
14. Ohlsson WTL. A study on oxygen toxicity at atmospheric pressure with special reference to the pathogenesis of pulmonary damage and clinical oxygen therapy. *Acta Med Scand* 1947;**128**:93.

15. Lambertsen CJ, Kough RH, Cooper DY, Emmel GL, Loeschcke HH, Schmidt CF. Oxygen toxicity. Effects in man of oxygen inhalation at 1 and 3.5 atmospheres upon blood gas transport, cerebral circulation, and cerebral metabolism. *J Appl Physiol* 1953;**5**:471.
16. DuBois AB. Oxygen toxicity. *Anesthesiology* 1962;**23**:473.
17. Allsopp CB. Theories of the biological actions of ionizing radiations. *Br J Radiol* 1948;**21**:72.
18. Park HF. Role of the unpaired electron in carcinogenesis. *J Phys Colloid Chem* 1950;**54**:1383.
19. Biesele JJ, Philips FS, Thiersch JB, Burchenal JH, Buckley SM, Stock CC. Chromosome alteration and tumor inhibition by nitrogen mustards. Hypothesis of cross-linking alkylation. *Nature* 1950;**166**:1112.
20. Hutchinson F. The distance that a radical formed by ionizing radiation can diffuse in a yeast cell. *Radiat Res* 1957;**7**:473.
21. Klebanoff SJ. The effect of x-radiation on the glutathione metabolism of intact erythrocytes in vitro. *J Gen Physiol* 1958;**41**:725.
22. Howard-Flanders P. Effect of oxygen on the radiosensitivity of bacteriophage in the presence of sulfhydryl compounds. *Nature* 1960;**186**:485.
23. Shields H, Gordy W. Electron-spin resonance studies of radiation damage to the nucleic acids and their constituents. *Proc Natl Acad Sci USA* 1959;**45**:269.
24. Hutchinson F, Arena J. Destruction of the activity of deoxyribonucleic acid in irradiated cells. *Radiat Res* 1960;**13**:137.
25. Ormerod MG, Alexander P. Repair of radiation damage in nucleoprotein by cysteamine. *Nature* 1962;**193**:290.
26. Dubinin NP, Sidorov BN, Sokolov NN. An experimental analysis of the initial mechanism responsible for the effect of radiation on the cell nucleus. *Dokl Akad Nauk SSSR* 1960;**133**:221.
27. Dalgliesh CE. Nonspecific formation of hydroxylated metabolites of the aromatic amino acids. *Arch. Biochem Biophys* 1955;**58**:214.
28. Weisburger JH, Weisburger EK, Morris HP. Orientation of biochemical hydroxylation in aromatic compounds. *Science* 1957;**125**:503.
29. Brodie BB, Axelrod J, Shore PA, Udenfriend S. Ascorbic acid in aromatic hydroxylation. II. Products formed by reaction of substrates with ascorbic acid, ferrous ion, and oxygen. *J Biol Chem* 1954;**208**:741.
30. Mason HS, Onoprienko I, Yasunobu K, Buhler D. Transfer of molecular oxygen by peroxidase. *J Am Chem Soc* 1957;**79**:5578.
31. Mason HS, Onopryenko I, Buhler D. Hyroxylation: activation of oxygen by peroxidase. *Biochim Biophys Acta* 1957;**24**:225.
32. Mitoma C, Posner HS, Reitz HC, Udenfriend S. Enzymic hydroxylation of aromatic compounds. *Arch Biochem Biophys* 1956;**61**:431.
33. Harman D. Aging: A Theory Based on Free Radical and Radiation Chemistry. *J Gerontol* 1956;**11**:298.
34. Harman D. Atherosclerosis: possible ill-effects of the use of highly unsaturated fats to lower serum-cholesterol levels. *Lancet* 1957;**273**:1116.
35. Harman D. Atherosclerosis. Effect of rate of growth. *Circ Res* 1962;**10**:851.
36. Harman D. Mutation, cancer, and ageing. *Lancet* 1961;**1**:200.
37. Harman D, Hendricks S, Eddy DE, Seibold J. Free radical theory of aging: effect of dietary fat on central nervous system function. *J Am Geriatr Soc* 1976;**24**:301.
38. Harman D. Free radical theory of aging: dietary implications. *Am J Clin Nutr* 1972;**25**:839.
39. Harman D. Nutritional implications of the free-radical theory of aging. *J Am Coll Nutr* 1982;**1**:27.
40. Commoner B, Hollocher Jr TC. Free radicals in heart muscle mitochondrial particles: general characteristics and localization in the electron transport system. *Proc Natl Acad Sci USA* 1960;**46**:405.
41. Hollocher Jr TC, Commoner B. Mechanism of formation. *Proc Natl Acad Sci USA* 1960;**46**:416.
42. Docampo R, Cruz FS, Boveris A, Muniz RPA, Esquivel DMS. β-Lapachone enhancement of lipid peroxidation and superoxide anion and hydrogen peroxide formation by Sarcoma 180 ascites tumor cells. *Biochem Pharmacol* 1979;**28**:723.

43. Nohl H, Breuninger V, Hegner D. Influence of mitochondrial radical formation on energy-linked respiration. *Eur J Biochem* 1978;**90**:385.
44. Nohl H, Jordan W, Hegner D. Identification of free hydroxyl radicals in respiring rat heart mitochondria by spin trapping with the nitrone DMPO. *FEBS Lett* 1981;**123**:241.
45. Neuman EW. Potassium superoxide and the three-electron bond. *J Chem Phys* 1934;**2**:31.
46. Meistrich ML. E.S.R. and optical studies of O2- in sodium fluoride. *J Phys Chem Solids* 1968;**29**:1111.
47. McCord JM, Fridovich I. Reduction of cytochrome c by milk xanthine oxidase. *J Biol Chem* 1968;**243**:5753.
48. Knowles PF, Gibson JF, Pick FM, Bray RC. Electron-spin-resonance evidence for enzymic reduction of oxygen to a free radical, the superoxide ion. *Biochem J* 1969;**111**:53.
49. Bray RC, Pick FM, Samuel D. Oxygen-17 hyperfine splitting in the electron paramagnetic resonance spectrum of enzymically generated superoxide. *Eur J Biochem* 1970;**15**:352.
50. Kastening B, Kazemifard G. Electrochemical reduction of oxygen to superoxide anions in aqueous solutions. *Ber Bunsenges Phys Chem* 1970;**74**:551.
51. Legg KD, Hercules DM. Electrochemically generated chemiluminescence of lucigenin. *J Amer Chem Soc* 1969;**91**:1902.
52. Arneson RM. Substrate-induced chemiluminescence of xanthine oxidase and aldehyde oxidase. *Arch Biochem Biophys* 1970;**136**:352.
53. Khan AU. Singlet molecular oxygen from superoxide anion and sensitized fluorescence of organic molecules. *Science* 1970;**168**:476.
54. Ballou D, Palmer G, Massey V. Direct demonstration of superoxide anion production during the oxidation of reduced flavin and of its catalytic decomposition by erythrocuprein. *Biochem Biophys Res Commun* 1969;**36**:898.
55. Massey V, Strickland S, Mayhew SG, Howell LG, Engel PC, Matthews RG, et al. Production of superoxide anion radicals in the reaction of reduced flavins and flavoproteins with molecular oxygen. *Biochem Biophys Res Commun* 1969;**36**:891.
56. Misra HP, Fridovich I. Univalent reduction of oxygen by reduced flavins and quinones. *J Biol Chem* 1971;**247**:188.
57. Orme-Johnson WH, Beinert H. Formation of the superoxide anion radical during the reaction of reduced iron-sulfur proteins with oxygen. *Biochem Biophys Res Commun* 1969;**36**:905.
58. McCord JM, Fridovich I. Superoxide dismutase. Enzymic function for erythrocuprein (hemocuprein). *J Biol Chem* 1969;**244**:6049.
59. McCord JM, Fridovich I. Utility of superoxide dismutase in studying free radical reactions. II. Mechanism of the mediation of cytochrome c reduction by a variety of electron carriers. *J Biol Chem* 1970;**245**:1374.
60. Keele Jr BB, McCord JM, Fridovich I. Further characterization of bovine superoxide dismutase in its isolation from bovine heart. *J Biol Chem* 1971;**246**:2875.
61. Arnold C, Mentzel C. Detection of hydrogen peroxide in milk. *Zeit Nahr Genussm* 1903;**6**:305.
62. Bach A. Detection of hydrogen peroxide in green plants. *Compt Rend* 1894;**119**:1218.
63. Darlington IT. Experimental data comparing the delicacy of different tests for hydrogen peroxide in milk. *J Ind Eng Chem* 1915;**7**:676.
64. Beckman AO, Gallaway WS, Kaye W, Ulrich WF. History of Spectrophotometry at Beckman Instruments, Inc. *Anal Chem* 1977;**49**:280A.
65. Fridovich I. Superoxide radical and superoxide dismutase. *Acc Chem Res* 1972;**5**:321.
66. Siegenthaler EJ, Kosikowski FV. Electrochemical method for detecting hydrogen peroxide-catalase-treated milk. *J Dairy Sci* 1969;**52**:1922.
67. Weetall HH, Weliky N. A new technique for the enzymic detection of hydrogen peroxide. *Anal Biochem* 1966;**14**:160.
68. Seitz WR. Chemiluminescence detection of enzymically generated peroxide. *Methods Enzymol* 1978;**57**:445.
69. Harbour JR, Bolton JR. Superoxide formation in spinach chloroplasts. Electron spin resonance detection by spin trapping. *Biochem Biophys Res Commun* 1975;**64**:803.

70. Repine JE, Eaton JW, Anders MW, Hoidal JR, Fox RB. Generation of hydroxyl radical by enzymes, chemicals, and human phagocytes in vitro. Detection with the anti-inflammatory agent, dimethyl sulfoxide. *J Clin Invest* 1979;**64**:1642.
71. Sagone Jr AL, Decker MA, Wells RM, Democko C. A new method for the detection of hydroxyl radical production by phagocytic cells. *Biochim Biophys Acta Gen Subj* 1980;**628**:90.
72. Richmond R, Halliwell B, Chauhan J, Darbre A. Superoxide-dependent formation of hydroxyl radicals: detection of hydroxyl radicals by the hydroxylation of aromatic compounds. *Anal Biochem* 1981;**118**:328.
73. Ignarro LJ, Buga GM, Byrns RE, Wood KS, Chaudhuri G. Endothelium-derived relaxing factor and nitric oxide possess identical pharmacologic properties as relaxants of bovine arterial and venous smooth muscle. *J Pharmacol Exp Ther* 1988;**246**:218.
74. Ignarro LJ, Buga GM, Wood KS, Byrns RE, Chaudhuri G. Endothelium-derived relaxing factor produced and released from artery and vein is nitric oxide. *Proc Natl Acad Sci USA* 1987;**84**:9265.
75. Angus JA, Cocks TM. The half-life of endothelium-derived relaxing factor released from bovine aortic endothelial cells in culture. *J Physiol* 1987;**388**:71.
76. Myers PR, Guerra Jr R, Harrison DG. Release of NO and EDRF from cultured bovine aortic endothelial cells. *Am J Physiol* 1989;**256**:H1030.
77. Myers PR, Minor Jr RL, Guerra Jr R, Bates JN, Harrison DG. Vasorelaxant properties of the endothelium-derived relaxing factor more closely resemble S-nitrosocysteine than nitric oxide. *Nature* 1990;**345**:161.
78. Bates JN, Harrison DG, Myers PR, Minor RL. EDRF: nitrosylated compound or authentic nitric oxide. *Basic Res Cardiol* 1991;**86**(Suppl 2):17.
79. Tracey WR, Linden J, Peach MJ, Johns RA. Comparison of spectrophotometric and biological assays for nitric oxide (NO) and endothelium-derived relaxing factor (EDRF): nonspecificity of the diazotization reaction for nitric oxide and failure to detect EDRF. *J Pharmacol Exp Ther* 1990;**252**:922.
80. Arroyo CM, Forray C, El-Fakahany EE, Rosen GM. Receptor-mediated generation of an EDRF-like intermediate in a neuronal cell line detected by spin trapping techniques. *Biochem Biophys Res Commun* 1990;**170**:1177.
81. Fasehun OA, Gross SS, Rubin LE, Jaffe EA, Griffith OW, Levi R. L-Arginine, but not Nα-benzoyl-L-arginine ethyl ester, is a precursor of endothelium-derived relaxing factor. *J Pharmacol Exp Ther* 1990;**255**:1348.
82. Greenberg SS, Wilcox DE, Rubanyi GM. Endothelium-derived relaxing factor released from canine femoral artery by acetylcholine cannot be identified as free nitric oxide by electron paramagnetic resonance spectroscopy. *Circ Res* 1990;**67**:1446.
83. Rubanyi GM, Johns A, Wilcox D, Bates FN, Harrison D. Evidence that a S-nitrosothiol, but not nitric oxide, may be identical with endothelium-derived relaxing factor. *J Cardiovasc Pharmacol* 1991;**17**:S41.
84. Ishii K, Sheng H, Warner TD, Forstermann U, Murad F. A simple and sensitive bioassay method for detection of EDRF with RFL-6 rat lung fibroblasts. *Am J Physiol* 1991;**261**:H598.
85. Baylis C, Harton P, Engels K. Endothelial derived relaxing factor controls renal hemodynamics in the normal rat kidney. *J Am Soc Nephrol* 1990;**1**:875.
86. Pagani ED, Buchholz RA, Silver PJ. Cardiovascular cyclic nucleotide phosphodiesterases and their role in regulating cardiovascular function. *Basic Res Cardiol* 1992;**87**:73.
87. McQueston JA, Cornfield DN, McMurtry IF, Abman SH. Effects of oxygen and exogenous L-arginine on EDRF activity in fetal pulmonary circulation. *Am J Physiol* 1993;**264**:H865.
88. Forstermann U, Pollock JS, Schmidt HHHW, Heller M, Murad F. Calmodulin-dependent endothelium-derived relaxing factor/nitric oxide synthase activity is present in the particulate and cytosolic fractions of bovine aortic endothelial cells. *Proc Natl Acad Sci USA* 1991;**88**:1788.
89. Broekman MJ, Eiroa AM, Marcus AJ. Inhibition of human platelet reactivity by endothelium-derived relaxing factor from human umbilical vein endothelial cells in suspension: blockade of aggregation and secretion by an aspirin-insensitive mechanism. *Blood* 1991;**78**:1033.

90. Wennmalm A, Petersson AS. Analysis of nitrite as a marker for endothelium-derived relaxing factor in biological fluids using electron paramagnetic resonance spectrometry. *J Cardiovasc Pharmacol* 1991;**17**:S34.
91. Gutzki FM, Tsikas D, Alheid U, Froelich JC. Determination of endothelium-derived nitrite/nitrate by gas chromatography/tandem mass spectrometry using (15N)sodium nitrite as internal standard. *Biol Mass Spectrom* 1992;**21**:97.
92. Beierwaltes WH, Carretero OA. Nonprostanoid endothelium-derived factors inhibit renin release. *Hypertension* 1992;**19**:II68.
93. Wang R. Hydrogen sulfide: a new EDRF. *Kidney Int* 2009;**76**:700.
94. Wardman P, Candeias LP. Fenton chemistry: an introduction. *Radiat Res* 1996;**145**:523.
95. Oda H, Nogami H, Kusumoto S, Nakajima T, Kurata A, Imai K. Long-term exposure of mice to nitric oxide. *Taiki Osen Kenkyu* 1976;**11**:150.
96. Li L, Kilbourn RG, Adams J, Fidler IJ. Role of nitric oxide in lysis of tumor cells by cytokine-activated endothelial cells. *Cancer Res* 1991;**51**:2531.
97. Kallmann B, Burkart V, Kroncke KD, Kolb-Bachofen V, Kolb H. Toxicity of chemically generated nitric oxide towards pancreatic islet cells can be prevented by nicotinamide. *Life Sci* 1992;**51**:671.
98. Oury TD, Ho YS, Piantadosi CA, Crapo JD. Extracellular superoxide dismutase, nitric oxide, and central nervous system oxygen toxicity. *Proc Natl Acad Sci USA* 1992;**89**:9715.
99. Koppenol WH, Moreno JJ, Pryor WA, Ischiropoulos H, Beckman JS. Peroxynitrite, a cloaked oxidant formed by nitric oxide and superoxide. *Chem Res Toxicol* 1992;**5**:834.
100. Ischiropoulos H, Zhu L, Beckman JS. Peroxynitrite formation from macrophage-derived nitric oxide. *Arch Biochem Biophys* 1992;**298**:446.
101. Jaeschke H, Schini VB, Farhood A. Role of nitric oxide in the oxidant stress during ischemia/reperfusion injury of the liver. *Life Sci* 1992;**50**:1797.
102. Beckman JS, Ischiropoulos H, Chen J, Zhu L, Smith CD. *Pergamon* 1991;251.
103. Crow JP, Beckman JS. The role of peroxynitrite in nitric oxide-mediated toxicity. *Curr Top Microbiol Immunol* 1995;**196**:57.
104. Crow JP, Beckman JS. The importance of superoxide in nitric oxide-dependent toxicity. Evidence for peroxynitrite-mediated injury. *Adv Exp Med Biol* 1996;**387**:147.
105. Varner PD, Beckman JS. *Academic* 1995;191.
106. Hogg N, Darley-Usmar VM, Graham A, Moncada S. Peroxynitrite and atherosclerosis. *Biochem Soc Trans* 1993;**21**:358.
107. White CR, Brock TA, Chang LY, Crapo J, Briscoe P, Ku D, et al. Superoxide and peroxynitrite in atherosclerosis. *Proc Natl Acad Sci USA* 1994;**91**:1044.
108. Darley-Usmar VM, Hogg N, O'Leary VJ, Wilson MT, Moncada S. The simultaneous generation of superoxide and nitric oxide can initiate lipid peroxidation in human low density lipoprotein. *Free Radical Res Commun* 1992;**17**:9.
109. Doyle MP, Pickering RA, Cook BR. Oxidation of oxymyoglobin by nitric oxide through dissociation from cobalt nitrosyls. *J Inorg Biochem* 1983;**19**:329.
110. Beckman JS, Beckman TW, Chen J, Marshall PA, Freeman BA. Apparent hydroxyl radical production by peroxynitrite: implications for endothelial injury from nitric oxide and superoxide. *Proc Natl Acad Sci USA* 1990;**87**:1620.
111. Radi R, Beckman JS, Bush KM, Freeman BA. Peroxynitrite oxidation of sulfhydryls. The cytotoxic potential of superoxide and nitric oxide. *J Biol Chem* 1991;**266**:4244.
112. Ischiropoulos H, Zhu L, Chen J, Tsai M, Martin JC, Smith CD, et al. Peroxynitrite-mediated tyrosine nitration catalyzed by superoxide dismutase. *Arch Biochem Biophys* 1992;**298**:431.
113. Radi R, Beckman JS, Bush KM, Freeman BA. Peroxynitrite-induced membrane lipid peroxidation: the cytotoxic potential of superoxide and nitric oxide. *Arch Biochem Biophys* 1991;**288**:481.
114. Douki T, Cadet J. Peroxynitrite mediated oxidation of purine bases of nucleosides and isolated DNA. *Free Radical Res* 1996;**24**:369.
115. Crow JP, Beckman JS. Quantitation of protein tyrosine, 3-nitrotyrosine, and 3-aminotyrosine utilizing HPLC and intrinsic ultraviolet absorbance. *Methods* 1995;**7**:116.

116. Van der Vliet A, O'Neill CA, Halliwell B, Cross CE, Kaur H. Aromatic hydroxylation and nitration of phenylalanine and tyrosine by peroxynitrite. Evidence for hydroxyl radical production from peroxynitrite. *FEBS Lett* 1994;**339**:89.
117. Augusto O, Gatti RM, Radi R. Spin-trapping studies of peroxynitrite decomposition and of 3-morpholinosydnonimine N-ethylcarbamide autoxidation: direct evidence for metal-independent formation of free radical intermediates. *Arch Biochem Biophys* 1994;**310**:118.
118. Schmidt K, Klatt P, Mayer B. Reaction of peroxynitrite with oxyhemoglobin: interference with photometrical determination of nitric oxide. *Biochem J* 1994;**301**:645.
119. Ischiropoulos H, Beckman JS, Crow JP, Ye YZ, Royall JA, Kooy NW. Detection of peroxynitrite. *Methods* 1995;**7**:109.
120. Crow JP, Ye YZ, Royall J, Kooy N, Beckman JS. Evidence of peroxynitrite formation in vivo: Detection of nitrated proteins with an anti-nitrotyrosine antibody. *Portland Press Proc* 1994;**8**:20.
121. Myatt L, Rosenfield RB, Eis ALW, Brockman DE, Greer I, Lyall F. Nitrotyrosine residues in placenta: Evidence of peroxynitrite formation and action. *Hypertension* 1996;**28**:488.
122. Malcolm S, Foust RIII, Hertkom C, Ischiropoulos H. Detection of peroxynitrite in biological fluids *Methods. Mol Med* 2000;**36**:171.
123. Crow JP, Ischiropoulos H. Detection and quantitation of nitrotyrosine residues in proteins: In vivo marker of peroxynitrite. *Methods Enzymol* 1996;**269**:185.
124. Kaur H, Lyras L, Jenner P, Halliwell B. Artifacts in HPLC detection of 3-nitrotyrosine in human brain tissue. *J Neurochem* 1998;**70**:2220.
125. Jiang H, Balazy M. Detection of 3-Nitrotyrosine in Human Platelets Exposed to Peroxynitrite by a New Gas Chromatography/Mass Spectrometry Assay. *Nitric Oxide* 1998;**2**:350.

CHAPTER 2

Chemistry of Reactive Species

2.1 INTRODUCTION

Molecules are composed of atoms that are held together by a sea of electrons through covalent bonding. The pairing of electrons between two atoms is the bedrock of molecular stability, but once such pairing breaks reactive intermediates can form as an ion pair (i.e., heterolytic cleavage):

$$A - B \rightarrow A^+ + B^-$$

If there is an equal partition of the electron pair between two atoms (i.e., homolytic cleavage), two neutral radical species are formed:

$$A - B \rightarrow A^\bullet + B^\bullet$$

Aside from being broken into component species, molecules can undergo transformation by accepting or losing electrons through energy or electron-transfer processes to form the reactive intermediates of a radical anion and a radical cation:

$$A - B \xrightarrow{+e^-} A - B^{\bullet -}$$
$$A - B \xrightarrow{-e^-} A - B^{\bullet +}$$

These ionic species (A^+/B^-) as well as neutral ($A^\bullet/B^\bullet$) or charged radicals ($AB^{\bullet -}/AB^{\bullet +}$) formed from stable molecules and are key contributors to biomolecular transformation; collectively, they are referred to as reactive species (RS) in biology. The driving force in most of these chemical transformations are biological oxidoreductants, mostly derived from oxygen, nitric oxide, sulfides, and halides. For examples, heterolytic cleavage of thiols (R-SHs) to form the nucleophile thiolate (RS^-) often serves as a basis for thiol nucleophilicity; homolytic cleavage of hydrogen peroxide (H_2O_2) forms the electrophilic ($HO^\bullet$) and is the basis of reactive oxygen species (ROS) oxidative properties; the electron transfer via one-electron reduction of oxygen (O_2) to form superoxide ($O_2^{\bullet -}$) is the precursor of some of the RS known to exist in biological systems. Superoxide radical is transformed into a variety of RS that come as neutral or charged species, either as radicals (e.g., $O_2^{\bullet -}$, $HO^\bullet$, $HO_2^\bullet$, $RO_2^\bullet$, $RO^\bullet$, $CO_3^{\bullet -}$, $RS^\bullet$, $GSSG^{\bullet -}$, and $NO_2^\bullet$) or as nonradicals (e.g., H_2O_2, HOCl, O_3, $O_2^1\Delta g$, ROOH, $ONOO^-$ and $ONOOCO_2^-$) as shown in Fig. 2.1. They are also classified as

Reactive Species Detection in Biology
DOI: http://dx.doi.org/10.1016/B978-0-12-420017-3.00005-0

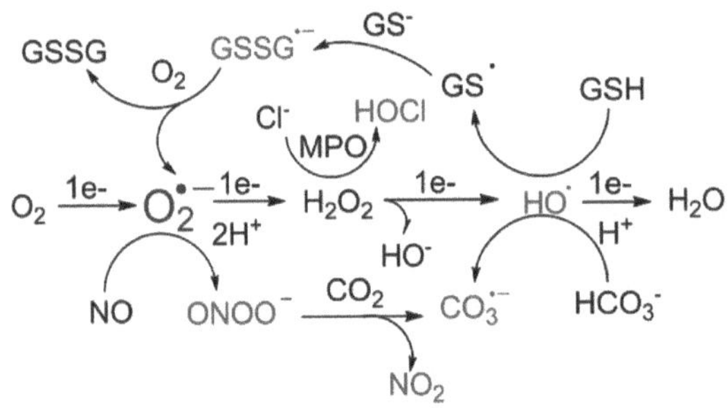

Figure 2.1 Formation of reactive species from oxygen via superoxide radical formation.

reactive oxygen-, nitrogen-, sulfur-, or halogen-species.[1,2] The formation of $O_2^{\bullet-}$ and its products signals the first sign of oxidative burst in a biological system. Due to their highly reactive nature, these species can be difficult to observe, but they are increasingly understood to play a key role in the regulation of vital cellular processes. For this reason, their direct or indirect detection is of critical importance.

2.2 REDOX CHEMISTRY

Reduction–oxidation (redox) processes are the major chemical reactions relevant to the modification of biomolecules involving RS. Redox reactions are characterized by changes in the oxidation state of particular atoms within a molecule either as the loss or gain of (1) a free electron or electrons or (2) electron carriers such as hydrogen, oxygen, and halogen atoms. Therefore, a redox reaction is mostly a dual process that involves electron donors (reducing agent) and acceptors (oxidizing agent). It is thus imperative to refer to the atom as being *oxidized* or *reduced* when two species combine to form one or more products. For example, peroxynitrite ($ONOO^-$) formation from $O_2^{\bullet-}$ and nitric oxide (NO) is shown in Eq. (2.1):

$$\underset{+2\ \ -2}{NO} + \underset{-0.5}{O_2^{\bullet-}} \rightarrow \underset{-2\ \ +3\ \ -1\ \ -1}{ON\text{-}OO^-} \qquad (2.1)$$

Here the entire reaction cannot be described as solely an oxidation or reduction reaction because several atoms are involved in the electron-transfer process, but what is clear is that the formation of $ONOO^-$ involves the formation of a $N_{\text{nitric oxide}}{-}O_{\text{superoxide}}$ bond. Knowing that the O atom is more electronegative than the N atom, the bond formed between N and O is polarized, giving more electron density to O than N and, since it is the electron being negatively charged, O assumes a more negative charge (-1) and N a more positive charge ($+3$). Hence, in this particular reaction, it is appropriate to describe NO as the reducing agent and $O_2^{\bullet-}$ as the oxidizing agent, where the N atom in NO underwent oxidation and reduction for the O atom in $O_2^{\bullet -}$. Traditionally, RS are referred to as reactive oxygen, nitrogen, sulfur, or halogen species, depending on which atom in that molecule has undergone a redox reaction. In the formation of $ONOO^-$, NO is the *reactive nitrogen species* while $O_2^{\bullet -}$ is the *reactive oxygen species* (ROS).

Moreover, $ONOO^-$ contains multiple atoms that can participate in redox reactions, so peroxynitrite cannot be described solely as a reactive nitrogen or oxygen species. For example, the conjugate acid of peroxynitrite, peroxynitrous acid (ONOOH), acts as an oxidizing agent that undergoes both hydroxylation or nitration reaction with phenols [Eq. (2.2)].[3] Homolytic cleavage of the O-O bond forms $NO_2^{\bullet}$ and $HO^{\bullet}$, so $ONOO^-$/ONOOH is considered both a reactive nitrogen and oxygen species. The oxidation state of N in $NO_2^{\bullet}$ is +4, while the O in $HO^{\bullet}$ is −1, which reduce to +3 and −2, respectively, on nitration and hydroxylation of phenols. Therefore, $ONOO^-$/ONOOH is an oxidizing agent whether it nitrates or hydroxylates the phenol. For most peroxynitrite reactions, however, nitration is the predominant reaction especially with tyrosine or guanine, which is why $ONOO^-$ is often referred to as a reactive nitrogen species.

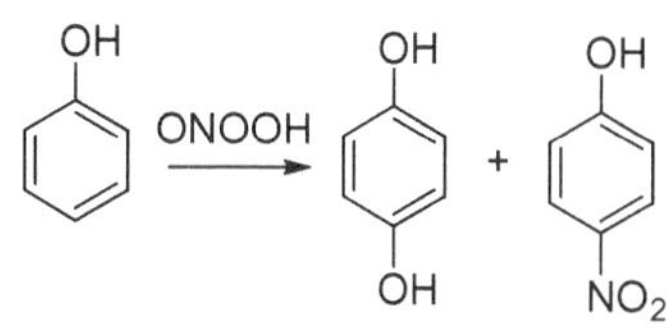

(2.2)

2.3 PROPERTIES OF REACTIVE SPECIES

2.3.1 Reactive Oxygen Species

Fig. 2.2 shows the various oxidation states of ROS originating from the one-, two-, and three-electron reductions of O_2 to form $O_2^{\bullet -}$, hydrogen peroxide (H_2O_2), and hydroxide (HO^-) with the concomitant release of a hydroxyl radical ($HO^{\bullet}$), respectively, and water (H_2O) being the final four-electron reduction product. The metabolism of O_2 typically occurs during aerobic respiration when it is converted to water via a series of electron-transfer processes in the mitochondrial electron-transport chain, although RS have also been shown to form through partial reduction of O_2 through a process commonly referred to as *electron leakage*, which is one of the main causes of ROS formation in cells during energy production. Other enzymatic processes for ROS production will be discussed in Chapter 3, Reactive Species Detection in Biology: From Fluorescence to Electron Paramagnetic Resonance Spectroscopy.

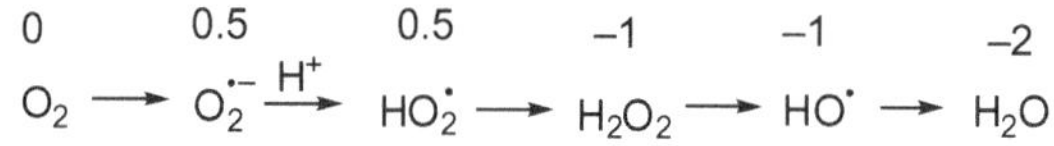

Figure 2.2 Oxidation states of oxygen atom in various oxygen species.

2.3.1.1 Oxygen (O_2)

Also referred to as *dioxygen* or *triplet* O_2 ($^3\Sigma_g^-$) owing to its electronic ground state in which both oxygen's two antibonding π_g-orbitals are occupied by an electron whose spin states are parallel with each other. The biradical (open-shell) nature of O_2 has significant implication for its reactivity such that O_2 has relatively higher reactivity toward paramagnetic species compared to other molecular gases such as N_2, CO_2, CH_4, and H_2S, which are all closed-shell systems. In fact, the two most favorable reactivities of O_2 are (1) electron-transfer reaction and (2) radical–radical addition, both involving O_2 reaction with species with unpaired electrons. For example, O_2 generates a $O_2^{\bullet -}/HOO^\bullet$ radical on reaction with a semiquinone radical via a one-electron–transfer reaction, according to Eq. (2.3).

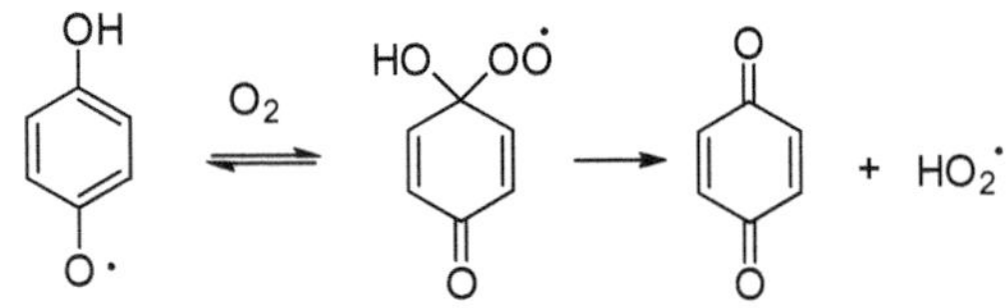

(2.3)

Iron (II) and other low-valence transition metals can catalyze the oxidation of biomolecules through activation of O_2. Iron, for example, could be loosely bound iron with small molecule chelators, free ionic iron, nonheme iron such as ferrtin or transferrin, and heme iron such as Hb, Mb, cytochrome *c* (cyt *c*), or catalase. Via the Haber–Weiss reaction, for example, Fe(II) reacts with O_2 to form $O_2^{\bullet -}$. The $O_2^{\bullet -}$ ultimately gives $HO^\bullet$, which can abstract an H atom from a polyunsaturated fatty acid (PUFA) which is an initial step in lipid peroxidation as shown in Eqs. (2.4)–(2.7):

$$Fe^{2+} + O_2 \rightarrow Fe^{3+} + O_2^{\bullet -} \quad (2.4)$$

$$2O_2^{\bullet -} + 2H^+ \rightarrow H_2O_2 + O_2 \quad (2.5)$$

$$H_2O_2 + Fe^{2+} \rightarrow Fe^{3+} + HO^\bullet + HO^- \quad (2.6)$$

$$LH + HO^\bullet \rightarrow H_2O + L^\bullet \quad (2.7)$$

Radical–radical reactions such as the addition reaction of O_2 to a carbon-centered and thiyl radical yields peroxyl and thiyl peroxyl radicals, respectively, according to Eqs. (2.8) and (2.9), making it relevant in lipid peroxidation and protein oxidation processes:

$$L^\bullet \overset{O_2}{\rightleftharpoons} LOO^\bullet \quad (2.8)$$

$$RS^\bullet \xrightarrow{O_2} RSOO^\bullet \quad (2.9)$$

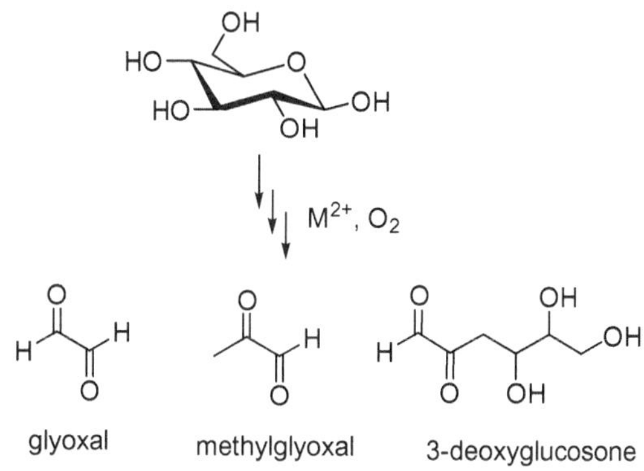

Figure 2.3 Dicarbonyl products of metal-catalyzed auto-oxidation of sugars by oxygen.

Glucose in the presence of oxygen undergoes a metal-catalyzed auto-oxidation to produce reactive dicarbonyls such as glyoxal, methylglyoxal, or 3-deoxyglucosone (Fig. 2.3). These dicarbonyls formed from the oxidation of glucose along with those formed from oxidized DNA and lipids yield advanced glycation end products (AGEs) via the Maillard reaction. This involves a condensation reaction of the carbonyl moiety with the amino groups of macromolecules to form Shiff base adducts. After a series of intramolecular rearrangements, dehydration and condensation reactions result in irreversible cross-linking of dicarbonyls with amino groups to form AGEs. Moreover, sugar oxidation not only gives dicarbonyl that can undergo Maillard reactions but also produces $O_2^{\bullet -}$, which could exacerbate ROS production.[5]

2.3.1.2 Superoxide Radical ($O_2^{\bullet -}$)

The term *superoxide radical anion* is redundant because the "-ide" ending already denotes the species' anionic nature therefore $O_2^{\bullet -}$ will be referred to as simply superoxide radical. The superoxide radical is described by the presence of an unpaired electron at its antibonding π orbital, which is delocalized between the two π^* orbitals, making $O_2^{\bullet -}$ less reactive compared to $CH_3^{\bullet}$, $RS^{\bullet}$, or $^{\bullet}OH$, which have unpaired electrons confined to only one π orbital. In fact, the rate of $O_2^{\bullet -}$ reaction with amino acids, lipids, nucleic acids, or sugars is especially slow if not unreactive.[6] Due to this poor reactivity, it can diffuse further from the site of its production, which is the reason for its selectivity. In general, since $O_2^{\bullet -}$ has an intermediate oxidation state, it can be reduced or oxidized to H_2O_2 or back to O_2, respectively. For example, in the case of a bimolecular reaction of two $O_2^{\bullet -}$ under acidic conditions, self-dismutation can result to produce H_2O_2 and O_2, according to Eq. (2.10).[7] However, dismutation of protonated $O_2^{\bullet -}$ in the form of $HO_2^{\bullet}$ is much more favorable due to a lessened repulsive effect between the negatively charged $O_2^{\bullet -}$ [Eqs. (2.11) and (2.12)].[7]

$$O_2^{\bullet -} + O_2^{\bullet -} + 2H^+ \rightarrow H_2O_2 + O_2\ (<0.3\ M^{-1}\ s^{-1}) \tag{2.10}$$

$$HO_2^{\bullet} + HO_2^{\bullet} \rightarrow H_2O_2 + O_2\ (8.3 \times 10^5\ M^{-1}\ s^{-1}) \tag{2.11}$$

$$HO_2^{\bullet} + O_2^{\bullet -} \rightarrow H_2O_2 + O_2\ (9.7 \times 10^7\ M^{-1}\ s^{-1}) \tag{2.12}$$

In a biological system, superoxide dismutation is enzymatically catalyzed via metal-mediated electron-transfer reactions such as those seen with superoxide dismutase (SOD). This bring us to another chemical property of $O_2^{\bullet -}$ that it is activated by metal centers to form the metal peroxo adducts. In the case of SOD,[8] SOD mimetics, or other high-valence transition metal complexes, $O_2^{\bullet -}$ forms H_2O_2 and O_2 (according to Fig. 2.4), or $O_2^{\bullet -}$ can be simply oxidized to O_2 by cytochrome *c* Fe^{3+}.[9] The same dismutation mechanism could be observed with stable radicals such as nitroxides as shown in Fig. 2.5.[10,11]

Aside from the direct oxidation and reduction of $O_2^{\bullet -}$ by itself through dismutation, metal centers, or nitroxides, $O_2^{\bullet -}$ is also capable of carrying out the following reactions: (1) Addition reaction, (2) proton–radical transfer, and (3) [Fe–S] cluster degradation. Addition of $O_2^{\bullet -}$ to a tyrosyl radical was observed to form tyrosyl [see Eq. (2.13)] or tryptophanyl hydroperoxide, which is relevant in the inactivation enzymes containing tyrosyl- or tryptophanyl radical active centers such as that found in ribonucleotide reductases, cyt *c* peroxidase, photosystem II, prostaglandin H synthase, amine- or galactose-oxidases, and methylamine dehydrogenase.[12]

$$M_a{}^{n}\text{-L} + O_2^{\bullet -} \longrightarrow M_b{}^{(n-1)}\text{-L} + O_2$$

$$M_b{}^{(n-1)}\text{-L} + O_2^{\bullet -} + 2H^+ \longrightarrow M_a{}^{n}\text{-L} + H_2O_2$$

$$\text{net reaction: } 2O_2^{\bullet -} + 2H^+ \longrightarrow H_2O_2 + O_2$$

where M_a/M_b = Cu(II)/Cu(I); Mn(III)/Mn(II); Fe(III)/Fe(II); Ni(II)/Ni(I); Mn(II)/Mn(I)

Figure 2.4 Superoxide radical dismutation by metal complexes.

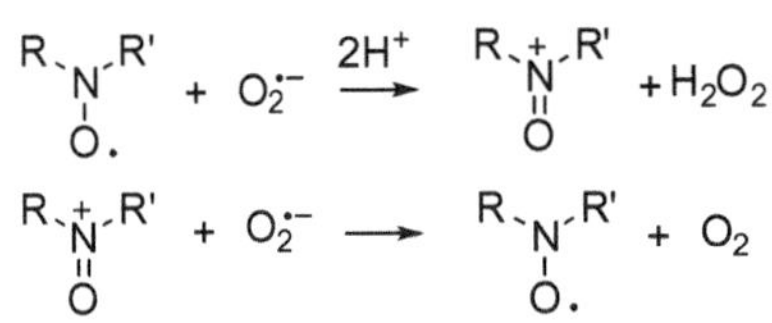

Figure 2.5 Superoxide radical dismutation by nitroxide.

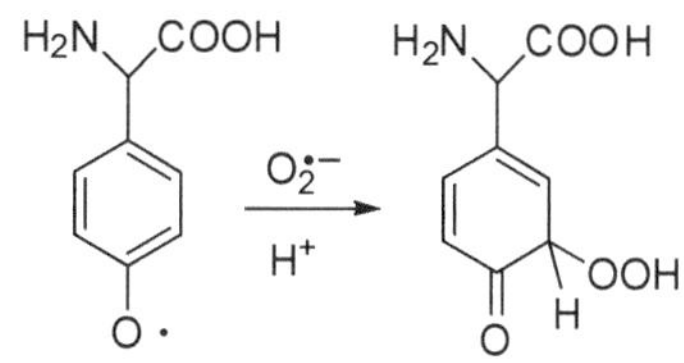

(2.13)

Perhaps the best known addition reaction of $O_2^{\bullet -}$ is its combination with NO to form $ONOO^-$, which occurs at rate constants approaching that of the diffusion-controlled limit (i.e., $4-6 \times 10^9\ M^{-1}\ s^{-1}$) compared to the enzymatic dismutation of $O_2^{\bullet -}$ by SOD with a rate constant that is on the order of $10^9\ M^{-1}\ s^{-1}$. Therefore, these two reactivities for $O_2^{\bullet -}$ are competitive and typically dictated by the cellular availability of either NO or SOD.[13,14]

The proton–radical transfer reaction of $O_2^{\bullet -}$ with catechols forms H_2O_2 and a semiquinone radical; the latter could further react with oxygen to form $O_2^{\bullet -}$ and *o*-quinone.[15] In a biological system, quinones can be reduced back to catechols (hydroquinones) or semiquinone by the flavoproteins nicotinamide adenine dinucleotide phosphate (NADPH):quinone oxidoreductase (NQOR) or NADPH-cyt P450 oxidoreductase (POR) via two- or one-electron reduction process, respectively, thus promulgating the production of ROS through what is known as a *redox cycling reaction (Fig. 2.6)*.[16,17]

Proton-radical reactions of $O_2^{\bullet -}$ with some thiols such as penicillamine or cysteine was also observed to form the oxidized disulfide (RSSR) and H_2O_2, according to Eqs. (2.14) and (2.15). However, with thiols such as GSH, cysteamine, *N*-acetylcysteine, and dithiothreitol (DTT), lower levels of H_2O_2 were formed along with the disulfide.[18]

$$O_2^{\bullet -} + R\text{-}SH + H^+ \rightarrow RS^{\bullet} + H_2O_2 \quad (2.14)$$

$$RS^{\bullet} + RS^{\bullet} \rightarrow RSSR \quad (2.15)$$

Finally, $O_2^{\bullet -}$ reacts with a [Fe–S] cluster in systems such as hydrolyases, mitochondrial aconitase, or ferredoxin-$NADP^+$ reductase, resulting in their degradation

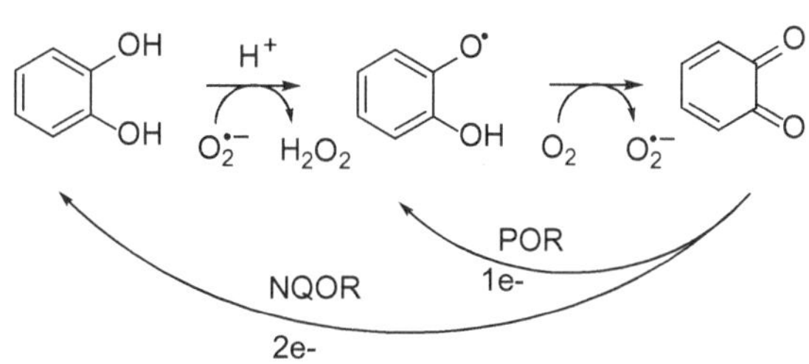

Figure 2.6 Redox cycling reaction of catechol with superoxide radical and oxidoreductases.

through the formation of unstable oxidation states of the [Fe–S] cluster, according to Eqs. (2.16) and (2.17), with rate constants $k = 10^6 - 10^7\ M^{-1}\ s^{-1}$.[19–21]

$$[4Fe\text{-}4S]^{2+} + O_2^{\bullet -} + 2H^+ \rightarrow [3Fe\text{-}4S]^{1+} + H_2O_2 + Fe^{2+} \tag{2.16}$$

$$[4Fe\text{-}4S]^{2+} + 2/3\ O_2^{\bullet -} + 8/3H^+ \rightarrow [3Fe\text{-}4S]^{1+} + 4/3H_2O_2 + Fe^{3+} \tag{2.17}$$

2.3.1.3 Hydroperoxyl/Peroxyl Radical ($HO_2^\bullet/RO_2^\bullet$)

The hydroperoxyl radical is the conjugate acid or protonated form of $O_2^{\bullet -}$ with a pK_a of 4.88 [Eq. (2.18)][22]:

$$HO^\bullet_{2(aq)} \overset{4.88}{\rightleftharpoons} O^{\bullet -}_{2(aq)} + H^+_{(aq)} \tag{2.18}$$

Hence, $O_2^{\bullet -}$ is mostly present in the unprotonated form at physiological pHs where only less than 0.3% of the $HO_2^\bullet$ was suggested to be present in the cytosol,[23] making the reactivity of $HO_2^\bullet$ mostly relevant to slightly acidic pHs such as during conditions of cellular acidosis. Although $HO_2^\bullet$ could be formed from the addition of a hydrogen atom to molecular oxygen [Eq. (2.19)], other mechanisms of $HO_2^\bullet$ formation in biological systems are more plausible—For example, the dehydrogenation of H_2O_2 by $HO^\bullet$ [Eq. (2.20)] or the oxidation of H_2O_2 by Fe(III) [Eqs. (2.21) and (2.22)].

$$O_2 + H^\bullet \rightarrow HO_2^\bullet \tag{2.19}$$

$$H_2O_2 + HO^\bullet \rightarrow HO_2^\bullet + H_2O \tag{2.20}$$

$$Fe^{3+} + H_2O_2 \rightleftharpoons FeOOH^{2+} + H^+ \tag{2.21}$$

$$FeOOH^{2+} \rightarrow Fe^{2+} + HO_2^\bullet \tag{2.22}$$

Due to protonation, $HO_2^\bullet$ has lower symmetry than $O_2^{\bullet -}$ making the electronic distribution in the former unequal between the two oxygen atoms where the spin density on the unprotonated oxygen atom is higher (73%) compared to the protonated one (27%). The implication of this electronic distribution on $HO_2^\bullet$ reactivity is significant compared to $O_2^{\bullet -}$ where the $HO_2^\bullet$ has been shown to be a stronger oxidizer than $O_2^{\bullet -}$ with $E^{\circ\prime} = 1.06$ V and 0.94 V, respectively.[24] Due to the less polar nature of $HO_2^\bullet$, it can penetrate the lipid bilayers, and its ability to abstract hydrogen atom of the methylene C–H in PUFA, $HO_2^\bullet$ plays a key role in the initiation of lipid peroxidation.

$HO_2^{\bullet}$ is capable not only of abstracting the hydrogen atom from a C–H bond, but also of abstracting a phenolic H-atom in alpha-tocopherol,[25] and this serves as a basis for the antioxidative property of alpha-tocopherol [Eq. (2.23)].

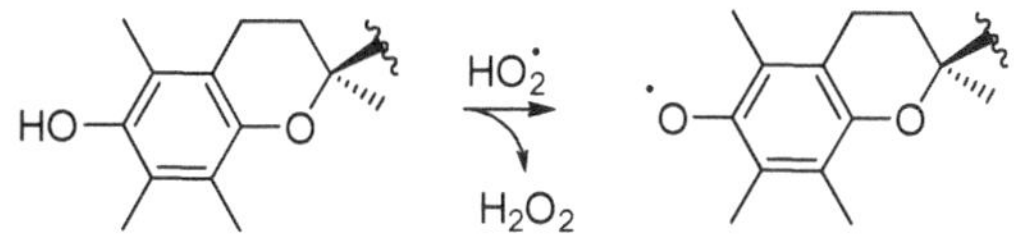

(2.23)

One relevant biological reaction of H-atom abstraction is in the termination of lipid peroxidation. The presence of antioxidants such as alpha-tocopherol or NO convert the reactive lipid peroxyl ($LOO^{\bullet}$) to the lipid hydroperoxide (LOOH) or the

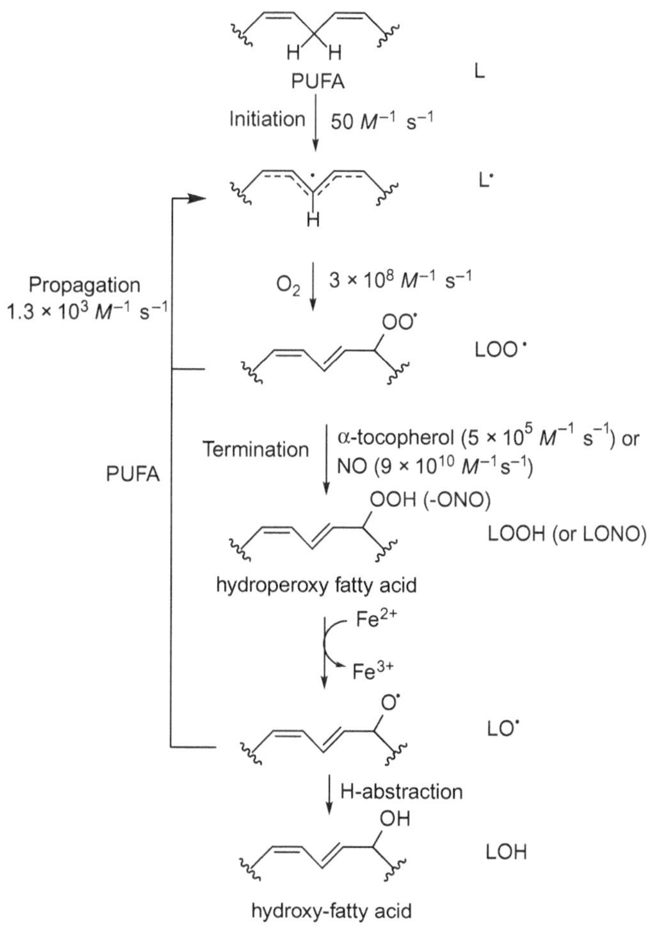

Figure 2.7 Classical auto-oxidation of PUFA or lipid peroxidation via free radical chain reaction involving initiation, propagation, and termination. *Adapted from Gardner, H. W., Free Radic. Biol. Med. 1989,* ***7*** *(1), 65–86.*

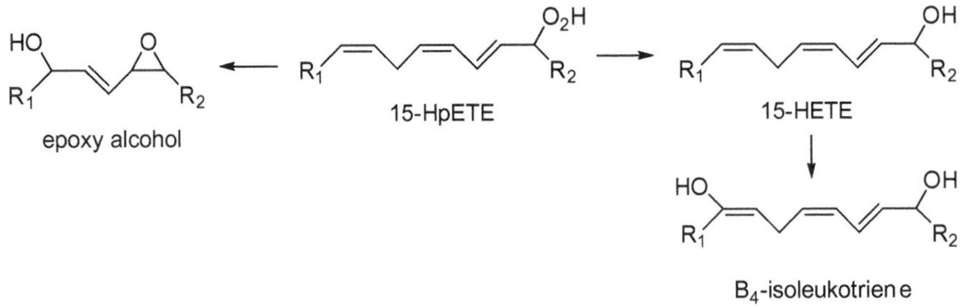

Figure 2.8 Decomposition of PUFA hydroperoxide to leukotriene via the lypoxygenase pathway.

nitritolipid (LONO) for NO (Fig. 2.7). However, LOOH can be converted to an alkoxyl radical ($LO^{\bullet}$) in the presence of a metal ion, which can further abstract an H atom from a PUFA, thereby catalyzing the peroxidation process, according to Fig. 2.7. Therefore, agents that can reduce Fe(III) back to Fe(II) also accelerate lipid peroxidation such as ascorbate (at low concentrations; at high concentrations it inhibits peroxidation), thiols, NADPH, catecholamines, and peroxynitrite.[26]

Not only that fatty acid hydroperoxides are generated from the chemical initiation of lipid peroxidation but they are also formed in great abundance by various lypoxygenases (LOs), which is one of the major pathways (i.e., LO pathway) for the production of PUFA hydroperoxides and hydroxides from arachidonic acid. Examples of these PUFA hydroperoxides and hydroxides are hydroperoxy eicosatetraenoic acid (15-HpETE) and hydroxy eicosatetraenoic acid (15-HETE), respectively, as shown in Fig. 2.8. Further transformation of the hydroxyl fatty acid could form isoleukotrine, or an internal rearrangement of the alcohol formed from the hydroperoxyl intermediate could form epoxy alcohol (Fig. 2.8). These leukotriene metabolites of the lypoxygenase-mediated peroxidation pathway are important chemotactic factors during acute inflammation and as potent vasoconstrictors in smooth muscles.[28]

Aside from hydroperoxides, endoperoxides are also formed as products of lipid peroxidation from another important metabolic pathway, the cyclooxygenase (COX) pathway to form optically pure prostaglandin (PG) metabolites[29,30] or via auto-oxidation of arachidonic acid to form racemic mixtures of isoprostanes (IsoPs) that are structurally similar to PGs.[31] One representative pathway is the auto-oxidation of arachidonic acid, which leads to the formation of G_2-isoprostanes (G_2-IsoPs), which are similar to the hydroperoxy endoperoxide prostaglandin (PGG_2) produced from the COX pathway. The G_2-IsoP is reduced to the hydroxyl analog (H_2-IsoP), and its complete reduction generates 64 regioisomers and stereoisomers of F_2-IsoPs (Fig. 2.9). Since IsoP formation is mediated by free radicals and they impart potent biological activity, measurements of F_2-IsoP concentrations in plasma and urine have been the gold standard for monitoring oxidative stress in vivo.[32]

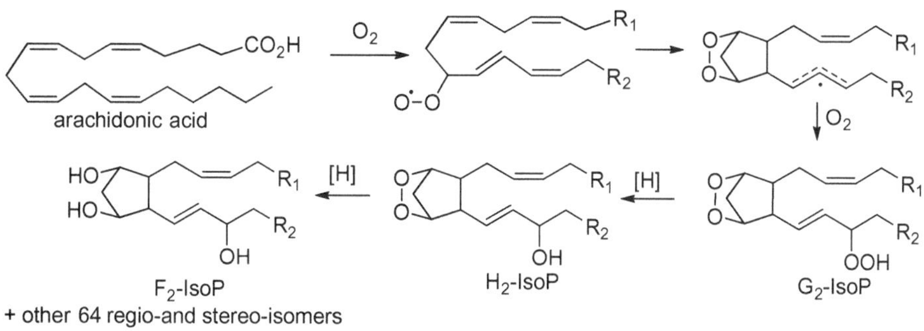

Figure 2.9 Free radical–mediated auto-oxidation of arachidonic acid to yield isoprostane.

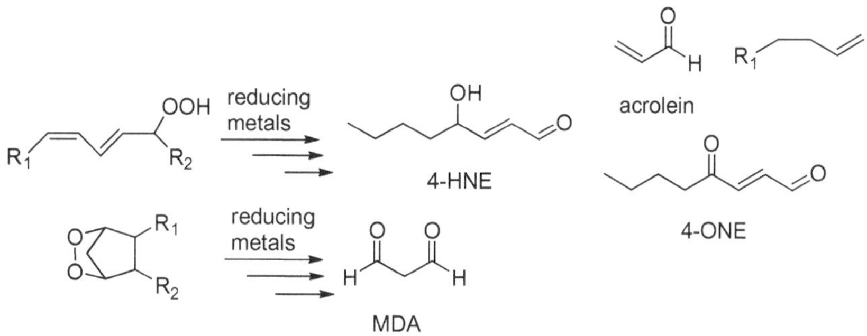

Figure 2.10 Lipid peroxidation-derived aldehydes.

In the presence of reducing metal ions, fatty acid hydroperoxides or endoperoxides ultimately lead to fragmentation of the acyl chain and the formation of reactive by-products such as malondialdehyde (MDA), acrolein, and the α,β-unsaturated aldehydes, 4-hydroxy-2-nonenal (4-HNE), and 4-oxononenal (4-ONE) all possess biological activity (Fig. 2.10).[33,34] These aldehydic electrophiles react with nucleophiles such as the cysteine thiol or, to a lesser extent, lysine and histidine to form a protein–carbonyl adduct via Michael addition, according to Eq. (2.24).[35] Protein carbonyls are well-studied and detectable using various analytical techniques. The formation of these adducts also has biological implications such as in-cell signaling, enzyme inactivation, and cell death, which are collectively known as electrophilic stress.[36]

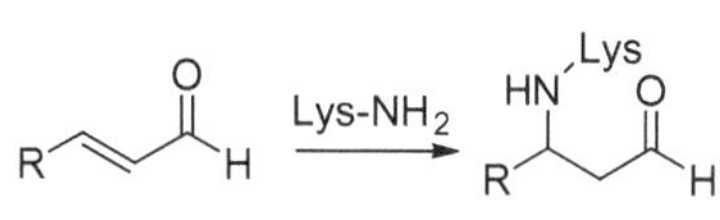

(2.24)

Cholesterol oxidation could yield products that are cytotoxic and mutagenic. The initial formation of cholesterol hydroperoxide occurs through H-atom abstraction at the C-7 position by $LOO^{\bullet}$ to form the cholesteryl radical, which rapidly reacts with O_2 to give the 7α- or 7β-cholesteryl peroxyl radicals and ultimately the cholesteryl hydroperoxides

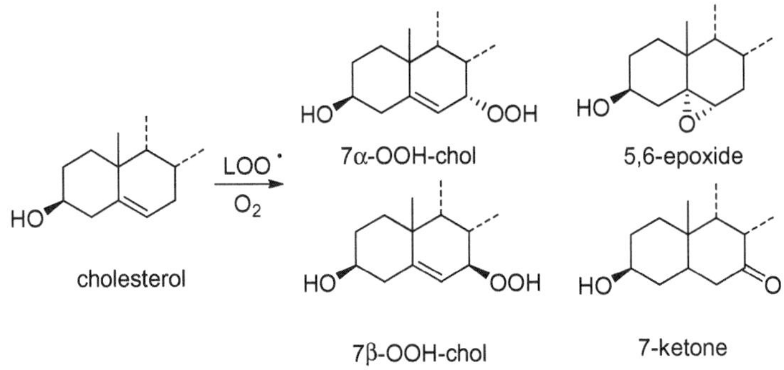

Figure 2.11 Free radical–mediated oxidation of cholesterol.

7α- or 7β-OOH-chol after subsequent abstraction of an H atom from allylic hydrogen of another PUFA (Fig. 2.11). Lipid peroxyl radicals ($LOO^{\bullet}$) can also attack the cholesterol double bond to give the epimeric 5,6-epoxides; or, in the presence of $LOO^{\bullet}$, 7α-/7β-oxyl radical can undergo β-hydrogen scission to give 7-ketone.[37]

2.3.1.4 Hydrogen Peroxide (H_2O_2)[38]

Hydrogen peroxide may be the most ubiquitous ROS present in biological systems due to its higher stability compared to other ROS such as $O_2^{\bullet-}$, $HO_2^{\bullet}$, and $HO^{\bullet}$ and reduction potential of $E^{o\prime} = 0.32$ V.[24] Hydrogen peroxide is formed from self-dismutation of $O_2^{\bullet-}$, enzymatic dismutation of $O_2^{\bullet-}$ by SOD, through H-atom abstraction by $HO_2^{\bullet}$, or via the radical coupling reaction of two $HO^{\bullet}$. At least three major reactions of H_2O_2 are implicated in oxidative modification: (1) formation of $HO^{\bullet}$ in the presence of low-valence transition metal ions such as iron, copper, molybdenum, or vanadium via a Fenton-type reaction[39] [Eq. (2.25)] but not from a metal-catalyzed Haber–Weiss reaction [Eqs. (2.26) and (2.27)] because toxicity arises from the Fenton reaction, which is the initiating process for $O_2^{\bullet-}$ production[40] [Eq. (2.26)]; (2) enzymatic formation of hypochlorous acid (HOCl) from myeloperoxidase and Cl^- [41] [Eq. (2.28)] or the ability to oxidize various substrates with other peroxidases[42] [Eq. (2.29)]; (3) oxidation of thiol resulting in the formation of disulfide[Eqs. (2.30) and (2.31)] that exhibits very low rate constants in the range of $k = 0.2–3.0\ M^{-1}\ s^{-1}$ for thiols such as glutathione (GSH), cysteine, *N*-acetylcysteine, or thioredoxin[43] but is much more efficient with peroxiredoxin cysteine residues of $k = 1.3 \times 10^7\ M^{-1}\ s^{-1}$ [44]. Fig. 2.12 shows the relative distances that each RS could diffuse to in the presence of GSH.[2] Note that the high reactivities of $HO^{\bullet}$, HOCl, and NO_2 (which will be discussed in the succeeding sections)make these RS the least diffusible RS in cellular systems by only diffusing at a short distance (compared to H_2O_2 or $ONOO^-$). In the presence of peroxiredoxin 2 (Prx2) , the diffusibility of H_2O_2 is lesser than H_2O_2 alone due to the high reactivity of H_2O_2 with Prx2.

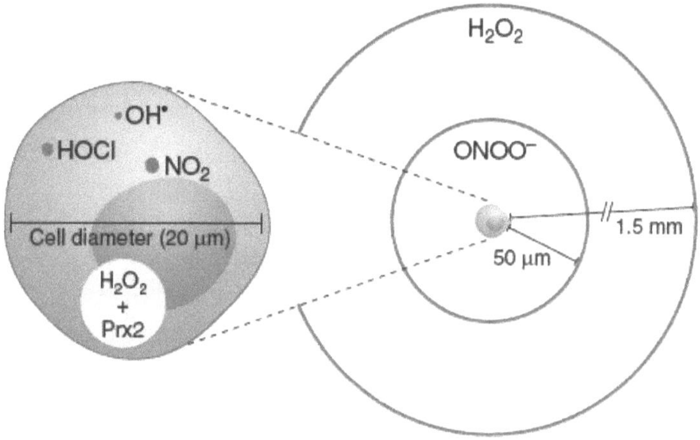

Figure 2.12 Relative diffusion distances (circles) of $HO^{\bullet}$, HOCl and NO_2 in the presence of 2 mM GSH relative to a cell with a diameter of 20 μm. The presence of Prx2 makes the H_2O_2 less diffusible compared to H_2O_2 alone of 1.5 mm. *Reprinted and adapted by permission from Macmillan Publishers Ltd: Winterbourn CC. Reconciling the chemistry and biology of reactive oxygen species,* Nat. Chem. Biol. *copyright © 2008;**4**:278–86.*

$$Fe^{2+} + H_2O_2 \rightarrow Fe^{3+} + HO^- + HO^{\bullet} \quad (2.25)$$

$$HO^{\bullet} + H_2O_2 \rightarrow H_2O + O_2^{\bullet-} + H^+ \quad (2.26)$$

$$O_2^{\bullet-} + H_2O_2 \rightarrow O_2 + HO^- + HO^{\bullet} \quad (2.27)$$

$$H_2O_2 + Cl^- \xrightarrow{\text{MPO}} HOCl + H_2O \quad (2.28)$$

$$AH_2 + H_2O_2 \xrightarrow{\text{peroxidases}} AH^{\bullet} + H_2O \quad (2.29)$$

$$RS^- + H_2O_2 \rightarrow RSOH + HO^- \quad (2.30)$$

$$RSOH + RSH \rightarrow RSSR + H_2O \quad (2.31)$$

2.3.1.5 Hydroxyl Radical ($HO^{\bullet}$)

The hydroxyl radical is one of the most reactive species known to form in a biological system, having an extremely short half-life of 10^{-9} s in aqueous solution compared to 10^{-6} s and 7 s for $RO^{\bullet}$ and $ROO^{\bullet}$, respectively.[45] A hydroxyl radical can be generated through various means such as chemically by the Fenton reaction or physically by photolysis or sonication of peroxides (e.g., ROOH or H_2O_2). The standard reduction potential for the $HO^{\bullet}_{aq}/HO^{-}_{aq}$ couple was determined to be $E^{\circ\prime} = 1.77$ V in neutral solution.[46] This high reactivity of $HO^{\bullet}$ translates into its indiscriminate reactivity with practically almost every biomolecule, including in (1) H-atom abstraction, (2) electrophilic addition, and (3) radical–radical, anion, metal cation reactions to name the most relevant ones.

For alcohols, H-atom abstraction by $HO^{\bullet}$ gives preference at the alpha–C–H bond but not at the O–H.[47] This is evident in the oxidation of glucose as initiated by $HO^{\bullet}$ abstraction of alpha-hydrogen,[48] which leads to oxidative degradation to malondialdehyde (Fig. 2.13).[49] With nucleotides, 5′-H atom abstraction in deoxyribose moiety (Fig. 2.14) or nucleic bases could lead to DNA cleavage.[50,51] With lipids, the abstraction of the hydrogen atom from allylic hydrogen adjacent to a double bond commonly present in a PUFA is energetically more favorable than the abstraction of an H atom adjacent to a single bond in saturated fatty acids, hence the rational for the susceptibility of PUFA to undergo lipid peroxidation (Fig. 2.15).[26]

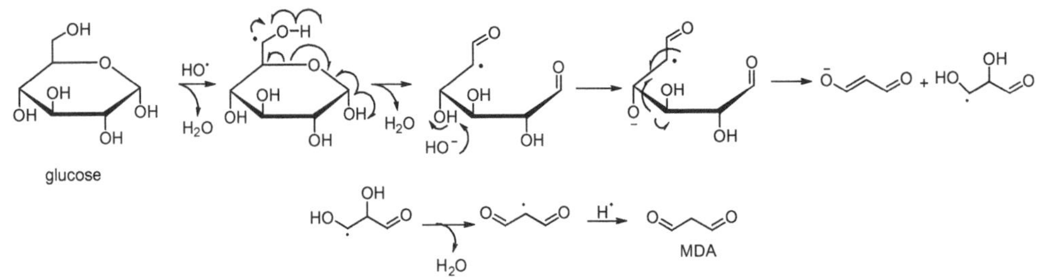

Figure 2.13 Degradation of glucose via hydroxyl radical abstraction of alpha-H to yield malondialdehyde.

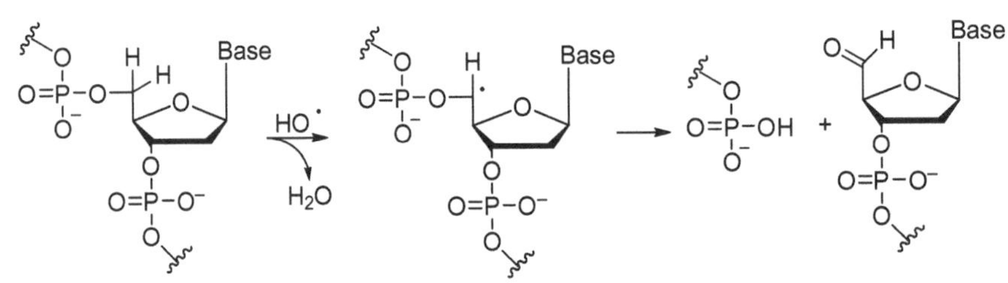

Figure 2.14 DNA cleavage via alpha-H abstraction by hydroxyl radical of the deoxyribose moiety.

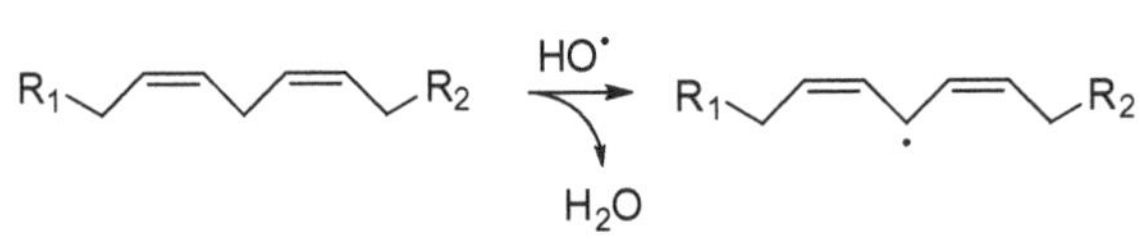

Figure 2.15 Allylic-H abstraction by hydroxyl radical in PUFA in the initiation of lipid peroxidation.

Electrophilic addition of HO• to the aromatic ring of Tyr, Phe, or Trp as well as the *N*-heterocyclic moiety of histidine (His) forms a variety of oxidation products such as hydroxylation in the case of the aromatic amino acids and formation of a urea functionality in the form of 2-oxohistidine. The formation of 3,3′-dityrosine is evident of H-atom abstraction to form the C-centered tyrosyl radicals and their subsequent bimolecular coupling (Fig. 2.16).[52] Oxidation of other nonaromatic amino

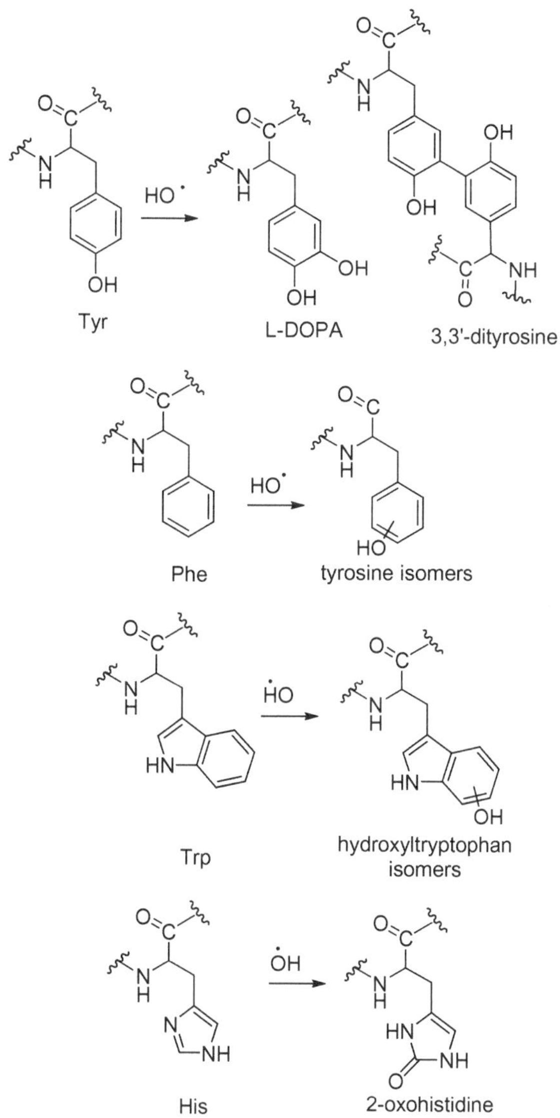

Figure 2.16 Hydroxyl radical–mediated oxidation of various amino acid side chains.

acids leads to the formation of reactive aldehydes from lysine, arginine, and proline; ketone from threonine; peptide backbone cleavage in the case of proline oxidation; and acrylamide formation from aspargine.[52] A comprehensive survey of oxidation products of various amino acids is shown elsewhere.[53]

The main cause of oxidative damage to DNA is through the generation of $HO^{\bullet}$ from H_2O_2 because the latter is more diffusible to the nucleus. Reaction of $HO^{\bullet}$ to guanine and thymine is the most common oxidative DNA damage due to the high oxidation potential of these nucleobases relative to other bases. The mechanisms of DNA oxidative damage by $HO^{\bullet}$ could proceed along two radical pathways as highlighted in Fig. 2.17. The first is through abstraction of an H-atom from the sugar moiety,[54] at the 4′ and 5′ position or at the methyl group of thymineto subsequently form the hydroxymethylated nucleobase.[55] The second mode of radical reaction is via addition to the electron-rich moieties of the purine and pyrimidine bases[56]. Fig. 2.18 shows the various $HO^{\bullet}$-mediated oxidation product of nucleobases. The $HO^{\bullet}$ addition to thymidine at the 5,6-double bond yields dihydroxythymine[57] while addition to the C-8 position of the purine ring of guanosine (dG) or adenosine (dA) forms 8-hydroxyguanosine (8-OH-dG) or 8-hydroxyadenosine (8-OH-dA), respectively. Further oxidation of 8-OH-dG yields 8-oxo-guanosine (8-oxo-dG), which is the most studied DNA lesion. Reduction could also occur for 8-OH-dG and result in ring opening and formation of 2,6-diamino-4-hydroxy-5-formamidopyrimidine (FapyGua), which could lead to base-pairing mismatch and adduction to a protein amino side chain.[58] A comprehensive survey of oxidized DNA base products is described elsewhere.[59]

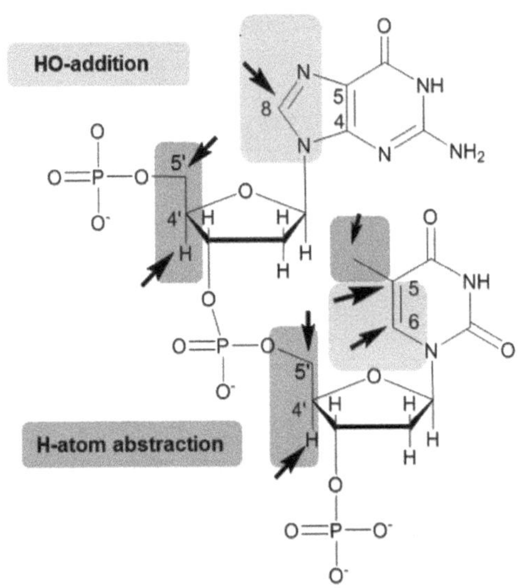

Figure 2.17 DNA sites of hydroxyl radical addition and H-atom abstraction reactions.

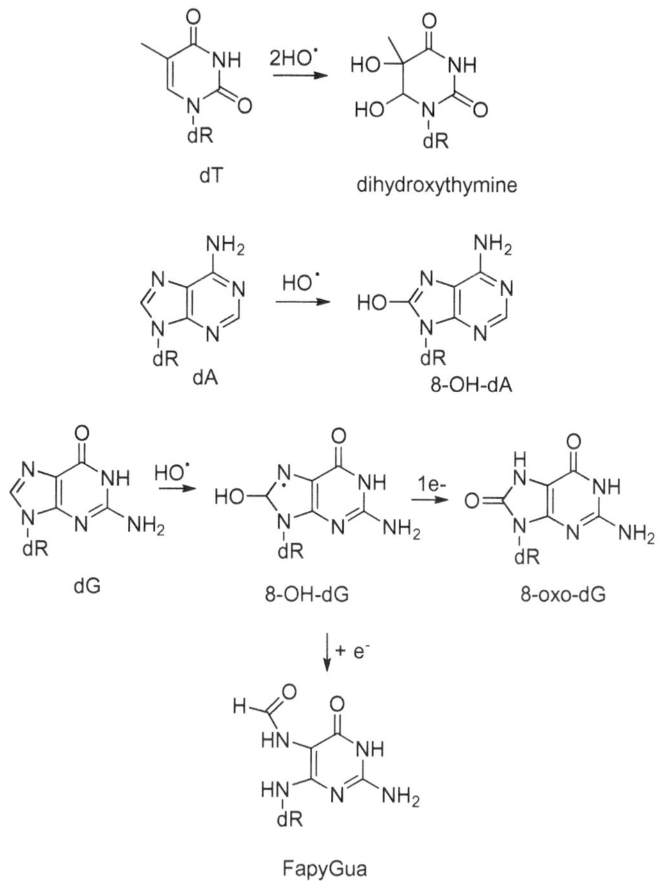

Figure 2.18 DNA sites of hydroxyl radical addition and H-atom abstraction reactions.

Oxidation of the DNA's sugar moiety could be extremely complex and involve H-atom abstraction from various C atoms of the molecules (Fig. 2.19). One pathway is the 4′ H-atom abstraction, which involves formation of an oxidized abasic site where the nucleotide is still connected but without the base and with the sugar ring open. The other pathway was recently shown to consist of two distinct reactions and, depending on the oxidizing agent, results in strand breakage. The first one consists of 3′-phosphoglycolate residue and 5′-phosphate with the base still attached, while the other pathway results in the formation of variable products that involve concomitant release of either base propenal or malondialdehyde and a free nucleobase. Similar to that of the lipid peroxidation carbonyl products, the aldehyde products from the strand breaking pathway are susceptible to nucleophilic attack by a lysine amino side chain to form DNA−protein adducts (i.e., protein−NH = C−DNA).[60,61]

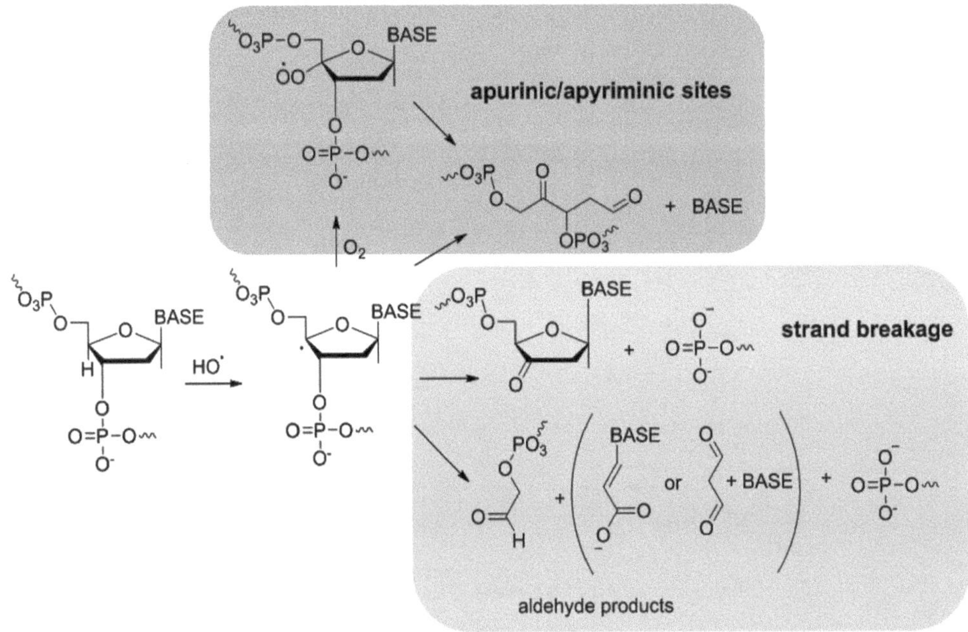

Figure 2.19 Formation of abasic sites and strand breakage on H-atom abstraction of the sugar moiety by a hydroxyl radical.

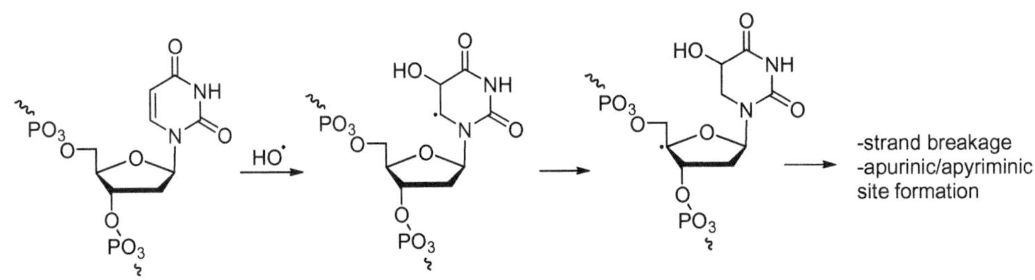

Figure 2.20 H-atom transfer from deoxyribose to the base leading to DNA damage.

Note that deoxyribose can also be a hydrogen atom donor—In this case, to the uracil radical to form the 4′-deoxyribosyl radical. Similar to the preceding, a sugar radical can undergo either strand breakage or the formation of abasic sites, demonstrating that damage to the base could be transferred to a sugar (Fig. 2.20).[62]

Finally, with small molecules, the role of $HO^{\bullet}$ is mainly as an oxidizing agent exhibiting high reactivity toward other ROS (e.g., $O_2^{\bullet -}$ and $HO_2^{\bullet}$), reactive nitrogen species (RNS) (e.g., NO and NO_2), anions (e.g., Cl^- and CO_3^{2-}), and metal ions (e.g., Fe^{2+} and Cu^{2+}) with rate constants approaching a diffusion-controlled limit to form their respective oxidized products according to Eqs. (2.32)–(2.39).[63]

$$HO^{\bullet} + O_2^{\bullet-} \rightarrow HO^- + O_2 \quad k = 7 \times 10^9 \tag{2.32}$$

$$HO^{\bullet} + HO_2^{\bullet} \rightarrow H_2O + O_2 \quad k = 6.6 \times 10^9 \tag{2.33}$$

$$HO^{\bullet} + NO \rightarrow H^+ + NO_2^- \quad k = 1 \times 10^{10} \tag{2.34}$$

$$HO^{\bullet} + NO_2 \rightarrow HO_2NO \quad k = 1 \times 10^{10} \tag{2.35}$$

$$HO^{\bullet} + Cl^- \rightarrow ClHO^{\bullet-} \quad k = 4.3 \times 10^9 \tag{2.36}$$

$$HO^{\bullet} + CO_3^{2-} \rightarrow HO^- + CO_3^{\bullet-} \quad k = 3.7 \times 10^8 \text{ (pH 11)} \tag{2.37}$$

$$HO^{\bullet} + Fe^{2+} \rightarrow FeOH^{2+} \quad k = 3.2 \times 10^8 \tag{2.38}$$

$$HO^{\bullet} + Cu^{2+} \rightarrow CuOH^{2+} \quad k = 3.5 \times 10^8 \tag{2.39}$$

2.3.1.6 Singlet Oxygen ($^1O_2$)

Singlet oxygen ($^1\Delta_g$) is the high-energy state of $^3O_2$ with an estimated energy separation of 22.5 kcal/mol or 7882 cm^{-1}. Although diamagnetic in nature, singlet oxygen has a short lifetime in H_2O of $\tau = 2$ μs.[64] There are actually two low-lying excited states for $^1O_2$, where the $^1\Delta_g$ is the first excited state and the second excited state is denoted by $^1\Sigma_g^+$, where the $^1\Sigma_g^+$ state has a shorter radiative lifetime in solution compared to $^1\Delta_g$ and has higher energy separation with the $^3O_2$ of 31.5 kcal/mol. Typically, $^1O_2$ is formed via sensitization of $^3O_2$ by light-activated photosensitizers such as rose of bengal, fluorescein, methylene blue, or porphyrins.[65,66] However, $^1O_2$ could also be directly formed from potassium superoxide on dissolution in water.[67] Other nonphotolytic generations of $^1O_2$ include reaction of H_2O_2 with NaOCl,[68] oxidation of lipid hydroperoxides[69] and their decomposition by cyt *c* and lactoperoxidase,[70] oxidation of NADPH by liver microsome[71] and an MPO-H_2O_2-halide system.[72] The cell-killing property of $^1O_2$ is due to its oxidizing property because it is more oxidizing than $^3O_2$ and with higher reactivity than $O_2^{\bullet-}$ but less reactive than $HO^{\bullet}$and is the basis of its therapeutic property in photodynamic therapy applications.

The most common reaction of $^1O_2$ is an addition reaction to double bonds to give a hydroperoxide via endoperoxide formation that is formed through the Diels–Alder mechanism or "-ene" reactions. This is exemplified by $^1O_2$ oxidation of unsaturated bonds in lipids such as a PUFA[73] to form lipid hydroperoxides (Fig. 2.21), amino acids such as

Tyr and Trp to form the hydroperoxide and Met to sulfoxide (Fig. 2.22),[37,74,75] and nucleic acids such as guanosine to produce 8-oxo-dG (Fig. 2.23).[76]

Singlet oxygen undergoes an ene-type reaction, a nonradical pathway for hydroperoxide formation, with cholesterol leading to the formation of 5a-OOH-Chol as the major product and 6α-/6β-OOH-Chol as the minor products (Fig. 2.24).[37]

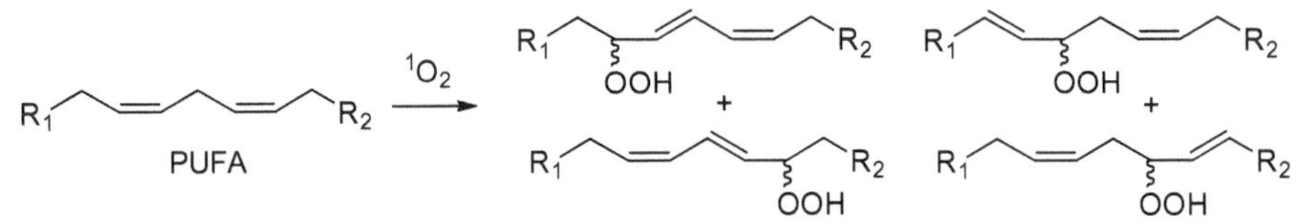

Figure 2.21 Singlet oxygen oxidation of PUFA.

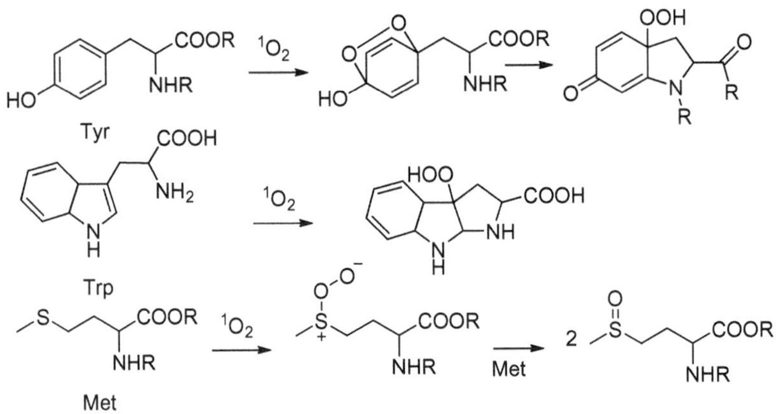

Figure 2.22 Oxidation of various amino acids by singlet oxygen.

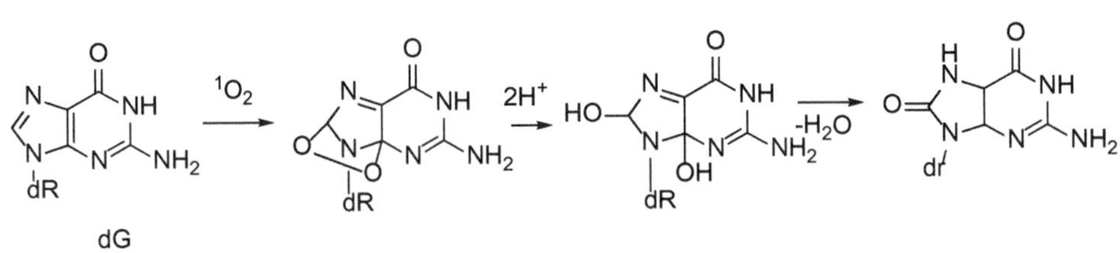

Figure 2.23 DNA base, 2-deoxyguanosine, oxidation by singlet oxygen.

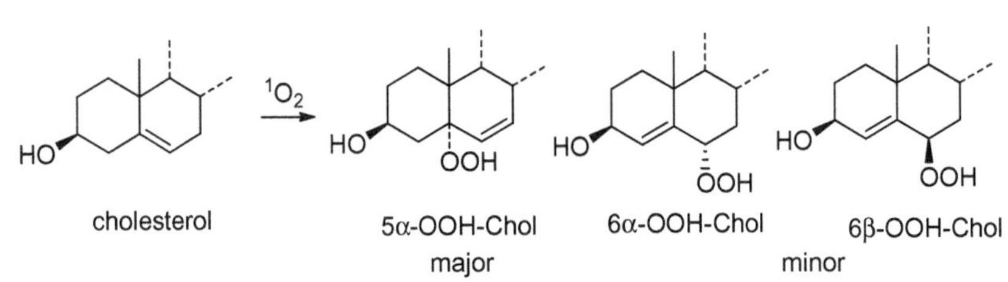

Figure 2.24 Oxidation of cholesterol by singlet oxygen.

2.3.2 Reactive Nitrogen Species

Reactive nitrogen species (RNS) comprises both paramagnetic and diamagnetic reactive intermediates whose nitrogen atoms participate in the redox reaction. In a biological system, RNS are mostly derived from NO where NO ultimately undergoes oxidation to nitrate or reduction to ammonia. Fig. 2.25 shows the various oxidation states of an RNS originating from the oxidation and reduction of NO. These redox products from NO are typically formed enzymatically or from reactions with oxygen, ROS, or metal ions. Two paramagnetic RNS are relevant in the initiation of oxidative damage to key biomolecular systems such as NO and NO_2;and the anionic, peroxynitrite ($ONOO^-$), which is also one of the most studied RNS and will be the subject of this section.

2.3.2.1 Nitric Oxide (NO)

Mainly produced enzymatically from oxidation of L-arginine, NO is one of the most ubiquitous paramagnetic gases in biological systems. Its molecular bonding orbital is characterized by an unpaired electron occupying an antibonding π orbital. As a small gas, NO has quite a few similarities with $^3O_2$ in terms of electronic nature (i.e., both are radicals) and reactivity with metal heme ions. However, there are obvious differences; for example, by virtue of molecular symmetry, NO is more polar than O_2, but their solubility[77] and diffusibility[78] in water and membranes[79] are similar, with NO just slightly higher than O_2. In terms of reactivity, the reduction potential for the $O_2/O_2^{\bullet-}$ couple is −0.16 V compared to the $NO/^3NO^-$ couple of −0.76 V, while reduction of NO to $^1NO^-$ is highly unfavorable with a predicted reduction potential of -1.67 V.[80] Therefore, NO is expected to be less reducible than O_2, which is consistent to that observed in physiological conditions as evidenced by the fact that NO is primarily oxidized by O_2 in biological system to NO_2^-, NO_2, and NO_3^-. Similar to O_2, the origin of NO toxicity is due to the formation of a more reactive species such as NO_2 from the oxidation of NO by O_2 or the formation of $ONOO^-$ from the reaction of NO and $O_2^{\bullet-}$. Due to its radical nature, NO not only reacts with other radicals such as O_2 or $O_2^{\bullet-}$ but also with other transient radicals, making it, in a sense, an efficient radical scavenger. For example, the reaction of NO with the oxidant carbonate radical $CO_3^{\bullet-}$ leads to the formation of NO_2^- and CO_2 with a rate constant of

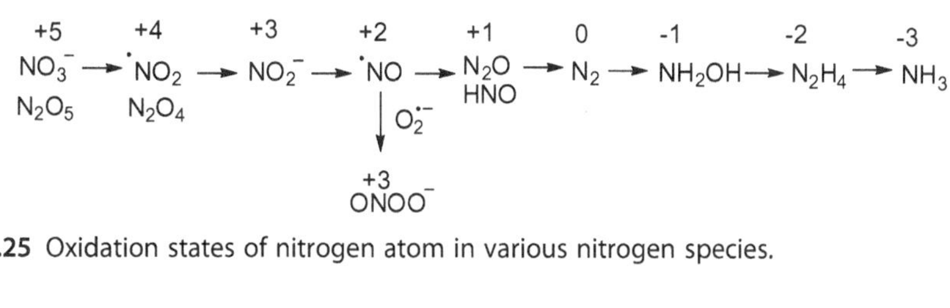

Figure 2.25 Oxidation states of nitrogen atom in various nitrogen species.

$k = 3.5 \times 10^9\ M^{-1}\ s^{-1}$ [Eq. (2.40)].[81] In the termination of lipid peroxidation [Eqs. (2.41)–(2.43)], a lipid peroxyl radical ($LOO^{\bullet}$) reacts with NO at a diffusion-controlled limit to form the lipid peroxynitrite (LOONO)[82,83] which further decomposes to NO_2 and lipid alkoxyl ($LO^{\bullet}$), where $LO^{\bullet}$ further reacts with NO to form the lipid nitrite (LONO), which is a relatively more stable product than the initially formed LOONO.[84]

$$NO^{\bullet} + CO_3^{\bullet-} \rightarrow NO_2^- + CO_2 \quad (k = 3.5 \times 10^9\ M^{-1}\ s^{-1}) \tag{2.40}$$

$$NO^{\bullet} + LOO^{\bullet} \rightarrow LOONO \quad (k = 3.5 \times 10^9\ M^{-1}\ s^{-1}) \tag{2.41}$$

$$LOONO \rightarrow LO^{\bullet} + NO_2^{\bullet} \tag{2.42}$$

$$NO^{\bullet} + LO^{\bullet} \rightarrow LONO \quad (k = 7 \times 10^4\ M^{-1}\ s^{-1}) \tag{2.43}$$

The ability of NO to form metal complexes—For example, with the heme active sites—Plays an important role in the regulation of enzyme function. One classical example of NO enzyme regulation via heme NO complexation is that with guanylate cyclase (sGC), which results in catalytic synthesis of cGMP from guanosine-5'-triphosphate (GTP) in response to calcium levels. In turn, this triggers a series of signal transudation events that ultimately lead to a physiological effect such as smooth muscle relaxation (Fig. 2.26).[85]

Other than NO complexation, NO binding to thiols such as the case of *S*-nitrosylation is also highly relevant in signal-transduction processes that regulate a variety of cellular functions and whose dysregulation has been implicated in the pathogenesis of various diseases.[86] This binding of NO to the reactive cysteine thiol, for example, is a form of reversible post-translational protein modification, similar to phosphorylation, that is capable of regulating enzyme activities via *S*-nitrosylation and denitrosylation processes [Eq. (2.44)]:

$$NO^{\bullet} + RS^{\bullet} \rightleftharpoons RS\text{-}NO \tag{2.44}$$

Metal-catalyzed *S*-nitrosylation is also believed to occur in vivo. Binding of NO to a metal ion can result in either its oxidation to nitrosonium (NO^+) or its reduction to nitroxyl (NO^-), both of which depend on the type of metal ion, oxidation state,

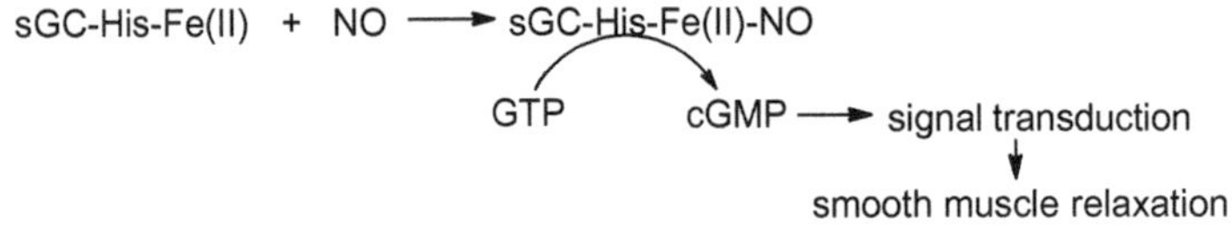

Figure 2.26 Nitric oxide binding to sGC heme resulting in physiological effect.

coordination number, and type of ligands. Due to the repulsive effect, NO^- is a poor nitrosylating agent while NO^+ has been shown to effectively *S*-nitrosylate solvent-accessible thiols.[87] Although NO^+'s lifetime in water is 3×10^{-10} s and is only present in highly acidic pH, *S*-nitrosylation by NO^+ is believed to occur in biological systems through transnitrosylation.[88] For example, dinitrosyl iron complexes (DNICs) coordinated by cysteine residues (Fig. 2.27) or by GSH are formed in biological systems from the reaction of NO with Fe–S cluster proteins, which has been shown to *S*-nitrosate bovine serum albumin (BSA) in the presence of excess low-mass thiols but is independent of trans-*S*-nitrosation reaction.[89] Others had shown RSNO formation from free NO via transnitrosation with DNIC in the cell.[90]

2.3.2.2 Nitrogen Dioxide (NO_2)[91]

Nitrogen Dioxide (NO_2)[91] Nitrogen dioxide is a radical-based RNS that is mainly derived from the reaction of NO with O_2 at a relatively fast rate, according to Eq. (2.45).[91] It is a powerful oxidizer with reduction potentials ranging from $E^{\circ\prime}$ $(NO_2/NO_2^-) = 0.6 - 0.9$ V.[92]

$$2NO + O_2 \rightarrow 2NO_2 \qquad (2.45)$$

NO's chemistry with O_2 is highly complex but ultimately leads to the formation of nitrate and nitrite via hydrolysis of various nitrogen oxides such as N_2O_3 and N_2O_4 as illustrated in Fig. 2.28.[13]

As mentioned previously, NO_2 is also formed from the decomposition of either organic peroxynitrites (RONOO) or peroxynitrous acid (ONOOH) through

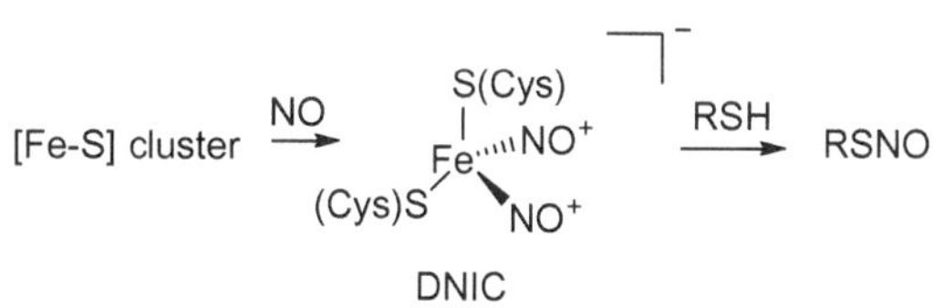

Figure 2.27 Trans-*S*-nitrosation of thiols by dinitrosyl iron complexes.

$$3NO_2 + H_2O \longrightarrow 2H^+ + 2NO_3^- + NO$$
$$2NO_2 \longrightarrow N_2O_4$$
$$N_2O_4 + H_2O \longrightarrow HNO_3 + HNO_2$$
$$2NO_2 + NO \longrightarrow N_2O_3$$
$$N_2O_3 + H_2O \longrightarrow 2HNO_2$$
$$3HNO_2 \longrightarrow HNO_3 + 2NO + H_2O$$

Figure 2.28 Oxidized congeners of nitrogen dioxide.

homolytic cleavage of the N–O bond to also form the $RO^{\bullet}$ and $HO^{\bullet}$, respectively, according to Eqs. (2.46) and (2.47):

$$RONOO \rightarrow RO^{\bullet} + NO_2 \tag{2.46}$$

$$HONOO \rightarrow HO^{\bullet} + NO_2 \tag{2.47}$$

The most relevant reaction of NO_2 in a biological system is nitration reaction, which occurs via a radical–radical reaction such as in the case of nitrotyrosine unit formation in peptides and proteins.[93] Tyrosine nitration proceeds through initial formation of tyrosyl radical, which is formed by oxidation of tyrosine by hydroxyl or carbonate radical, according to Fig. 2.29. A radical–radical coupling reaction then follows in which NO_2 adds to the C-centered tautomer of the phenoxyl radical. A similar nitration reaction of aromatic amino acids such as Phe and Trp was also observed as shown in Fig. 2.29. Similar to lipid hydroperoxide formation from O_2, radical–radical reaction was proposed for the nitration of a PUFA (linoleic acid), which involves initial H-atom abstraction and subsequent NO_2 addition to the PUFA radical (Fig. 2.30).[94]

Due to the electron-withdrawing nature of the nitro group, nitroalkenes formed from lipid nitration can undergo a Michael addition reaction with nucleophiles such

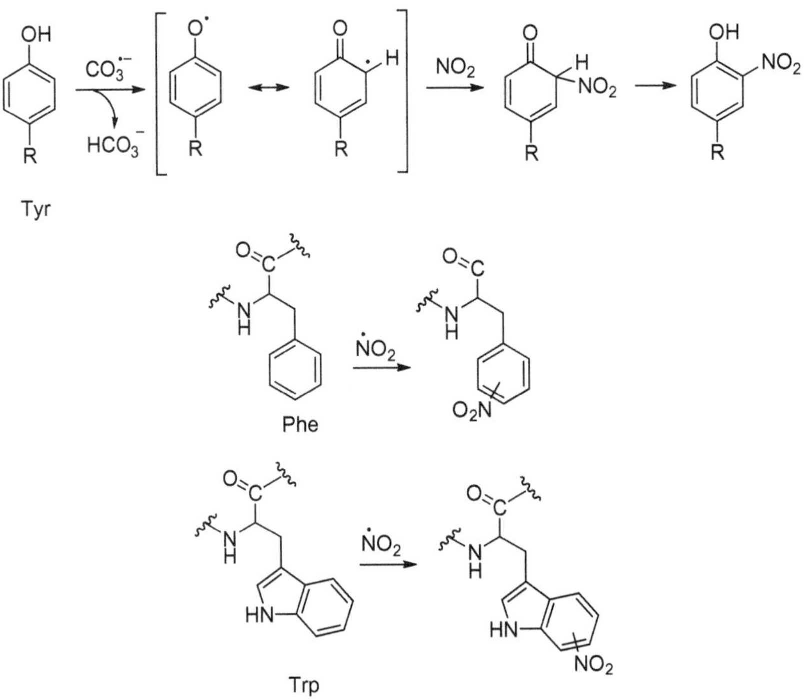

Figure 2.29 Nitration of aromatic amino acids by nitrogen dioxide.

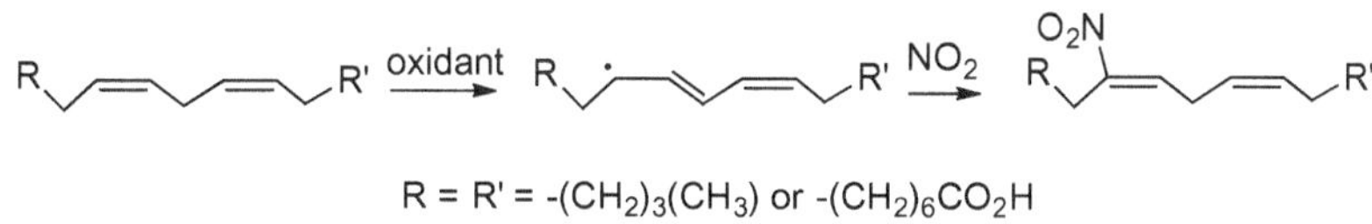

Figure 2.30 Lipid nitration by nitrogen dioxide addition to linoleic acid radical.

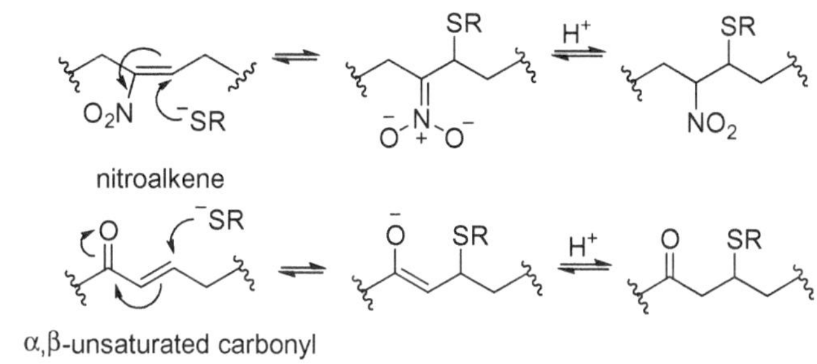

Figure 2.31 Michael addition reaction of nitroalkene with thiolate to form the nitro–fatty acid–protein adducts. Also shown is the analogous reaction with α,β-unsaturated carbonyl.

as thiols, which is analogous to the Michael addition reaction exhibited by α,β-unsaturated carbonyl with thiolates (Fig. 2.31). Carbonyl and nitro–fatty acid adducts of proteins and glutathione are labile and reversible. These nitro–fatty acids adducts were identified in healthy human blood and were shown to be responsible for some reversible post-translational regulation of protein functions such as signaling, which could lead to the induction of gene expression.[95]

Aside from the radical–radical reaction of thiyl radical with NO as previously mentioned, another route for *S*-nitrosylation is the oxidative pathway that involves N_2O_3 formed from NO_2 and NO. The N_2O_3 produced from activated macrophage was observed to undergo *S*-nitrosylation with glutathione at a rate constant of $4.4 \times 10^7\ M^{-1}\ s^{-1}$ (Eq. 2.48). Although N_2O_3 also reacts with water at a rate of $k = 1600\ M^{-1}\ s^{-1}$ [Eq. 2.49), the hydrolysis is considerably slower than in *S*-nitrosylation, so the formation of *S*-nitrosothiol from N_2O_3 is a relevant mechanism under nitrosative stress conditions.[96]

$$N_2O_3 + GSH \rightarrow NO_2^- + GSNO \tag{2.48}$$

$$N_2O_3 + H_2O \rightarrow 2NO_2^- + H^+ \tag{2.49}$$

2.3.2.3 Peroxynitrite ($ONOO^-$)[97]

Production of peroxynitrite–peroxynitrous acid ($ONOO^-/ONOOH$)[97] results from the radical–radical coupling reaction of NO and $O_2^{\bullet-}$,[98] or $HO^\bullet$ and NO_2,[99] where both pathways proceed at a diffusion-controlled limit according to Eqs. (2.50) and (2.51), respectively:

$$NO + O_2^{\bullet -} \xrightarrow{7\times 10^9\ -1\times 10^{10}\ M^{-1}\ s^{-1}} ONOO^- \qquad (2.50)$$

$$HO^{\bullet} + NO_2^{\bullet} \underset{10^{-8}-10^{-6}\ s^{-1}}{\overset{5\times 10^9\ M^{-1}\ s^{-1}}{\rightleftharpoons}} ONOOH \xrightarrow{1.2\ s^{-1}} HNO_3 \qquad (2.51)$$

Peroxynitrite could also be formed from the reaction of NO_2^- and H_2O_2, according to Eq. (2.52), which is a common method for peroxynitrite synthesis.[100,101]

$$H_2O_2 + NO_2^- \rightarrow ONOO^- + H_2O \qquad (2.52)$$

The conjugate acid of $ONOO^-$, ONOOH, has a pK_a of 6.6[102,103] and therefore in physiological pH, a significant amount of $ONOO^-$ is protonated to form ONOOH, which is known to undergo ONO–OH cleavage to form NO_2 and $HO^{\bullet}$. Hence, $ONOO^-$ is a vital source of NO_2 and $HO^{\bullet}$, known oxidizing and nitrating agents, respectively, as discussed in previous sections. Other than the reactions originating from NO_2 and $HO^{\bullet}$, peroxynitrite is a strong nucleophile, one capable of reacting with CO_2 to form the nitrosoperoxycarbonate ($ONOOCO_2^-$),[104] which is a two-electron oxidant that further decomposes to form $CO_3^{\bullet -}$ and NO_2, according to Eq. (2.53), and has been implicated in a variety of biochemical oxidative modification.[105–108]

$$ONOO^- + CO_2 \xrightarrow{\sim 10^4\ M^{-1}\ s^{-1}} ONO - OCO_2^- \rightarrow NO_2 + CO_3^{\bullet -} \qquad (2.53)$$

The homolytic cleavage of ONO–OX (X = H or CO_2, in the case of peroxynitrous acid and nitrosoperoxycarbonate) also occurs when for example $ONOO^-$ reacts with oxidized metal centers as in the case of myeloperoxidase resulting in the formation of NO_2 and metal–oxo complexes as shown in Eq. (2.54).[109,110]

$$ONOO^- + M^nX \rightarrow NO_2 + O = M^{n+1}X \qquad (2.54)$$

Three potential modes of redox processes can be ascribed to $ONOO^-$: (1) direct reduction or oxidation by $ONOO^-$, (2) oxidation by NO_2 or $HO^{\bullet}$, and (3) oxidation by $CO_3^{\bullet -}$ or NO_2. Direct reduction of free radical species such as $CO_3^{\bullet -}$, $N_3^{\bullet}$, $HO^{\bullet}$, and $ClO_2^{\bullet}$ by $ONOO^-$ occurs through the one-electron-transfer process from $ONOO^-$ to the radical species, yielding the $ONOO^{\bullet}$ and corresponding anions CO_3^{2-}, N_3^-, HO^-, and ClO_2^-, with rate constants ranging from 10^{-4} to $10^9\ M^{-1}\ s^{-1}$ (Eqs. (2.55)–(2.58)). Also, $ONOO^-$ could also undergo direct oxidation by two-electron nucleophilic displacement or one-electron oxidation such as in the case of its highly exoergic reaction with NO to produce NO_2 and NO_2^-

[Eq. (2.59)][111], while oxidation of H_2O_2 by ONOOH produces NO_2^- and O_2 [Eq. (2.60)][112]:

$$ONOO^- + CO_3^{\bullet -} \rightarrow ONOO^{\bullet} + CO_3^{2-} \quad (2.55)$$

$$ONOO^- + N_3^{\bullet} \rightarrow ONOO^{\bullet} + N_3^- \quad (2.56)$$

$$ONOO^- + HO^{\bullet} \rightarrow ONOO^{\bullet} + HO^- \quad (2.57)$$

$$ONOO^- + ClO_2^{\bullet} \rightarrow ONOO^{\bullet} + ClO_2^- \quad (2.58)$$

$$ONOO^- + NO \rightarrow NO_2^{\bullet} + NO_2^- \quad (2.59)$$

$$ONOOH + H_2O_2 \rightarrow NO_2^- + O_2 + H^+ + H_2O \quad (2.60)$$

Peroxynitrite has been implicated in the oxidation and nitration of DNA, proteins, and lipids. DNA modification by $ONOO^-$ results in the formation of 8-nitroguanosine and 8-hydroxyguanosine (Fig. 2.32) as well as single-strand breakage.[113–116]

With proteins, $ONOO^-$ also acts as a two-electron oxidant such as in the case of thiol oxidation to sulfenic acid (RSOH), which ultimately forms the disulfide [Eq. (2.61)],[112] or with methionine to produce methionine sulfoxide and NO_2^- (Fig. 2.33).[117]

$$RS^- + ONOOH \rightarrow RSOH + NO_2^- \quad (2.61)$$

Peroxynitrite is also reactive against aromatic amino acids such as Trp,[118] Tyr,[119] and Phe[120] whose reaction typically involves prior formation of the aromatic radical

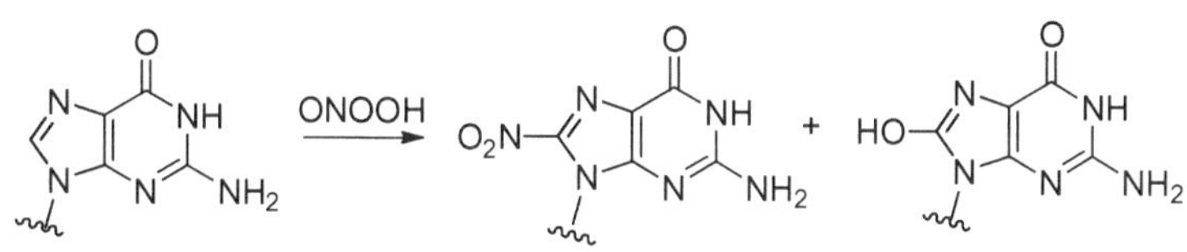

Figure 2.32 DNA base oxidative modification by peroxynitrite.

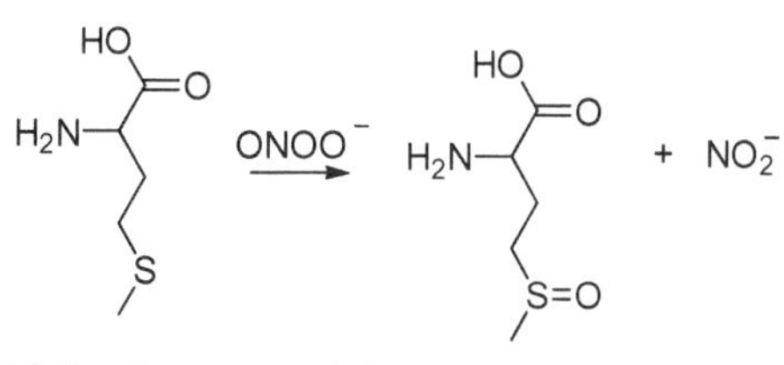

Figure 2.33 Methionine oxidation by peroxynitrite.

species perhaps via H-atom abstraction by peroxynitrous acid–derived $HO^{\bullet}$, $CO_3^{\bullet -}$, or the oxo–iron complex formed from the oxidation of iron heme by peroxynitrite. Subsequent addition of NO_2 to these aromatic radicals forms nitrated (Fig. 2.29) and hydroxylated products (Fig. 2.16). Peroxynitrite also induces lipid peroxidation [121] through H-atom abstraction by $HO^{\bullet}$ or NO_2 without the requirement of an iron ion. Termination reactions occur by addition of NO_2 to the conjugated diene radical to form the nitrated lipid (Figure 2.30).

2.3.3 Reactive Sulfur Species

Thiols (R-SHs) are integral part of biological systems and have significant roles in protein structure, detoxification, cell signaling, and enzyme activities as well as being cofactors. For these reasons, their oxidative modification has profound implications in cellular function and viability. As shown in Fig. 2.34, thiols are easily oxidized to sulfur species with higher oxidation states such as disulfides (RSSR), thiyl radicals ($RS^{\bullet}$), and disulfide radical anions ($RSSR^{\bullet -}$) as well as the organosulfur oxoacid family such as sulfenic (RSOH), sulfinic (RSO_2H), and sulfonic acids (RSO_3H). Sulfides (RSR) are also oxidizable to sulfoxides ($R_2S=O$) and sulfones [$R_2S(=O)_2$]. Some of the oxidized species are reversibly converted back to the more reduced thiol through chemical or enzymatic means but not for sulfinic and sulfonic acid, which typically require highly reducing conditions that are not typically possible in a biological system.[122] It is worth noting that the nucleophilicity and oxidizability of thiols highly depends on accessibility of the thiolate. Thiols are chemically analogous to alcohols, but because S is less electronegative than O, S–H bond in thiols are weaker than O–H bonds in alcohols and therefore thiol protons are more acidic than alcohol protons. Thiols' acidity, as characterized by low pK_a, enhances their reactivity, and while the pK_a varies greatly for both low-molecular-weight and protein thiols—That is, $pK_a = 3-11$,[43,123–125]—It is worth noting that some protein cysteine thiols such as in human peroxiredoxin-5 and glutathioredoxin-1 are more acidic ($pK_a = 3.7-4.6$) and nucleophilic compared to low-molecular-weight thiols with pK_a of 8–10.[126] In the succeeding section, notice that thiol's diverse reactivity involves the one-electron oxidized form, the thiyl radical ($RS^{\bullet}$), and the two-electron sulfenic acid (RS-OH), both showing a myriad of biochemical transformation with biological consequences (Fig. 2.34).

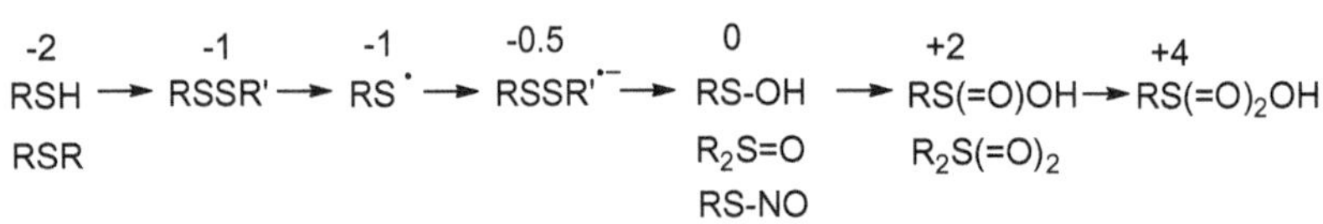

Figure 2.34 Oxidation states of the sulfur atom in various sulfur species.

2.3.3.1 Thiyl Radical ($RS^{\bullet}$)

Also referred to as a *sulfuhydryl radical*, the thiyl radical is more electrophilic than the chemically analogous alkoxyl radicals due to the lower electronegativity of the S atom compared to O. In fact, comparison of $RS^{\bullet}$, $RO^{\bullet}$, and $RR^{\bullet}$ addition reactions indicates low activation barriers for $RS^{\bullet}$ addition to alkenes due to the ability of the unpaired electron on the $RS^{\bullet}$ to undergo resonance stabilization with the transition state π^* orbitals of the substrate.[127] The thiyl radical is formed from one-electron oxidation of thiols (RSH)[128] as catalyzed by high-valence transition metal ions,[129] oxyhemoglobin,[130] horseradish peroxidase,[131] $HO^{\bullet}$,[132] NO_2,[133] from $ONOO^-$ (as $HO^{\bullet}$ and NO_2),[106,134,135] or from $ONOO^-$ as catalyzed by CO_2 as $CO_3^{\bullet -}$ and NO_2.[136] This facile oxidation of thiols is the basis of its antioxidant activity and repair mechanisms in cellular systems. One predominant intracellular antioxidant, GSH, prevents the formation of protein disulfides in a competitive manner via *S*-glutathionylation thus regulating protein function Fig. 2.35.

GSH also reacts with $Tyr^{\bullet}$ to form $GS^{\bullet}$ and TyrOH as a repair mechanism, but it is not as competitive as ascorbate reaction with $Tyr^{\bullet}$.[137]

One important property of $RS^{\bullet}$ is their ability to abstract H atoms from C—H bonds with rate constant of $k = 10^7 M^{-1} s^{-1}$, which has relevance in initiating oxidation[138] and isomerization of fatty acids (Fig. 2.36),[139] catalysis of protein backbone fragmentation or epimerization (Fig. 2.37),[140–143] catalytic formation of

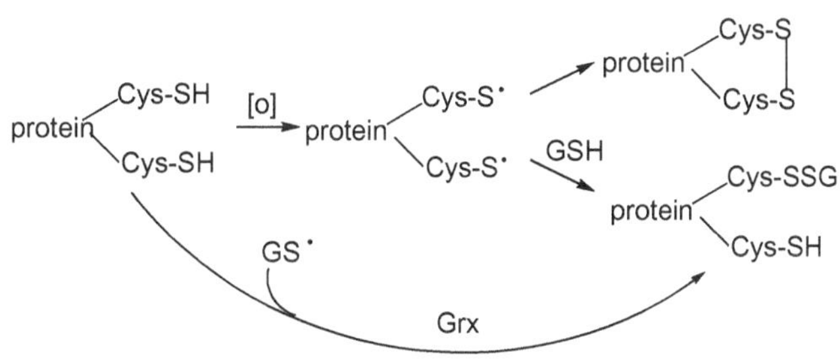

Figure 2.35 Role of thiyl radical in the regulation of disulfide formation.

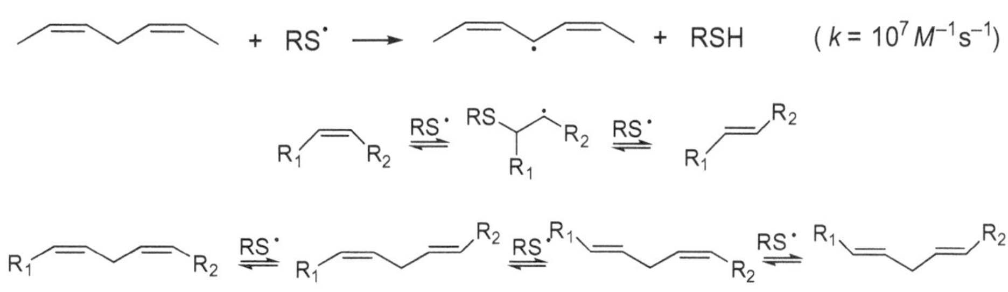

Figure 2.36 Abstraction of H atom from PUFA by thiyl radical leading to isomerization.

deoxyribonucleic acid building blocks,[144] as well as degradation of carbohydrates.[145] The generation of C-centered radicals in peptides and protein systems in the presence of $RS^{\bullet}$ occurs by H-atom transfer from the peptidyl C—H bond, which could lead to protein degradation.[146,147] The Cys-$S^{\bullet}$ residue not only reacts with adjacent amino acids within the peptide or protein but also can catalyze the oxidation of remote amino acids. Fig. 2.37 shows H-atom abstraction of the amino acid C_{α}—H by $RS^{\bullet}$ to form the amino acid C_{α}—radical, which subsequently adds to O_2 at a diffusion-controlled limit to ultimately yield the fragmentation products. Similar to $RS^{\bullet}$-mediated isomerization of mono- and polyunsaturated fatty acids and in the absence of O_2, $RS^{\bullet}$ could also cause epimerization at the C_{α}—H position.

Analogous to H-atom transfer in peptidyl C—H, $RS^{\bullet}$—Specifically, $GS^{\bullet}$—Can cause epimerization at the C(4′) position of the DNA sugar moiety during a repair mechanism involving GSH (see Fig. 2.38), so this "repair" mechanism is considered a form of DNA damage.[148] The thiyl radical has also been shown to react with the DNA base moiety by abstraction of the C5-CH_3 hydrogen atom in thymine and thymidine monophosphate (Fig. 2.39) with rate constants that are in the order of $10^4\ M^{-1}\ s^{-1}$, which is similar to the rate constants observed for the H-atom abstraction from the sugar moiety.[149]

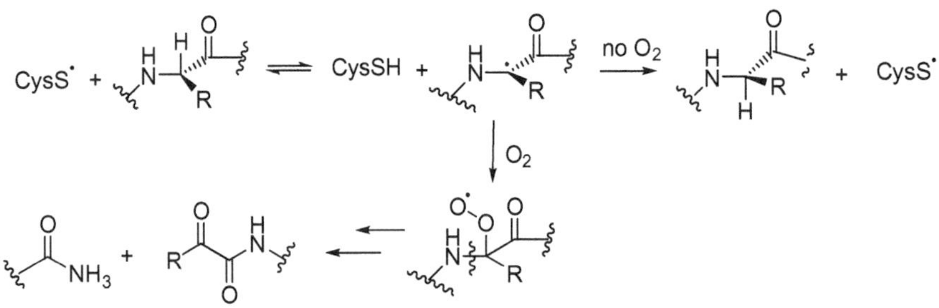

Figure 2.37 Epimerization and fragmentation of peptide by thiyl radical via H-atom abstraction at the C_{α}—H position.

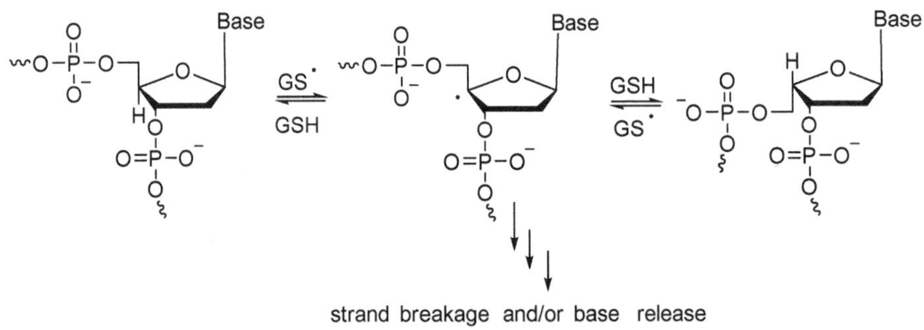

Figure 2.38 Epimerization of DNA sugar by H atom by thiyl radical.

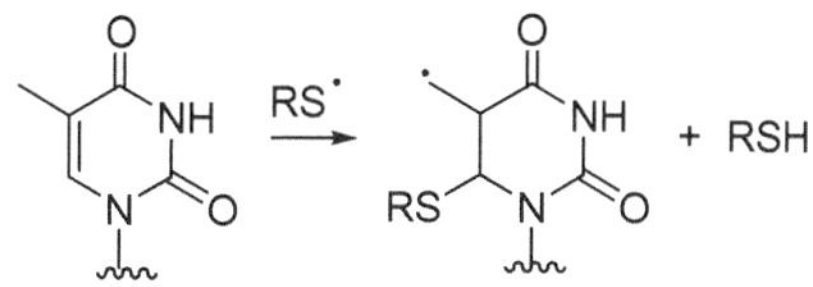

Figure 2.39 H-atom abstraction at the thymine C5-CH_3 by thiyl radical.

Similar to that observed for thiyl radical—mediated H-atom abstraction in DNA sugar moieties, H-atom abstraction by $RS^\bullet$ in selected carbohydrates such as 2-deoxy-D-ribose, 2-deoxy-D-glucose, α-D-glucose, and inositol gave rate constants that are on the order of $10^4\ M^{-1}\ s^{-1}$ with preference to the C1—H bond.[145]

Other biologically relevant $RS^\bullet$ reactions include that of NO to form the *S*-nitrosothiol as described previously [see Eq. (2.44)] with the rate approaching that of a diffusion-controlled limit (e.g., $2.7 \times 10^9\ M^{-1}\ s^{-1}$ for $GS^\bullet$). Reaction of $GS^\bullet$ with GS^- to form the oxidized glutathione radical disulfide anion $GSSG^{\bullet-}$ with $k = 4.65 \times 10^8\ M^{-1}\ s^{-1}$ [Eq. (2.62)][150,151] could occur at a wide pH range of 7—12 with rate constants of 10^8—$10^9\ M^{-1}\ s^{-1}$, which indicates that even protonated thiols are highly reactive to $RS^\bullet$.[152] The relevance of $GSSG^{\bullet-}$ formation is that it serves as a reducing agent for the formation of $O_2^{\bullet-}$ from O_2, according to Eq. (2.62).[153]

$$GS^\bullet + GS^- \rightarrow GSSG^{\bullet-} \xrightarrow{O_2} O_2^{\bullet-} + GSSG \tag{2.62}$$

Lastly, the reaction of $RS^\bullet$ with O_2 is fast, with $k = 10^7$—$10^8\ M^{-1}\ s^{-1}$ [154] according to Eq. (2.63), to form the thiylperoxy radical ($RSOO^\bullet$),[155] which further reacts with another thiol molecule, transferring an oxygen atom to form the sulfenic acid and sulfinyl radical[156] or undergo slow rearrangement to form the sulfonyl radical,[147] which is a nonreducible thiol species.

$$RS^\bullet + O_2 \rightarrow RS\text{-}O\text{-}O^\bullet \xrightarrow{RSH} R\dot{S}O + RSOH$$
$$RS\text{-}O\text{-}O^\bullet \rightarrow RS(O)O^\bullet \tag{2.63}$$

2.3.3.2 *Sulfenic Acid (RSOH)*

Sulfenic acid arises from the two-electron oxidation of thiolates just as $RS^\bullet$ arises from one-electron oxidation. Like $RS^\bullet$, RSOHs are considered transient reactive intermediates that play a key role in a variety biomolecular transformations such as in redox signaling leading to induction gene expression as well as in the control of protein functions for enzyme reactivity and catalysis.[157] Sulfenic acids are short-lived due their ability to undergo condensation with another thiol to form the disulfide, although analytical techniques are available to detect and identify their formation in vivo systems (*Note:* Detection of RSOH's was not covered in detail in this book but readers are encouraged to consult the references herein).[158,159] The S—O bond on the acid is basically a heteroatomic version found in the analogous hydroperoxides (ROOH) and hydropersulfides (RSSH)[160,161] compounds. The H–OSR bond disassociation energies (BDEs) for the relatively stable small molecule RSOHs were reported to be in the range of 68–73 kcal/mol based on experimental and theoretical data[162,163] which are less than those determined for the H–OOR in hydroperoxides (84–111 kcal/mol)[164] but higher than the H-OSR in hydropersulfides (70 kcal/mol).[165] Depending on the substituent electronic and steric effect on the RSOH, the pK_a for various low-molecular-weight and protein RSOHs varies significantly and could range from 5–12,[166–169] with protein RSOH exhibiting higher acidity. Moreover, the reduction potential for the $RSO^{\bullet}/RSO^-$ couple was determined to be 0.74 V which suggests that $RSO^{\bullet}$ is less oxidizing than the $ROO^{\bullet}$ analog of 1.05 V versus normal hydrogen electrode (NHE).[163]

The formation of RSOH arises from the oxidation of thiolates with H_2O_2, ROOH, $ONOO^-$, HOCl, or chloramines via a two-electron process (Fig. 2.40).[170] The previous paradigm for RSOH formation from H_2O_2 is via SN_2 nucleophilic attack by the thiolate as characterized by the presence of $RSOH-^{-}OH$ as an intermediate. However, new insight on this transformation shows a barrierless reaction in which the unprotonated RSO^- and H_2O are produced directly.[171]

Shown in Fig. 2.41 are the biologically relevant reactions of RSOH involve further oxidation to RSSR′/RSSH[172] and thiosulfinate [RSS(=O)R′] via an electrophilic addition reaction with $R'SH/H_2S$ and another R′SOH, respectively. An electrophilic addition reaction of RSOH to amines ($R'NH_2$) results in the formation sulfonamide (RSNHR′), while nucleophilic reaction of RSOH with H_2O_2 yields sulfinic acid [RS(=O)OH].[173]

The kinetics of thiolate transformation by H_2O_2 is slow, with rate constants ranging from 10 to 26 $M^{-1}\ s^{-1}$ [18,174–176] for low-molecular-weight thiols, while protein

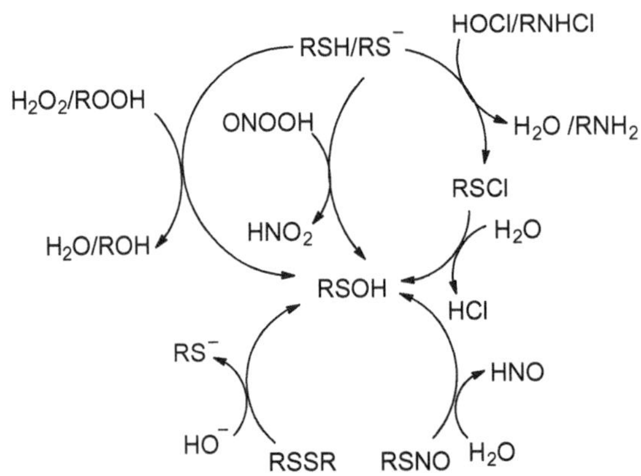

Figure 2.40 Sulfenic acid formation from (1) two-electron thiol oxidation pathway by peroxides (H_2O_2/ROOH), peroxynitrite ($ONOO^-$), hypochlorous acid (HOCl), and chloramine (RNHCl); and (2) hydrolysis pathway from disulfide (RSSR) and nitrosothiol (RSNO). *Adapted from V. Gupta, K.S. Carroll,* Biochim. Biophys. Acta, *2008,* 1840, *847–875.*

thiols such as catalase and peroxiredoxin (Prx) 2 oxidation by H_2O_2 (see Fig. 2.42) proceeds at orders of magnitude faster, with $k = 10^7 - 10^8\ M^{-1}\ s^{-1}$.[44,177] Other protein systems have varying rate constants (in $M^{-1}\ s^{-1}$) such as glutathione peroxidases ($10^5 - 10^8$), the bacterial sensor OxyR (10^5), glyceraldehyde-3-phosphate dehydrogenase (10^3), papain (62), and protein tyrosine phosphatase 1B (9), but all are mostly lower compared to that observed for Prx 2.[173] This wide array of reactivity for protein systems suggests that other factors other than low pK_a have been proposed to be responsible for higher reactivity to H_2O_2. Fig. 2.42 shows the Prx catalytic cycle where the Prx thiol in the form of a thiolate is oxidized by peroxides (H_2O_2 or ROOH) to yield the RSOH, where the RSOH is transformed to disulfide through reaction with another thiol from the same subunit or exogenous sources. This disulfide is recycled back to the thiol through a disulfide exchange reaction with thioredoxin (Trx) or other protein thiols (see the following discussion for Trx).[179,180]

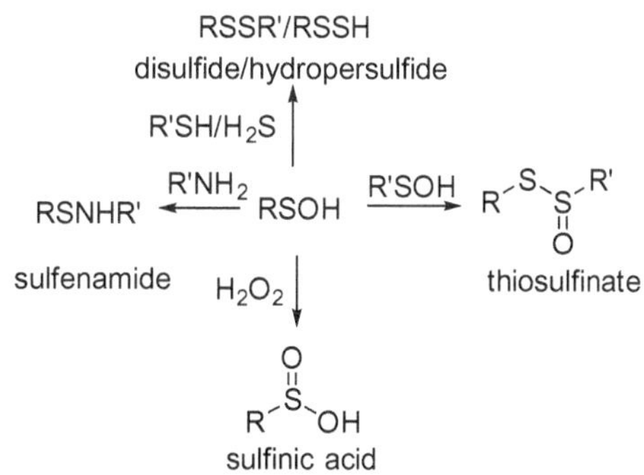

Figure 2.41 Biologically relevant electrophilic and nucleophilic transformations of sulfenic acid. *Adapted from V. Gupta, K.S. Carroll,* Biochim. Biophys. Acta, *2008,* 1840, *847–875.*

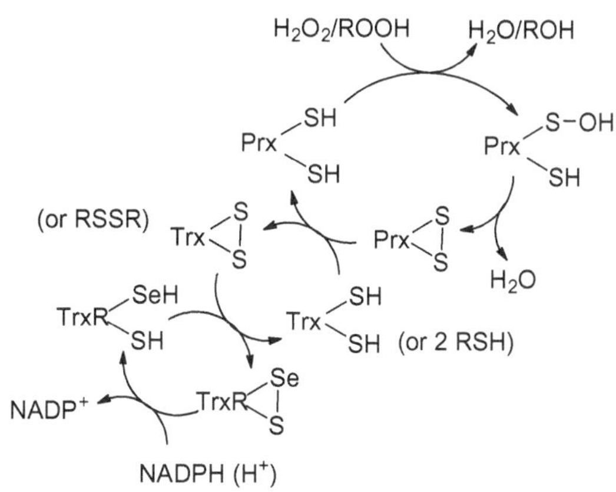

Figure 2.42 Peroxiredoxin (Prx) catalytic cycle of the peroxides involving thioredoxin (Trx), thioredoxin reductase (TrxR), and NADPH.

2.3.3.3 Disulfide (RSSR')

Disulfide bond formation is one of the most important biochemical processes to play a far-reaching role in cell-signaling processes as redox switches, in protein structure, and protein folding.[181–183] The disulfide S–S bond has a higher BDE compared to its oxygen analog, peroxides (ROOR′). The BDE_{S-S} for MeS–SMe was reported to be 74 kcal/mol compared to 38 kcal/mol for MeO–OMe, therefore the former is much more difficult to cleave than the peroxides. As discussed previously, RSSR′ are formed from one-electron reduction of RSH to form the $RS^{\bullet}$, then a subsequent radical–radical coupling reaction forms the disulfide with rate constants of $1.5\times10^{9}\ M^{-1}\ s^{-1}$ in the case of GSSG formation [Eq. (2.64)],[184] or two-electron oxidation of GSH by H_2O_2 to GSSG via formation of GSOH [Eqs. (2.65) and (2.66)], for example, which occurs at a much slower rate of $\geq 15\ M^{-1}\ s^{-1}$.[176]

$$GS^{\bullet} + GS^{\bullet} \rightarrow GSSG \tag{2.64}$$

$$H_2O_2 + GSH \rightarrow GSOH + H_2O \tag{2.65}$$

$$GSOH + GSH \rightarrow GSSG + H_2O \tag{2.66}$$

Rates of reaction of RSH with R′SOH to give the RSSR′ are greatly varied from small molecule RSOH and macomolecular RSOH. For example, the reported rate constant for the reaction of CysSOH with CysSH to form CysS–SCys is $>10^{5}\ M^{-1}\ s^{-1}$ [Eq. (2.67)],[185] while reaction of human serum albumin–sulfenic acid with Cys was determined to be $21.6\ M^{-1}\ s^{-1}$ [Eq. (2.68)].[186] Oxidation of GSH to GSSG in biological systems is therefore carried out enzymatically by mammalian glutathione peroxidases (GPX) [Eq. (2.69)].[187]

$$\text{Cys-SOH} + \text{Cys} \rightarrow \text{Cys-S-S-Cys} \tag{2.67}$$

$$\text{HSA-SOH} + \text{Cys} \rightarrow \text{HSA-S-S-Cys} \tag{2.68}$$

$$2GSH + H_2O_2 \xrightarrow{GPx} GSSG + 2H_2O \tag{2.69}$$

The thiol–disulfide exchange interaction is also one of the most relevant biological processes because it highlights the role disulfide plays in protein-folding processes, which are governed by nucleophilic substitution reaction of the reduced thiol with the disulfide, resulting in mixed disulfide exchange [Eq. (2.70)]. The rate constants for the exchange reaction for some thiols such as GSH, Cys, and homocysteine were determined to be in the range of $12–60\ M^{-1}\ s^{-1}$ at $pH > 10$, which are slow reactions. However, oxidoreductases such as thioredoxin (Trx), glutaredoxin, protein disulfide isomerase, and other organelle-residing enzymes could make this exchange processes proceed at a much faster rate—That is, $k = 10^4 - 10^6\ M^{-1}\ s^{-1}$ [Eq. (2.71)].[180]

$$\begin{array}{lcl} RSH + GSSG & \rightleftharpoons & RSSG + GSH \\ RSH + RSSG & \rightleftharpoons & RSSR + GSH \\ \hline \text{Net: } 2\ RSH + GSSG & \rightleftharpoons & RSSR + 2\ GSH \end{array} \quad (2.70)$$

$$RSH + GSSG \overset{Grx}{\rightleftharpoons} RSSG + GSH \quad (2.71)$$

Another pathway for disulfide-bond scrambling is via initial oxidation of the disulfide to S-monoxides [RS(=O)SR] or S-dioxides [$RS(=O)_2SR$] ,[188] wherein the S(O)−S bond weakens and becomes susceptible to nucleophilic attack by thiol to form the mixed disulfide and the corresponding sulfenic (RSOH) and sulfinic acids (RSO_2H) (Fig. 2.43). Therefore, both thiol−disulfide exchange interaction and thiol redox reactions are two important thiol reactions regulating protein structure and function.

As mentioned in the section titled "Sulfenic Acid (RSOH)" and as shown in Fig. 2.42, reduction of disulfide to thiols through thiol−disulfide exchange can be carried out by either monothiol reductants or dithiol reductants such as GSH or thioredoxin, respectively, the latter being more efficient due to the proximity of the second vicinal cysteine, which drives the reduction reaction with higher efficiency. Oxidized Trx can then be recycled back by thioredoxin reductase (TrxR), with NAPDH supplying the reducing electron (Fig. 2.42).[189] Other relevant disulfide reductant include glutathione reductase (GSR), which catalyzes the reduction of GSSG to GSH using the GSR's flavin adenine dinucleotide (FAD)prosthetic group and NADPH. The reduced GSH is also used concomitantly by GPX to detoxify the cells with harmful peroxides, according to Fig. 2.44.[190]

Aside from the reaction of $RS^{\bullet}$ with RS^- to form the disulfide radical anion ($RSSR^{\bullet -}$), direct one-electron reduction of the RSSR by hydrated electrons (e^-_{aq}) or

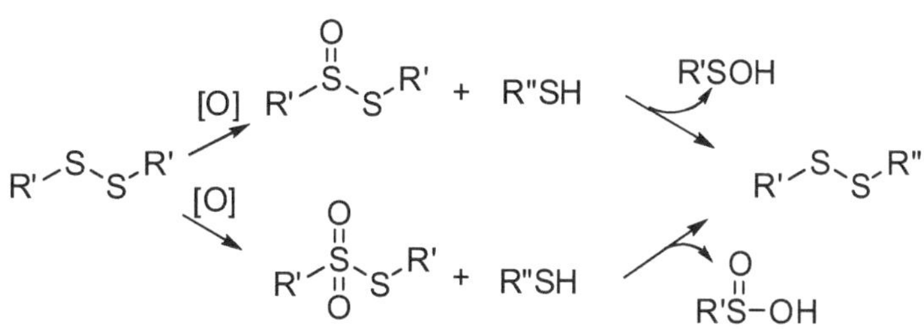

Figure 2.43 Disulfide-bond exchange via formation of oxidized thiols.

H atoms ($H^{\bullet}$) from the ionizing radiation of water yields $RSSR^{\bullet -}$, which further decomposes to RSH/RS^{-} and $RS^{\bullet}$ with rate constants approaching that of the diffusion-controlled limit, thus giving a rationale for the rupture of disulfide linkage via one-electron transfer under ionizing conditions (Fig. 2.45).[151] Proton-abstraction α or β protons of the Cys residue in proteins using a strong base such as hydroxide can lead to C—S or S—S bond breakage, respectively, as shown in Fig. 2.46.[191]

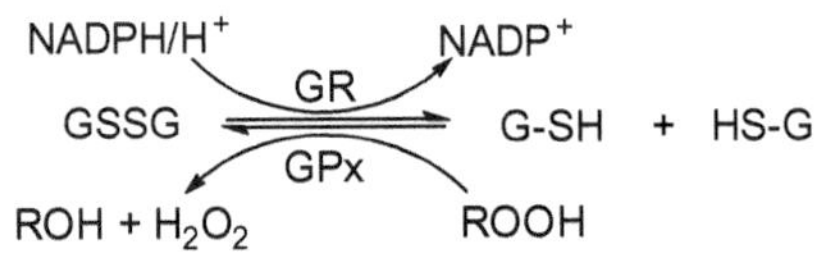

Figure 2.44 Catalytic reduction of GSSG to RSH by glutathione reductases.

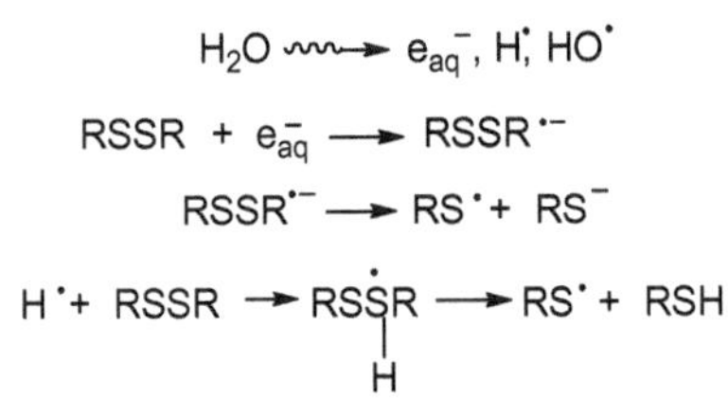

Figure 2.45 Formation of disulfide radical anion ($RSSR^{\bullet -}$) from ionizing radiation of water.

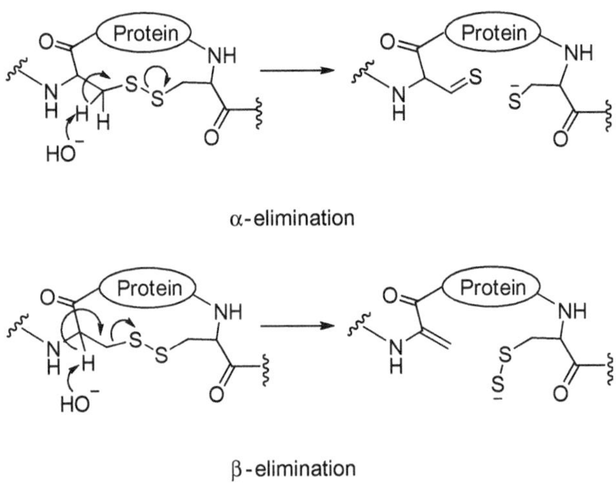

Figure 2.46 Carbon—sulfur and disulfide bond breakage by proton abstraction of the α- or β-protons of Cys.

2.3.4 Reactive Halogen Species

Hypochlorous acid (HOCl). Formed from the activation of phagocytes through myeloperoxidase (MPO)-mediated peroxidation of Cl^-using H_2O_2. Although a weak acid ($pK_a = 7.5$), HOCl is considered a strong oxidizing agent responsible for the killing action of phagocytes against a variety of pathogens.[41,192–194] Other than Cl^-, other halides such as Br^- and I^- as well as the anion isothiocyanate (SCN^-) are oxidized by the MPO/H_2O_2 system to generate the hypohalous acids, HOBr and HOCl, or hypothiocyanous acid (HOSCN), respectively. Hypochlorous acid reacts with various ROS such as $O_2^{\bullet -}$, H_2O_2, and $HO^\bullet$ to yield $HO^\bullet$, $^1O_2$, and $ClO^\bullet$, respectively, according to Eqs. (2.72)–(2.74). Similar to the reaction of HOCl with H_2O_2, fatty acid hydroperoxide (LOOH) also yield singlet oxygen along with the respective alcohol and ketone.[195]

$$HOCl \xrightarrow{O_2^{\bullet -}} HO^\bullet + Cl^- + O_2 \tag{2.72}$$

$$HOCl \xrightarrow{H_2O_2} H_2O + HCl + {}^1O_2 \tag{2.73}$$

$$HOCl \xrightarrow{HO^\bullet} ClO^\bullet + H_2O \tag{2.74}$$

Hypochlorous acid also directly oxidizes ionic substances such as NO_2^- to yield a NO_2 radical as an intermediate [Eq. (2.75)] that can nitrate the tyrosine residue of proteins.[196]

$$HOCl \xrightarrow[HO^-]{NO_2^-} NO_2\text{-}Cl \longrightarrow Cl^\bullet + NO_2^\bullet \tag{2.75}$$

In the presence of metal ions such as Fe^{2+} in free[197] or in protein-bound heme form,[198] HOCl was shown to generate $HO^\bullet$(Eq. 2.76):

$$HOCl \xrightarrow[\text{or } Fe^{2+}\text{-cyt c}]{Fe^{2+}} Cl^- + HO^\bullet + Fe^{3+}\ (Fe^{3+}\text{-cyt } c) \tag{2.76}$$

Reaction of HOCl with thiols yields RSOH via the radical–radical coupling reaction of the intermediates, $RS^\bullet$ and $HO^\bullet$, with an estimated rate constant of $>10^7\ M^{-1}\ s^{-1}$, according to Eqs. (2.77)–(2.79).[199]

$$HOCl + GSH \rightarrow GSCl + H_2O \rightarrow GS^\bullet + Cl^\bullet \tag{2.77}$$

$$Cl^\bullet + H_2O \rightarrow HO^\bullet + Cl + H^+ \tag{2.78}$$

$$GS^\bullet + HO^\bullet \rightarrow GSOH \tag{2.79}$$

The reactivity of HO-Cl mimics those of the two-electron oxidation reaction exhibited by peroxynitrite (HO–NO_2) or hydrogen peroxide (HO–OH) whereby the homolytic cleavage of O–X (X = N, O, or Cl) to form the $HO^{\bullet}$ and $Cl^{\bullet}$ pair could impart a diversity of chemical reactions with a variety of biomolecules. As shown in Fig. 2.47, HOCl gives a variety of unique oxidation products at varying rates. For example, reaction of HOCl with protein amino acid residues are highest for Cys and Met thiols with rate constants of $k = 3–3.8 \times 10^7\ M^{-1}\ s^{-1}$. These are followed by His and a terminal α-amino group both with $k = 10^5\ M^{-1}\ s^{-1}$ and Lys amino with $k = 5 \times 10^3\ M^{-1}\ s^{-1}$. Another amino group that is prone to oxidation by HOCl is Trp with $k = 1.1 \times 10^4\ M^{-1}\ s^{-1}$ and Tyr being the least reactive with $k = 44\ M^{-1}\ s^{-1}$.[200]

The diversity of HOCl chemistry with free amino acids, peptides, and proteins is also associated with complexity due to the occurrence of secondary reactions arising from decomposition or reactivity with other amino acids' side chains, terminal amino, or carboxyl groups.[198] Thiols such as Met could be oxidized by HOCl to methionine sulfoxide in high yield, which is one of the main contributors for the killing action of neutrophils against bacteria (Fig. 2.47).[201] The peptide methionine sulfoxide reductase exhibits high specificity for the reduction of free or protein-bound methionine sulfoxide back to methionine, which serves as a basis for the reversibility of methionine oxidation.[202] Cysteine oxidation by HOCl leads to the formation of cysteic acid (RSO_3H) and cystine (RSSR) via the sulfenyl chloride intermediate (RSCl) reaction with water and additional cysteine residue, respectively (Fig. 2.47).[203] Other more complex oxidation motifs have also been detected and involve intramolecular reaction of the RSCl intermediate with adjacent primary or secondary amines to form sulfonamide linkages (Fig. 2.48).[204]

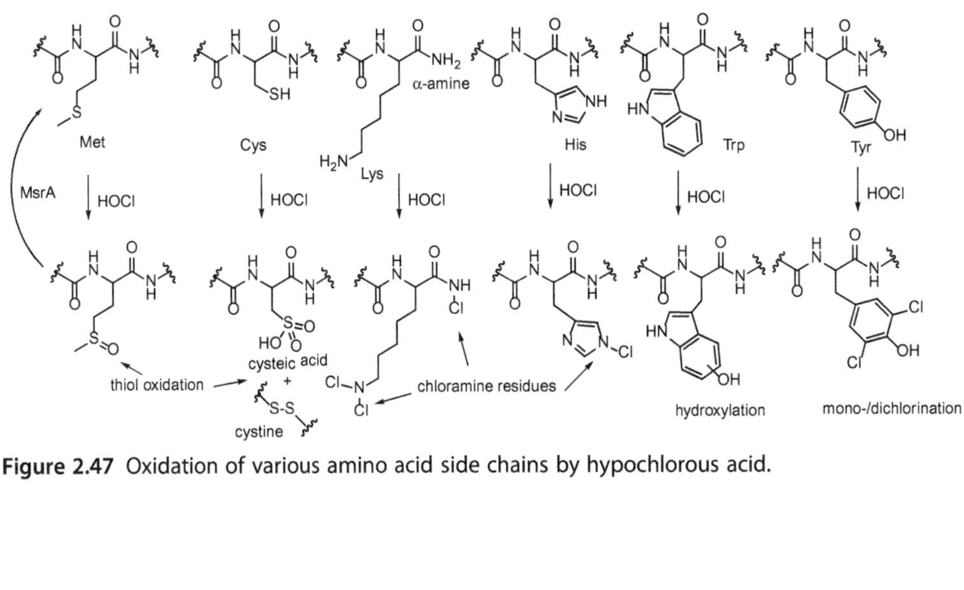

Figure 2.47 Oxidation of various amino acid side chains by hypochlorous acid.

N-Chloramine formation is perhaps one of the most relevant reactivities of HOCl with the amino groups of the Lys[205] amino acid side chain and His[206] leading to the formation of $RNHCl/RNCl_2$ and R_2NCl, respectively (see Fig. 2.47). The Lys-ε-haloamines could also serve as halogenating agents where they can promote halogen transfer to Tyr, resulting in the formation of monochlorinated Tyr residues[207] or act as sacrificial antioxidants against HOCl to ultimately form Lys nitrile .[208] Cross-linking of *N*-chloramines could also be observed with other amino acids such as oxidized Met to form sulfilimine covalent adducts.[209] The terminal α-amino group of the peptide or protein are also susceptible to oxidation leading to the formation of terminal *N*-α-chloramines (Fig. 2.47).[210] Also, the formation of *N*-chloramine has been implicated in protein and nucleoside degradation[211,212] because they serve as precursors to the formation of *N*-centered radical and $Cl^{\bullet}$ when activated, for example, by metal ions or when subject to further degradation via nucleophilic substitution, decarboxylation, H-atom transfer, fragmentation, or dehydrohalogentaion and elimination reactions (Fig. 2.49).[213]

Oxidation of Trp by HOCl yields two types of products—Oxygenated or chlorinated Trp—Depending on the type of amino acid that is adjacent to it. When Trp is adjacent to Gly, only the formation of oxygenated Trp (Fig. 2.47) was observed but when Trp is adjacent to other amino acids it produced high yields of a chloroindolenine derivative.[214,215]

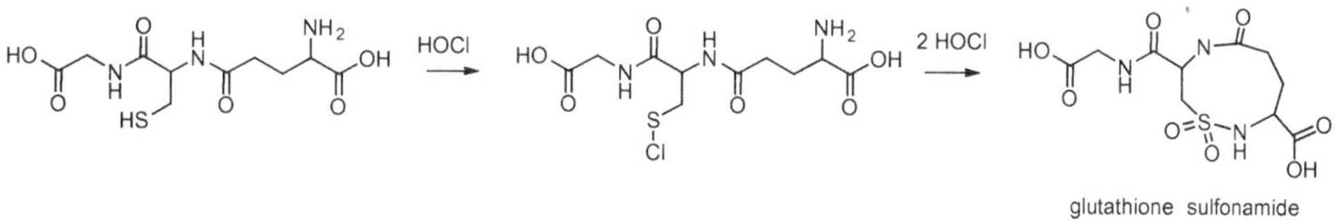

Figure 2.48 Intramolecular cyclization of RSCl intermediate with terminal amino group on oxidation by hypochlorous acid.

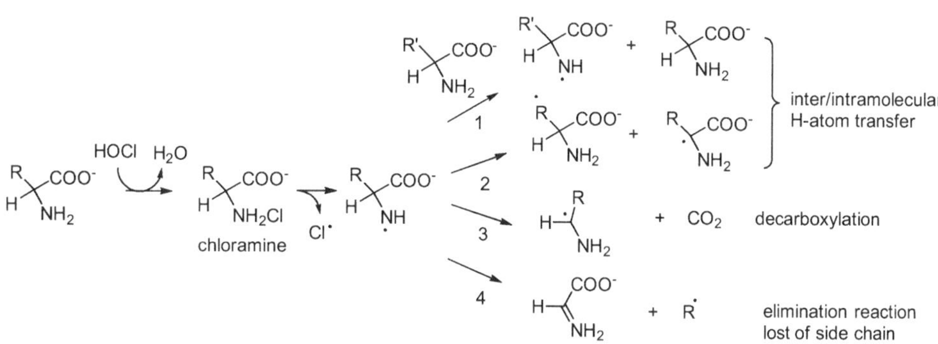

Figure 2.49 Decomposition pathways of chloramine intermediate from the reaction of hypochlorous acid with terminal amino group. *Adapted from Hawkins, Clare L. and Davies, Michael J.* J. Chem. Soc., Perkin Trans. 2, *1998, 1937–1946.*

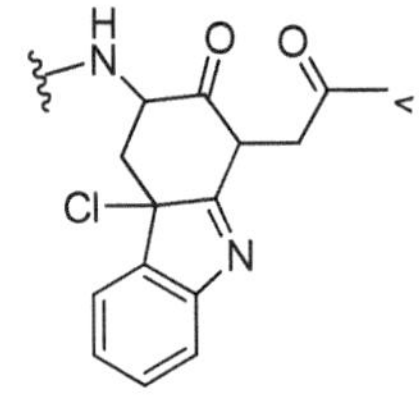

3-chloroindolenine

Mono- and dechlorinated Tyr are also products of tyrosine oxidation in HOCl-modified proteins (Fig. 2.47),[216] while free tyrosine was found to be oxidized to p-hydroxyphenylacetaldehyde, which serves as a rationale for the protective effect of Tyr against HOCl-induced modification of LDL.[217] Moreover, this p-hydroxyphenylacetaldehyde has been shown to covalently modify the ε-amino groups of protein lysine residues to form Schiff base adducts.[218]

Oxidation of DNA as mediated by HOCl has been implicated in mutagenesis[219] and could occur via two major pathways: (1) Direct oxidation by HOCl and (2) through chloramine-mediated oxidation.[220] Regardless of the pathway for HOCl-mediated DNA damage, the main oxidation occurs through chlorination or oxygenation of the aromatic ring or the amino or hetero NH groups of the DNA bases. This oxidative events lead to double-strand dissociation through loss of H bonding and ultimately to DNA denaturation[221,222] or strand breakage.[220] It has been demonstrated that reaction of dG with HOCl yields 8-oxo-dG and 8-chloro-2′-deoxyguanosine (8-Cl-dG), where the former is further oxidized to spiroiminodihydantoin (dS) nucleoside (Fig. 2.50).[223]

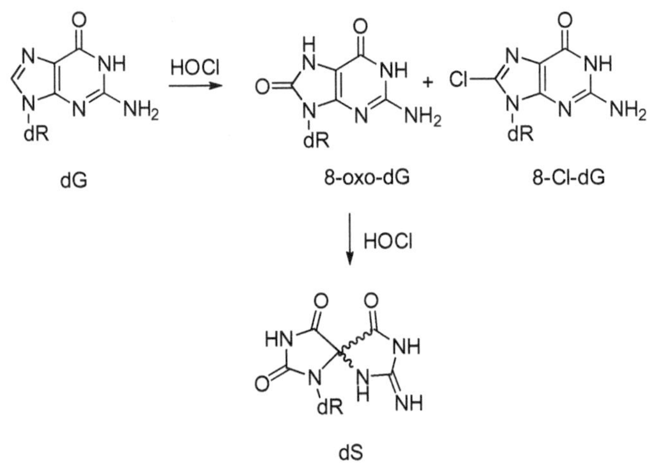

Figure 2.50 Oxidation of dG by HOCl to form oxidation and chlorinated products.

Protein chloramine was shown to form adducts with DNA as exemplified by addition reaction of the Lys-chloramine radical to the thymidine base at the C-5 or C-6 position.[220] Exposure of calf thymus DNA to HOCl yielded 8-Cl adenine and, to a lesser extent, 5-Cl uracil via 5-Cl cytosine formation with 5-Cl uracil to be a more reliable biomarker than 8-Cl adenine due to the susceptibility of the latter to degradation by other RNS or ROS.[224]

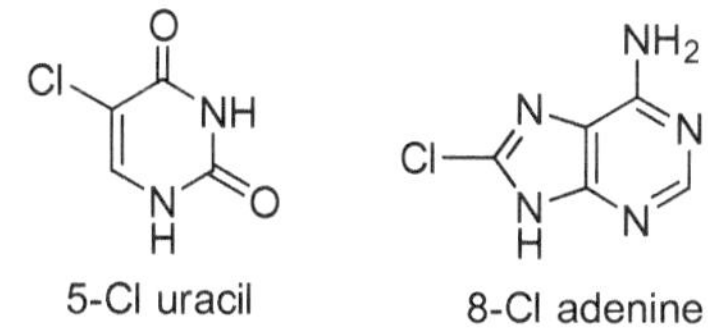

Formation of *N*-centered radicals from DNA−chloramine at the exocylic $-NH_2$ groups of cytidine, adenosine, and guanosine as well as the ring NH groups of all the bases were also observed.[212] Nucleoside reaction with HOCl generated from human neutrophils gave various chlorinated (2′-deoxy)nucleosides, including 8-chloro (2-deoxy)guanosine (8-Cl-dG), 5-chloro(2-deoxy)cytidine (5-Cl-dCyd), and 8-chloro (2-deoxy)adenosine (8-CldA).[225]

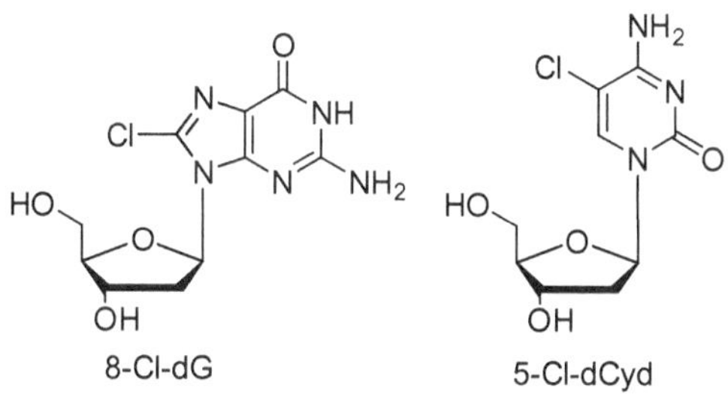

Lipid peroxidation initiation by HOCl is widely considered as one of the main contributors for the pathogenesis of atherosclerosis due to the role of phagocytes in generating HOCl upon stimulation and the presence of oxidizable lipoproteins in the blood (both LDL and very-low-density lipoproteins).[226−228] The HOCl reaction with a PUFA or phosphatidylcholine occurs at the double bonds to form the chlorohydrin as observed in linoleic acid, arachidonic acid, and oleic acid[229] as well as in LDL[230] (Fig. 2.51). With cholesterol, the same formation of chlorhydrins was observed along

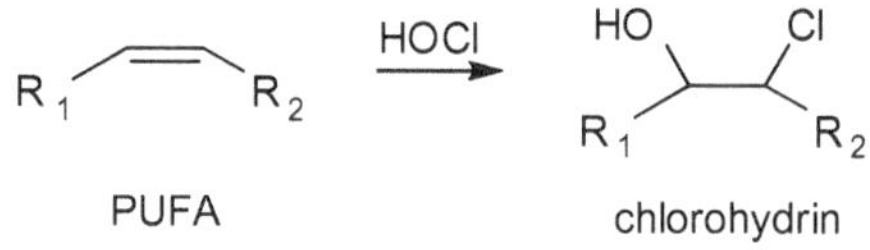

Figure 2.51 Formation of chlorohydrin from the oxidation of PUFA by hypochlorous acid.

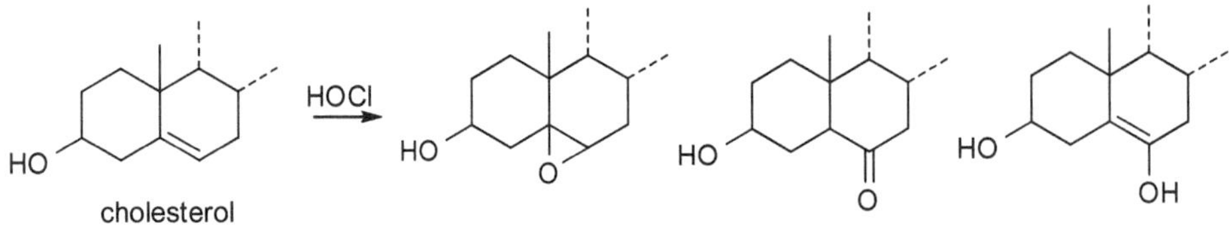

Figure 2.52 Oxidation products of cholesterol with hypochlorous acid.

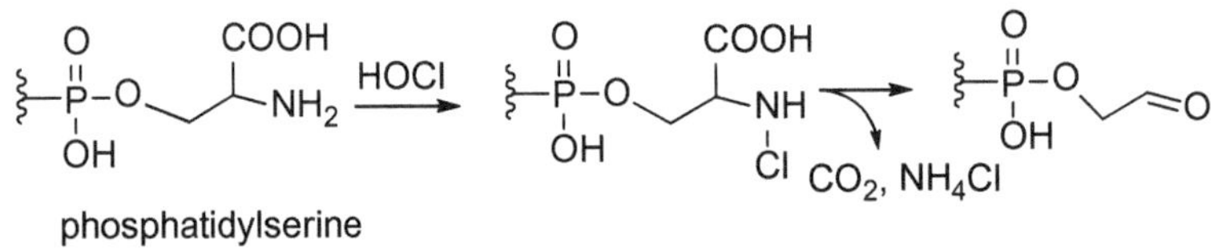

Figure 2.53 Formation of chloramine from the oxidation of phospholipid by hypochlorous acid.

with the formation of dichlorinated products,[231] although cholesterol chlorohydrin can undergo further dehydrohalogenation to form epoxide, ketone, or hydroxyl, according to Fig. 2.52.[232]

Reaction of HOCl with aminophospholipids generates both mono- and dichloramines that could also decompose to form the phosphatydylglycoaldehyde as in the case of phosphatyldserine (Fig. 2.53).[233]

REFERENCES

1. Halliwell B, Gutteridge JMC. *Free Radicals in Biology and Medicine*. 4th ed Oxford: Oxford University Press; 2007
2. Winterbourn CC. Reconciling the chemistry and biology of reactive oxygen species. *Nat. Chem. Biol.* 2008;**4**(5):278–86.
3. Pryor WA, Squadrito GL. The chemistry of peroxynitrite: a product from the reaction of nitric oxide with superoxide. *Am. J. Physiol.* 1995;**268**(5 Pt 1):L699–722.
4. Nursten HE. *The Maillard reaction: Chemistry, biochemistry, and implications*. Cambridge, UK: Royal Society of Chemistry; 2005
5. Bierhaus A, Hofmann MA, Ziegler R, Nawroth PP. AGEs and their interaction with AGE-receptors in vascular disease and diabetes mellitus. I. The AGE concept. *Cardiovasc. Res* 1998;**37**(3):586–600.
6. Bielski BHJ, Cabelli DE, Arudi RL, Ross AB. Reactivity of HO_2/O_2^- Radicals in Aqueous Solution. *J. Phys. Chem. Ref. Data* 1985;**1041-1100** 1041-1041
7. Sawyer DT, Valentine JS. How super is superoxide? *Acc. Chem. Res.* 1981;**14**(12):393–400.
8. Fridovich I. Superoxide anion radical (O_2^-.), superoxide dismutases, and related matters. *J. Biol. Chem.* 1997;**272**:18515–17.
9. Butler J, Jayson GG, Swallow AJ. , The reaction between the superoxide anion radical and cytochrome *c*. *Biochim. Biophys. Acta* 1975;**408**(3):215–22.
10. Krishna MC, Russo A, Mitchell JB, Goldstein S, Dafni H, Samuni A. Do nitroxide antioxidants act as scavengers of O_2^-. or as SOD mimics?. *J. Biol. Chem.* 1996;**271**(42):26026–31.
11. Krishna MC, Grahame DA, Samuni A, Mitchell JB, Russo A. Oxoammonium cation intermediate in the nitroxide-catalyzed dismutation of superoxide. *Proc. Natl. Acad. Sci. U.S.A* 1992;**89**(12): 5537–41.
12. Stubbe J, van Der Donk WA. Protein Radicals in Enzyme Catalysis. *Chem. Rev.* 1998;**98**:705–62.

13. Lancaster Jr. JR. Nitroxidative, Nitrosative, and Nitrative Stress: Kinetic Predictions of Reactive Nitrogen Species Chemistry Under Biological Conditions. *Chem. Res. Toxicol.* 2006;**19**(9):1160–74.
14. Ferrer-Sueta G, Radi R. Chemical Biology of Peroxynitrite: Kinetics, Diffusion, and Radicals. *ACS Chem. Biol.* 2009;**4**(3):161–77.
15. Valgimigli L, Amorati R, Fumo MG, DiLabio GA, Pedulli GF, Ingold KU, et al. The unusual reaction of semiquinone radicals with molecular oxygen. *J. Org. Chem.* 2008;**73**(5):1830–41.
16. Siegel D, Reigan P, Ross D. One- and two-electron-mediated reduction of quinones: enzymology and toxicological implications. *Biotechnol.: Pharm. Aspects* 2008;**9**:169–97 (Advances in Bioactivation Research)
17. Vella F, Ferry G, Delagrange P, Boutin JA. NRH:quinone reductase 2: An enzyme of surprises and mysteries. *Biochem. Pharmacol.* 2005;**71**(1-2):1–12.
18. Winterbourn CC, Metodiewa D. Reactivity of biologically important thiol compounds with superoxide and hydrogen peroxide. *Free Radic. Biol. Med.* 1999;**27**(3-4):322–8.
19. Flint DH, Tuminello JF, Emptage MH. The inactivation of Fe-S cluster containing hydro-lyases by superoxide. *J. Biol. Chem.* 1993;**268**(30):22369–76.
20. Sutton VR, Stubna A, Patschkowski T, Munck E, Beinert H, Kiley PJ. Superoxide destroys the $[2Fe\text{-}2S]^{2+}$ cluster of FNR from Escherichia coli. *Biochemistry* 2004;**43**(3):791–8.
21. Liang L-P, Patel M. Iron-sulfur enzyme mediated mitochondrial superoxide toxicity in experimental Parkinson's disease. *J. Neurochem.* 2004;**90**(5):1076–84.
22. Behar D, Czapski G, Rabani J, Dorfman LM, Schwarz HA. Acid dissociation constant and decay kinetics of the perhydroxyl radical. *J. Phys. Chem.* 1970;**74**(17):3209–13.
23. De Grey AD. HO_2^*: the forgotten radical. *DNA Cell Biol.* 2002;**21**:251–7.
24. Buettner GR. The Pecking Order of Free Radicals and Antioxidants: Lipid Peroxidation, α-Tocopherol, and Ascorbate. *Arch. Biochem. Biophys.* 1993;**300**(2):535–43.
25. Navarrete M, Rangel C, Espinosa-García J, Corchado JC. Theoretical Study of the Antioxidant Activity of Vitamin E: Reactions of α-Tocopherol with the Hydroperoxy Radical. *J. Chem. Theory Comput.* 2005;**1**(2):337–44.
26. Davies SS, Guo L. *Lipid peroxidation and nitration.* John Wiley & Sons, Inc; 2013p. 49–70.
27. Gardner HW. Oxygen radical chemistry of polyunsaturated fatty acids. *Free Radic. Biol. Med.* 1989; **7**(1):65–86.
28. Haeggström JZ, Funk CD. Lipoxygenase and Leukotriene Pathways: Biochemistry, Biology, and Roles in Disease. *Chem. Rev.* 2011;**111**(10):5866–98.
29. Rouzer CA, Marnett LJ. Mechanism of Free Radical Oxygenation of Polyunsaturated Fatty Acids by Cyclooxygenases. *Chem. Rev.* 2003;**103**(6):2239–304.
30. Smith WL, Urade Y, Jakobsson P-J. Enzymes of the Cyclooxygenase Pathways of Prostanoid Biosynthesis. *Chem. Rev.* 2011;**111**(10):5821–65.
31. Milne GL, Yin H, Hardy KD, Davies SS, Roberts LJ. Isoprostane Generation and Function. *Chem. Rev.* 2011;**111**(10):5973–96.
32. Milne GL, Sanchez SC, Musiek ES, Morrow JD. Quantification of F2-isoprostanes as a biomarker of oxidative stress. *Nat. Protoc.* 2007;**2**(1):221–6.
33. Esterbauer H, Schaur RJ, Zollner H. Chemistry and biochemistry of 4-hydroxynonenal, malonaldehyde and related aldehydes. *Free Radic. Biol. Med.* 1991;**11**:81–128.
34. Lee SH, Blair IA. Characterization of 4-Oxo-2-nonenal as a Novel Product of Lipid Peroxidation. *Chem. Res. Toxicol.* 2000;**13**(8):698–702.
35. Doorn JA, Petersen DR. Covalent Modification of Amino Acid Nucleophiles by the Lipid Peroxidation Products 4-Hydroxy-2-nonenal and 4-Oxo-2-nonenal. *Chem. Res. Toxicol.* 2002; **15**(11):1445–50.
36. Zimniak P. Relationship of electrophilic stress to aging. *Free Radic. Biol. Med.* 2011;**51**(6):1087–105.
37. Xu L, Porter NA. Free radical oxidation of cholesterol and its precursors: Implications in cholesterol biosynthesis disorders. *Free Radic. Res.* 2015;**49**(7):835–49.
38. Winterbourn CC. The biological chemistry of hydrogen peroxide. *Methods Enzymol.* 2013;**528**:3–25.

39. Winterbourn CC. Toxicity of iron and hydrogen peroxide: the Fenton reaction. *Toxicol. Lett.* 1995;**82-83**:969–74.
40. Koppenol WH. The Haber-Weiss cycle–70 years later. *Redox Rep* 2001;**6**(4):229–34.
41. Winterbourn CC. Biological reactivity and biomarkers of the neutrophil oxidant, hypochlorous acid. *Toxicology* 2002;**181–182**:223–7.
42. Davies MJ, Hawkins CL, Pattison DI, Rees MD. Mammalian heme peroxidases: from molecular mechanisms to health implications. *Antioxid. Redox Signal.* 2008;**10**(7):1199–234.
43. Winterbourn CC, Hampton MB. Thiol chemistry and specificity in redox signaling. *Free Radic. Biol. Med.* 2008;**45**(5):549–61.
44. Peskin AV, Low FM, Paton LN, Maghzal GJ, Hampton MB, Winterbourn CC. The High Reactivity of Peroxiredoxin 2 with H_2O_2 Is Not Reflected in Its Reaction with Other Oxidants and Thiol Reagents. *J. Biol. Chem.* 2007;**282**(16):11885–92.
45. Pryor WA. Oxy-radicals and related species: their formation, lifetimes, and reactions. *Annu. Rev. Physiol.* 1986;**48**:657–67.
46. Koppenol WH, Liebman JF. The oxidizing nature of the hydroxyl radical. A comparison with the ferryl ion (FeO^{2+}). *J. Phys. Chem.* 1984;**88**(1):99–101.
47. Seal P, Oyedepo G, Truhlar DG. Kinetics of the Hydrogen Atom Abstraction Reactions from 1-Butanol by Hydroxyl Radical: Theory Matches Experiment and More. *J. Phys. Chem.* 2013; **117**(2):275–82.
48. Gilbert BC, King DM, Thomas CB. Radical reactions of carbohydrates. Part 2. An electron spin resonance study of the oxidation of D-glucose and related compounds with the hydroxyl radical. *J. Chem. Soc., Perkin Trans.* 1981;**2**(8):1186–99.
49. Cheeseman KH, Beavis A, Esterbauer H. Hydroxyl-radical-induced iron-catalysed degradation of 2-deoxyribose. Quantitative determination of malondialdehyde. *Biochem. J* 1988;**252**(3):649–53.
50. Jain SS, Tullius TD. Footprinting protein-DNA complexes using the hydroxyl radical. *Nat. Protoc* 2008;**3**(6):1092–100.
51. Nemera DB, Jones AR, Merino EJ. *DNA oxidation*. John Wiley & Sons, Inc; 2013p. 93–112.
52. Hougland JL, Darling J, Flynn S. *Protein posttranslational modification*. John Wiley & Sons, Inc.; 2013p. 71–92.
53. Winyard PG, Ryan B, Eggleton P, Nissim A, Taylor E, Lo Faro ML, et al. Measurement and meaning of markers of reactive species of oxygen, nitrogen and sulfur in healthy human subjects and patients with inflammatory joint disease. *Biochem. Soc. Trans.* 2011;**39**(5):1226–32.
54. Regulus P, Duroux B, Bayle P-A, Favier A, Cadet J, Ravanat J-L. Oxidation of the sugar moiety of DNA by ionizing radiation or bleomycin could induce the formation of a cluster DNA lesion. *Proc. Natl. Acad. Sci. USA* 2007;**104**(35):14032–7.
55. Kriukiene E, Liutkeviciute Z, Klimasauskas S. 5-Hydroxymethylcytosine - the elusive epigenetic mark in mammalian DNA. *Chem. Soc. Rev.* 2012;**41**(21):6916–30.
56. Kanvah S, Joseph J, Schuster GB, Barnett RN, Cleveland CL, Landman U. Oxidation of DNA: Damage to Nucleobases. *Acc. Chem. Res.* 2010;**43**(2):280–7.
57. Joseph J, Schuster GB. One-electron oxidation of DNA: reaction at thymine. *Chem. Commun.* 2010;**46**(42):7872–8.
58. Cadet J, Douki T, Ravanat J-L, Di Mascio P. Sensitized formation of oxidatively generated damage to cellular DNA by UVA radiation. *Photochem. Photobiol. Sci.* 2009;**8**(7):903–11.
59. Cooke MS, Evans MD, Dizdaroglu M, Lunec J. Oxidative DNA damage: mechanisms, mutation, and disease. *FASEB J.* 2003;**17**(10):1195–214.
60. Dedon PC. The Chemical Toxicology of 2-Deoxyribose Oxidation in DNA. *Chem. Res. Toxicol.* 2008;**21**(1):206–19.
61. Chan SW, Dedon PC. The biological and metabolic fates of endogenous DNA damage products. *J. Nucleic Acids* 2010. 929047, 13 pp
62. von Sonntag C. *The Chemical Basis of Radiation Biology*. London; Philadelphia, PA: Taylor & Francis; 1989
63. Buxton GV, Greenstock CL, Helman WP, Ross AB. Critical Review of rate constants for reactions of hydrated electrons, hydrogen atoms and hydroxyl radicals (OH/O – in Aqueous Solution. *J. Phys. Chem. Ref. Data* 1988;**17**(2):513–886.

64. Khan AU. Singlet molecular oxygen. A new kind of oxygen. *J. Phys. Chem.* 1976;**80**(20):2219–28.
65. Schweitzer C, Mehrdad Z, Noll A, Grabner E-W, Schmidt R. Mechanism of Photosensitized Generation of Singlet Oxygen during Oxygen Quenching of Triplet States and the General Dependence of the Rate Constants and Efficiencies of $O_2(^1\Sigma_g^+)$, $O_2(^1\Delta_g)$, and $O_2(^3\Sigma_g^-)$ Formation on Sensitizer Triplet State Energy and Oxidation Potential. *J. Phys. Chem.* 2003;**107**(13):2192–8.
66. DeRosa MC, Crutchley RJ. Photosensitized singlet oxygen and its applications. *Coord. Chem. Rev.* 2002;**233–234**:351–71.
67. Khan AU. Direct spectral evidence of the generation of singlet molecular oxygen ($^1\Delta_g$) in the reaction of potassium superoxide with water. *J. Am. Chem. Soc.* 1981;**103**(21):6516–17.
68. Greer A. Christopher Foote's Discovery of the Role of Singlet Oxygen [$^1O_2$ ($^1\Delta_g$)] in Photosensitized Oxidation Reactions. *Acc. Chem. Res.* 2006;**39**(11):797–804.
69. Miyamoto S, Ronsein GE, Prado FM, Uemi M, Corrêa TC, Toma IN, et al. Biological hydroperoxides and singlet molecular oxygen generation. *IUBMB Life* 2007;**59**(4-5):322–31.
70. Sun S, Bao Z, Ma H, Zhang D, Zheng X. Singlet oxygen generation from the decomposition of alpha-linolenic acid hydroperoxide by cytochrome c and lactoperoxidase. *Biochemistry* 2007; **46**(22):6668–73.
71. Sugioka K, Nakano M. A possible mechanism of the generation of singlet molecular oxygen in nadph-dependent microsomal lipid peroxidation. *Biochim. Biophys. Acta* 1976;**423**(2):203–16.
72. Rosen H, Klebanoff SJ. Formation of singlet oxygen by the myeloperoxidase-mediated antimicrobial system. *J. Biol. Chem.* 1977;**252**(14):4803–10.
73. Stratton SP, Liebler DC. Determination of Singlet Oxygen-Specific versus Radical-Mediated Lipid Peroxidation in Photosensitized Oxidation of Lipid Bilayers: Effect of β-Carotene and α-Tocopherol. *Biochemistry* 1997;**36**(42):12911–20.
74. Xu L, Porter NA. Free radical oxidation of cholesterol and its precursors: Implications in cholesterol biosynthesis disorders. *Free Radical Res* 2015;**49**(7):835–49.
75. Davies MJ. Protein oxidation and peroxidation. *Biochem. J* 2016;**473**(7):805.
76. Klotz L-O, Kroncke K-D, Sies H. Singlet oxygen-induced signaling effects in mammalian cells. *Photochem. Photobiol. Sci.* 2003;**2**(2):88–94.
77. Ravanat J-L, Martinez GR, Medeiros MHG, Di Mascio P, Cadet J. Singlet oxygen oxidation of 2′-deoxyguanosine. Formation and mechanistic insights. *Tetrahedron* 2006;**62**(46):10709–15.
78. Gevantman LH. CRC Handbook of Chemistry and Physics , 8-87-8-90, 85th Edition. 85th ed. Boca Raton, FL: CRC Press; 2004
79. Zacharia IG, Deen WM. Diffusivity and Solubility of Nitric Oxide in Water and Saline. *Ann. Biomed. Eng* 2005;**33**(2):214–22.
80. Möller MN, Lancaster Jr JR, Denicola A. Chapter 2: The Interaction of Reactive Oxygen and Nitrogen Species with Membranes. In: Sadis M, editor. *Curr. Top. Membr.*, Vol. 61. Academic Press; 2008. p. 23–42.
81. Bartberger MD, Liu W, Ford E, Miranda KM, Switzer C, Fukuto JM, et al. The reduction potential of nitric oxide (NO) and its importance to NO biochemistry. *Proc. Natl. Acad. Sci. U.S.A.* 2002; **99**(17):10958–63.
82. Czapski G, Holcman J, Bielski BHJ. Reactivity of Nitric Oxide with Simple Short-Lived Radicals in Aqueous Solutions. *J. Am. Chem. Soc.* 1994;**116**(25):11465–9.
83. O'Donnell VB, Chumley PH, Hogg N, Bloodsworth A, Darley-Usmar VM, Freeman BA. Nitric Oxide Inhibition of Lipid Peroxidation: Kinetics of Reaction with Lipid Peroxyl Radicals and Comparison with α-Tocopherol. *Biochemistry* 1997;**36**(49):15216–23.
84. O'Donnell VB, Eiserich JP, Bloodsworth A, Chumley PH, Kirk M, Barnes S, et al. Nitration of unsaturated fatty acids by nitric oxide-derived reactive species. *Methods Enzymol.* 1999;**301**:454–70.
85. Morabito P, Heicklen J. Disproportionation to combination ratios of alkoxy radicals with nitric oxide. *J. Phys. Chem.* 1985;**89**(13):2914–16.
86. Denninger JW, Marletta MA. Guanylate cyclase and the ·NO/cGMP signaling pathway. *Biochim. Biophys. Acta* 1999;**1411**(2–3):334–50.
87. Hess DT, Matsumoto A, Kim S-O, Marshall HE, Stamler JS. Protein S-nitrosylation: purview and parameters. *Nat. Rev. Mol. Cell Biol.* 2005;**6**(2):150–66.

88. Foster MW, Stamler JS. New Insights into Protein S-Nitrosylation: Mitochondria as a Model System. *J. Biol. Chem.* 2004;**279**(24):25891–7.
89. Hughes MN. Relationships between nitric oxide, nitroxyl ion, nitrosonium cation and peroxynitrite. *Biochim. Biophys. Acta* 1999;**1411**(2–3):263–72.
90. Boese M, Mordvintcev PI, Vanin AF, Busse R, Mülsch A. S-Nitrosation of Serum Albumin by Dinitrosyl-Iron Complex. *J. Biol. Chem.* 1995;**270**(49):29244–9.
91. Bosworth CA, Toledo JC, Zmijewski JW, Li Q, Lancaster JR. Dinitrosyliron complexes and the mechanism(s) of cellular protein nitrosothiol formation from nitric oxide. *Proc. Natl. Acad. Sci. USA* 2009;**106**(12):4671–6.
92. Augusto O, Bonini MG, Amanso AM, Linares E, Santos CCX, De Menezes S l L. Nitrogen dioxide and carbonate radical anion: two emerging radicals in biology. *Free Radic. Biol. Med.* 2002; **32**(9):841–59.
93. Dutton AS, Fukuto JM, Houk KN. Theoretical Reduction Potentials for Nitrogen Oxides from CBS-QB3 Energetics and (C)PCM Solvation Calculations. *Inorg. Chem.* 2005;**44**(11):4024–8.
94. Lancaster JR. Nitroxidative, Nitrosative, and Nitrative Stress: Kinetic Predictions of Reactive Nitrogen Species Chemistry Under Biological Conditions. *Chem. Res. Toxicol.* 2006; **19**(9):1160–74.
95. Prutz WA, Monig H, Butler J, Land EJ. Reactions of nitrogen dioxide in aqueous model systems: oxidation of tyrosine units in peptides and proteins. *Arch. Biochem. Biophys.* 1985;**243**(1):125–34.
96. O'Donnell VB, Eiserich JP, Chumley PH, Jablonsky MJ, Krishna NR, Kirk M, et al. Nitration of unsaturated fatty acids by nitric oxide-derived reactive nitrogen species peroxynitrite, nitrous acid, nitrogen dioxide, and nitronium ion. *Chem. Res. Toxicol.* 1999;**12**(1):83–92.
97. Baker LM, Baker PR, Golin-Bisello F, Schopfer FJ, Fink M, Woodcock SR, et al. Nitro-fatty acid reaction with glutathione and cysteine. Kinetic analysis of thiol alkylation by a Michael addition reaction. *J. Biol. Chem.* 2007;**282**(42):31085–93.
98. Gobert AP, Vincendeau P, Mossalayi D, Veyret B. Mechanism of Extracellular Thiol Nitrosylation by N2O3 Produced by Activated Macrophages. *Nitric Oxide* 1999;**3**(6):467–72.
99. Szabo C, Ischiropoulos H, Radi R. Peroxynitrite: biochemistry, pathophysiology and development of therapeutics. *Nat. Rev. Drug Discov* 2007;**6**(8):662–80.
100. Huie RE, Padmaja S. The Reaction of no With Superoxide. *Free Radic. Res. Commun.* 1993; **18**(4):195–9.
101. Loegager T, Sehested K. Formation and decay of peroxynitric acid: a pulse radiolysis study. *J. Phys. Chem.* 1993;**97**(39):10047–52.
102. Carmichael AJ, Steel-Goodwin L, Gray B, Arroyo CM. Reactions of active oxygen and nitrogen species studied by EPR and spin trapping. *Free Radical Res. Commun.* 1993;**19**(Suppl. 1):S1–16.
103. Robinson KM, Beckman JS. Synthesis of peroxynitrite from nitrite and hydrogen peroxide. *Methods Enzymol.* 2005;**396**:207–14 (Nitric Oxide, Part E)
104. Keith WG, Powell RE. Kinetics of decomposition of peroxynitrous acid. *J. Chem. Soc. A* 1969; (1):90.
105. Goldstein S, Meyerstein D, Van Eldik R, Czapski G. Peroxynitrous acid decomposes via homolysis: Evidence from high-pressure pulse radiolysis. *J. Phys. Chem. A* 1999;**103**(33):6587–90.
106. Uppu RM, Squadrito GL, Pryor WA. Acceleration of peroxynitrite oxidations by carbon dioxide. *Arch. Biochem. Biophys.* 1996;**327**(2):335–43.
107. Padmaja S, Squadrito GL, Pryor WA. Inactivation of glutathione peroxidase by peroxynitrite. *Arch. Biochem. Biophys.* 1998;**349**(1):1–6.
108. Scorza G, Minetti M. One-electron oxidation pathway of thiols by peroxynitrite in biological fluids: bicarbonate and ascorbate promote the formation of albumin disulfide dimers in human blood plasma. *Biochem. J.* 1998;**329**(2):405–13.
109. Lee YA, Yun BH, Kim SK, Margolin Y, Dedon PC, Geacintov NE, et al. Mechanisms of oxidation of guanine in DNA by carbonate radical anion, a decomposition product of nitrosoperoxycarbonate. *Chem. - Eur. J.* 2007;**13**(16):4571–81.
110. Yun BH, Geacintov NE, Shafirovich V. Generation of Guanine-Thymidine Cross-Links in DNA by Peroxynitrite/Carbon Dioxide. *Chem. Res. Toxicol.* 2011;**24**(7):1144–52.

111. Sampson JB, Rosen H, Beckman JS. Peroxynitrite-dependent tyrosine nitration catalyzed by superoxide dismutase, myeloperoxidase, and horseradish peroxidase. *Methods Enzymol* 1996;**269**:210–18.
112. Ferrer-Sueta G, Quijano C, Alvarez B, Radi R. Reactions of manganese porphyrins and manganese-superoxide dismutase with peroxynitrite. *Methods Enzymol.* 2002;**349**:23–37.
113. Koppenol WH. Thermodynamics of reactions involving nitrogen-oxygen compounds. *Methods Enzymol.* 1996;**268**:7–12.
114. Alvarez B, Denicola A, Radi R. Reaction between Peroxynitrite and Hydrogen Peroxide: Formation of Oxygen and Slowing of Peroxynitrite Decomposition. *Chem. Res. Toxicol.* 1995; **8**(6):859–64.
115. Douki T, Cadet J. Peroxynitrite mediated oxidation of purine bases of nucleosides and isolated DNA. *Free Radic. Res.* 1996;**24**(5):369–80.
116. Douki T, Cadet J, Ames BN. An adduct between peroxynitrite and 2′-deoxyguanosine: 4,5-dihydro-5-hydroxy-4-(nitrosooxy)-2′-deoxyguanosine. *Chem. Res. Toxicol.* 1996;**9**(1):3–7.
117. Spencer JP, Wong J, Jenner A, Aruoma OI, Cross CE, Halliwell B. Base modification and strand breakage in isolated calf thymus DNA and in DNA from human skin epidermal keratinocytes exposed to peroxynitrite or 3-morpholinosydnonimine. *Chem. Res. Toxicol.* 1996;**9**(7):1152–8.
118. Szabo C, Ohshima H. DNA damage induced by peroxynitrite: subsequent biological effects. *Nitric Oxide* 1997;**1**(5):373–85.
119. Perrin D, Koppenol WH. The quantitative oxidation of methionine to methionine sulfoxide by peroxynitrite. *Arch. Biochem. Biophys.* 2000;**377**(2):266–72.
120. Alvarez B, Rubbo H, Kirk M, Barnes S, Freeman BA, Radi R. Peroxynitrite-dependent tryptophan nitration. *Chem. Res. Toxicol.* 1996;**9**(2):390–6.
121. Radi R. Nitric oxide, oxidants, and protein tyrosine nitration. *Proc. Natl. Acad. Sci. U.S.A.* 2004;**101**(12):4003–8.
122. van der Vliet A, O'Neill CA, Halliwell B, Cross CE, Kaur H. Aromatic hydroxylation and nitration of phenylalanine and tyrosine by peroxynitrite. Evidence for hydroxyl radical production from peroxynitrite. *FEBS Lett.* 1994;**339**(1-2):89–92.
123. Radi R, Beckman JS, Bush KM, Freeman BA. Peroxynitrite-induced membrane lipid peroxidation: the cytotoxic potential of superoxide and nitric oxide. *Arch. Biochem. Biophys.* 1991; **288**(2):481–7.
124. Poole LB. The basics of thiols and cysteines in redox biology and chemistry. *Free Radic. Biol. Med.* 2015;**80**:148–57.
125. Thapa B, Schlegel HB. Density Functional Theory Calculation of pKa's of Thiols in Aqueous Solution Using Explicit Water Molecules and the Polarizable Continuum Model. *J. Phys. Chem. A* 2016;**120**:5726–35.
126. Tolbert BS, Tajc SG, Webb H, Snyder J, Nielsen JE, Miller BL, et al. The active site cysteine of ubiquitin-conjugating enzymes has a significantly elevated pKa: functional implications. *Biochemistry* 2005;**44**(50):16385–91.
127. Nelson JW, Creighton TE. Reactivity and ionization of the active site cysteine residues of DsbA, a protein required for disulfide bond formation in vivo. *Biochemistry* 1994;**33**(19):5974–83.
128. Sardi F, Manta B, Portillo-Ledesma S, Knoops B, Comini MA, Ferrer-Sueta G. Determination of acidity and nucleophilicity in thiols by reaction with monobromobimane and fluorescence detection. *Anal. Biochem.* 2013;**435**(1):74–82.
129. Degirmenci I, Coote ML. Comparison of Thiyl, Alkoxyl, and Alkyl Radical Addition to Double Bonds: The Unusual Contrasting Behavior of Sulfur and Oxygen Radical Chemistry. *J. Phys. Chem.* 2016;**120**(10):1750–5.
130. Munday R. Bioactivation of thiols by one-electron oxidation. *Adv. Pharmacol.* 1994;**27**:237–70.
131. Cullis CF, Trimm DL. Homogeneous catalysis of the oxidation of thiols by metal ions. *Discuss. Faraday Soc.* 1968;**46**(0):144–9.
132. Munday R. In vivo toxicity of thiols: Relationship to rate of one-electron oxidation by oxyhemoglobin. *Methods Enzymol.* 1995;**251**:117–20.
133. Burner U, Obinger C. Transient-state and steady-state kinetics of the oxidation of aliphatic and aromatic thiols by horseradish peroxidase. *FEBS Lett.* 1997;**411**(2,3):269–74.

134. Enescu M, Cardey B. Mechanism of cysteine oxidation by a hydroxyl radical: a theoretical study. *ChemPhysChem* 2006;**7**(4):912–19.
135. Jourd'heuil D, Jourd'heuil FL, Feelisch M. Oxidation and Nitrosation of Thiols at Low Micromolar Exposure to Nitric Oxide. *J. Biol. Chem.* 2003;**278**(18):15720–6.
136. Scorza G, Minetti M. One-electron oxidation pathway of thiols by peroxynitrite in biological fluids: bicarbonate and ascorbate promote the formation of albumin disulphide dimers in human blood plasma. *Biochem. J* 1998;**329**(Pt 2):405–13.
137. Radi R, Beckman JS, Bush KM, Freeman BA. Peroxynitrite oxidation of sulfhydryls. The cytotoxic potential of superoxide and nitric oxide. *J. Biol. Chem.* 1991;**266**(7):4244–50.
138. Quijano C, Alvarez B, Gatti RM, Augusto O, Radi R. Pathways of peroxynitrite oxidation of thiol groups. *Biochem. J.* 1997;**322**(1):167–73.
139. Bonini MG, Augusto O. Carbon dioxide stimulates the production of Thiyl, sulfinyl, and disulfide radical anion from thiol oxidation by peroxynitrite. *J. Biol. Chem.* 2001;**276**(13):9749–54.
140. Folkes LK, Trujillo M, Bartesaghi S, Radi R, Wardman P. Kinetics of reduction of tyrosine phenoxyl radicals by glutathione. *Arch. Biochem. Biophys.* 2011;**506**(2):242–9.
141. Schoneich C, Dillinger U, von Bruchhausen F, Asmus KD. Oxidation of polyunsaturated fatty acids and lipids through thiyl and sulfonyl radicals: reaction kinetics, and influence of oxygen and structure of thiyl radicals. *Arch. Biochem. Biophys.* 1992;**292**(2):456–67.
142. Chatgilialoglu C, Ferreri C, Melchiorre M, Sansone A, Torreggiani A. Lipid geometrical isomerism: from chemistry to biology and diagnostics. *Chem. Rev.* 2014;**114**(1):255–84.
143. Mozziconacci O, Sharov V, Williams TD, Kerwin BA, Schoneich C. Peptide cysteine thiyl radicals abstract hydrogen atoms from surrounding amino acids: the photolysis of a cystine containing model peptide. *J. Phys. Chem. B* 2008;**112**(30):9250–7.
144. Zhao R, Lind J, Merenyi G, Eriksen TE. Kinetics of One-Electron Oxidation of Thiols and Hydrogen Abstraction by Thiyl Radicals from .alpha.-Amino C-H Bonds. *J. Am. Chem. Soc.* 1994;**116**(26):12010–15.
145. Schöneich C. Thiyl radicals and induction of protein degradation. *Free Radical Res.* 2016; **50**(2):143–9.
146. Nauser T, Schöneich C. Thiyl Radicals Abstract Hydrogen Atoms from the αC − H Bonds in Model Peptides: Absolute Rate Constants and Effect of Amino Acid Structure. *J. Am. Chem. Soc.* 2003;**125**(8):2042–3.
147. Follmann H. Deoxyribonucleotides: the unusual chemistry and biochemistry of DNA precursors. *Chem. Soc. Rev.* 2004;**33**(4):225–33.
148. Pogocki D, Schoneich C. Thiyl radicals abstract hydrogen atoms from carbohydrates: reactivity and selectivity. *Free Radical Biol. Med.* 2001;**31**(1):98–107.
149. Nauser T, Pelling J, Schoneich C. Thiyl radical reaction with amino acid side chains: rate constants for hydrogen transfer and relevance for posttranslational protein modification. *Chem. Res. Toxicol.* 2004;**17**(10):1323–8.
150. Schoneich C. Mechanisms of protein damage induced by cysteine thiyl radical formation. *Chem. Res. Toxicol.* 2008;**21**(6):1175–9.
151. von Sonntag C. The Chemistry of Free-Radical-Mediated DNA Damage. In: Glass WA, Varma MN, editors. *Physical and Chemical Mechanisms in Molecular Radiation Biology.* Boston, MA: Springer US; 1991. p. 287–321.
152. Nauser T, Schoneich C. Thiyl radical reaction with thymine: absolute rate constant for hydrogen abstraction and comparison to benzylic C–H bonds. *Chem. Res. Toxicol.* 2003;**16**(9):1056–61.
153. Pogocki D, Schoneich C. Thiyl radicals abstract hydrogen atoms from carbohydrates: reactivity and selectivity. *Free Radic. Biol. Med.* 2001;**31**(1):98–107.
154. Mezyk SP. Rate Constant Determination for the Reaction of Hydroxyl and Glutathione Thiyl Radicals with Glutathione in Aqueous Solution. *J. Phys. Chem.* 1996;**100**(21):8861–6.
155. Hoffman MZ, Hayon E. One-electron reduction of the disulfide linkage in aqueous solution. Formation, protonation, and decay kinetics of the RSSR- radical. *J. Am. Chem. Soc.* 1972; **94**(23):7950–7.

156. Mezyk SP. Direct rate constant measurement of radical disulphide anion formation for cysteine and cysteamine in aqueous solution. *Chem. Phys. Lett.* 1995;**235**(1–2):89–93.
157. Fang X, Wu J, Wei G, Schuchmann HP, von Sonntag C. Generation and reactions of the disulphide radical anion derived from metallothionein: a pulse radiolytic study. *Int. J. Radiat Biol.* 1995;**68**(4):459–66.
158. Monig J, Asmus KD, Forni LG, Willson RL. On the reaction of molecular oxygen with thiyl radicals: a re-examination. *Int. J. Radiat. Biol. Relat. Stud. Phys. Chem. Med.* 1987;**52**(4):589–602.
159. Sevilla MD, Yan MY, Becker D. Thiol peroxyl radical formation from the reaction of cysteine thiyl radical with molecular oxygen: an ESR investigation. *Biochem. Biophys. Res. Commun.* 1988; **155**(1):405–10.
160. Dénès F, Pichowicz M, Povie G, Renaud P. Thiyl Radicals in Organic Synthesis. *Chem. Rev.* 2014;**114**(5):2587–693.
161. Poole LB, Karplus PA, Claiborne A. Protein sulfenic acids in redox signaling. *Annu. Rev. Pharmacol. Toxicol.* 2004;**44**:325–47.
162. Poole LB, Nelson KJ. Discovering mechanisms of signaling-mediated cysteine oxidation. *Curr. Opin. Chem. Biol.* 2008;**12**(1):18–24.
163. Furdui CM, Poole LB. Chemical approaches to detect and analyze protein sulfenic acids. *Mass Spectrom. Rev.* 2014;**33**(2):126–46.
164. Saund SS, Sosa V, Henriquez S, Nguyen QNN, Bianco CL, Soeda S, et al. The chemical biology of hydropersulfides (RSSH): Chemical stability, reactivity and redox roles. *Arch. Biochem. Biophys.* 2015;**588**:15–24.
165. Ono K, Akaike T, Sawa T, Kumagai Y, Wink DA, Tantillo DJ, et al. Redox chemistry and chemical biology of H2S, hydropersulfides, and derived species: implications of their possible biological activity and utility. *Free Radic. Biol. Med.* 2014;**77**:82–94.
166. Vaidya V, Ingold KU, Pratt DA. Garlic: Source of the Ultimate Antioxidants—Sulfenic Acids. *Angew. Chem. Int. Ed.* 2009;**48**(1):157–60.
167. McGrath AJ, Garrett GE, Valgimigli L, Pratt DA. The Redox Chemistry of Sulfenic Acids. *J. Am. Chem. Soc.* 2010;**132**(47):16759–61.
168. Blanksby SJ, Ellison GB. Bond Dissociation Energies of Organic Molecules. *Acc. Chem. Res.* 2003;**36**(4):255–63.
169. Benson SW. Thermochemistry and kinetics of sulfur-containing molecules and radicals. *Chem. Rev.* 1978;**78**(1):23–35.
170. Hugo M, Turell L, Manta B, Botti H, Monteiro G, Netto LES, et al. Thiol and Sulfenic Acid Oxidation of AhpE, the One-Cysteine Peroxiredoxin from Mycobacterium tuberculosis: Kinetics, Acidity Constants, and Conformational Dynamics. *Biochemistry* 2009;**48**(40):9416–26.
171. Kice JL, Weclas-Henderson L, Kewan A. Equilibrium and kinetic studies of some reactions of 1-anthraquinonesulfenic acid and its methyl ester. *J. Org. Chem.* 1989;**54**(17):4198–203.
172. Okuyama T, Miyake K, Fueno T, Yoshimura T, Soga S, Tsukurimichi E. Equilibrium and kinetic studies of reactions of 2-methyl-2-propanesulfenic acid *Heteroat. Chem.* 1992;**3**(5-6):577–83.
173. Poole LB, Ellis HR. Identification of cysteine sulfenic acid in AhpC of alkyl hydroperoxide reductase. *Methods Enzymol.* 2002;**348**:122–36.
174. Zeida A, Guardia CM, Lichtig P, Perissinotti LL, Defelipe LA, Turjanski A, et al. Thiol redox biochemistry: insights from computer simulations. *Biophys. Rev.* 2014;**6**(1):27–46.
175. Zeida A, Babbush R, Gonzalez Lebrero MC, Trujillo M, Radi R, Estrin DA. Molecular Basis of the Mechanism of Thiol Oxidation by Hydrogen Peroxide in Aqueous Solution: Challenging the SN2 Paradigm. *Chem. Res. Toxicol.* 2012;**25**(3):741–6.
176. Cuevasanta E, Lange M, Bonanata J, Coitiño EL, Ferrer-Sueta G, Filipovic MR, et al. Reaction of Hydrogen Sulfide with Disulfide and Sulfenic Acid to Form the Strongly Nucleophilic Persulfide. *J. Biol. Chem.* 2015;**290**(45):26866–80.
177. Gupta V, Carroll KS. Sulfenic acid chemistry, detection and cellular lifetime. *Biochim. Biophys. Acta - General Subjects* 2014;**1840**(2):847–75.
178. Luo D, Smith SW, Anderson BD. Kinetics and mechanism of the reaction of cysteine and hydrogen peroxide in aqueous solution. *J. Pharm. Sci.* 2005;**94**(2):304–16.

179. Barton JP, Packer JE, Sims RJ. Kinetics of the reaction of hydrogen peroxide with cysteine and cysteamine. *J. Chem. Soc., Perkin Trans.* 1973;**2**(11):1547–9.
180. Abedinzadeh Z, Gardes-Albert M, Ferradini C. Kinetic study of the oxidation mechanism of glutathione by hydrogen peroxide in neutral aqueous medium. *Can. J. Chem.* 1989;**67**(7): 1247–55.
181. Winterbourn CC, Metodiewa D. Reactivity of biologically important thiol compounds with superoxide and hydrogen peroxide. *Free Radical Biol. Med.* 1999;**27**(3/4):322–8.
182. Peskin AV, Low FM, Paton LN, Maghzal GJ, Hampton MB, Winterbourn CC. The High Reactivity of Peroxiredoxin 2 with H2O2 Is Not Reflected in Its Reaction with Other Oxidants and Thiol Reagents. *J. Biol. Chem.* 2007;**282**(16):11885–92.
183. Manta B, Hugo M, Ortiz C, Ferrer-Sueta G, Trujillo M, Denicola A. The peroxidase and peroxynitrite reductase activity of human erythrocyte peroxiredoxin 2. *Arch. Biochem. Biophys.* 2009; **484**(2):146–54.
184. Bindoli A, Fukuto JM, Forman HJ. Thiol chemistry in peroxidase catalysis and redox signaling. *Antioxid. Redox Signal.* 2008;**10**(9):1549–64.
185. Poole LB, Nelson KJ, Karplus PA. Sulfenic Acids and Peroxiredoxins in Oxidant Defense and Signaling. In: Jakob U, Reichmann D, editors. *Oxidative Stress and Redox Regulation.* Netherlands: Dordrecht: Springer; 2013. p. 85–118.
186. Nagy P. Kinetics and mechanisms of thiol-disulfide exchange covering direct substitution and thiol oxidation-mediated pathways. *Antioxid. Redox Signal.* 2013;**18**(13):1623–41.
187. Forman HJ, Maiorino M, Ursini F. Signaling functions of reactive oxygen species. *Biochemistry* 2010;**49**(5):835–42.
188. Hatahet F, Boyd D, Beckwith J. Disulfide bond formation in prokaryotes: history, diversity and design. *Biochim. Biophys. Acta* 2014;**1844**(8):1402–14.
189. Braakman I, Helenius J, Helenius A. Manipulating disulfide bond formation and protein folding in the endoplasmic reticulum. *EMBO J.* 1992;**11**(5):1717–22.
190. Hoffman MZ, Hayon E. Pulse radiolysis study of sulfhydryl compounds in aqueous solution. *J. Phys. Chem.* 1973;**77**(8):990–6.
191. Nagy P, Ashby MT. Reactive Sulfur Species: Kinetics and Mechanisms of the Oxidation of Cysteine by Hypohalous Acid to Give Cysteine Sulfenic Acid. *J. Am. Chem. Soc.* 2007; **129**(45):14082–91.
192. Turell L, Botti H, Carballal S, Ferrer-Sueta G, Souza JM, Duran R, et al. Reactivity of sulfenic acid in human serum albumin. *Biochemistry* 2008;**47**(1):358–67.
193. Brigelius-Flohe R, Maiorino M. Glutathione peroxidases. *Biochim. Biophys. Acta* 2013; **1830**(5):3289–303.
194. Giles GI, Tasker KM, Jacob C. Hypothesis: the role of reactive sulfur species in oxidative stress. *Free Radic. Biol. Med.* 2001;**31**(10):1279–83.
195. Danon A. Redox reactions of regulatory proteins: do kinetics promote specificity? *Trends Biochem. Sci* 2002;**27**(4):197–203.
196. Couto N, Wood J, Barber J. The role of glutathione reductase and related enzymes on cellular redox homoeostasis network. *Free Radic. Biol. Med.* 2016;**95**:27–42.
197. Trivedi MV, Laurence JS, Siahaan TJ. The role of thiols and disulfides on protein stability. *Curr Protein Pept Sci* 2009;**10**(6):614–25.
198. Winterbourn CC. Biological reactivity and biomarkers of the neutrophil oxidant, hypochlorous acid. *Toxicology* 2002;**181-182**:223–7.
199. Davies MJ. Myeloperoxidase-derived oxidation: mechanisms of biological damage and its prevention. *J. Clin. Biochem. Nutr.* 2011;**48**(1):8–19.
200. Klebanoff SJ, Kettle AJ, Rosen H, Winterbourn CC, Nauseef WM. Myeloperoxidase: a front-line defender against phagocytosed microorganisms. *J. Leukoc. Biol.* 2013;**93**(2):185–98.
201. Winterbourn CC, Vissers MC, Kettle AJ. Myeloperoxidase. *Curr. Opin. Hematol.* 2000;**7**(1):53–8.
202. Miyamoto S, Martinez GR, Rettori D, Augusto O, Medeiros MH, Di Mascio P. Linoleic acid hydroperoxide reacts with hypochlorous acid, generating peroxyl radical intermediates and singlet molecular oxygen. *Proc. Natl. Acad. Sci. U. S. A.* 2006;**103**(2):293–8.

203. Eiserich JP, Cross CE, Jones AD, Halliwell B, van der Vliet A. Formation of nitrating and chlorinating species by reaction of nitrite with hypochlorous acid. A novel mechanism for nitric oxide-mediated protein modification. *J. Biol. Chem.* 1996;**271**(32):19199–208.
204. Folkes LK, Candeias LP, Wardman P. Kinetics and mechanisms of hypochlorous acid reactions. *Arch. Biochem. Biophys.* 1995;**323**(1):120–6.
205. Prutz WA, Kissner R, Nauser T, Koppenol WH. On the oxidation of cytochrome c by hypohalous acids. *Arch. Biochem. Biophys.* 2001;**389**(1):110–22.
206. Peskin AV, Winterbourn CC. Kinetics of the reactions of hypochlorous acid and amino acid chloramines with thiols, methionine, and ascorbate. *Free Radic. Biol. Med.* 2001;**30**(5):572–9.
207. Pattison DI, Davies MJ. Absolute rate constants for the reaction of hypochlorous acid with protein side chains and peptide bonds. *Chem. Res. Toxicol.* 2001;**14**(10):1453–64.
208. Rosen H, Klebanoff SJ, Wang Y, Brot N, Heinecke JW, Fu X. Methionine oxidation contributes to bacterial killing by the myeloperoxidase system of neutrophils. *Proc. Natl. Acad. Sci. U.S.A.* 2009;**106**(44):18686–91.
209. Boschi-Muller S, Azza S, Sanglier-Cianferani S, Talfournier F, Van Dorsselear A, Branlant G. A sulfenic acid enzyme intermediate is involved in the catalytic mechanism of peptide methionine sulfoxide reductase from Escherichia coli. *J. Biol. Chem.* 2000;**275**(46):35908–13.
210. Hawkins CL, Pattison DI, Davies MJ. Hypochlorite-induced oxidation of amino acids, peptides and proteins. *Amino Acids* 2003;**25**(3-4):259–74.
211. Winterbourn CC, Brennan SO. Characterization of the oxidation products of the reaction between reduced glutathione and hypochlorous acid. *Biochem. J* 1997;**326**(1):87.
212. Hawkins CL, Davies MJ. Hypochlorite-induced damage to proteins: formation of nitrogen-centred radicals from lysine residues and their role in protein fragmentation. *Biochem. J* 1998;**332** (Pt 3):617–25.
213. Pattison DI, Davies MJ. Kinetic analysis of the role of histidine chloramines in hypochlorous acid mediated protein oxidation. *Biochemistry* 2005;**44**(19):7378–87.
214. Bergt C, Fu X, Huq NP, Kao J, Heinecke JW. Lysine residues direct the chlorination of tyrosines in YXXK motifs of apolipoprotein A-I when hypochlorous acid oxidizes high density lipoprotein. *J. Biol. Chem.* 2004;**279**(9):7856–66.
215. Sivey JD, Howell SC, Bean DJ, McCurry DL, Mitch WA, Wilson CJ. Role of lysine during protein modification by HOCl and HOBr: halogen-transfer agent or sacrificial antioxidant? *Biochemistry* 2013;**52**(7):1260–71.
216. Ronsein GE, Winterbourn CC, Di Mascio P, Kettle AJ. Cross-linking methionine and amine residues with reactive halogen species. *Free Radic. Biol. Med.* 2014;**70**:278–87.
217. Raftery MJ. Detection and characterization of N-alpha-chloramines by electrospray tandem mass spectrometry. *Anal. Biochem.* 2007;**366**(2):218–27.
218. Hawkins CL, Davies MJ. Hypochlorite-induced oxidation of proteins in plasma: formation of chloramines and nitrogen-centred radicals and their role in protein fragmentation. *Biochem. J* 1999;**340**(Pt 2):539–48.
219. Hawkins CL, Davies MJ. Hypochlorite-induced damage to nucleosides: formation of chloramines and nitrogen-centered radicals. *Chem. Res. Toxicol.* 2001;**14**(8):1071–81.
220. Hawkins CL, Davies M. l., Reaction of HOCl with amino acids and peptides: EPR evidence for rapid rearrangement and fragmentation reactions of nitrogen-centred radicals. *J. Chem. Soc., Perkin Trans.* 1998;**2**(9):1937–46.
221. Fu X, Wang Y, Kao J, Irwin A, d'Avignon A, Mecham RP, et al. Specific sequence motifs direct the oxygenation and chlorination of tryptophan by myeloperoxidase. *Biochemistry* 2006; **45**(12):3961–71.
222. Fu X, Kao JL, Bergt C, Kassim SY, Huq NP, d'Avignon A, et al. Oxidative cross-linking of tryptophan to glycine restrains matrix metalloproteinase activity: specific structural motifs control protein oxidation. *J. Biol. Chem.* 2004;**279**(8):6209–12.
223. Chapman AL, Senthilmohan R, Winterbourn CC, Kettle AJ. Comparison of mono- and dichlorinated tyrosines with carbonyls for detection of hypochlorous acid modified proteins. *Arch. Biochem. Biophys.* 2000;**377**(1):95–100.

224. Exner M, Alt E, Hermann M, Hofbauer R, Kapiotis S, Quehenberger P, et al. p-Hydroxyphenylacetaldehyde, the major product of tyrosine oxidation by the activated myeloperoxidase system can act as an antioxidant in LDL. *FEBS Lett.* 2001;**490**(1-2):28–31.
225. Hazen SL, Gaut JP, Hsu FF, Crowley JR, d'Avignon A, Heinecke JW. p-Hydroxyphenylacetaldehyde, the major product of L-tyrosine oxidation by the myeloperoxidase-H2O2-chloride system of phagocytes, covalently modifies epsilon-amino groups of protein lysine residues. *J. Biol. Chem.* 1997;**272**(27):16990–8.
226. Sassa A, Kamoshita N, Matsuda T, Ishii Y, Kuraoka I, Nohmi T, et al. Miscoding properties of 8-chloro-2′-deoxyguanosine, a hypochlorous acid-induced DNA adduct, catalysed by human DNA polymerases. *Mutagenesis* 2013;**28**(1):81–8.
227. Hawkins CL, Pattison DI, Davies MJ. Reaction of protein chloramines with DNA and nucleosides: evidence for the formation of radicals, protein-DNA cross-links and DNA fragmentation. *Biochem. J* 2002;**365**(Pt 3):605–15.
228. Prutz WA. Interactions of hypochlorous acid with pyrimidine nucleotides, and secondary reactions of chlorinated pyrimidines with GSH, NADH, and other substrates. *Arch. Biochem. Biophys.* 1998;**349**(1):183–91.
229. Prutz WA. Hypochlorous acid interactions with thiols, nucleotides, DNA, and other biological substrates. *Arch. Biochem. Biophys.* 1996;**332**(1):110–20.
230. Suzuki T, Ohshima H. Nicotine-modulated formation of spiroiminodihydantoin nucleoside via 8-oxo-7,8-dihydro-2′-deoxyguanosine in 2′-deoxyguanosine-hypochlorous acid reaction. *FEBS Lett* 2002;**516**(1-3):67–70.
231. Whiteman M, Hong HS, Jenner A, Halliwell B. Loss of oxidized and chlorinated bases in DNA treated with reactive oxygen species: implications for assessment of oxidative damage in vivo. *Biochem. Biophys. Res. Commun.* 2002;**296**(4):883–9.
232. Masuda M, Suzuki T, Friesen MD, Ravanat JL, Cadet J, Pignatelli B, et al. Chlorination of guanosine and other nucleosides by hypochlorous acid and myeloperoxidase of activated human neutrophils. Catalysis by nicotine and trimethylamine. *J. Biol. Chem.* 2001;**276**(44):40486–96.
233. Stocker R, Keaney JF. Role of Oxidative Modifications in Atherosclerosis. *Physiol. Rev.* 2004; **84**(4):1381.
234. Sugiyama S, Kugiyama K, Aikawa M, Nakamura S, Ogawa H, Libby P. Hypochlorous Acid, a Macrophage Product, Induces Endothelial Apoptosis and Tissue Factor Expression. *Arterioscler. Thromb. Vasc. Biol.* 2004;**24**(7):1309.
235. Carr AC, McCall MR, Frei B. Oxidation of LDL by Myeloperoxidase and Reactive Nitrogen Species. *Arterioscler. Thromb. Vasc. Biol.* 2000;**20**(7):1716.
236. Winterbourn CC, van den Berg JJ, Roitman E, Kuypers FA. Chlorohydrin formation from unsaturated fatty acids reacted with hypochlorous acid. *Arch. Biochem. Biophys.* 1992;**296**(2):547–55.
237. Jerlich A, Fabjan JS, Tschabuschnig S, Smirnova AV, Horakova L, Hayn M, et al. Human low density lipoprotein as a target of hypochlorite generated by myeloperoxidase. *Free Radic. Biol. Med.* 1998;**24**(7-8):1139–48.
238. Hazen SL, Hsu FF, Duffin K, Heinecke JW. Molecular chlorine generated by the myeloperoxidase-hydrogen peroxide-chloride system of phagocytes converts low density lipoprotein cholesterol into a family of chlorinated sterols. *J. Biol. Chem.* 1996;**271**(38):23080–8.
239. van den Berg JJ, Winterbourn CC, Kuypers FA. Hypochlorous acid-mediated modification of cholesterol and phospholipid: analysis of reaction products by gas chromatography-mass spectrometry. *J. Lipid Res.* 1993;**34**(11):2005–12.
240. Kawai Y, Kiyokawa H, Kimura Y, Kato Y, Tsuchiya K, Terao J. Hypochlorous acid-derived modification of phospholipids: characterization of aminophospholipids as regulatory molecules for lipid peroxidation. *Biochemistry* 2006;**45**(47):14201–11.

CHAPTER 3

Reactive Species in Biological Systems

3.1 INTRODUCTION

The biomolecular basis of aging has confounded many scientists over the years and remains one of the most intriguing questions in biology. Several biological theories have been proposed—from the classical evolutionary perspective to telomeres and reproductive cell cycles. The genetic theory of aging is probably one of the most compelling theories and suggests the role of free radicals in the initiation of oxidative damage to key biomolecular systems such as DNA. In the late 1950s, Harman[1] postulated that production of free radicals from living things through irradiation chemistry can be implicated in the development of radiation-induced mutation, cancer, and aging. Furthermore, direct detection of free radicals from lyophilized microorganisms have been demonstrated using electron paramagnetic resonance (EPR) spectroscopy[2] and showed the dependence of radical formation in the presence of air. Hence, the role of molecular oxygen in the production of free radicals had become apparent.

Backtrack to the late 19th century. Fenton in his seminal work proposed that oxidation of organic acid substrates can be achieved in the presence of H_2O_2 and Fe^{2+} salt.[3] Although purely chemical, this important observation suggests that an oxygen radical intermediate can be formed. Several decades later, scientists were beginning to discover that radical production can occur in biological systems. Yamasaki and Piette[4] showed that free radical formation can be catalyzed by peroxidase in the presence of substrates such as dihydroxyfumarate, triose reductone or indoleacetic acid, and hydrogen peroxide. These free radicals can reduce oxygen to form hydroperoxyl radical and the peroxidase itself. Oxygen can also be metabolically incorporated into aliphatic amine substrates via *N*-oxidation to hydroxylamine where the oxygen radical or flavoprotein formation was implicated to be a key intermediate.[5] Further evidence also corroborated the idea that enzymes play a direct role in the one-electron reduction of the xenobiotics, nitrofurazone, by the oxygen-sensitive nitroreductase purified from *Escherichia coli*,[6] forming the nitro radical anion intermediate that stimulates the formation of superoxide radical and oxygen consumption. Although it was demonstrated that oxygen radicals can indeed be formed metabolically, the biochemistry of their formations remained unclear.

Through rapid-freezing EPR, the direct detection of superoxide radical produced from the enzymatic reduction of oxygen was demonstrated by Knowles et al.[7] using

Reactive Species Detection in Biology
DOI: http://dx.doi.org/10.1016/B978-0-12-420017-3.00002-5

xanthine oxidase at elevated pH (pH = 10.5). With the EPR signal intensity being dependent on the oxygen rather than the enzyme concentration, it was further revealed that radicals can be endogenously generated from oxygen. However, the active enzymatic site of superoxide radical formation had not been identified until Massey et al.[8] proposed this site in the native xanthine oxidase to be the reduced flavin moiety rather than the non-heme iron and was further confirmed by EPR using reduced flavin.[9] More evidence followed showing superoxide radicals as precursors of hydrogen peroxide generated from mitochondrial respiration.[10] In the several decades that followed, it has become clear that radicals can be generated in biological systems. In this chapter, the various cellular sites of radical production will be further discussed and more detailed mechanisms of their generation will be presented.

3.2 EXTRACELLULAR MILIEU

3.2.1 Exogenous agents

Extracellular radical generation from an exogenous agent (EA) can be mediated through oxidation by oxygen (Eq. 3.1), oxidation by high-valent metal ions (Eq. 3.2) (also through reduction by low-valent metal ions), oxidation by oxygen of reduced EA (Eq. 3.3), or from direct reduction of oxygen by metal ions (Eq. 3.4).

$$\mathrm{EA} \xrightarrow{O_2 \;\; O_2^{\bullet-}} \mathrm{EA}^{\bullet} \tag{3.1}$$

$$\mathrm{EA} \xrightarrow{M^{n} \;\; M^{n-1}} \mathrm{EA}^{\bullet} \tag{3.2}$$

$$\mathrm{EA}^{\bullet} \xrightarrow{O_2 \;\; O_2^{\bullet-}} \mathrm{EA(ox)} \tag{3.3}$$

$$M^{n} \xrightarrow{O_2 \;\; O_2^{\bullet-}} M^{n+1} \tag{3.4}$$

Examples of EAs are the circulating catechols such as the catecholamines (e.g., dopamine, norepinepherine, and epinephrine) and estrogens, as well as the xenobiotics such as quinones (e.g., doxorubicin, menadione) and bipyridinium (e.g., paraquat) (Fig. 3.1).

For catechols, their one-electron oxidation by metal ions (or oxygen in basic conditions[11]) to the semiquinone form leads to the reduction of oxygen to form superoxide radicals and quinone. Quinones in biological systems can be reduced back to the catechol via a two-electron reduction process by NADPH-quinone

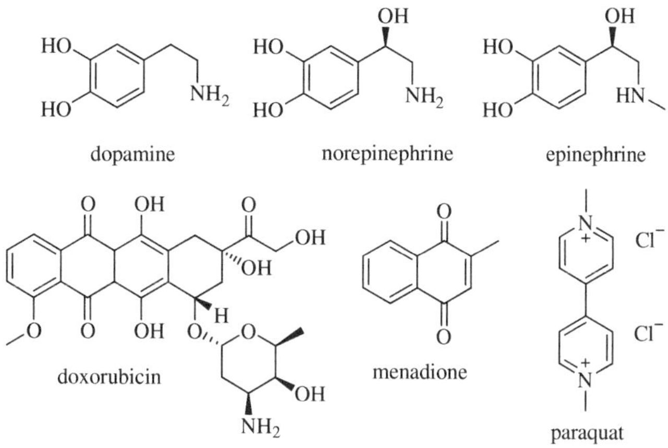

Figure 3.1 Exogenous agents capable of generating radicals.

oxidoreductase 1, other NADPH-dependent oxidoreductases, or vitamin C, hence allowing for the continuous production of radicals from oxygen (Eq. 3.5).[12]

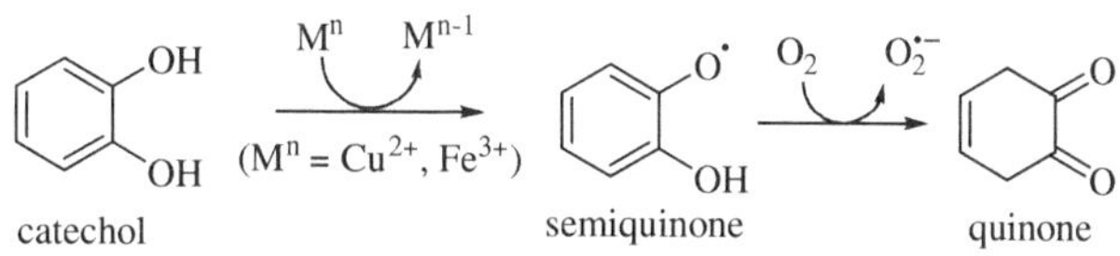

(3.5)

In the case of doxorubicin, paraquinones undergo a one-electron reduction by enzymes such as NADPH cytochrome P450 reductase or xanthine oxidoreductase to yield the semiquinone radical that can subsequently reduce oxygen to superoxide radical to form the quinone once again (Eq. 3.6). This ability of doxorubicin to redox cycle oxygen to radicals—resulting in lipid peroxidation and DNA damage of cancer cells—is one of the main mechanisms of its antitumor property[13] but also the main basis of its cardiotoxicity.[14]

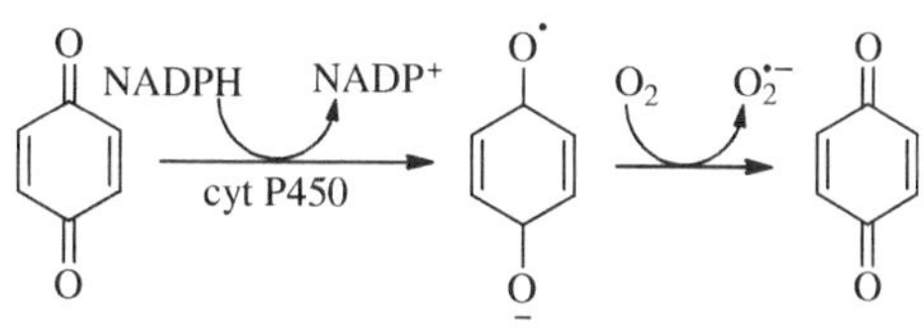

(3.6)

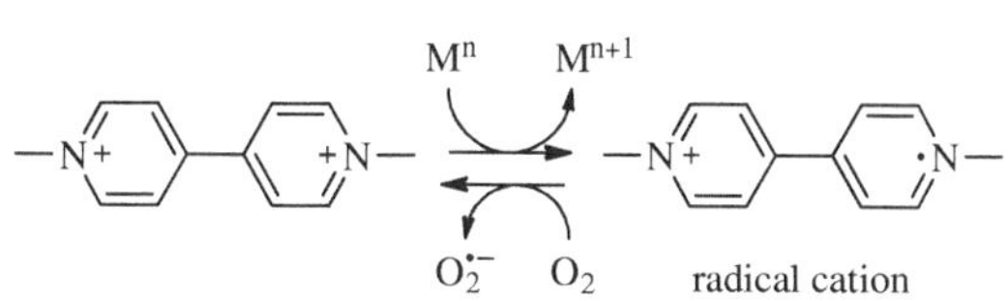

(3.7)

Paraquat can be one-electron reduced extracellularly by metal ions or in the presence of NADPH or NADPH-generating systems to the radical cation form which it can subsequently reduce oxygen to a superoxide radical that regenerates the paraquat (Eq. 3.7).[15] Anaerobically, the paraquat radical has been shown to be generated by electron transfer from xanthine using xanthine oxidase and can react with hydrogen peroxide to form hydroxyl radical.[16]

Although most of these EAs are metabolized intracellularly by phase I and II enzymes, which results in their activation and further modification, respectively, their ability to extracellularly form radicals could have important pathophysiological consequences such as in the development of inflammation, atherosclerosis, or membrane damage leading to other cardiovascular and neurodegenerative diseases. Exoenzymes can play a role in the generation of extracellular radicals. Exoenzymes are extracellular macromolecules that are secreted by cells and exhibit functional roles such as virulence factors for the destruction of other cells and tissues, as clotting and anticlotting agents, and as digestive enzymes. In acute pancreatitis, elevated plasma concentrations of xanthine oxidase has been observed and can be responsible for the circulation of free radicals, triggering an inflammatory response to the lungs.[17]

3.2.2 Photochemical and ionizing radiation

Photochemical radical formation in biological systems can be initiated by ultraviolet (100–400 nm), visible (400–700 nm), and infrared (760–1440 nm) lights. The mechanism of radical generation can vary from one system to another. These radicals are formed directly through homolytic cleavage of O–O or S–S bonds from peroxides or disulfides, respectively, or through initiation by sensitizing agents from exogenous sources such as riboflavin (vitamin B2) or photodynamic agents (e.g., phthalocyanines, porphyrin, and nanoparticles) via formation of singlet oxygen or superoxide radical.

Hydrogen peroxide is intracellularly generated but can also cross cell membranes into the extracellular matrix to extracellularly form radicals. Hydrogen peroxide had been measured in the extracellular space of the brain using microamperometric microsensors in vivo[18] and from the extracellular pool as generated from phagocytes.[19] Although ultraviolet-A irradiation at 320–400 nm of cells kept in cell culture media results in the activation of stress kinases and the generation of hydrogen peroxide,[20] the exact mechanism by which these reactive oxygen species (ROS) are extracellularly produced is not clear but it may be partly the photochemical sensitization of light-sensitive moieties (e.g., flavins) that lead to the formation of ROS such as singlet oxygen and superoxide radicals, both of which can initiate further signaling events leading to more radical production.

Radical production through photosensitization of riboflavin or photodynamic agents such as phorphyrins occurs via several mechanisms. For example, light sensitization of riboflavin could lead to the formation of its excited triplet state and allow the transfer of energy from the excited riboflavin to oxygen and forming the ground state riboflavin and triple oxygen. Singlet oxygen can be quenched by electron-rich substrates such as aromatic amines to produce superoxide radicals[21] or olefins to produce peroxyl radicals.[22] Photosensitized phorphyrins lead to the formation of its triplet excited state; in the presence of a substrate and oxygen, they can form superoxide radicals via type I charge transfer mechanism and subsequently form other reactive oxygen species. For example, ROS formation from the photobleaching of hematoporphyrins has been shown to mediate the destruction of tumor cells during photodynamic therapy.[23]

Ionizing radiation produces radicals differently from photolysis. In the former, absorption of high-energy particles by an atom or molecule leads to the release of atomic particles such as electrons, protons, and neutrons. In general, x-rays (0.01−10 nm) and gamma rays (<10 pm) are ionizing, although the far ultraviolet (UV) wavelength between 170 and 182 nm have also been shown to yield superoxide radicals in oxygen-saturated water via homolysis of the water molecules to H and HO radicals and reaction of H with oxygen to form hydroperoxyl or superoxide radicals.[24] Ionizing radiation exhibits either direct or indirect effects on radical generation. Direct energy deposition on biomolecules results in the formation of radical cations plus electrons as well as excited states of the biomolecules, while indirect effects results in absorption of energy by the surrounding water, producing a hydrogen atom, a hydroxyl radical, and solvated electrons that can further react with oxygen to form superoxide and other biomolecules forming peroxyl radicals.[25] The effects of ionizing radiation play a crucial role in carcinogenesis,[26] but radiotherapy also have been exploited for the treatment of cancer.[27] Like photochemistry, the effects of persistent irradiation can induce metabolic production of reactive oxygen species and adaptive responses as well as exacerbate the degeneration of tissues and the development of some diseases, including cancer, inflammation, and fibrosis.[28]

3.2.3 Metal ions or heme metals

Radical generation from low-valent transition metal ions such as Fe(II), Cr(II), Ti(III), or Cu(I) occurs mostly through electron-transfer processes from the metal ion to the substrate. Transition metals are good electron stores by virtue of their atomic size and accessibility of electrons through their *d* orbitals. However, for iron, e.g., the high-valent ferric form, Fe(III), is predominantly present in biological systems in chelated form. Chelates determine Fe(III)/Fe(II) redox potential such that in the case of iron-transferrin or desferrioxamine, the ferric state is nonreducible by any reducing agents

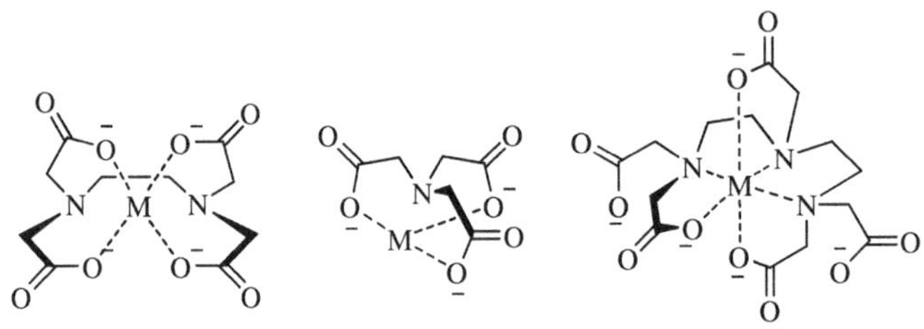

Figure 3.2 Metal complexes with various chelators. *EDTA*, ethylenediaminetetraacetate; *NTA*, nitrilotriacetate; *DTPA*, diethylene triamine pentaacetate.

while ferric along with small molecule chelators such as ethylenediaminetetraacetate (EDTA), nitrilotriacetate (NTA), or diethylene triamine pentaacetate (DTPA), is reduced by superoxide to the ferrous form (Eq. 3.8)[29] (Fig. 3.2).

$$Fe(III) + O_2^{\bullet -} \rightarrow Fe(II) + O_2 \quad (3.8)$$

The reduction of ferric to ferrous iron is perhaps one the most important life processes because the ferrous state is involved in the activation and reduction of oxygen, the decomposition of peroxides, and the transport and storage of oxygen and electrons.[29] The ability to reduce ferric to ferrous depends not only on the type of chelator it is associated with but also on the type of reducing agents, the most common of which are thiols, ascorbate, and ferric reductases. Chelates such as DTPA and EDTA can significantly slow the ferric- and cupric-catalyzed autoxidation of ascorbate in neutral pH,[30] while metmyoglobin can be reduced to oxymyoglobin by dihydrolipoic acid but not by glutathione, cysteine, or *N*-acetyl cysteine. However, only the high oxidation state ferrylmyoglobin, Fe(IV) = O, is capable of oxidizing all the thiols mentioned previously to thiyl radical.[31]

Likewise, the ability of metal ions to reduce hydrogen peroxide to yield hydroxyl radical via Fenton's reaction depends on the type and oxidation state of the metal ion, the type of

$$Fe(III) + H_2O_2 \rightarrow Fe(III) + HO^- + HO^\bullet \quad (3.9)$$

chelator, as well as the polarity and pH of the medium (Eq. 3.9).[29] For example, Mn (II), Co(II), or Ag(I) only react slowly to hydrogen peroxide but fast with Ti(III), Cr (II), and Cu(I) and intermediate rate with Fe(II).[32] The accessibility of the ferrous ion, e.g., to oxidize hydrogen peroxide reaction is determined by the type of chelator. Larger and bulkier intact iron heme systems such as hemoglobin, ferritin, and transferrin are not effective catalysts for HO radical production from hydrogen peroxide unless free ferrous is released from these systems through oxidation of hemoglobin by peroxides or by reduction of the ferric form in ferritin or transferrin by ascorbate or superoxide due to the weak binding of ferrous. Ferrous complexes with lower-molecular-weight ligands such as DTPA and EDTA directly undergo Fenton's reaction.[33]

Other non-Fenton's type reactions that are known to generate free radicals are the reaction of Fe(III) with hydrogen peroxide to yield hydroperoxyl and subsequently HO radicals as shown in Eqs. (3.10)–(3.12).[34]

$$\mathrm{Fe(III)} + \mathrm{H_2O_2} \rightleftharpoons \mathrm{Fe(II)OOH} + \mathrm{H^+} \quad (3.10)$$

$$\mathrm{Fe(II)OOH} \overset{\text{slow}}{\rightarrow} \mathrm{Fe(II)} + \mathrm{HOO^\bullet} \quad (3.11)$$

$$\mathrm{Fe(II)} + \mathrm{H_2O_2} \rightarrow \mathrm{Fe(III)} + \mathrm{HO^-} + \mathrm{HO^\bullet} \quad (3.12)$$

Although a very slow reaction (i.e., over a 24-h period), superoxide radicals have also been shown to be generated in partially oxygenated Fe(II)hemoglobin producing Fe(III)hemoglobin (Eq. 3.13), while the reaction of hydrogen peroxide (produced from the dismutation reaction of the generated superoxide) with Fe(II)hemoglobin yields the Fe(IV)ferrylhemoglobin and water (Eqs. 3.14 and 3.15), where the former further reacts with hydrogen peroxide to form superoxide radicals and Fe(III)hemoglobin (Eq. 3.16).[35]

$$\mathrm{Hb(II)} + \mathrm{O_2} \rightarrow \mathrm{Hb(III)} + \mathrm{O_2^{\bullet -}} \quad (3.13)$$

$$2\,\mathrm{O_2^{\bullet -}} + \mathrm{H^+} \rightarrow \mathrm{H_2O_2} + \mathrm{O_2} \quad (3.14)$$

$$\mathrm{Hb(II)} + \mathrm{H_2O_2} \rightarrow \mathrm{Hb(IV)} = \mathrm{O} + \mathrm{H_2O} \quad (3.15)$$

$$\mathrm{Hb(IV)} = \mathrm{O} + \mathrm{H_2O_2} \rightarrow \mathrm{Hb(III)} + \mathrm{O_2^{\bullet -}} + 2\mathrm{H^+} \quad (3.16)$$

3.3 MEMBRANE-BOUND ENZYMES

3.3.1 NADPH oxidases (NOXs)

NOX enzymes are transmembrane proteins that are essential in signal transduction processes and in innate immunity systems, where they are used by the cells as a first line of defense against invading organisms. Phagocytic NOX (or *phox*) includes the NOX2 (formerly gp91^{phox}) that catalyzes the NADPH-dependent reduction of oxygen to superoxide radical and is composed of various components that are located in the cytosol and the membrane. The membrane-bound flavocytochrome b_{558} is composed of the gp91^{phox} and p22^{phox}, while the cytosolic components are p47^{phox}, p67^{phox}, p40^{phox}, and small G-protein Rac1/2,[36] which upon activation catalyze the electron flow from the electron donor (NADPH) to the electron acceptor (oxygen) to form the superoxide based on Eq. (3.17).

$$NADPH + 2O_2 \rightarrow NADP^+ \rightarrow 2O_2^{\bullet -} + H^+ \quad (3.17)$$

Superoxide production can be influenced through NOX activation by various cellular stimuli such as exogenous inorganic and organic chemical agents, endogenously formed cytotoxic molecules such as oxidized lipoproteins, ROS itself, ionizing radiation, shear stress, low temperature, osmotic pressure, inflammatory factors such as LPS, cytokines, angiotensin II, and lipids.[37] Upon activation, the cytosolic components assemble and translocate into the membrane to generate superoxide radical.

Various nonphagocytic cells also exhibit NOX activity though they produce lesser ROS than the NOX2.[36] So far, nonphagocyte NADPH oxidases include the NOX isoforms, NOX1, NOX3, NOX4, NOX5, DUOX1, DUOX2, and the NOXO1 (p41^{nox}) and NOXA1 (p67^{nox}) as the p47^{phox} and p67^{phox} homologs, respectively. While these NOXs can also be found in other cell types, they produce much less superoxide and by different mechanisms.[38] Measurement of oxygen consumption along with free radical formation has been detected in LPS-activated alveolar macrophages [39] and rat aortic homogenates using an EPR spin-trapping technique,[40] but the amount of superoxide can be overestimated spectrophotometrically using lucigenin.[41]

3.3.2 Endothelial nitric oxide synthase (eNOS, NOS3)

Endothelial nitric oxide synthase (eNOS) resides in the plasmalemmal caveolae but is also known to co-localize intracellularly in the Golgi region,[42] and its intracellular translocation as a result of external stimulation has been associated with decrease nitric oxide production.[43] eNOS catalyzes the production of nitric oxide from L-arginine as substrate via reaction with oxygen and reducing equivalents of NADPH according to Eq. (3.18).

$$\text{L-arginine} + 3/2\ NADPH + H^+ + 2O_2 \rightarrow \text{citrulline} + NO + 3/2\ NADP^+ \quad (3.18)$$

The enzyme eNOS is composed of two noncovalently bound protein monomers. Each monomer consists of the *C*-terminus reductase domain that is homologous to cytochrome P450 reductase[44] and contains the flavin mononucleotide (FMN), flavin adenine dinucleotide (FAD), and the reduced NADPH binding site, while the *N*-terminal oxygenase domain contains the heme prosthetic group, tetrahydrobiopterin (H_4B), and the arginine binding site.[45] The reductase domain is linked to the oxygenase domain by Ca^{2+}-calmodulin to form a dimer. The electron flows from the NADPH to FAD to FMN and finally to the ferric heme, which then activates oxygen that is responsible for the hydroxylation of L-arginine and further leads to its oxidation to form citrulline and NO. The presence of the enzyme cofactor, H_4B, is essential during the course of L-arginine oxidation processes because NO production from NOS is highly regulated H_4B; in the absence of H_4B, NOS has been shown to produce hydrogen peroxide[46] from superoxide. Hence, reduced H_4B bioavailability or destruction of the heme due to its susceptibility to reaction with peroxynitrite or reduced synthesis in the case of H_4B have been implicated in the development of oxidative stress-mediated cardiovascular diseases.[47] Furthermore, eNOS can also switch from producing normal levels of nitric oxide to higher superoxide levels via denitrosylation and S-glutathionylation of its cysteine residues[48] or via inhibition of the Akt-mediated phosphorylation of eNOS at Ser-1177.[49]

Detection of nitric oxide is usually employed to assess the enzyme viability using EPR spin-trapping technique.[50] Both in in-vitro and in-vivo systems, the uncoupled eNOS is characterized by lower nitric oxide and higher superoxide levels than normal as detected using fluorescence and EPR spin-trapping techniques,[51,52] and these results were correlated with decreased H_4B levels using HPLC techniques.[53] Detection of both nitric oxide and superoxide can be accomplished through cell stimulation with Ca^{2+}/calmodulin.[52] Complementary studies can also be carried out by analyzing nitrite or nitrate ions concentrations as secondary products from the reaction of nitric oxide with oxygen or superoxide.[54]

3.4 CYTOSOLIC ENZYMES

3.4.1 Xanthine oxidase (XO, XAO)

Xanthine oxidase (XO) belongs to the molybdenum hydroxylase family within the oxidoreductase class of enzymes. Hypoxanthine is oxidized as part of its metabolic degradation at the molybdenum center where hypoxanthine is converted ultimately to uric acid via xanthine. Electrons are then passed on to FAD via the iron−sulfur clusters and finally to oxygen as the final electron acceptor to form hydrogen peroxide or superoxide.[55] The stoichiometry of ROS production depends on the kinetics of XO reduction of oxygen, which exhibits two initial fast-phase two-electron reduction and two final slower-phase two one-electron reduction of oxygen. In a fully six-electron reduced XO, XO can sequentially dump the first four electrons to each molecule of oxygen to form two equivalents of hydrogen peroxide. The final two electrons then leave the enzyme to form two equivalents of superoxide according to Eq. (3.19).[56]

$$XO_{6e-} \xrightarrow[35\ s^{-1}]{O_2 \to H_2O_2} XO_{4e-} \xrightarrow[33.5\ s^{-1}]{O_2 \to H_2O_2} XO_{2e-} \xrightarrow[20\ s^{-1}]{O_2 \to O_2^{\bullet-}} XO_{1e-} \xrightarrow[0.9\ s^{-1}]{O_2 \to O_2^{\bullet-}} XO_{ox} \quad (3.19)$$

EPR techniques have been used to detect superoxide radical from pure isolated bovine xanthine oxidase using xanthine or hypoxanthine as substrates to initiate radical production.[57,58] Chemiluminescence techniques using lucigenin have also been employed.[59] In intact cells, reactive species production was measured by fluorescence spectroscopy using dichlorofluorescin diacetate and Amplex Red fluorescenct probes[60] or dihydroethidium.[61] In both in vitro and in vivo systems, xanthine oxidase–dependent radical production can be confirmed using an XO inhibitor such as allopurinol or oxypurinol.[62–64]

3.4.2 Aldehyde oxidase (AO, AOX)

Aldehyde oxidase (AO) belongs to the same family of xanthine oxidoreductases as XO. Although AO has structural similarities to XO, they differ in the substrate specificity and inhibitors.[55,65] AO can effectively hydroxylate most aldehydes to carboxylic acid but is unreactive toward xanthine; its oxidative and reductive half-reactions of AO do not involve biphasic kinetic properties as observed with XO.[65] Using purified AO and *p*-dimethyl aminocinnamaldehyde as reducing substrate, superoxide and hydrogen peroxide have been detected using spectroscopic and EPR spin techniques.[66] Nitrite reduction by AO has also been observed to form nitric oxide[67,68] and have been shown to function as superoxide-generating NADH oxidase in which the oxygen reduction to superoxide occurs at the FAD site.[69] Being a metabolizing enzyme, AO is highly ubiquitous in the liver but can also be found in other tissues such as lung,[70] spleen, and heart[71] and therefore are implicated in the development of some oxidative stress-mediated diseases such as increased risk of breast cancer from alcohol consumption.[72]

3.4.3 Inducible and neuronal nitric oxide synthase (iNOS, NOS2, and nNOS, NOS1)

In cardiomyocytes, nNOS is localized in the cardiac sarcoplasmic reticulum and regulates cytosolic-SR Ca^{2+} homeostasis, compared to eNOS in the calveolae, whose role is to regulate extracellular-cytosolic Ca^{2+} via L-type Ca^{2+} channel,[73] while iNOS localizes in the cytoplasm in adipocytes.[74] However, NOS intacellular compartmentalization varies from one cell type to another and cellular response to external stimulation.[75] All the three NOS (i.e., eNOS, iNOS, and nNOS) are encoded by distinct genes and have similar genomic structures. Similar to eNOS, iNOS and nNOS are homodimers with the reductase *C*-terminus and oxygenase *N*-terminus, which are linked together by Ca^{2+}/calmodulin and associated with tightly bound H_4B, FAD, FMN and the iron heme.[76] Similar to eNOS, nNOS, is a constitutively expressed enzyme whose function is highly regulated by the Ca^{2+} fluxes and binding to calmodulin and is expressed, though not exclusively, in peripheral neurons. On the other hand, iNOS is ubiquitous in the immune and cardiovascular systems, and its expression is highly regulated by inflammatory agents such as cytokines and LPS, playing a crucial role in immune response as well as in the maintenance of smooth muscle tone. In contrast to the constitutive NOSs, iNOS is insensitive to Ca^{2+} due probably to the tight and noncovalent nature of calmodulin binding to both the oxygenase and reductase domains.[77]

Superoxide generation has been detected in purified iNOS and nNOS using EPR spin trapping. In the case of iNOS, superoxide has been shown to be produced from the flavin-binding sites of the reductase domain as opposed to the heme-oxygenase domain in eNOS.[78,79] In the cells, nNOS-mediated nitric oxide production as activated by signal transduction pathways[80] and iNOS-mediated superoxide and peroxynitrite generation in macrophages have been detected using EPR spin trapping and immuno-staining and chemiluminescence techniques in the case of peroxynitrite.[81] In an in vivo system, iNOS-mediated free radical production as induced by LPS or by long-term exposure to acetone was demonstrated using immunological and EPR spin trapping,[64,82] and nNOS-mediated ROS production has been monitored in the mouse brain using dihydroethidium microfluorography.[83]

3.4.4 Cyclooxygenases or prostaglandin-endoperoxide synthase (COX, PTGS)

Prostanoids—which include prostaglandins, prostacyclin, and thromboxane—are important mediators of inflammation and vessel constriction. They are synthesized by cyclooxygenases (COXs) from arachidonic acid as the main precursor. There are two major isoforms of COX—COX-1 and COX-2—and both function as prostaglandin H synthase catalyzing the bis-dioxygenation of arachidonic acid to form hydroxyperoxy endoperoxide (PGG_2), a prostaglandin. Its two-electron reduction to hydroxyperoxy endoperoxide (PGH_2) occurs through the COX peroxidase activity. PGH_2 is then subsequently metabolized by various enzymes to form a variety of prostanoids.[84] COX-1 is a constitutively expressed enzyme important in the normal functions of the gut and vascular system and is highly expressed in blood vessels, interstitial cells, smooth muscles cells, platelets, and mesothelial cells, whereas COX-2 is an inducible enzyme that is upregulated during inflammatory conditions but is also ubiquitous in parenchymal cells with the exception of the heart.[85]

Oxygenation of arachidonic acid by COX-1 via free radical mechanism has been described as producing peroxyl radicals.[86] Even with the formation of hydroperoxides, one can envision their homolytic cleavage to form the alkoxyl and hydroxyl radicals. Radical production from COX-2–catalyzed metabolism of arachidonic acid is evidenced by a significant increase in the amount of oxidized DNA and nucleosides by detecting the 8-oxodG lesion using LC/UV-MS-MS.[87] Arachidonic acid–derived free radicals have been detected using a combination of LC-EPR and LC-MS through a spin-trapping agent from purified COX and human prostate cancer cells.[88] Using EPR spin trapping or the fluorogenic probe hydroethidium, COX-2 transfected PC12 cells also show arachidonic acid–dependent production of carbon-centered radicals while no oxygen or thiyl radicals were formed; in the presence of ascorbic acid, ascobate radical was observed.[89] In in vivo, it was proposed that prostanoids and not ROS are implicated in the development of COX-2–mediated neurotoxicity in mice on brain activation of *N*-methyl-D-aspartate receptors as detected by fluoromicrography using hydroethidine as a superoxide probe.[90]

3.5 ORGANELLE ENZYMES

3.5.1 Mitochondria

Mitochondrial dysfunction has been implicated in a variety of diseases, including neurodegeneration,[91] cancer,[92] and cardiovascular disease.[93] Mitochondria are perhaps the most studied organelles in biological system for the generation of reactive species due to the susceptibility of oxygen to be partially reduced during the electron transfer process as a result of electron leakage from the various electron transport enzyme complexes under normal physiological conditions. However, during pathophysiological conditions, ROS production could intensify. Oxidation of pyruvate through the Krebs cycle produces reducing equivalents of NADH. Electrons from the NADH coenzyme are shuttled through the mitochondrial electron transport chain, which is composed of complexes I (NADH:ubiquinone oxidoreductase), II (succinate dehydrogenase), III (cytochrome bc_1 complex), and IV (cytochrome *c* oxidase). Via a two one–electron transfer process, an electron in complex I is passed on to the ubiquinone via flavin and the Fe–S cluster forming a semiquinone radical with the second electron reducing the semiquinone further to ubiquinol. Complex I is considered to be the major source of superoxide in which superoxide has been proposed to be formed from the fully reduced flavin to oxygen.[94]

In parallel to complex I, electrons from complex II are also transferred to the ubiquinone from $FADH_2$ as the electron donor. Although complex II is not known to be a major source of superoxide, superoxide generation has been proposed to be mediated via the FAD cofactor and ubiquinone binding sites.[95] For example, mutated complex II exhibited reduced succinate-ubiquinone oxidoreductase activities and show high sensitivity to oxygen.[96] Complex III through the Q cycle catalyzes the electron transfer from ubiquinol to cytochrome c. The semiquinone formed at the cytoplasmic site or Q_o pocket of complex III has been implicated as another source of superoxide production.[97,98] Finally, complex IV catalyzes the sequential transfer of four electrons from cytochrome c to oxygen to form water. Like complex II, complex IV does not typically produce superoxide under normal physiological conditions, although production of superoxide can be mediated by these enzymes under pathophysiological conditions such as during ischemia and reperfusion injury.[99]

Using EPR spin trapping, superoxide generation was observed from purified complex 1, and the sites of superoxide generation were confirmed by inhibition experiments using diphenyleneiodinium chloride specific to the FMN-binding site.[100] Reactive oxygen species production from purified complex II was also measured using the cytochrome c reduction method for superoxide. In a horseradish peroxidase assay for hydrogen peroxide, the rate of superoxide production on flavin auto-oxidation was found to be proportional to oxygen concentration; the hydrogen peroxide was produced as a dismutation product of superoxide.[101] Superoxide generation from purified complex III at the Q_o site have been detected using spectrophotometric[102] and EPR spin-trapping[103] techniques as supported by inhibition studies using antimycin A, myothiazol, and stigmatellin. In intact cells, mitochondrially derived reactive species could be detected by confocal microscopy using conventional ROS-specific dyes or protein-based ROS reporter molecules that can be selectively expressed in different organelles.[104]

3.5.2 Lysosomes

Along with mitochondria, these membrane-bound organelles are the main generators of reactive species in cells. Known to be responsible for the breakdown of unwanted biomolecules in the cytosol, they are also involved in such secretory mechanisms as endocytosis, autophagy, and phagocytosis,[105] as well as in cell signaling, membrane repair, and energy metabolism.[106,107] All of the lysosomal enzymes are acid hydrolases that mediate in the digestion of cellular components and other exogenous phagocytosed materials. During oxidative stress conditions, these enzymes are susceptible to oxidation and could lead to altered catabolic processes inside the cell. ROS formation from lysosomes originates from the hydrogen peroxide that is generated from other organelles such as mitochondria, peroxisome, the endoplasmic reticulum (ER), and cytosolic oxidases that can diffuse into the lysosome.[106] Inside the lysosome, hydrogen peroxide can react with labile iron (from degraded iron-containing biomolecules) to form hydroxyl radicals, which could result in lysosomal rupture and subsequently more ROS production and mitochondrial-mediated pro-apoptotic cascade. Release of amyloidogenic oligomers upon radical-mediated disruption of liposomal membrane have been implicated in the accumulation of oxidatively damaged proteins such as those associated with Alzheimer's, hence there is no other disease conditions in which lysosomal function is more studied than in age-related diseases including Alzheimer's disease.[107]

In neutrophils and monocytes, lysosomes contain a peroxidase enzyme called myeloperoxidase (MPO) with a heme prosthetic group. In the presence of hydrogen peroxide and chloride ion, hypochlorous acid (HOCl) is produced as a product of a neutrophil's respiratory burst from the assembled NADPH oxidase complex at the phagosomal membrane. Due to the high toxicity of HOCl, it is the main chemical responsible for the killing capability of neutrophils against bacteria and other pathogens.[108] HOCl generation was visualized by fluorescence microscopy in neutrophils using a green fluorescent protein-expressing *E. coli* as a selective probe for HOCl[109] and was indirectly detected by mass spectrometric analysis of HOCl biomarkers.[110] Peroxynitrite can also be produced from neutrophils—although not directly from MPO but from the NOX-derived superoxide reaction with NOS-derived NO.[111] However, nitrite can be converted to nitryl chloride (NO_2Cl) and nitrogen dioxide (NO_2) through MPO-dependent pathways as evidenced by tyrosine nitration. Moreover, MPO can potentiate the nitrosation reactions in the presence of NO and H_2O_2 via NO^+ and nitrogen oxides–like species formation.[112] Also, hyprobromous acid and hypothiocyanite can also be formed in the presence of other anions such as bromide and isocyanate.[113,114] Other bactericidal substances that have been shown to be formed by MPO are chloramines as by-products of HOCl reaction with amines.[115] MPO deficiency has been implicated in a variety of diseases such as infection from species of *Candida* in patients with diabetes mellitus or increased incidence of malignant tumors.[116]

3.5.3 Endoplasmic reticulum

There are three main classes of endoplasmic reticulum: the rough endoplasmic reticulum (RER), smooth endoplasmic reticulum, and the sarcoplasmic reticulum (SR). They have different functions depending on the type of cells in which they are present. Aside from being calcium stores that are involved in the active intracellular transport of Ca^{2+}, one of the ER's functions—specifically that of RER—is the manufacture of secreted proteins or integral membrane proteins, and this process generates disulfide bonds as catalyzed by the thiol oxidase enzyme by transferring electrons from the dithiols to oxygen. Disulfide formation is important in protein folding, which in turn is required for their proper transport within the cell to the Golgi. Increases in intracellular unfolded proteins is a manifestation of endoplasmic reticulum stress response or ER stress. Disturbances in cellular redox regulation is one major cause of ER stress, which is apparent in several diseases such as ischemia-reperfusion injury, neurodegeneration, and diabetes. Hence, ER stress plays a major role in the initiation of pathophysiological conditions.[117] ER stress activates unfolded protein response (UPR) as a way to restore normal cell function by inducing expression of genes that can enhance the production of ER-related proteins to increase the protein-folding capacity of ER as well as stop translation of misfolded proteins and their removal. However, when UPR fails, ER stress can lead to apoptosis.[118] Growing evidence has shown that ROS are generated during UPR, and ROS production is highly integrated with ROS production from the mitochondria and NOX4.[119] Major source of

ER-derived ROS generation was proposed to be from the flavoenzyme Ero1p, which produces disulfide bonds during protein folding.[120] Ero1p is a flavin-containing oxidase, whose activity is strongly sensitive to FAD concentration in the ER, where the bound FAD cofactor can oxidize thiol substrates to disulfides. The reduced $FADH_2$ can transfer two electrons to oxygen to form equimolar amounts of hydrogen peroxide, according to Eqs. (3.20) and (3.21).[120] Hence, ER stress is a promising therapeutic target for the treatment of certain diseases such as target cardiovascular diseases,[121] cancer,[122] and diabetes and insulin resistance.[123]

$$\text{E-FAD} + 2\,\text{RSH} \rightarrow \text{E-FADH}_2 + \text{RSSR} \tag{3.20}$$

$$\text{E-FADH}_2 + \text{O}_2 \rightarrow \text{E-FAD} + \text{H}_2\text{O}_2 \tag{3.21}$$

Using a combination of oxygen consumption and hydrogen peroxide assays, ROS formation have been demonstrated from purified Ero1p enzyme.[120] Also, by analyzing markers of oxidative damage from liver tissue extracts and fluorescence analysis by flow cytometry of Chines hamster ovary (CHO) cells, ROS generated from misfolded proteins have been shown to cause UPR activation and cell death. The effects of ER stress can be reduced by antioxidants through improving protein secretion.[124]

3.5.4 Peroxisomes

Peroxisomes are organelles found in almost all eukaryotic cells. They are responsible for the metabolism of hydrophobic substrates such as fatty acids, steroids, and amino acids. As the name implies, a peroxisome contains a variety of enzymes that produce hydrogen peroxide as well as superoxides that are essential to metabolic processes. Among such reactive species-producing enzymes are acyl-CoA oxidases, D-amino acid oxidase, α-hydroxyacid oxidase, polyamine oxidase, urate oxidase, xanthine oxidase, and NO synthase; these are important in the catabolism of various substrates such as long-chain fatty acids, D-proline, lactate, *N*-acetyl spermine, uric acid, xanthine, and arginine, respectively. The reactive species produced from these oxidases are intracellularly counteracted by antioxidant enzymes, including catalase, glutathione peroxidase, superoxide dismutases, epoxide hydrolase, and peroxiredoxin. This delicate balance inside peroxisomes between reactive species production and their sequestration ensures normal catabolic processes. However, during oxidative stress conditions triggered by various factors such as exposure to UV or xenobiotics or during ischemia reperfusion, the unregulated peroxisomal production of reactive species can occur.[125] The ability of peroxisomes to cause uncontrolled production of reactive species results from their abnormal biogenesis, their downregulation of peroxisomal antioxidant enzymes, and their ability to undergo massive proliferation under oxidative stress conditions have been observed in a variety of disease conditions such as in inflammation, carcinogenesis, heart failure from alcoholism, and aging.[126,127] Peroxisomal reactive species production can be assessed using various techniques such as chemiluminescence with luminol,[128] peroxisome-targeted fluorescent probe,[129] and proteinaceous reporter molecule.[130]

3.5.5 Golgi apparatus

Proteins excreted by the RER are further modified by the Golgi apparatus and play a crucial role in their processing and sorting, as well as the delivery of proteins and other macromolecules to the extracellular matrix as well as within the cell. Although there is no evidence thus far for direct radical generation from Golgi, the Golgi can cause increased ROS production indirectly due to the crucial role it plays in Ca^{2+} dyshomeostasis.[131] Under oxidative stress conditions, the Golgi's Ca^{2+}/Mn^{2+} ATPases (SPCA1) activity, being important ion regulations, is diminished, leading to Ca^{2+} dyshomeostasis as characterized by reduced Ca^{2+} uptake and cytosolic Ca^{2+} overload. Therefore, increased intacellular Ca^{2+} can trigger a series of Ca^{2+}-dependent metabolic disturbances in the mitochondria, phospholipases, NOS, or xanthine oxidase, leading to indirect increases in ROS production.[131]

REFERENCES

1. Harman D. Aging: a theory based on free radical and radiation chemistry. *J Gerontol* 1956;**11**:298.
2. Heckly RJ, Dimmick RL, Windle JJ. Free radical formation and survival of lyophilized microorganisms. *J Bacteriol* 1963;**85**:961.
3. Fenton HJH, Jones HO. The oxidation of organic acids in presence of ferrous iron. Part I. *J Chem Soc Trans* 1900;**77**:69.
4. Yamazaki I, Piette LH. The Mechanism of Aerobic Oxidase Reaction Catalyzed by Peroxidase. *Biochim Biophys Acta* 1963;**77**:47.
5. Beckett AH, Belanger PM. Metabolic incorporation of oxygen into primary and secondary aliphatic amines and the consequences in carbon-nitrogen bond cleavage. *J Pharm Pharmacol* 1975;**27**:547.
6. Peterson FJ, Mason RP, Hovsepian J, Holtzman JL. Oxygen-sensitive and -insensitive nitroreduction by Escherichia coli and rat hepatic microsomes. *J Biol Chem* 1979;**254**:4009.
7. Knowles PF, Gibson JF, Pick FM, Bray RC. Electron-spin-resonance evidence for enzymic reduction of oxygen to a free radical, the superoxide ion. *Biochem J* 1969;**111**:53.
8. Massey V, Strickland S, Mayhew SG, Howell LG, Engel PC, Matthews RG, et al. The production of superoxide anion radicals in the reaction of reduced flavins and flavoproteins with molecular oxygen. *Biochem Biophys Res Commun* 1969;**36**:891.
9. Ballou D, Palmer G, Massey V. Direct demonstration of superoxide anion production during the oxidation of reduced flavin and of its catalytic decomposition by erythrocuprein. *Biochem Biophys Res Commun* 1969;**36**:898.
10. Loschen G, Azzi A, Richter C, Flohe L. Superoxide radicals as precursors of mitochondrial hydrogen peroxide. *FEBS Lett* 1974;**42**:68.
11. Klegeris A, Korkina LG, Greenfield SA. Autoxidation of dopamine: a comparison of luminescent and spectrophotometric detection in basic solutions. *Free Radic Biol Med* 1995;**18**:215.
12. Liehr JG, Roy D. Free radical generation by redox cycling of estrogens. *Free Radic Biol Med* 1990;**8**:415.
13. Cummings J, Willmott N, Hoey BM, Marley ES, Smyth JF. The consequences of doxorubicin quinone reduction in vivo in tumour tissue. *Biochem Pharmacol* 1992;**44**:2165.

14. Marcillat O, Zhang Y, Davies KJ. Oxidative and non-oxidative mechanisms in the inactivation of cardiac mitochondrial electron transport chain components by doxorubicin. *Biochem J* 1989;**259**:181.
15. Hassan HM. Exacerbation of superoxide radical formation by paraquat. *Methods Enzymol* 1984;**105**:523.
16. Winterbourn CC. Production of hydroxyl radicals from paraquat radicals and H2O2. *FEBS Lett* 1981;**128**:339.
17. Granell S, Gironella M, Bulbena O, Panes J, Mauri M, Sabater L, et al. Heparin mobilizes xanthine oxidase and induces lung inflammation in acute pancreatitis. *Crit Care Med* 2003;**31**:525.
18. Kulagina NV, Michael AC. Monitoring hydrogen peroxide in the extracellular space of the brain with amperometric microsensors. *Anal Chem* 2003;**75**:4875.
19. Test ST, Weiss SJ. Assay of the extracellular hydrogen peroxide pool generated by phagocytes. *Methods Enzymol* 1986;**132**:401.
20. Mahns A, Melchheier I, Suschek CV, Sies H, Klotz LO. Irradiation of cells with ultraviolet-A (320-400 nm) in the presence of cell culture medium elicits biological effects due to extracellular generation of hydrogen peroxide. *Free Radic Res* 2003;**37**:391.
21. Saito I, Matsuura T, Inoue K. Formation of superoxide ion from singlet oxygen. Use of a water-soluble singlet oxygen source. *J Am Chem Soc* 1981;**103**:188.
22. Frimer AA. The reaction of singlet oxygen with olefins: the question of mechanism. *Chem Rev* 1979;**79**:359.
23. Chekulayeva LV, Chekulayev VA, Shevchuk IN. Active oxygen intermediates in the degradation of hematoporphyrin derivative in tumor cells subjected to photodynamic therapy. *J Photochem Photobiol B* 2008;**93**:94.
24. Holroyd RA, Bielski BHJ. Photochemical generation of superoxide radicals in aqueous solutions. *J Am Chem Soc* 1978;**100**:5796.
25. Von Sonntag C. Some aspects of radiation-induced free-radical chemistry of biologically important molecules. *Radiat Phys Chem* 1992;**39**:477.
26. Shah DJ, Sachs RK, Wilson DJ. Radiation-induced cancer: a modern view. *Br J Radiol* 2012;**85**: e1166.
27. Lomax ME, Folkes LK, O'Neill P. Biological consequences of radiation-induced DNA damage: relevance to radiotherapy. *Clin Oncol (R Coll Radiol)* 2013;**25**:578.
28. Spitz DR, Hauer-Jensen M. Ionizing radiation-induced responses: where free radical chemistry meets redox biology and medicine. *Antioxid Redox Signal* 2014;**20**:1407.
29. Pierre JL, Fontecave M, Crichton RR. Chemistry for an essential biological process: the reduction of ferric iron. *BioMetals* 2002;**15**:341.
30. Buettner GR. Ascorbate autoxidation in the presence of iron and copper chelates. *Free Radic Res Commun* 1986;**1**:349.
31. Romero FJ, Ordonez I, Arduini A, Cadenas E. The reactivity of thiols and disulfides with different redox states of myoglobin. Redox and addition reactions and formation of thiyl radical intermediates. *J Biol Chem* 1680;**1992**:267.
32. Edwards JO, Curci R. *Fenton Type Activation and Chemistry of Hydroxyl Radical*. Dordrecht: Kluwer Academic Publisher; 1992.
33. Buettner GR. Activation of oxygen by metal complexes and its relevance to autoxidative processes in living systems *Bioelectrochem. Bioenergetics* 1987;**18**:29.
34. Perez-Benito JF. Iron(III) – Hydrogen Peroxide Reaction: Kinetic Evidence of a Hydroxyl-Mediated Chain Mechanism. *J Phys Chem A* 2004;**108**:4853.
35. Rifkind JM, Ramasamy S, Manoharan PT, Nagababu E, Mohanty JG. Redox reactions of hemoglobin. *Antioxid Redox Signal* 2004;**6**:657.
36. Quinn MT. In: Villamena FA, editor. *Molecular Basis of Oxidative Stress: Chemistry, Mechnisms, and Disease Pathogenesis*. Hoboken, NJ: John Wiley & Sons, Inc; 2013. p. 137.
37. Jiang F, Zhang Y, Dusting GJ. NADPH oxidase-mediated redox signaling: roles in cellular stress response, stress tolerance, and tissue repair. *Pharmacol Rev* 2011;**63**:218.

38. Bedard K, Krause KH. The NOX family of ROS-generating NADPH oxidases: physiology and pathophysiology. *Physiol Rev* 2007;**87**:245.
39. Sato K, Kadiiska MB, Ghio AJ, Corbett J, Fann YC, Holland SM, et al. In vivo lipid-derived free radical formation by NADPH oxidase in acute lung injury induced by lipopolysaccharide: a model for ARDS. *FASEB J* 2002;**16**:1713.
40. Souza HP, Liu X, Samouilov A, Kuppusamy P, Laurindo FR, Zweier JL. Quantitation of superoxide generation and substrate utilization by vascular NAD(P)H oxidase. *Am J Physiol Heart Circ Physiol* 2002;**282**:H466.
41. Janiszewski M, Souza HP, Liu X, Pedro MA, Zweier JL, Laurindo FR. Overestimation of NADH-driven vascular oxidase activity due to lucigenin artifacts. *Free Radic Biol Med* 2002;**32**:446.
42. Fulton D, Fontana J, Sowa G, Gratton JP, Lin M, Li KX, et al. Localization of endothelial nitric-oxide synthase phosphorylated on serine 1179 and nitric oxide in Golgi and plasma membrane defines the existence of two pools of active enzyme. *J Biol Chem* 2002;**277**:4277.
43. Blair A, Shaul PW, Yuhanna IS, Conrad PA, Smart EJ. Oxidized low density lipoprotein displaces endothelial nitric-oxide synthase (eNOS) from plasmalemmal caveolae and impairs eNOS activation. *J Biol Chem* 1999;**274**:32512.
44. Marletta MA. Nitric oxide synthase: aspects concerning structure and catalysis. *Cell* 1994;**78**:927.
45. Wei CC, Crane BR, Stuehr DJ. Tetrahydrobiopterin radical enzymology. *Chem Rev* 2003;**103**:2365.
46. Scott-Burden T. Regulation of nitric oxide production by tetrahydrobiopterin. *Circulation* 1995;**91**:248.
47. Forstermann U, Sessa WC. Nitric oxide synthases: regulation and function. *Eur Heart J* 2012;**33**:829.
48. Chen CA, Wang TY, Varadharaj S, Reyes LA, Hemann C, Talukder MA, et al. S-glutathionylation uncouples eNOS and regulates its cellular and vascular function. *Nature* 2010;**468**:1115.
49. Chen C, Druhan L, Varadharaj S, Chen Y, Zweier J. Phosphorylation of endothelial nitric-oxide synthase regulates superoxide generation from the enzyme. *J Biol Chem* 2008;**283**:27038.
50. Gopalakrishnan B, Nash K, Velayutham M, Villamena F. Detection of nitric oxide and superoxide radical anion by electron paramagnetic resonance spectroscopy from cells using spin traps. *J Vis Exp* 2012;e2810.
51. Das A, Gopalakrishnan B, Druhan LJ, Wang TY, De Pascali F, Rockenbauer A, et al. Reversal of SIN-1-induced eNOS dysfunction by the spin trap, DMPO, in bovine aortic endothelial cells via eNOS phosphorylation. *Br J Pharmacol* 2014.
52. Xia Y, Tsai A, Berka V, Zweier J. Superoxide generation from endothelial nitric-oxide synthase. A Ca2 + /calmodulin-dependent and tetrahydrobiopterin regulatory process. *J Biol Chem* 1998;**273**:25804.
53. Dumitrescu C, Biondi R, Xia Y, Cardounel A, Druhan L, Ambrosio G, et al. Myocardial ischemia results in tetrahydrobiopterin (BH4) oxidation with impaired endothelial function ameliorated by BH4. *Proc Natl Acad Sci USA* 2007;**104**:15081.
54. Joshi MS, Ferguson Jr TB, Johnson FK, Johnson RA, Parthasarathy S, Lancaster Jr JR. Receptor-mediated activation of nitric oxide synthesis by arginine in endothelial cells. *Proc Natl Acad Sci USA* 2007;**104**:9982.
55. Hille R, Nishino T, Bittner F. Molybdenum enzymes in higher organisms. *Coord Chem Rev* 2011;**255**:1179.
56. Hille R, Nishino T. Flavoprotein structure and mechanism. 4. Xanthine oxidase and xanthine dehydrogenase. *FASEB J* 1995;**9**:995.
57. Britigan BE, Pou S, Rosen GM, Lilleg DM, Buettner GR. Hydroxyl radical is not a product of the reaction of xanthine oxidase and xanthine. The confounding problem of adventitious iron bound to xanthine oxidase. *J Biol Chem* 1990;**265**:17533.
58. Liu Y, Song Y, De Pascali F, Liu X, Villamena FA, Zweier JL. Tetrathiatriarylmethyl radical with a single aromatic hydrogen as a highly sensitive and specific superoxide probe. *Free Radic Biol Med* 2012;**53**:2081.

59. Li Y, Zhu H, Kuppusamy P, Roubaud V, Zweier JL, Trush MA. Validation of lucigenin (bis-N-methylacridinium) as a chemilumigenic probe for detecting superoxide anion radical production by enzymatic and cellular systems. *J Biol Chem* 2015;**1998**:273.
60. McNally JS, Davis ME, Giddens DP, Saha A, Hwang J, Dikalov S, et al. Role of xanthine oxidoreductase and NAD(P)H oxidase in endothelial superoxide production in response to oscillatory shear stress. *Am J Physiol Heart Circ Physiol* 2003;**285**:H2290.
61. Beetsch JW, Park TS, Dugan LL, Shah AR, Gidday JM. Xanthine oxidase-derived superoxide causes reoxygenation injury of ischemic cerebral endothelial cells. *Brain Res* 1998;**786**:89.
62. Arroyo CM, Carmichael AJ, Bouscarel B, Liang JH, Weglicki WB. Endothelial cells as a source of oxygen-free radicals. An ESR study. *Free Radic Res Commun* 1990;**9**:287.
63. Sato K, Corbett J, Mason RP, Kadiiska MB. In vivo evidence of free radical generation in the mouse lung after exposure to Pseudomonas aeruginosa bacterium: an ESR spin-trapping investigation. *Free Radic Res* 2012;**46**:645.
64. Nakai K, Kadiiska MB, Jiang JJ, Stadler K, Mason RP. Free radical production requires both inducible nitric oxide synthase and xanthine oxidase in LPS-treated skin. *Proc Natl Acad Sci USA* 2006;**103**:4616.
65. Hille R. The Mononuclear Molybdenum Enzymes. *Chem Rev* 1996;**96**:2757.
66. Kundu TK, Hille R, Velayutham M, Zweier JL. Characterization of superoxide production from aldehyde oxidase: an important source of oxidants in biological tissues. *Arch Biochem Biophys* 2007;**460**:113.
67. Li H, Cui H, Kundu TK, Alzawahra W, Zweier JL. Nitric oxide production from nitrite occurs primarily in tissues not in the blood: critical role of xanthine oxidase and aldehyde oxidase. *J Biol Chem* 2008;**283**:17855.
68. Li H, Kundu TK, Zweier JL. Characterization of the magnitude and mechanism of aldehyde oxidase-mediated nitric oxide production from nitrite. *J Biol Chem* 2009;**284**:33850.
69. Kundu TK, Velayutham M, Zweier JL. Aldehyde oxidase functions as a superoxide generating NADH oxidase: an important redox regulated pathway of cellular oxygen radical formation. *Biochemistry* 2012;**51**:2930.
70. Neumeier M, Weigert J, Schaffler A, Weiss TS, Schmidl C, Buttner R, et al. Aldehyde oxidase 1 is highly abundant in hepatic steatosis and is downregulated by adiponectin and fenofibric acid in hepatocytes in vitro. *Biochem Biophys Res Commun* 2006;**350**:731.
71. Calzi ML, Raviolo C, Ghibaudi E, de Gioia L, Salmona M, Cazzaniga G, et al. Purification, cDNA cloning, and tissue distribution of bovine liver aldehyde oxidase. *J Biol Chem* 1995;**270**:31037.
72. Wright RM, McManaman JL, Repine JE. Alcohol-induced breast cancer: a proposed mechanism. *Free Radic Biol Med* 1999;**26**:348.
73. Barouch LA, Harrison RW, Skaf MW, Rosas GO, Cappola TP, Kobeissi ZA, et al. Nitric oxide regulates the heart by spatial confinement of nitric oxide synthase isoforms. *Nature* 2002;**416**:337.
74. Giordano A, Tonello C, Bulbarelli A, Cozzi V, Cinti S, Carruba MO, et al. Evidence for a functional nitric oxide synthase system in brown adipocyte nucleus. *FEBS Lett* 2002;**514**:135.
75. Villanueva C, Giulivi C. Subcellular and cellular locations of nitric oxide synthase isoforms as determinants of health and disease. *Free Radic Biol Med* 2010;**49**:307.
76. Alderton WK, Cooper CE, Knowles RG. Nitric oxide synthases: structure, function and inhibition. *Biochem J* 2001;**357**:593.
77. Spratt DE, Taiakina V, Palmer M, Guillemette JG. Differential binding of calmodulin domains to constitutive and inducible nitric oxide synthase enzymes. *Biochemistry* 2007;**46**:8288.
78. Xia Y, Roman LJ, Masters BS, Zweier JL. Inducible nitric-oxide synthase generates superoxide from the reductase domain. *J Biol Chem* 1998;**273**:22635.
79. Song Y, Cardounel AJ, Zweier JL, Xia Y. Inhibition of superoxide generation from neuronal nitric oxide synthase by heat shock protein 90: implications in NOS regulation. *Biochemistry* 2002;**41**:10616.
80. Raines KW, Cao GL, Porsuphatana S, Tsai P, Rosen GM, Shapiro P. Nitric oxide inhibition of ERK1/2 activity in cells expressing neuronal nitric-oxide synthase. *J Biol Chem* 2004;**279**:3933.

81. Xia Y, Zweier JL. Superoxide and peroxynitrite generation from inducible nitric oxide synthase in macrophages. *Proc Natl Acad Sci USA* 1997;**94**:6954.
82. Stadler K, Bonini MG, Dallas S, Duma D, Mason RP, Kadiiska MB. Direct evidence of iNOS-mediated in vivo free radical production and protein oxidation in acetone-induced ketosis. *Am J Physiol Endocrinol Metab* 2008;**295**:E456.
83. Girouard H, Wang G, Gallo EF, Anrather J, Zhou P, Pickel VM, et al. NMDA receptor activation increases free radical production through nitric oxide and NOX2. *J Neurosci* 2009;**29**:2545.
84. Rouzer CA, Marnett LJ. Mechanism of free radical oxygenation of polyunsaturated fatty acids by cyclooxygenases. *Chem Rev* 2003;**103**:2239.
85. Zidar N, Odar K, Glavac D, Jerse M, Zupanc T, Stajer D. Cyclooxygenase in normal human tissues–is COX-1 really a constitutive isoform, and COX-2 an inducible isoform? *J Cell Mol Med* 2009;**13**:3753.
86. Mukherjee A, Brinkley DW, Chang K-M, Roth JP. Molecular Oxygen Dependent Steps in Fatty Acid Oxidation by Cyclooxygenase-1. *Biochemistry* 2007;**46**:3975.
87. Nikolic D, van Breemen RB. DNA Oxidation Induced by Cyclooxygenase-2. *Chem Res Toxicol* 2001;**14**:351.
88. Yu Q, Purwaha P, Ni K, Sun C, Mallik S, Qian SY. Characterization of novel radicals from COX-catalyzed arachidonic acid peroxidation. *Free Radical Biol Med* 2009;**47**:568.
89. Jiang J, Borisenko GG, Osipov A, Martin I, Chen R, Shvedova AA, et al. Arachidonic acid-induced carbon-centered radicals and phospholipid peroxidation in cyclo-oxygenase-2-transfected PC12 cells. *J Neurochem* 2004;**90**:1036.
90. Manabe Y, Anrather J, Kawano T, Niwa K, Zhou P, Ross ME, et al. Prostanoids, not reactive oxygen species, mediate COX-2-dependent neurotoxicity. *Ann Neurol* 2004;**55**:668.
91. Lin MT, Beal MF. Mitochondrial dysfunction and oxidative stress in neurodegenerative diseases. *Nature* 2006;**443**:787.
92. Wallace DC. Mitochondria and cancer. *Nat Rev Cancer* 2012;**12**:685.
93. Ballinger SW. Mitochondrial dysfunction in cardiovascular disease. *Free Radic Biol Med* 2005;**38**:1278.
94. Kussmaul L, Hirst J. The mechanism of superoxide production by NADH:ubiquinone oxidoreductase (complex I) from bovine heart mitochondria. *Proc Natl Acad Sci USA* 2006;**103**:7607.
95. Yankovskaya V, Horsefield R, Tornroth S, Luna-Chavez C, Miyoshi H, Leger C, et al. Architecture of succinate dehydrogenase and reactive oxygen species generation. *Science* 2003;**299**:700.
96. Guo J, Lemire BD. The ubiquinone-binding site of the Saccharomyces cerevisiae succinate-ubiquinone oxidoreductase is a source of superoxide. *J Biol Chem* 2003;**278**:47629.
97. Bleier L, Drose S. Superoxide generation by complex III: from mechanistic rationales to functional consequences. *Biochim Biophys. Acta* 2013;**1827**:1320.
98. Cape JL, Bowman MK, Kramer DM. A semiquinone intermediate generated at the Qo site of the cytochrome bc1 complex: importance for the Q-cycle and superoxide production. *Proc Natl Acad Sci USA*. 2007;**104**:7887.
99. Chen Y-R. In: Villamena FA, editor. *Molecular Basis of Oxidative Stress: Chemistry, Mechnisms, and Disease Pathogenesis*. Hoboken, NJ: John Wiley & Sons, Inc; 2013. p. 123.
100. Chen YR, Chen CL, Zhang L, Green-Church KB, Zweier JL. Superoxide generation from mitochondrial NADH dehydrogenase induces self-inactivation with specific protein radical formation. *J Biol Chem* 2005;**280**:37339.
101. Messner KR, Imlay JA. Mechanism of superoxide and hydrogen peroxide formation by fumarate reductase, succinate dehydrogenase, and aspartate oxidase. *J Biol Chem* 2002;**277**:42563.
102. Muller FL, Roberts AG, Bowman MK, Kramer DM. Architecture of the Qo site of the cytochrome bc1 complex probed by superoxide production. *Biochemistry* 2003;**42**:6493.
103. Chen YR, Chen CL, Yeh A, Liu X, Zweier JL. Direct and indirect roles of cytochrome b in the mediation of superoxide generation and NO catabolism by mitochondrial succinate-cytochrome c reductase. *J Biol Chem* 2006;**281**:13159.

104. Forkink M, Smeitink JA, Brock R, Willems PH, Koopman WJ. Detection and manipulation of mitochondrial reactive oxygen species in mammalian cells. *Biochim Biophys Acta* 2010;**1797**:1034.
105. Luzio JP, Pryor PR, Bright NA. Lysosomes: fusion and function. *Nat Rev Mol Cell Biol* 2007;**8**:622.
106. Kurz T, Terman A, Gustafsson B, Brunk UT. Lysosomes and oxidative stress in aging and apoptosis. *Biochim Biophys Acta* 2008;**1780**:1291.
107. Butler D, Bahr BA. Oxidative stress and lysosomes: CNS-related consequences and implications for lysosomal enhancement strategies and induction of autophagy. *Antioxid Redox Signal* 2006;**8**:185.
108. Hampton MB, Kettle AJ, Winterbourn CC. Inside the neutrophil phagosome: oxidants, myeloperoxidase, and bacterial killing. *Blood* 1998;**92**:3007.
109. Palazzolo AM, Suquet C, Konkel ME, Hurst JK. Green fluorescent protein-expressing Escherichia coli as a selective probe for HOCl generation within neutrophils. *Biochemistry* 2005;**44**:6910.
110. Kettle AJ, Albrett AM, Chapman AL, Dickerhof N, Forbes LV, Khalilova I, et al. Measuring chlorine bleach in biology and medicine. *Biochim Biophys Acta* 2014;**1840**:781.
111. Evans TJ, Buttery LD, Carpenter A, Springall DR, Polak JM, Cohen J. Cytokine-treated human neutrophils contain inducible nitric oxide synthase that produces nitration of ingested bacteria. *Proc Natl Acad Sci USA* 1996;**93**:9553.
112. Lakshmi VM, Nauseef WM, Zenser TV. Myeloperoxidase potentiates nitric oxide-mediated nitrosation. *J Biol Chem* 1746;**2005**:280.
113. Chapman AL, Skaff O, Senthilmohan R, Kettle AJ, Davies MJ. Hypobromous acid and bromamine production by neutrophils and modulation by superoxide. *Biochem J* 2009;**417**:773.
114. van Dalen CJ, Whitehouse MW, Winterbourn CC, Kettle AJ. Thiocyanate and chloride as competing substrates for myeloperoxidase. *Biochem J* 1997;**327**(Pt 2):487.
115. Coker MS, Hu WP, Senthilmohan ST, Kettle AJ. Pathways for the decay of organic dichloramines and liberation of antimicrobial chloramine gases. *Chem Res Toxicol* 2008;**21**:2334.
116. Lanza F. Clinical manifestation of myeloperoxidase deficiency. *J Mol Med* 1998;**76**:676.
117. Xu C, Bailly-Maitre B, Reed JC. Endoplasmic reticulum stress: cell life and death decisions. *J Clin Invest* 2005;**115**:2656.
118. Hetz C. The unfolded protein response: controlling cell fate decisions under ER stress and beyond. *Nat Rev Mol Cell Biol* 2012;**13**:89.
119. Santos CX, Tanaka LY, Wosniak J, Laurindo FR. Mechanisms and implications of reactive oxygen species generation during the unfolded protein response: roles of endoplasmic reticulum oxidoreductases, mitochondrial electron transport, and NADPH oxidase. *Antioxid Redox Signal* 2009;**11**:2409.
120. Gross E, Sevier CS, Heldman N, Vitu E, Bentzur M, Kaiser CA, et al. Generating disulfides enzymatically: reaction products and electron acceptors of the endoplasmic reticulum thiol oxidase Ero1p. *Proc Natl Acad Sci USA* 2006;**103**:299.
121. Minamino T, Komuro I, Kitakaze M. Endoplasmic reticulum stress as a therapeutic target in cardiovascular disease. *Circ Res* 2010;**107**:1071.
122. Verfaillie T, Garg AD, Agostinis P. Targeting ER stress induced apoptosis and inflammation in cancer. *Cancer Lett* 2013;**332**:249.
123. Nakatani Y, Kaneto H, Kawamori D, Yoshiuchi K, Hatazaki M, Matsuoka TA, et al. Involvement of endoplasmic reticulum stress in insulin resistance and diabetes. *J Biol Chem* 2005;**280**:847.
124. Malhotra JD, Miao H, Zhang K, Wolfson A, Pennathur S, Pipe SW, et al. Antioxidants reduce endoplasmic reticulum stress and improve protein secretion. *Proc Natl Acad Sci USA* 2008;**105**:18525.
125. Schrader M, Fahimi HD. Peroxisomes and oxidative stress. *Biochim Biophys Acta* 2006;**1763**:1755.
126. Antonenkov VD, Grunau S, Ohlmeier S, Hiltunen JK. Peroxisomes are oxidative organelles. *Antioxid Redox Signal* 2010;**13**:525.
127. Fransen M, Nordgren M, Wang B, Apanasets O. Role of peroxisomes in ROS/RNS-metabolism: implications for human disease. *Biochim Biophys Acta* 2012;**1822**:1363.
128. Mueller S, Weber A, Fritz R, Mutze S, Rost D, Walczak H, et al. Sensitive and real-time determination of H2O2 release from intact peroxisomes. *Biochem J* 2002;**363**:483.

129. Yano T, Oku M, Akeyama N, Itoyama A, Yurimoto H, Kuge S, et al. A novel fluorescent sensor protein for visualization of redox states in the cytoplasm and in peroxisomes. *Mol Cell Biol* 2010;**30**:3758.
130. Gehrmann W, Elsner M. A specific fluorescence probe for hydrogen peroxide detection in peroxisomes. *Free Radic Res* 2011;**45**:501.
131. Jiang Z, Hu Z, Zeng L, Lu W, Zhang H, Li T, et al. The role of the Golgi apparatus in oxidative stress: is this organelle less significant than mitochondria? *Free Radic Biol Med* 2011;**50**:907.

CHAPTER 4

Fluorescence Technique

4.1 INTRODUCTION

The recorded history of photoluminescence dates back to the 16th century when the Aztecs exploited bluish extract from the wood of the medicinal plant *Lignum nephriticum* to detect counterfeited varieties used for medicinal purposes.[1] It was not until the mid-1800s that the word *fluorescence* was coined by Sir George Gabriel Stokes, who made the seminal observation of dispersive reflection in which the wavelength of the emitted light is longer than the wavelength of the exciting light. Using a solution of quinine sulfate, Stoke observed that the quinine solution emits blue light when irradiated with ultraviolet (UV) light but not with visible light.[1,2] The earliest and most successful dye, fluorescein, was synthesized by Adolph von Baeyer in 1871 and was the first dye ever applied as histological stain.[3,4] Currently, there is a myriad of fluorophores that are organically, small peptide-, or protein-based and used in diverse applications such as cell structure imaging microscopy, flow cytometry, cell viability, proliferation and function, nucleic acid and protein labeling, genotyping and genomic profiling, and drug discovery. Over the past several decades, fluorescent dyes have been employed to assess overall cellular redox states through direct detection of reactive species generation as well as biomarkers of oxidative stress such as lipid peroxidation or oxidized and reduced glutathione levels. Recently, protein-based sensors that are genetically encoded are employed to measure H_2O_2 or glutathione redox potentials in live cells in a reversible and reproducible manner.[5] Organically based dyes can be used in concert with a protein sensor, but the latter needs to be transduced or transfected into the cells. The fluorescence technique is attractive since it offers less tedious preparation, high detection sensitivity, and the ability to give spectrotemporal as well as spectrospatial imaging of cells, which allows real-time monitoring of reactive species production in live cells. Detection techniques using fluorescent probes are employed through the aid of optical microscopes, spectrofluorimetric microplate readers, flow cytometric systems allowing high-throughput studies, and high-performance liquid chromatography (HPLC) analysis providing a more specific detection.

4.2 FLUORESCENCE SPECTROSCOPY AND MICROSCOPY

There are several other mechanisms by which substances can emit light—by chemical reaction (chemiluminescence), electric current passing through (electroluminescence),

Reactive Species Detection in Biology
DOI: http://dx.doi.org/10.1016/B978-0-12-420017-3.00003-7

or mechanical stress (mechanoluminescence)—but the emission of light through absorption of an electromagnetic radiation is a natural phenomenon called *photoluminescence*. There are two major modes of photoluminescent emission—fluorescence or phosphorescence—as shown in Fig. 4.1 in the form of a Jablonski diagram. Molecules occupy the lowest possible vibrational energy level, which is the singlet ground state, S_0 (i.e., two electrons with opposite spins occupying the lowest energy level). Upon absorption of electromagnetic radiation, the singlet ground state is transformed into the first or second singlet excited state, S_1 or S_2, respectively (i.e., one of the electrons is promoted to a higher energy level and assumes a spin opposite that of the other electron occupying the lowest energy level). The S_2 state can decay to S_1 via internal conversion (IC); in some cases, S_1 can undergo intersystem crossing (ISC) to assume a triplet excited state, T_1 (i.e., one of the electrons that was promoted to the higher energy level changes its spin, which is similar to the other electron occupying the lowest energy level), where both IC and ISC do not emit light. The excited energy

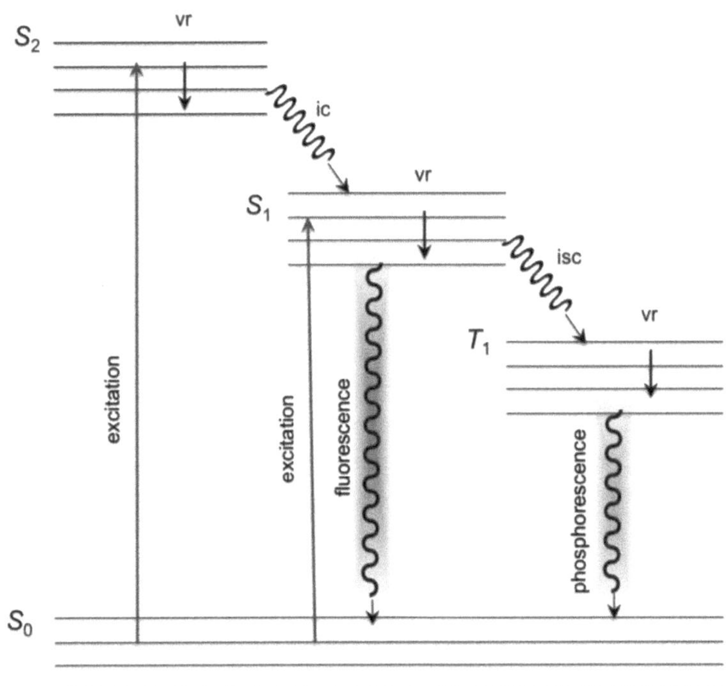

Figure 4.1 Photoluminescence showing various radiative (fluorescence, phosphorescence) and radiationless (vibrational relaxation (VR), internal conversion (IC), and intersystem crossing (ISC)) mechanisms.

states—namely, S_1, and T_1—undergo relaxation to the S_0 with light emission (i.e., relaxation by fluorescence or phosphorescence, respectively) or indirectly without light emission (i.e., radiationless deactivation via vibrational relaxation (VR), IC, ISC, or external conversion due to loss of energy to the sourroundings). The lifetime for these modes of relaxation are VR or IC, 10^{-14} to 10^{-11} s; fluorescence, 10^{-9} to 10^{-7} s; and phosphorescence, 10^{-3} to 10^{2} s. Both absorbed and emitted lights can be measured as an extinction coefficient and quantum yield, respectively.

The extinction coefficient (ε) is described by how strongly or efficiently the substance absorbs the excitation light; the quantum yield (Φ) is a measure of the efficiency of photon emission as defined by the ratio of the number of photons emitted to the number of photons absorbed. The intensity of the emitted light compared to the absorbed light is typically lower and is shifted toward a longer wavelength, a phenomenon typically referred to as the Stokes shift. The λ_{max}'s of the excitation and emission are used for maximum absorption and the fluorescence signal. Although there is usually a significant spectral overlap between the absorption and fluorescence bands, this is rectified by properly chosen filters, which will be discussed later.

Two major types of instrumentation measure fluorescence—spectrofluorometers and filter fluorometers. In a spectrofluorometer, monochromator is used to select both the excitation and emission wavelengths but in some cases a combination of both a monochromator and filter is used for the selection of exciting and analysis of emission wavelengths, respectively. The emitted light is measured at a 90° angle relative to the exciting light to prevent interference of the two wavelengths. In a filter fluorometer, optical filters provide an opportunity to choose excitation as well as emission wavelengths that are relevant to the probe, which is the main principle employed in fluorescence microscopy. An epifluorescence microscope has a similar optical layout as a compound microscope; as the name implies, epi-illumination has the incident light shone from above the specimen. This light source is typically a high-intensity lamp with wavelengths that range from the visible to the UV region of the electromagnetic spectrum. Light passes through an excitation filter with a narrow range of wavelengths, where it is reflected by a dichroic mirror toward the objective lens before it is absorbed by the sample. The sample absorbs the light and emits florescent light where it then again passes through the objective, then to the dichroic mirror and to the barrier filter, which also blocks the incident radiation. Thereby only light that is emitted—not the excitation light—is observed. Hence, epifluorescence microscope has three major filters: excitation filter, dichroic mirror, and barrier filter.

4.3 CHEMISTRY OF REDOX DETECTION BY FLUORESCENCE

The chemistry of reactive species (RS) detection mostly relies on the redox reaction of the RS with a nonfluorescent to weakly fluorescent probe and a yield of products with enhanced fluorescence. These fluorophores (light-emitting chromophores) typically possess highly conjugated systems that are a series of double bonds separated by single bonds. In conjugated double bonds, electrons residing in the π orbitals are less bound to the nucleus, making them highly delocalized along the π system. Due to the diminished electrostatic effect by the nucleus, these π electrons are excitable at lower-energy wavelengths. The energy required for the transition from $S_0 \rightarrow S_1$ is the energy difference between the highest occupied molecular orbital (HOMO) and lowest occupied molecular orbital (LOMO) and, for fluorophores, this requires less energy and falls between the near UV (300 nm) to visible region (700 nm). Therefore, application of fluorescent probes for redox measurement must involve formation of a conjugated π system from a stable reduced form. Strategies to achieve formation of a conjugated system, depending on the reactive species, involves both redox and nonredox mechanisms. For example, conversion of phenolic,[6] boronate,[7] sulfonate,[8] and phenoxyl[9] compounds to the respective quinonoid-like product via a two-electron process was achieved as shown in Fig. 4.2. In the succeeding section, mechanisms of these reactions will be categorized based on their reactivities to various reactive species.

As shown in Fig. 4.3, the formation of highly conjugated aromatic rings such as eosin, rhodamine, and dichlorodehydrofluorescein involves two-electron oxidation that leads to the conversion of the phenol group to ketone (or iminium in the case of

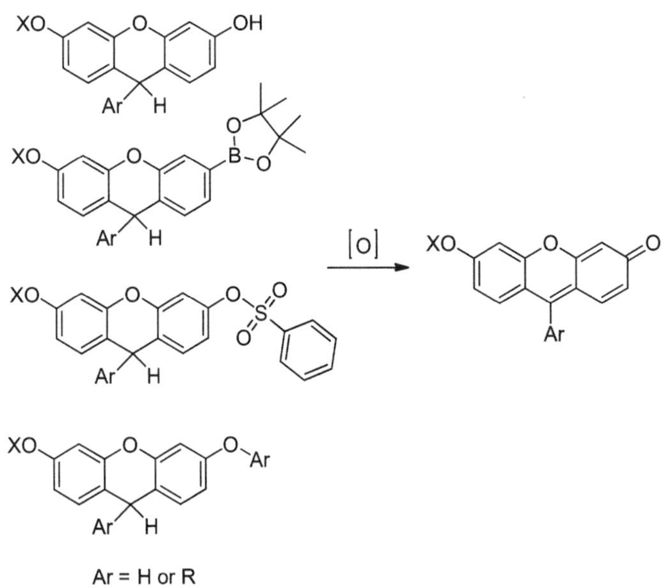

Figure 4.2 Chemical conversion of various nonconjugated reduced probes (FLH_2) to an oxidized conjugated system exhibiting fluorescence (FL).

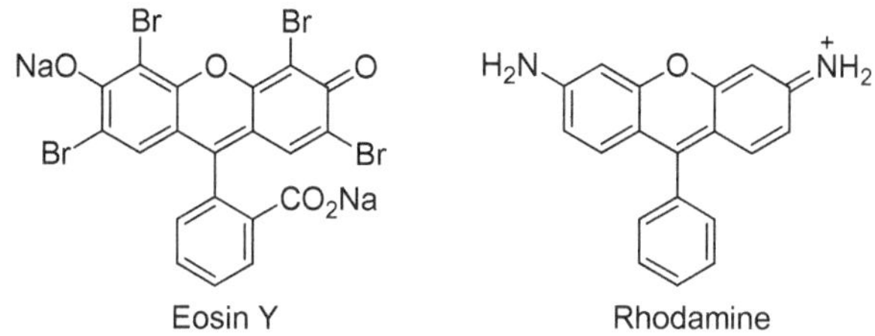

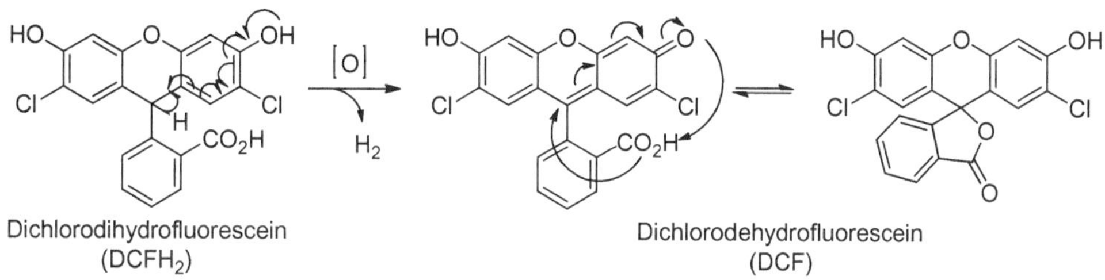

Figure 4.3 Oxidation of the $DCFH_2$ to yield the highly conjugated fluorescent molecule, DCF.

rhodamine) with a quinonoid-like structure facilitating π conjugation along the xanthene moiety and is the basis of their fluorescent properties.

Probes, therefore, must have at least two main components: the reporter molecule (i.e., the fluorophore) and the trigger moiety (i.e., the RS-reactive moiety). Several fluorescent probes were designed offering various modes of fluorescence formation from direct reaction with the RS of interest. For example, in the case of the turn-on (or off–on) modality, the formation of the emission spectrum is monitored. The on–off–on strategy involves multiple analytes with the fluorescence probe quenched, e.g., by H_2O_2; the resulting product can be reduced back to the nonfluorescent product such as shown in Fig. 4.4.[10] The rarely employed off–on–off modality uses metal ion to induce fluorescence formation and is reversed to the nonfluorescent form through metal displacement by an analyte such as in the sequential detection of Zn^{2+} and H_2O_2.[11] Indirect formation of fluorescence can be achieved through molecular designs that involve a total or partial quenching of the chromophore's fluorescence by an electron acceptor exhibiting a shift in or formation of emission spectrum.

While the off–on and on–off modalities have been the most popular approach to probe design, ratiometric detection has been gaining widespread application. Ratiometric methods compare the ratio of intensities of one or two fluorescence signals at the same or two different wavelengths, respectively, hence the name *ratiometric* as shown in Fig. 4.4. The intensity ratio of the emission signal of the RS-modified and RS-unmodified fluorophore is measured at single or dual excitation wavelengths. Advantages of using ratiometric detection over the turn-on and turn-off modality are that it avoids problems encountered when reporting absolute fluorescence values such

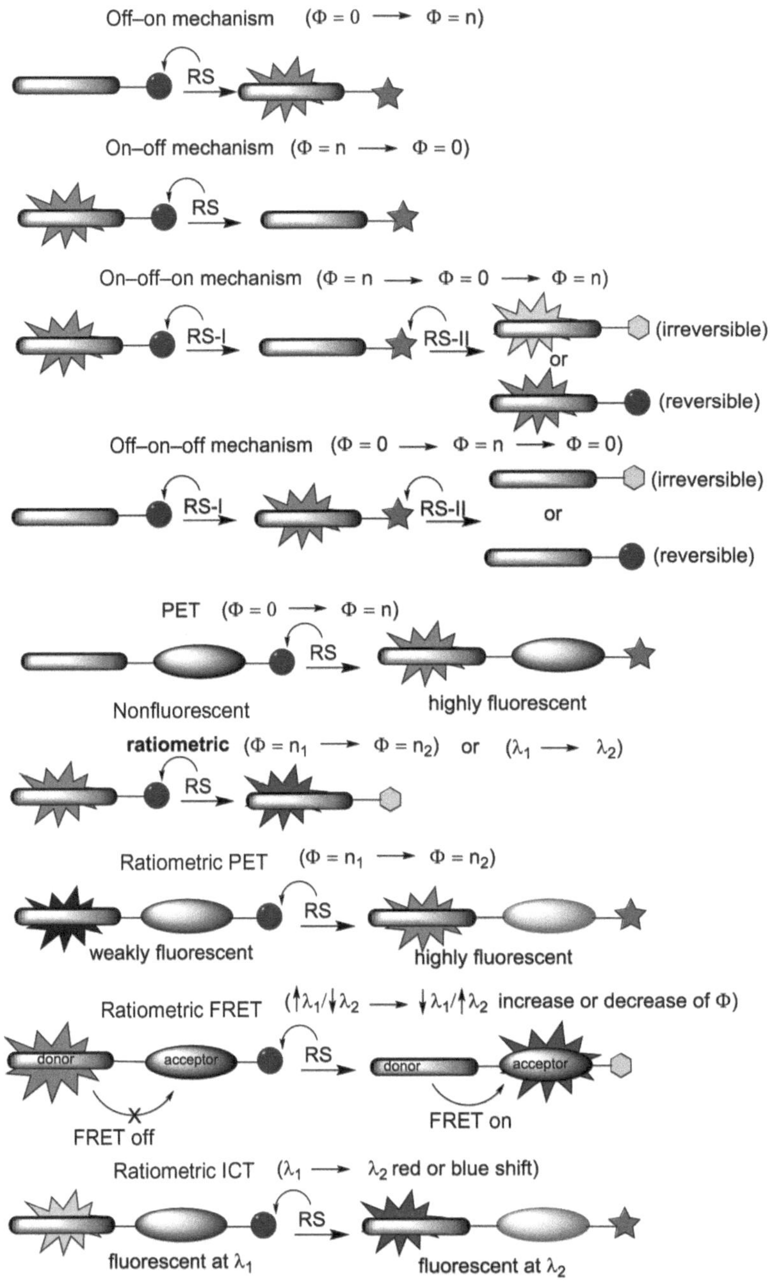

Figure 4.4 Fluorescent probe designs showing various modalities for fluorescence signal formation or shifts.

as probe loading concentration, photobleaching, and probe distribution, as well as instrumental considerations such as optical path length and illumination intensity.

A photo-induced electron-transfer (PET) mechanism is typically employed in RS probe applications to quench fluorescence from an excited fluorophore through reductive or oxidative process. A photo-induced fluorophore promotes an electron from a ground state orbital to the excited state. In the absence of an electron donor (electron-rich moieties) or an electron acceptor (electron-poor moieties are typically those with electron withdrawing groups), the excited electron can undergo normal relaxation giving off fluorescence. However, in the presence of an electron donor (D), the excited molecule (*A) can act as an oxidizing agent, allowing the electron from the donor (D) to occupy the ground state orbital left by the excited electron (i.e., $^*A + D \rightarrow A^{\bullet -} + D^{\bullet +}$). Conversely, in the presence of an electron acceptor (A), the excited electron (*D) can be transferred to the lowest unoccupied molecular orbital (LUMO) of the acceptor, hence the fluorophore acting as a reducing agent according to the equation, $^*D + A \rightarrow A^{\bullet -} + D^{\bullet +}$. These electron-transfer processes, either oxidatively or reductively, create a charge separation, which is the underlying mechanism of fluorescence quenching as shown in Fig. 4.5.

Fluorescence (or Förster) resonance energy transfer (FRET) process involves two light-sensitive chromophores with differing excited energy states in which the chromophore with the higher excited energy assumes the donor chromophore; the one with the lower excited energy state is the acceptor. When excited, the energy from the donor chromophore may transfer its energy to an acceptor chromophore through radiative or nonradiative pathway. Typical FRET-based fluorescent probes for RS detection involves nonradiative energy transfer from the excited donor to the acceptor (see Fig. 4.6) through dipole interaction, which is highly dependent on the distance and relative orientation of the dipole moments of the donor emission and the

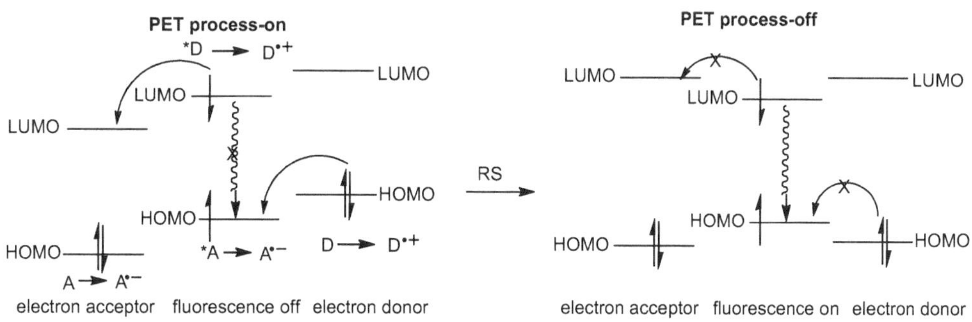

Figure 4.5 Reaction of the RS to the reactive moiety of the donor or acceptor molecule causing changes in the donor or acceptor electronic properties, hence the HOMO–LUMO energy gap widens, making the PET processes unfavorable. The PET process can also be terminated through a bond-breaking process that separates the fluorophore and the donor or acceptor molecule.

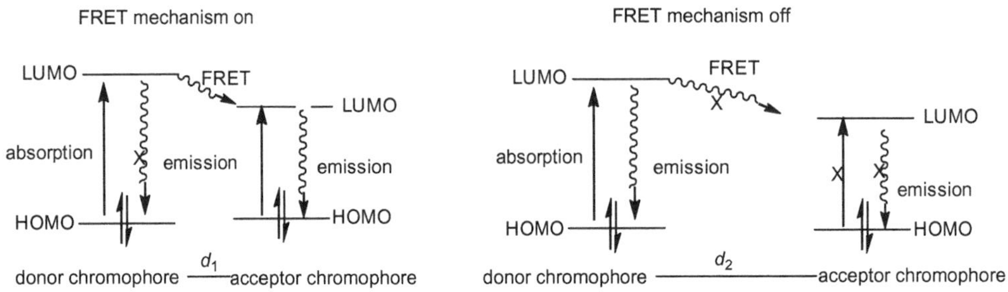

Figure 4.6 FRET process showing the dependence of this process on the distance of two light-sensitive chromophores (donor–acceptor) through initial transfer of energy via a nonradiative dipole–dipole coupling pathway.

acceptor's absorption. FRET therefore causes fluorescence of the acceptor chromophore; in the absence of the FRET process, this causes fluorescence of the donor's chromophore. Due to the sensitivity of the efficiency of energy transfer to the distance between the donor and acceptor, this process is widely employed in the field of biophysics to measure intermolecular distances (1–10 nm) in macromolecules.

The internal charge-transfer (ICT) or photo-induced charge-transfer (PCT) mechanism allows for the observation of changes in the ratio of the absorption intensities or changes in the emission wavelengths upon excitation by light. Electron transfer occurs at the excited state when the vacated orbital (in this case, that of the acceptor fluorophore, A) is filled by the electron from an electron-rich donor substituent (D). Due to the relatively low ionization potential of D and the high electron affinity of A, partial polarization of charges occurs at the excited state to form the stabilized charge transfer complex, $A^{\delta-}$–$D^{\delta+}$ (Fig. 4.7). This intramolecular charge transfer state is stabilized by conformational change through molecular distortion or by increased solvent polarity or by both. This electron push–pull molecular design was exploited to exhibit an emission wavelength (λ_1) that is highly dependent on the charge-transfer state. The extent of this charge-transfer separation can be affected on reaction of the probe with a reactive species; the respective dipole moments of D or A could be altered and hence favor a more efficient charge separation to form the radical ion pairs A^{-}–D^{+}. resulting in increase in HOMO–LUMO energy separation, which may translate to a blue shift in the absorption and emission spectra (λ_2). Depending on the type of linker group (e.g., coupled by a σ or π bond), conformational changes such as twisting or planarization could occur at the excited state and exhibit dual fluorescence such as in the case of a twisted ICT mechanism.

4.4 CLASSIFICATION OF FLUORESCENT RS PROBES BY SPECIFICITY

Fluorescent probes will be categorized according to their specificity to RS. Although these probes' selectivity may not be exclusive, the classification will present their

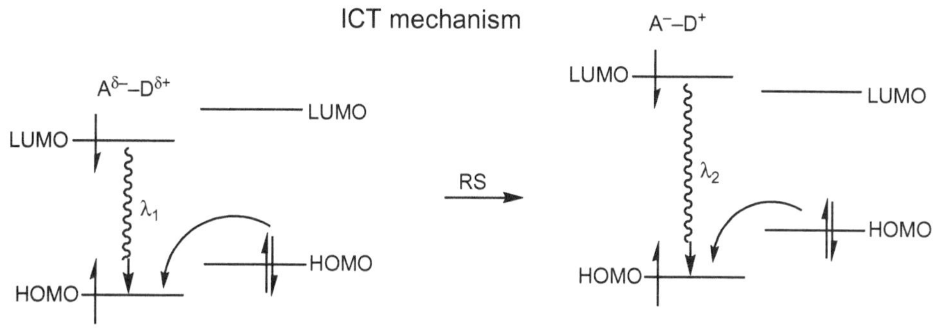

Figure 4.7 ICT process in weakly coupled acceptor and donor ($A^{\delta-}$–$D^{\delta+}$) with an emission wavelength λ_1; on reaction with RS, this forms a more pronounced charge separation (A^-–D^+) with emission wavelength λ_2.

mechanism of action, thus providing users a better understanding of their appropriateness for a particular application and their limitations.

4.4.1 Global redox state or total ROS detection

4.4.1.1 Dichlorodihydrofluorescein ($DCFH_2$) and derivatives

Dichlorodihydrofluorescein ($DCFH_2$) probes are employed to assess the global redox state of cells and tissues. In cellular systems, the exact mechanism of $DCFH_2$ conversion to DCF is not clear, but evidence shows an initial one-electron oxidation to the semiquinone radical form ($DCFH^{\bullet}/DCF^{\bullet-}$) by $CO_3^{\bullet-}$ and $NO_2^{\bullet}$ via $ONOO^-/ONOOCO_2^{\bullet-}$ generation as measured using pulse radiolytic and spectrophotometric kinetic techniques.[6] Or it can also be formed from horseradish peroxidase (HRP) (P-Cpd I and II).[12] Semiquinone radical intermediate reacts with molecular oxygen to form the $O_2^{\bullet-}$ as detected by EPR spin trapping and the final fluorescent product, DCF (see Fig. 4.8).[13]

Therefore, when using fluorescein, one should consider the formation of $O_2^{\bullet-}$ as a by-product of DCF. In spite of this limitation, several $DCFH_2$-derivative probes were designed to improve their cellular permeability and specificity. For example, 2′,7′-dichlorodihydrofluorescein diacetate ($DCFH_2$-DA) was designed to have improved cellular permeability due to the presence of an ester group. Once internalized, esterases cleave the ester moieties of $DCFH_2$-DA to form the $DCFH_2$, according to Fig. 4.9 and see Fig. 4.10 for other $DCFH_2$ derivatives.

Several factors must be considered in interpreting fluorescence signal from the use of $DCFH_2$ and its analog, $DCFH_2$-DA. In the study of respiratory burst activity in mononuclear phagocytes, $DCFH_2$ is preferred over $DCFH_2$-DA due to insufficient probe-esterase activity in mammalian phagocytes. Also, reaction buffers or cultured media alone causes $DCFH_2$ or $DCFH_2$-DA oxidation, which can be exacerbated by the addition of

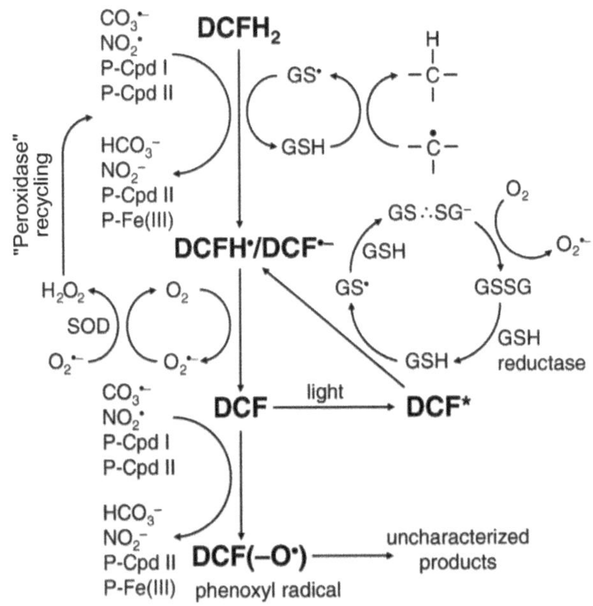

Figure 4.8 Various one-electron oxidizing agents toward the formation DCF from $DCFH_2$. P stands for peroxidase-like catalysts in the cell. Source: *Reprinted with permission from Wardman P. Fluorescent and luminescent probes for measurement of oxidative and nitrosative species in cells and tissues: Progress, pitfalls, and prospects. Free Radic Biol Med 2007;***43***(7):995–1022.*[14] *Copyright © 2007 American Chemical Society.*

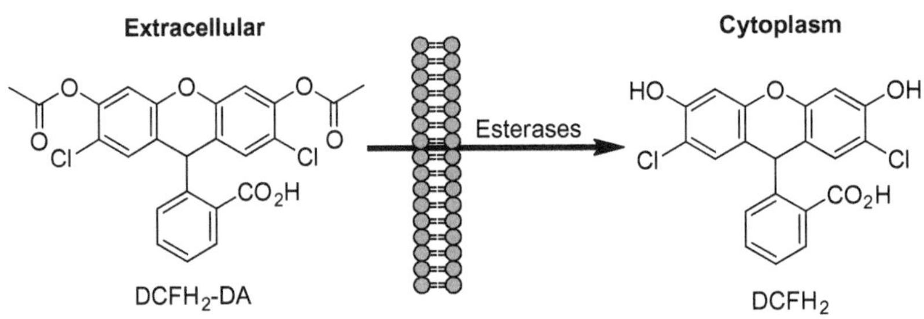

Figure 4.9 Cellular internalization of $DCFH_2$-DA to form $DCFH_2$ on esterase hydrolysis.

tyrosine combined with metal ions contamination.[15] In the study of the toxicology of xenobiotics, reactive oxygen species (ROS) generation in both in vitro and in vivo systems are typically monitored using $DCFH_2$ dyes, although the toxicants used could directly oxidize $DCFH_2$ to the fluorescent DCF. Artifactual data was observed from direct interaction of $DCFH_2$-DA with a toxicant such as tetrabromobisphenol A (TBBPA) in serum-free and cell-free Hank's buffered salt solution media; it is dependent on the dye concentrations typically used for ROS detection studies.[16] Exposure of $DCFH_2$ to xenobiotics such as pyocyanin, mitoxantrone, and ametantrone (but not menadione, paraquat,

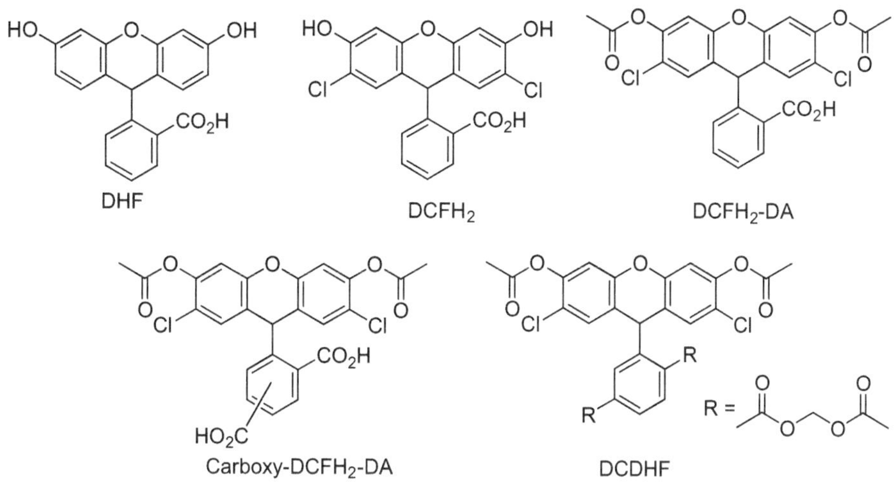

Figure 4.10 $DCFH_2$ derivatives for global ROS detection.

plumbagin, streptonigrin, doxorubicin, daunorubicin, or 5-iminodaunorubicin) could also result in the direct oxidation of $DCFH_2$ to DCF in cell-free media, and this oxidation is not inhibitable by superoxide dismutase (SOD) or catalase, which indicates that ROS are not involved in the oxidation process.[17] This interaction by $DCFH_2$ with pyocyanin produces superoxide radicals under aerobic conditions. Reagents used as inhibitor of cellular metabolism such as that of the mitochondrial electron transport chain (mETC) were proven to directly interact with the probes themselves.

Antimycin and 2-heptyl-4-hydroxy-quinoline-*N*-oxide are inhibitors of complex III of the mETC and showed time-dependent increases in fluorescence with $DCFH_2$ as well as with dihydrorhodamine (DHR) or dihydroethidium (DHE) at concentrations typically employed for inhibition experiments. This inhibitor—probe interaction is oxygen dependent and is abolished by the addition of serum or albumin to the media, indicating possible competition between the proteins and the probe.[18] Catalase and Cu/Zn-SOD were shown to directly increase fluorescence of DCFH2-DA and also act as cofactors for $DCFH_2$-DA oxidation by H_2O_2.[19] Also, H_2O_2-dependent oxidation of $DCFH_2$-DA requires the presence of redox active transition metal ions or simply oxidation by mitochondrial cytochrome c alone in the cytosol or both.[20] While exogenous chemical agents could have a direct positive effect on probes causing fluorescence, physical agents such as ultraviolet A (UVA) irradiation, which is commonly employed to study radiation-mediated oxidative stress in biological systems, can have a direct negative effect on the probes themselves independently of the presence of cells. For example, in Dulbecco's modified eagle's medium, irradiation of DHR123 or $DCFH_2$-DA diminishes the probes' ability to fluoresce after ROS generation in the presence or absence of cells, which indicates that the probe's molecular structure could be altered on direct irradiation.[21] Exogenously added Fe ions or the presence of an endogenous labile iron pool can cause direct oxidation of $DCFH_2$-DA. Cell lines with varying labile iron pools such as murine lymphoma L5178Y(LY) sublines (LY-R, LY-S) gave higher fluorescence signals in the LY-R subline, which has higher iron pool levels and can be inhibited by cell-permeable lipophilic iron chelator salicylaldehyde isonicotinoyl hydrazine but is not inhibitable by peroxidase or SOD, indicating that iron plays a direct role in $DCFH_2$-DA oxidation.[22] This parallels previous study on a cell-free system that shows that the slow rate of $DCFH_2$ oxidation by H_2O_2 can be catalyzed by exogenously added Fe^{2+}.[23] Endogenous and exogenous agents not only affect the probe's ability to impart fluorescence but also probes may participate in biological processes. For example, in HeLA cells, $DCFH_2$-DA was shown to block heme oxygenase (HO-1) expression at the gene level by abolishing phosphorylation of extracellular signal-regulated kinases, which in turn inhibit the Nrf2 translocation and thereby suppresses HO-1 gene activation by oxidants such as hemin, arsenite, or cadmium and thereby preventing their cytotoxicity.[24] However, though these studies were performed under 1–5 h of $DCFH_2$-DA incubation, the effect of $DCFH_2$-DA on early gene formation within the typical recommended time of $DCFH_2$-DA incubation for imaging purposes must be considered.

Using nuclear magnetic resonance chemical shift–polarity correlation, it was demonstrated that $DCFH_2$-DA and its hydrolysis product, $DCFH_2$, compartmentalize within the lipid bilayer; it was proposed that intracellularly generated H_2O_2 diffuses into the membrane and oxidizes probes at the lipid bilayer.[25] However, other studies show otherwise as demonstrated by confocal imaging in which intracellular compartmentalization of $DCFH_2$-DA can detect mitochondrial ROS production under oxidative insult via hypoxia–reoxygenation and pro-oxidant treatments. In this case, mitochondrial ROS are released in to the cytosol, and ROS detection can be achieved by combining $DCFH_2$-DA detection with mitochondrial staining using MitoSOX and MitoTracker Red CM-H2XRos.[26] The $DCFH_2$-DA probe was also shown to detect ROS in the mitochondria when counterstained with tetramethylrhodamine methyl ester perchlorate, a mitochondrial membrane potential probe, and showed sharper and more superior confocal images compared to the acetoxymethoxy analog DCDHF.[27] In the study of kinetics of respiratory burst in phagocytes, $DCFH_2$-DA data did not mirror the cytochrome c assay, but $DCFH_2$ did, which indicates that $DCFH_2$ is appropriate for the investigation of extracellular ROS generation.[15] In endothelial cells, although control cells already show some fluorescence, the presence of menadione alone increases probe fluorescence intensity (Fig. 4.11), with DHF giving the brightest image compared to $DCFH_2$, 5,6-carboxy-2′,7′-dichlorodihydrofluorescein diacetate and DHR123 under similar experiment conditions and instrumental settings.[19] In an independent study using a cell-free system, DHF gave the lowest fluorescence response to various oxidants or oxidant-generating systems (e.g., H_2O_2,

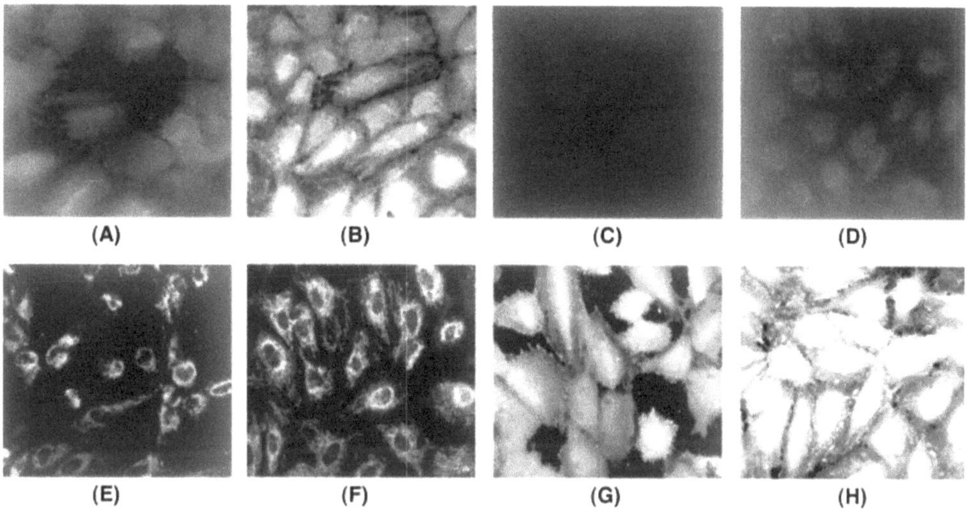

Figure 4.11 Confocal micrograph of human umbilical vein endothelial cells loaded with 20 μM of various probes: (A) $DCFH_2$, control; (B) $DCFH_2$, menadione; (C) 5,6-carboxy-2′,7′-dichlorodihydrofluorescein diacetate, control; (D) 5,6-carboxy-2′,7′-dichlorodihydrofluorescein diacetate, menadione; (E) DHR123, control; (F) DHR123, menadione; (G) DHF, control; and (H) DHF, menadione. Source: *Adapted from Hempel et al.,* Free Radic Biol Med *1999;**27**:146–59.*[19]

KO_2, $ONOO^-$, NO, horseradish peroxidase, ferric iron, xanthine oxidase, cytochrome c, and lipoxygenase) compared to the other probes studied. DHF-DA showed the highest cell-loading property, which indicates that the quality of fluorescence image is determined by the probes' ability to localize intracellularly and exhibit high molar fluorescence more than the probes' ease of oxidation.[19]

4.4.1.2 Dihydrorhodamine 123 (DHR 123) and reduced MitoTracker probe derivatives

Dihydrorhodamine is one of the most commonly used reduced probes. DHR 123 enters the cell and gets oxidized by ROS to rhodamine 123 (RH 123) and accumulates in the mitochondria. The ability of RH 123 to compartmentalize in the mitochondria on oxidation is due to the formation of the cationic rhodamine 123 iminium cation (in resonance with xanthylium salt), which exhibits green fluorescence (Fig. 4.12). Mitochondrial respiration involves a series of redox reactions whereby the mETC that is present in the inner mitochondrial membrane creates an electrochemical gradient that is characterized by a difference in proton concentrations and electrical transmembrane potential between the intermembrane space and mitochondrial matrix as a result of O_2 consumption and synthesis of adenosine triphosphate (ATP). Hence, under normal conditions, mitochondrial potential is more negative than the extracellular matrix. This membrane potential, $\Delta\Psi_m$, can be measured using

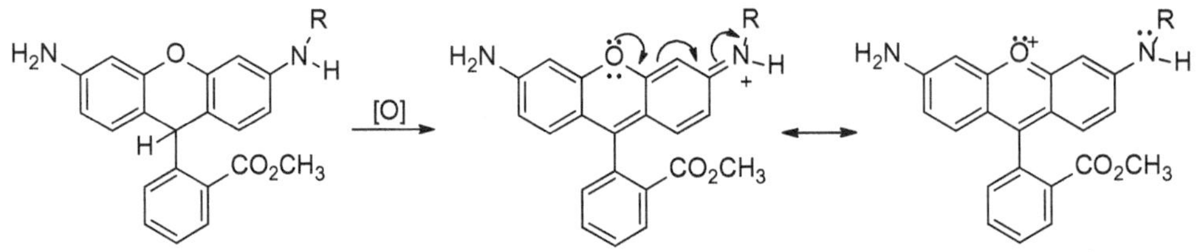

DHR 123 (R = H)
DHR 6G (R =CH_3)

rhodamine 123 (RH-123)
rhodamine 6G (RH-6G)

xanthyllium form

Figure 4.12 Oxidation of reduced DHR to the fluorescent RH.

lipophilic cationic probes such as DHR 123 in which the ester group provides some degree of lipophilicity and the oxidized iminium cation provides the positive charge needed to facilitate mitochondrial membrane permeability and accumulation within the mitochondria. Since RH 123 accumulates inside the mitochondria, initial loading concentration of DHR 123 will be orders of magnitude lower than the final accumulated concentration of RH 123 in the mitochondria, causing inhibition of the oxidative phosphorylation of the ATP synthase and quenching of the RH 123.[28]

Similar in selectivity to $DCHF_2$-DA, as shown in Fig. 4.13, a steady-state radiolytic study shows formation of fluorescent products from DHR 123 on oxidation by $HO^{\bullet}$, $CO_3^{\bullet-}$, and $NO_2^{\bullet}$ but poor reactivity with $O_2^{\bullet-}$, which only resulted in low fluorescence formation in a cell-free system.[29] However, reaction of $NO_2^{\bullet}$ to DHR 123 is >1000-fold slower than the corresponding oxidation by $CO_3^{\bullet-}$.[30]

Oxidation of DHR 123 to RH 123 by H_2O_2 depends on the presence of cytochrome c or Fe^{2+} as shown in endothelial cells,[31] and DHR 123 oxidation is further enhanced in the presence of peroxidase.[32] Neither $O_2^{\bullet-}$[32] nor NO alone was able to oxidize DHR 123 to RH 123, whereas $ONOO^-$ and hypochlorous acid (HOCl) cause DHR 123 oxidation.[33] Labile plasma iron is essential for the ROS-mediated oxidation of the DHR 123 in blood samples of healthy and β thalassemic patients.[34] However, DHR reaction with $ONOO^-$ is not metal-ion dependent, and DHR 123 oxidation is a direct reaction with ONOO- through an initial one-electron oxidation to form the $DHR^{\bullet-}$ and its subsequent disproportionation reaction to yield the recycled DHR 123 and the fluorescent RH 123.[35] The $DHR^{\bullet-}$ intermediate can also form the RH 123 via one-electron oxidation by molecular O_2 with a rate constant of $7 \times 10^8\ M^{-1}\ s^{-1}$ using a pulse radiolytic technique,[30] while oxidation of DHR 123 to RH 123 by NO is oxygen dependent, suggesting that NO_2 is the oxidizing agent.[35] Another suggested mechanism was that $NO_2^{\bullet}$ or $HO^{\bullet}$ as decomposition products of $ONOO^-$ or both directly react with DHR 123. A thorough examination of the kinetics of oxidation of DHR 123 revealed a zero-order kinetics, which involves formation of the $[\text{DHR 123–ONOO}]^-$ adduct at pH>7 followed by protonation and subsequent two-electron oxidation to yield NO_2^- and the fluorescent RH 123; the same mechanism

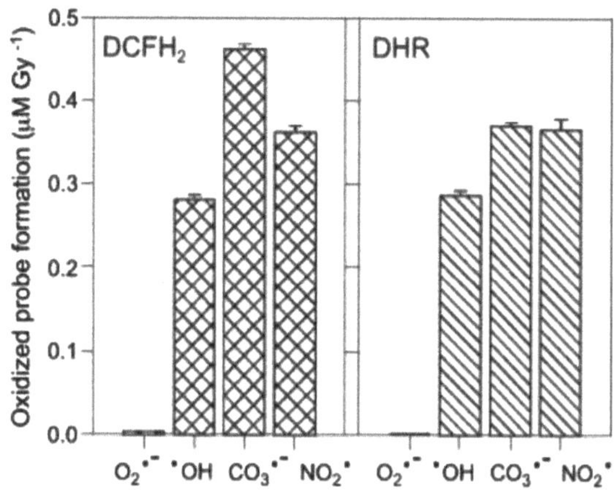

Figure 4.13 Yield of fluorescent products formed from the reaction of $DCFH_2$ and DHR with superoxide, hydroxyl, carbonate, and nitrogen dioxide radicals. Source: *Adopted from Wrona et al.,* Free Radic Biol Med *2005;**38**:262–70.*[29]

was proposed for the $ONOO^-$-mediated oxidation of $DCFH_2$ to DCF.[36] Therefore, for in vitro studies requiring $ONOO^-$-mediated oxidative insult, either from direct addition of commercially available ONOOH or by employing 3-morpholinosydnonimine (SIN-1) as the $ONOO^-$ source, careful controls should be carried that consider the direct oxidation of DHR 123 by $ONOO^-$.[37] Although $ONOO^-$ was found to be a more potent oxidant than nitroxyl, NO^- from Angeli's salt, DHR 123 oxidation by NO^- was found to be oxygen dependent,[38] which mirrors previous studies on the oxidation of DHR 123 by NO^- and O_2 producing $ONOO^-$.[39] These studies show the nonspecificity of DHR 123 reaction to several oxidants.

Similar to $DCFH_2$, DHR 123 reacts directly with reagents such as the mETC complex III inhibitor and is oxygen dependent.[18] Fluorometric assay of high-density lipoprotein based on DHR 123 oxidation was found to be independent of ROS generation.[40,41] ROS-independent oxidation of DHR 123 or $DCFH_2$ by Cr(V) in Cr (VI)-treated lung carcinoma A549 cells were reported, which cautions the use of chromate-induced ROS production in cellular systems.[42] Moreover, in the study of UVA-mediated oxidative stress, UVA induces DHR 123 oxidation to RH 123 independently of cellular ROS generation.[21]

Although $DCFH_2$-DA gave advantages over other probes, it is cell type–dependent. For example, comparison of the fluorescence intensities between DHR 123 and $DCFH_2$-DA resulting from exogenously added H_2O_2 to tumor SPC-A-1 cells shows that DHR 123 is superior to $DCFH_2$-DA.[43] Comparison of the appropriateness of using DHR 123 versus $DCFH_2$-DA or hydroethidine (HE) for flow cytometric analysis of oxidative burst from neutrophils revealed that while all probes are responsive to the overall ROS production on activation by phorbol myristate acetate

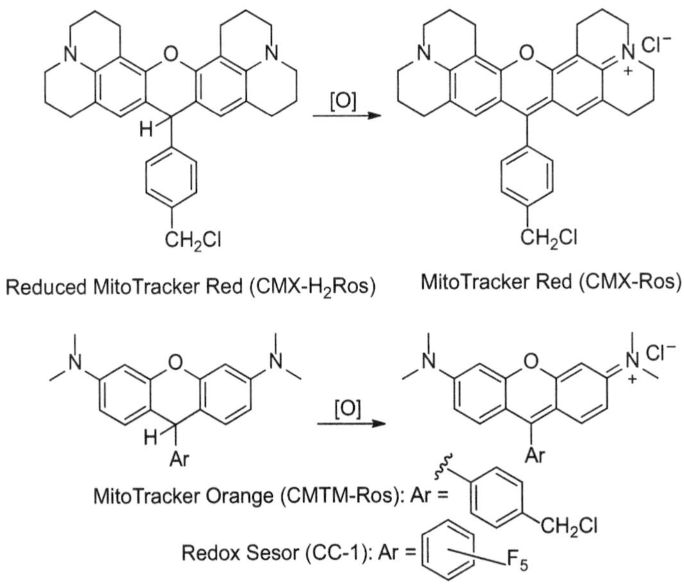

Figure 4.14 Oxidation of CMX-H_2Ros, CMTM-H_2Ros, and CC-1 to xanthyllium salt (iminium cation) forms.

(PMA), each has different requirements for oxidation.[44] Probe response to ROS production from alveolar macrophages is also dependent on the type of stimulant used. For example, macrophage stimulation by TiO_2 or quartz results in DHR 123 and HE oxidation but not with DCFH_2-DA, while residual oil fly ash stimulation oxidizes DCFH_2-DA more efficiently than DHR 123 or HE.[45]

Similarly, reduced MitoTracker Red (CMX-H_2Ros), reduced MitoTracker Orange (CMTM-H_2Ros) and Redox Sensor (Red CC-1) are oxidized to the fluorescent xanthyllium salt forms (Fig. 4.14). These DHR analogs are commonly used to stain mitochondria and their intra-mitochondrial accumulation, like DHR 123, is driven by the membrane potential. One has to consider that they only accumulate once oxidized—i.e., they do not accumulate in the mitochondria in the reduced form since it is the lypophilic positive character that drives the probe's migration inside the mitochondria. Therefore, one cannot unequivocally determine the site of ROS production from using these probes other than to say the ROS were formed prior to mitochondrial compartmentalization.

4.4.1.3 3′-(p-Aminophenyl) fluorescein (APF) and hydroxyphenyl fluorescein (HPF)

Unlike the DCFH_2 and DHR 123 types of fluorescent probes in which the probes are in the reduced state and yield a highly fluorescent conjugated system on oxidation; with aminophenyl fluorescein (APF) and HPF, these probe molecules are already

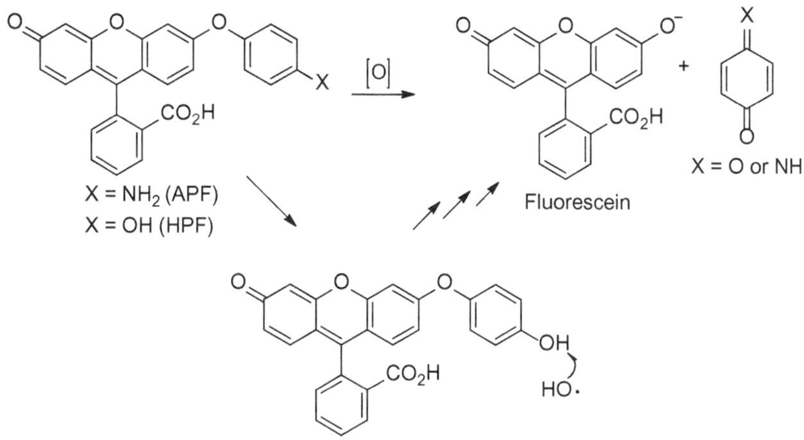

Figure 4.15 Oxidation of weakly fluorescing O-arylated fluoresceins APF and HPF, yielding the highly fluorescent fluorescein and *p*-benzoquinone imine or *p*-benzoquinone.

Table 4.1 Fluorescence increase of HPF, APF, and $DCFH_2$ in the presence of various ROS generating systems

ROS	HPF	APF	DCFH
$HO^{\bullet}$	730	1200	7400
$ONOO^{-}$	120	560	6600
^{-}OCl	6	3600	86
$^{1}O_2$	5	9	26
$O_2^{\bullet -}$	8	6	67
H_2O_2	2	<1	190
NO	6	<1	150
$ROO^{\bullet}$	17	2	710
Auto-oxidation	<1	<1	2000

Source: Adapted from Setsukinai et al., *J Biol Chem* **2003**;278:3170–5.[46]

oxidized with highly conjugated aromatic rings, but we can presume these are highly fluorescing already, making them useless analytical probes for ROS. However, the presence of a phenyl-amino or phenolic group at the 6′-position in APF and HPF, respectively, suppresses the fluorescence of the fluorescein moiety. Therefore, O-dearylation of APF or HPF yields the strongly fluorescent fluorescein and *p*-benzoquinone imine or *p*-benzoquinone according to Fig. 4.15.

Table 4.1 shows the relative reactivity of APF and HPF with various oxidants compared to $DCFH_2$. Data show that $DCFH_2$ exhibited the highest reactivity to $HO^{\bullet}$, $ONOO^{-}$, H_2O_2, NO, and $ROO^{\bullet}$ and the least reactivity to HOCl, $^{1}O_2$, and $O_2^{\bullet -}$, while APF exhibited high reactivity to HOCl compared to HPF and $DCFH_2$. This remarkable sensitivity of APF to HOCl makes APF a selective probe for

HOCl.[46] Due to the relatively low reactivity of HPF and APF with NO and H_2O_2, when used along with $DCFH_2$ these reagents are ideal in distinguishing between NO or H_2O_2 from $HO^•$ radicals since $DCFH_2$ gives fluorescence to all of the radicals. Distinguishing between NO or $O_2^{•-}$ from $ONOO^-$ could be achieved using HPF or APF. Another advantage from using APF and HPF is that, unlike $DCHF_2$, they are least susceptible to light-induced auto-oxidation in vitro. Also, APF was employed in the detection of HOBr from the peroxidase activity of eosinophil granulocytes in the presence of bromide ions with equal sensitivity as with HOCl detection using a flow cytometric technique.[47,48]

The differences in the selectivity by HPF, APF, and $DCFH_2$ toward various oxidants are due to the initial chemistry of oxidation. In HPF and APF, the oxidants may initially act on the phenoxy-H or amino-H through H atom abstraction or abstraction of electron from the O or N atoms, respectively, while for $DCFH_2$, the initial mode of attack is on the *ipso*-H, which is more facile for most oxidants. The higher reactivity of APF than HPF to HOCl is due to the fact that Cl atom addition to the amino-N is much faster compared to the addition of a phenolic aromatic ring of HPF. For example, *N*-chlorination of primary amines by HOCl is on the order of $10^7-10^8\ M^{-1}\ s^{-1}$ compared to aromatic ring chlorination in phenol of $\sim 250\ M^{-1}\ s^{-1}$.[49]

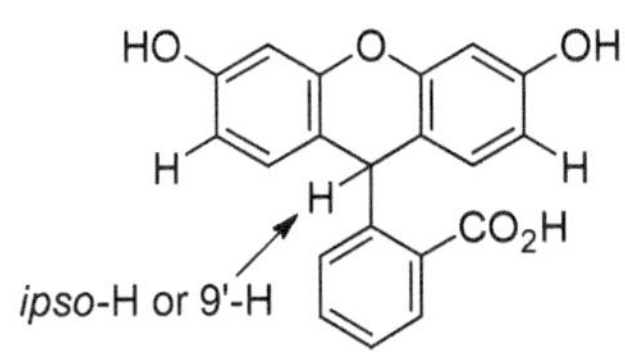

4.4.1.4 Dihydrocalcein acetoxymethylester (AM)

Intracellular detection of ROS is desirable due obviously to its provision of mechanistic insights into the oxidative processes that occur in the cytoplasm. By protecting the carboxylate and phenolic moieties with an ester group such as in the case of dihydrocalcein acetoxymethylester (AM), probe internalization is facilitated through passive diffusion of the hydrophobic molecule across the plasma membrane; the esters are then cleaved by esterases to give the phenolic molecule. Ester hydrolysis allows the now hydrophilic molecule trapped in the cytosol to be further oxidized to its fluorescent form (see Fig. 4.16). Compared to DCF, the intracellular half-life of the fluorescent calcein is several hours compared to 5 min for DCF and is oxidized nonspecifically by $ONOO^-$, $HO^•$, and peroxidase but not by H_2O_2 or NO. Furthermore, calcein accumulates in the mitochondria of the vascular smooth muscle cells and its oxidation is most likely due to the direct electron transfer to the complex I of the mETC.[50] However, using breast cancer MCF-7 cells, calcein distributes throughout the cytoplasm and cell membrane. Fluorescence can be intracellularly

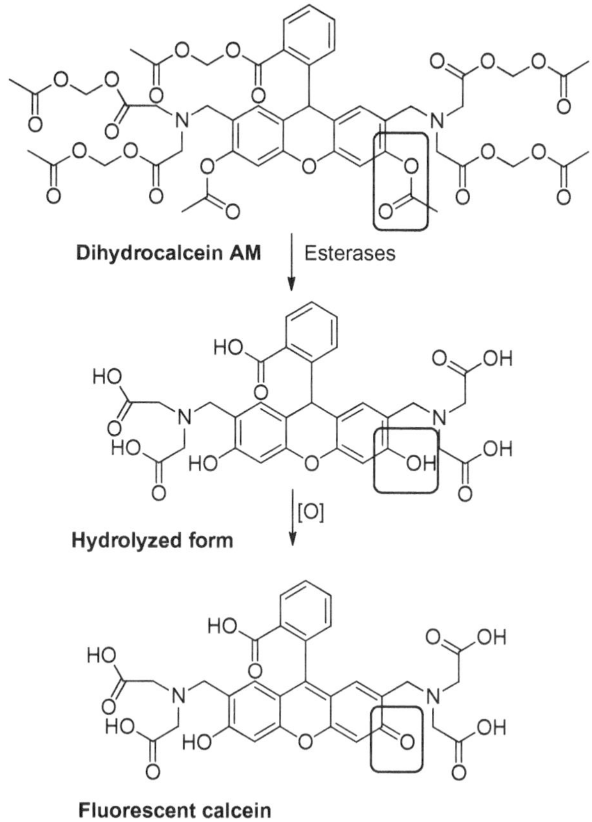

Figure 4.16 Ester hydrolysis of dihydrocalcein by esterases and its oxidation to its fluorescent form.

generated from *t*-butylhydroperoxide–derived ROS or through visible light illumination to generate singlet oxygen.[51] While the use of an acetoxymethyl group demonstrated effective intracellular compartmentalization of the probes in in vitro settings, their applicability to in vivo ROS detection could be limited due to extracellular esterase activity that can cleavage the esters before they enter the cells.[52]

4.4.2 Selective RS detection

Superoxide radical ($O_2^{\bullet -}$)

Dihydroethidium (DHE) or simply hydroethidine (HE) is a fluorescent-based probe that is employed to directly detect $O_2^{\bullet -}$ via formation of a four-electron oxidation product, 2-HO-E^+. However, the formation of 2-HO-E^+ (excitation, 480 nm; emission, 567 nm) is not an exclusive product where other by-products such as the two-electron product E^+ (excitation, 500–530 nm; emission, 590–620 nm) and other products that are nonspecific to $O_2^{\bullet-}$ are also formed and can interfere with the fluorescence measurement.[53] In biological systems, formation of 2-HO-E^+ may also occur via enzymatic means. HPLC analysis of the reaction of HE with $O_2^{\bullet-}$ gave other by-products such as the iminium cation, E^+, and several dimeric products such as HE–HE, HE–E^+, and E^+–E^+ as a result of the bimolecular radical–radical addition reaction. While 2-HO-E^+ and all other by-products originate from a single intermediate, $HE^{\bullet+}$, the formation of this one-electron oxidation product of HE in the presence of $O_2^{\bullet-}$ was proposed, and several oxidants such as $HO^{\bullet}$ and $Br^{\bullet-}$ were shown to oxidize HE to $HE^{\bullet+}$ but only $O_2^{\bullet-}$ gave 2-HO-E^+.[54] Cytochrome c–mediated oxidation of HE or MitoSox by the mitochondria were reported and showed formation of the red fluorescent E^+ or Mito-E^+ and other oxidation products, but not 2-HO-E^+ or 2-OH-Mito-E^+.[55] Since 2-HO-E^+ and E^+ as well as 2-OH-Mito-E^+ and Mito-E^+ impart red fluorescence, it is therefore imperative to discern one from the other by using an HPLC separation technique[56,57] Fig. 4.17).

HE was also found to react with oxidants derived from myeloperoxidase (MPO), HOCl or chloromaine, forming 2-Cl-E^+. Therefore, caution should be observed when detecting $O_2^{\bullet-}$ from inflammatory responses, especially when MPO is involved.[58] Moreover, exogenous agents (physical or chemical) that are typically employed to study oxidative stress such as visible light could cause photo-oxidation of HE and Mn(III) TBAP could directly oxidize HE, yielding E^+ but not 2-OH-E^+ as the final product, while sonication employed to lyse the cells could lead to the formation of 2-OH-E^+ via formation of $O_2^{\bullet-}$ during cavitation.[59] HE was also shown to be oxidized by H_2O_2 via nonspecific peroxidase catalysts such as HRP and MPO and by heme proteins such as mitochondrial cytochromes, hemoglobin, and myoglobin, although these studies did not distinguish between E^+ and 2-OH-E^+ using other methods.[60]

Using capillary electrophoresis and laser-induced fluorescence detection, MitoSox was employed for the detection of $O_2^{\bullet-}$ in both sides of the mitochondrial inner membrane as 2-OH-Mito-E^+[61] or by HPLC methods using UV visible absorption, fluorescence, and electrochemical and mass spectrometry detection.[62] Using fluorescence microscopy, monitoring MitoHE and HE oxidation at 396- and 510-nm excitation wavelengths allows for selective detection of $O_2^{\bullet-}$ in living cells.[63] Superoxide detection by fluorescence microscopy and flow cytometry using HE from various cell lines such as K562 (human leukemia cells), A431 (human epidermoid carcinoma cells), and SCE2304 (human mesenchymal stem cells derived from endometrium) revealed that oxidized HE accumulates in the mitochondria and altered mitochondrial metabolic function.[64] This parallels earlier studies showing that oxidized product of HE is toxic to *Escherichia coli* under both aerobic and anaerobic conditions, implicating the formation of E^+ as the cytotoxic agent.[65]

Nonredox-based superoxide-specific probe such as the bis(2,4-dinitrobenzenesulfonyl) fluorescein (DBSF) (Fig. 4.18) was employed, although a similar

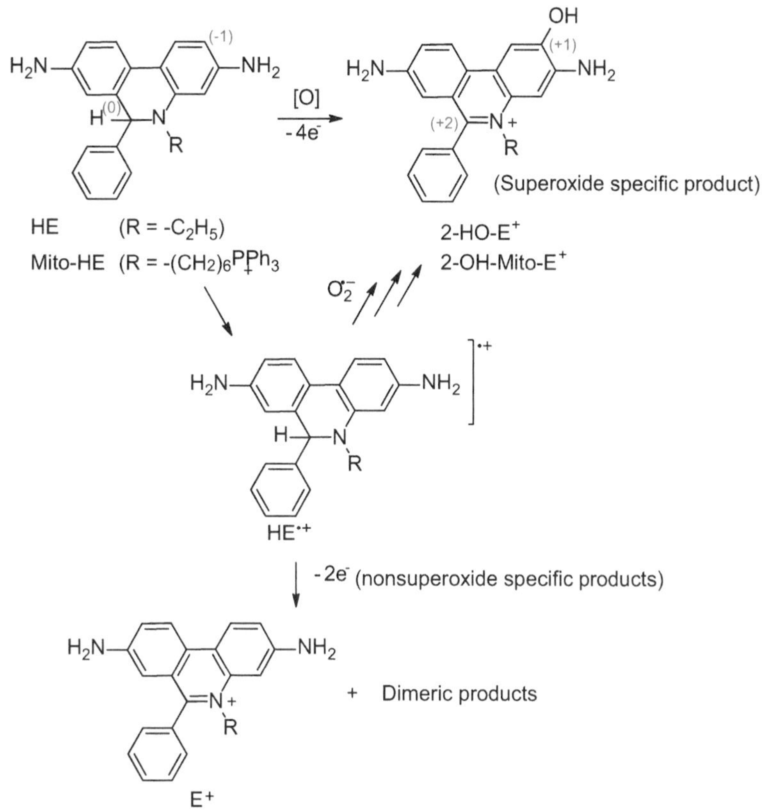

Figure 4.17 Four-electron oxidation of HE to 2HO-E^+ by $O_2^{\bullet -}$ via the $HE^{\bullet +}$ intermediate.

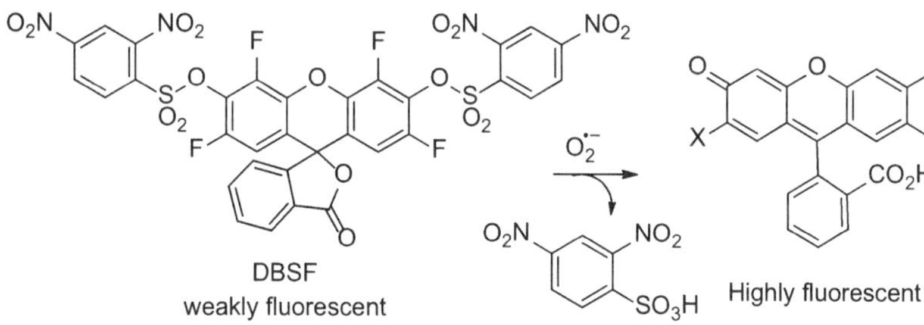

Figure 4.18 Nonredox mode of $O_2^{\bullet -}$ detection using bis(2,4-dinitrobenzenesulfonyl) fluorescein (DBSF).

perfluorobenzylsulofonyl-fluorscein analog was previously used to specifically detect H_2O_2. This dinitrobenzyl analog exhibited high specificity to $O_2^{\bullet -}$ with relative intensity ratio for superoxide to nonsuperoxide ROS of 5–50.[8]

The mitochondrial circularly permuted yellow fluorescent protein (mt-cpYFP was serendipitously discovered as an $O_2^{\bullet -}$ sensor and is a ratiometric probe. This protein-based $O_2^{\bullet-}$ sensor was expressed in cardiomyocytes in conditions of hypoxia and reoxygenation and imparted $O_2^{\bullet-}$ "flashes" in the mitochondria that can be inhibited by adenosine.[66] However, further studies revealed that these "mitoflashes" from cpYFB only respond slightly to the $O_2^{\bullet-}$−generating system, xanthine (X)−xanthine oxidase (XO), and instead were mainly caused by increased pH due to the addition of KOH to the xanthine stock solution.[67] This controversy was later clarified and suggests that the cpYFP must be in a "fully reduced" state by incubation with dithioreitol, thus exhibiting a twofold increase in fluorescence on oxygenation and a twofold increase on introduction with X−XO. Therefore, cpYFP must be used in its reduced form and expressed in cells that have gluthathione reductase.[68]

4.4.2.2 Hydrogen peroxide (H_2O_2)

Boronate reactivity to hydrogen peroxide was first proposed in the early 1930s when phenylboric acid reacting with H_2O_2 yielded phenol and $B(OH)_3$.[69] In the 1950s, kinetic studies of H_2O_2 reacting with phenylboric acid showed that this reaction is base catalyzed, indicating involvement of hydroperoxide.[70] On reaction with H_2O_2 (as deprotonated HOO^-), boronate is cleaved via nucleophilic addition of the hydroperoxide to the boron atom, which forms the hydroperoxyl-boronate complex and loses a hydroxide ion before undergoing a 1, 2 insertion reaction to form the boronate ester that is hydrolyzable to the phenolic form (see Fig. 4.19).

Boronate-based H_2O_2 probes that are conjugated to a fluorophore (nitroaniline, **1a** or 7-amino-4-methylcoumarin, **1b**) via a carbamate linker group were first designed in the early 2000s.[71] Here the boronate reaction with H_2O_2 in alkaline conditions forms the phenol with subsequent release of the fluorophore via the intermolecular bond-breaking process as shown in Fig. 4.20.

When tethered to the already oxidized fluorescent fluorescein or rhodamine, boronate could mask the fluorescence and give a low quantum yield such as in the case of MitoPY1,[72] PY1,[73] PF1,[74,75] and the nonfluorescein-based boronate probes such as PR-1 and PX-1[75] (Fig. 4.21). This unmasking of fluorescence of fluorescein- and rhodamine-based probes on reaction with H_2O_2 is the basis of H_2O_2 detection.

The apparent second-order rate constant for the boronic acid reaction with H_2O_2 to form the phenol and boric acid was calculated to be $k_2 = 3 \times 10^{-3}\ M^{-1}\ s^{-1}$ at pH 4.44[70] (Eq. 4.1), and this slow reactivity was also observed with PG1 and PC1 boronate dyes (Fig. 4.21) at neutral pH with pseudo−first-order rate constants of $k_{obs} = 1.1(1) \times 10^{-3}\ s^{-1}$ and $k_{obs} = 1.0(1) \times 10^{-3}\ s^{-1}$, respectively, with a maximum turn-on response time of 2−3 h.[76] This slow H_2O_2 reactivity toward boronates is due to the fact that the concentration of hydroperoxide (HOO^-) at neutral pH is very low (pK_a for H_2O_2 is 11.8).

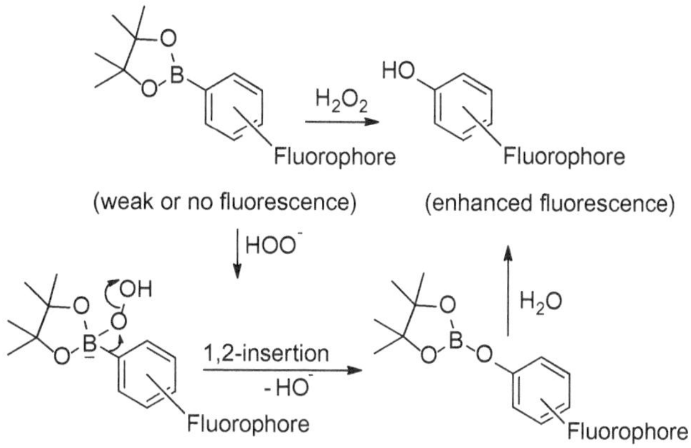

Figure 4.19 Mechanism of H_2O_2 reaction to phenyl-boronate.

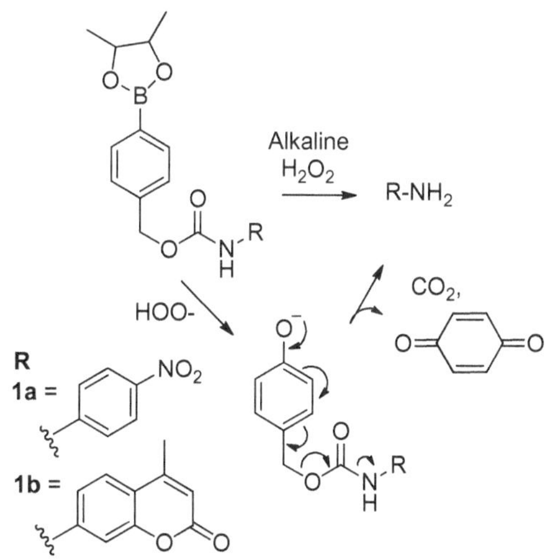

Figure 4.20 Reaction of H_2O_2 in alkaline condition showing release of the fluorophores, nitroaniline, **1a** and 7-amino-4-methylcoumarin, **1b**.

$$C_6H_5OH + H_2O_2 \overset{k_2}{\rightarrow} C_6H_5OH + B(OH)_3 \tag{4.1}$$

The rate constants of oxidation of various boronic acid and esters by $ONOO^-$ and HOCl as compared to H_2O_2 were previously determined.[77] Second-order rate constants for boronate oxidation by $ONOO^-$, HOCl, and H_2O_2 were calculated to be on the order of 10^{5-6}, 10^{3-4}, and $\sim 2\ M^{-1}\ s^{-1}$, respectively. Therefore, in systems where $ONOO^-$ is generated—such as where there is concomitant generation of $O_2^{\bullet -}$ and NO, activation of phagocytic cells, or either $ONOO^-$ or HOCl are introduced exogenously to induce oxidative stress—proper controls or inhibitory studies

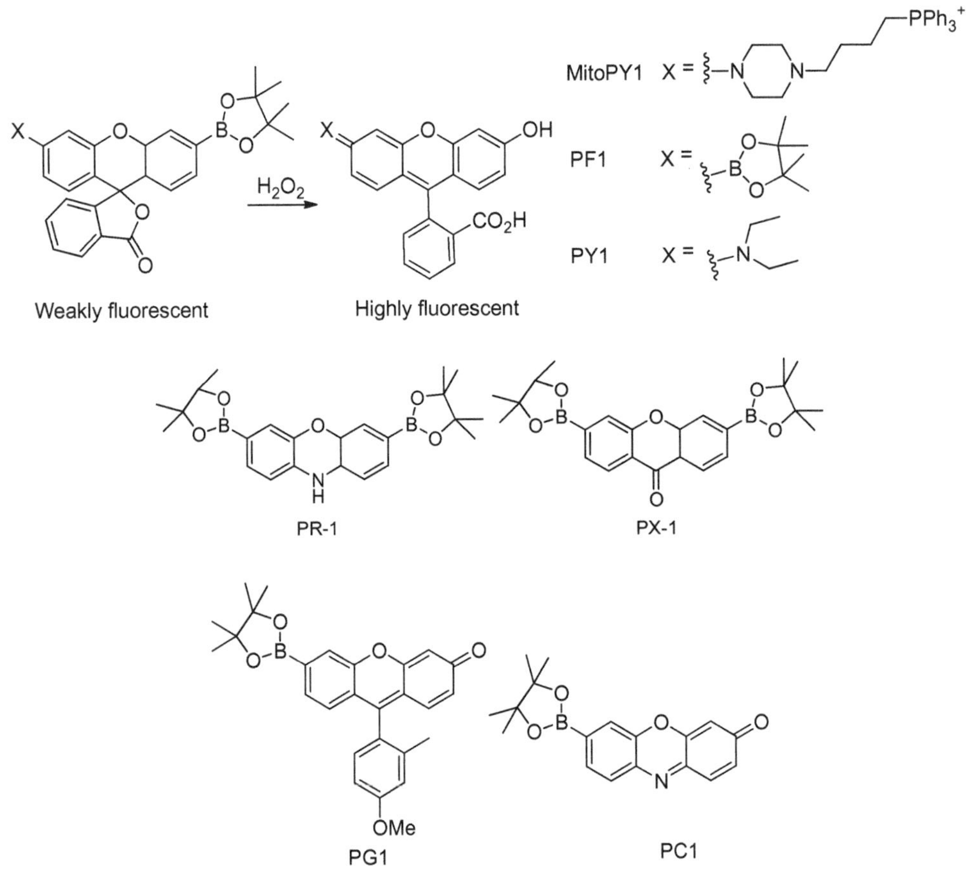

Figure 4.21 Various types of boronate-fluorescein or rhodamine probes.

should be implemented to differentiate H_2O_2 production from other reactive species that could lead to wrong interpretation of the type of oxidant generated.

Detection of H_2O_2 in human spermatozoa using boronate-based probes was also shown to be give positive results with $ONOO^-$.[78] Other nonconventional fluorophore boronate probes were developed based on 1,8-napthalimide such as the cytoplasm-targeted NP1 and the nuclei-targeted pep-NP1 with high selectivity to H_2O_2, although reactivity to $ONOO^-$ was not reported[79] (Fig. 4.22).

As previously discussed earlier, emission intensity can be affected by various factors such as probe concentration and microenvironment distribution, as well as by instrumental settings such as incident laser power. In the absence and presence of the analyte, the observation of emission intensities at two wavelengths, which is known as *ratiometric measurement*, has become attractive since it avoids errors in analyte measurement and improves sensitivity. By reporting fluorescence in terms of ratiometric

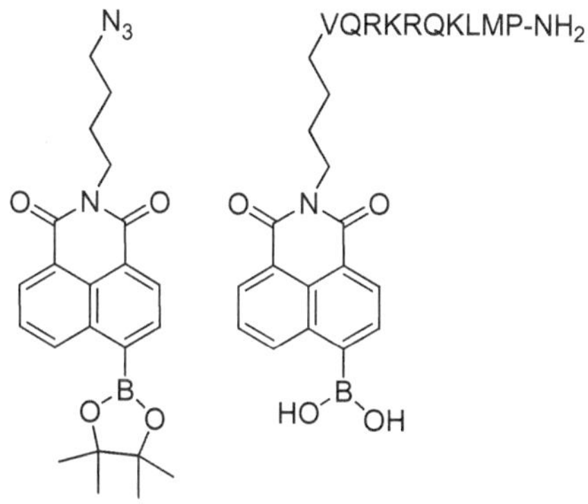

Figure 4.22 1,8-Napthalimide-boronate probes such as the cytoplasm-targeted **NP1** and the nuclei-targeted **pep-NP1**.

values, one could eliminate gross variations in fluorescence reading. Boronate-based two-photon probes that are ICT modulated and requiring short-wavelength excitations such as PL1[80] and PN1[81] were developed with ratiometric capabilities and were employed for the visualization of H_2O_2 in live cells and in tissue with PN1using two-photon confocal microscopy (Fig. 4.23).

Other molecules that exhibit the ICT mechanism were developed that allowed spectral shift as a function of H_2O_2 concentration. For example, *N,N*-dimethylamino-cinnamaldehyde boronic ester (DACB) with low emission energy due to twisted ICT. On reaction with H_2O_2, DACB is converted to a protonated form as an ammonium cation to give a high-energy emission band thus allowing for the monitoring of changes in the emission intensities at two wavelengths—i.e., disappearance of emission wavelength at 566 nm and formation of emission at 484 nm[82] (Fig. 4.24).

Boronic ester moiety may not necessarily always be the site of attack by H_2O_2. Hydrogen peroxide was shown to react with the carbamate linker group such as in the case of SHP-Mito (Fig. 4.25), which cleaves the fluorophore from the phenyl-boronate group to allow a two-photon florescence detection of H_2O_2 production in the mitochondria and intact tissue at a depth of 100−180 μm. Perhaps one advantage of this type of probe, unlike the rhodamine-based ones, is that it does not need to be oxidized in order to compartmentalize into the mitochondria, hence allowing more accurate H_2O_2 monitoring in the mitochondria such as in RAW 264.7 cells coupled with Mitotracker Red FM staining. Moreover, since carbamate is the main target of H_2O_2 reaction, the probe is not specific to $ONOO^-$ and ^-OCl, which is common in boronate-based probes.[83]

A metal-based fluorescent probe (MBFh1) was also developed to exploit the reactivity of Fe^{3+} with H_2O_2 and form secondary oxidants that could release the nonfluorescent moiety—in this case, 3,7-dihydroxyphenoxazine—via the C−N bond-breaking

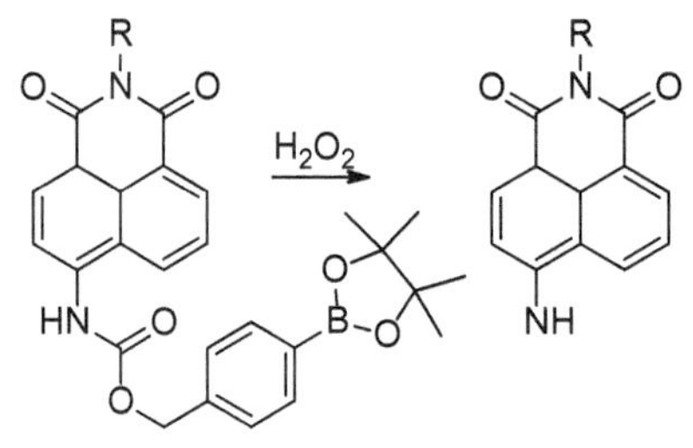

PL-1

$\lambda_{ex}\backslash\lambda_{em}$ = 375\475 nm $\lambda_{ex}\backslash\lambda_{em}$ = 435\540 nm

R = -(CH2)2O(CH2)O(C=O)CH3

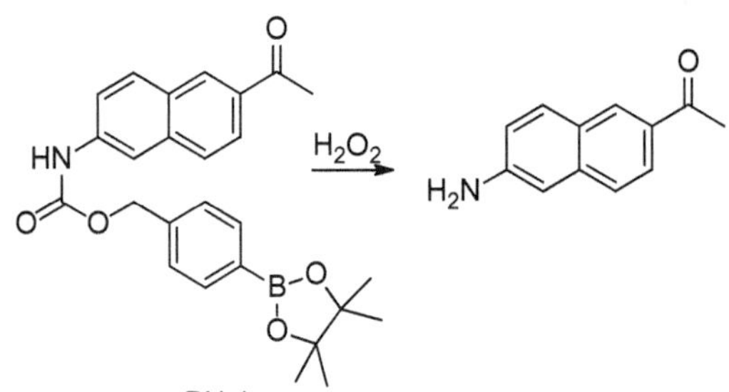

PN-1

$\lambda_{ex}\backslash\lambda_{em}$ = 321\328 nm $\lambda_{ex}\backslash\lambda_{em}$ = 338\358 nm

Figure 4.23 Examples of probes with ratiometric capabilities employed using two-photon confocal microscopy.

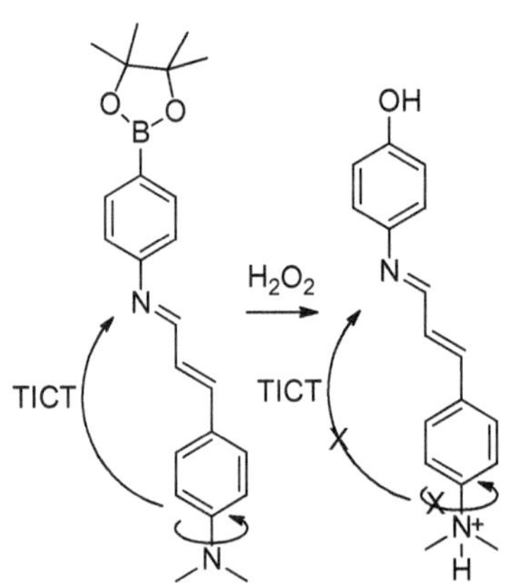

DACB   DAC-H

$\lambda_{ex}\backslash\lambda_{em}$ = 400\566 nm $\lambda_{ex}\backslash\lambda_{em}$ = 400\484 nm

Figure 4.24 Twisted-ICT approach for the detection of H_2O_2 using DACB.

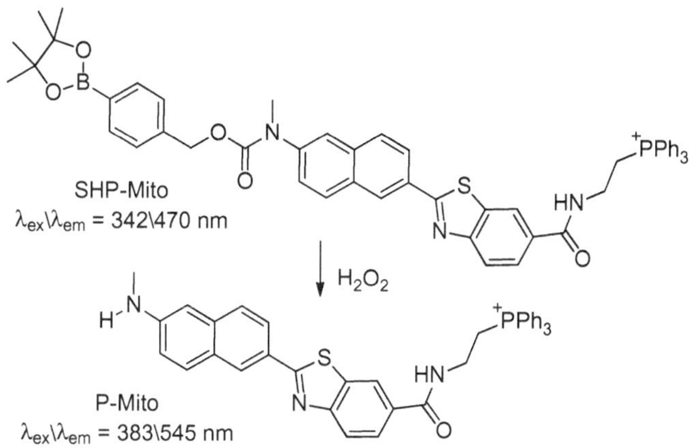

Figure 4.25 Formation of P-Mito from SHP-Mito from the carbamate cleavage by H_2O_2 showing a Stoke shift in P-Mito due to the stabilization of its ICT excited state.

process to form the fluorescent resorufin, according to Fig. 4.26. The C—N bond-breaking process was proposed as being mediated by highly oxidizing iron-oxo species (Fe^{IV} or $Fe^{V}=O$) generated from the initial formation of Fe^{3+}-OOH to form the cationic amide group, which subsequently hydrolyzes to form resorufin and the Fe^{III}-amino-carboxylate complex. MBFh1 has a limit of detection of 3.2 μM with a reaction rate constant of 38 M^{-1} s^{-1} and is only slightly specific since it displayed significant fluorescence formation with *t*-butylperoxide but moderate increase with $O_2^{\bullet -}$.[84]

As shown in Fig. 4.27, the nonredox-based approach for detecting H_2O_2 was also demonstrated through hydrolysis of the sulfonate linker group connecting fluorescein and the pentafluorobenzyl moieties. Hydrolysis of the sulfonate by H_2O_2 unmasks the fluorescence of the fluorescein via removal of the photo-induced electron transfer process. However, the degree of fluorescence response with H_2O_2 was comparable to that of $O_2^{\bullet -}$ and to a lesser but still significant degree to $ONOO^-$, *t*-BuOOH, $HO^{\bullet}$, and NO.[85]

Fig. 4.28 shows a nonboronate-based turn-on probe (D-HMSe) with high selectivity to H_2O_2. Unlike the boronate ones, this selenium-based probe does not react with $ONOO^-$, but its sensitivity for H_2O_2 was low with no notable changes in fluorescence below 50 μM.[86]

Quenching of fluorescence can also be exploited in order to assess H_2O_2 and thiol formation such as in the case of DA-Cy with catechol, which is conjugated to a cyanine scaffold (see Fig. 4.29). Here an on—off—on modality for H_2O_2/RSH detection was applied where the DA-Cy fluorescence is quenched via the donor-excited photo-induced electron transfer process on reaction of DA-Cy with H_2O_2—i.e., from the cyanine to the ortho-quinone moiety. Intracellular thiols (mostly GSH) undergoes the Michael addition reaction to form the Michael adduct, which can regenerate the fluorescence signal and thereby allow the detection of intracellular reduced thiol levels in HL-7702 and HepG2 cell lines as well as in rat hippocampus tissues.[87]

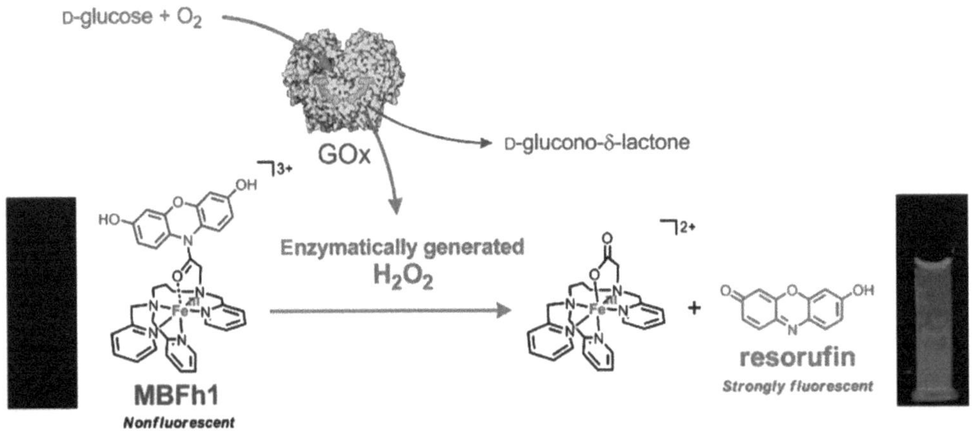

Figure 4.26 Detection of H_2O_2 using the metal complex-based fluorescent probe, MBFh1, showing the C—N bond cleavage and release of the fluorescent resorufin. Source: *Reprinted with permission from Hitomi Y, Takeyasu T, Funabiki T, Kodera M. Anal Chem 2011;**83**:9213.*[84] *Copyright © 2011 American Chemical Society.*

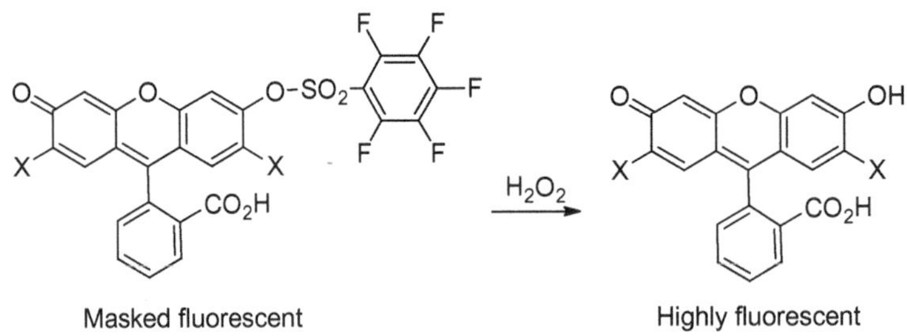

Figure 4.27 Nonoxidative mechanism for the detection of H_2O_2 via hydrolysis of sulfonate linker group leading to the removal of PET-induced masking of the probe's fluorescence.

Also, by exploiting the unique reactivity of benzil with H_2O_2, one can restore fluorescence from benzil-based probes (NBzF) exhibiting quenched fluorescence ($\Phi = 0.004$) through the donor-excited photo induced electron transfer (d-PeT) process as shown in Fig. 4.30. In the presence of H_2O_2, 5-carboxyfluorscein ($\Phi = 0.8$) is formed within 15 min from NBzF as visualized in live RAW 264.7 macrophages and A431 carcinoma cells.[88]

The development of genetically encoded probes could pave the way to a more specific, faster response time, tissue specificity, and real-time reversible oxidation—reduction dynamics for H_2O_2 detection in living cells.[89] Among these H_2O_2-sensitive probes were the HyPer, Hyper-2, Hyper-3, and roGFP2-Orp1. The Hyper redox probes exploit the dependence of fluorescence of the circularly permuted yellow fluorescent protein (cpYFP) on conformational changes of Oxy-R. Hydrogen peroxide can cause conformational changes of the regulatory domain (OxyR-RD) of

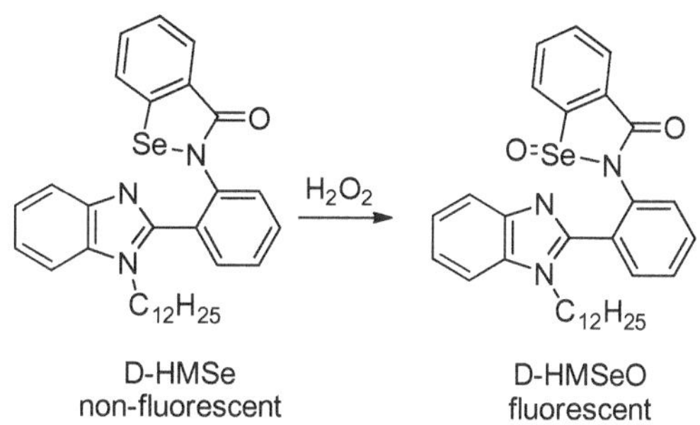

Figure 4.28 Selenium-based probe for the selective detection of H_2O_2.

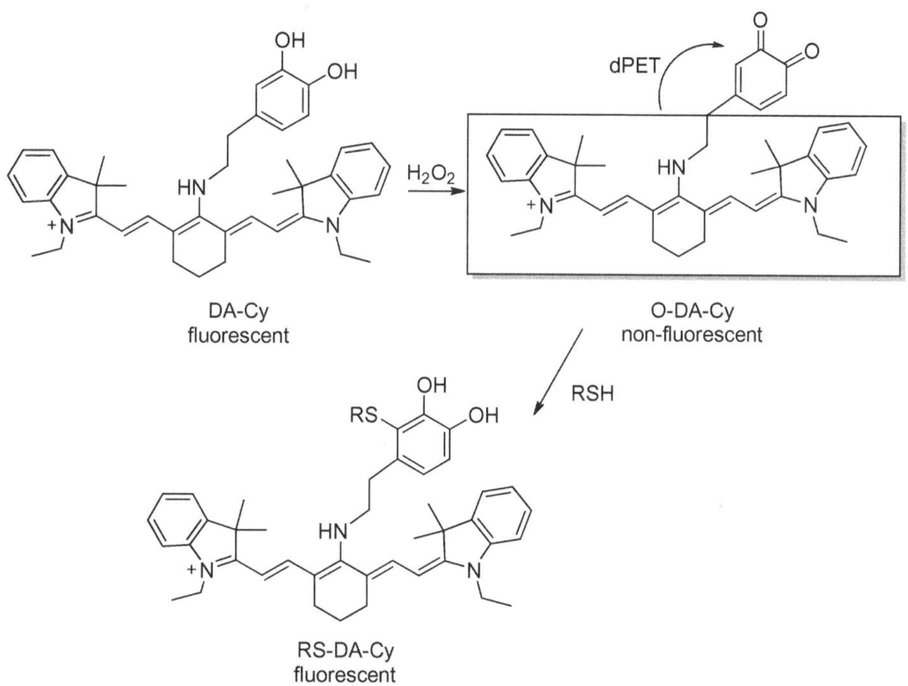

Figure 4.29 On–off–on fluorescent probe, DA-Cy, for the detection of H_2O_2 and thiols.

OxyR via oxidation of Cys199 to sulfenic acid, which is located in the hydrophobic packet of OxyR-RD. This sulfenic acid can then be released from the hydrophobic packet to form a disulfide bond with Cys 208 and cause conformational changes that allow DNA binding of OxyR. With cpYFP being integrated into OxyR-RD as cpYFP-OxyR-RD, one would be able to detect conformational changes via change in fluorescence as a function of H_2O_2 levels with emission at 516 nm and two excitation maxima at 420 and 500 nm corresponding to the protonated and deprotonated chromophores, respectively, and therefore allow for ratiometric excitation at two wavelengths where the shift in fluorescence intensities are measured as F500/F420,[90] thereby eliminating gross errors in the interpretation of fluorescence image caused by unsteady images due to cell movement and differences in protein expression. Hper-2 is a second-generation Hyper probe with an expanded dynamic range. By mutating A233V, the dimer formed from H_2O_2 oxidation is stabilized, resulting in an increased dynamic ratio of F500/F420 by sixfold compared to threefold in Hyper.[91] Hyper-3 is the latest generation of Hyper probes with H114Y mutations that produce a variant with a high signal-to-noise ratio and with improved response time and speed than Hyper as well as with faster redox dynamics compared to HyPer-2.[92] Moreover, the oxidation lifetime (disulfide formation) was faster for Hyper-3 than Hyper-2, with the former having the shortest reduction half-time (disulfide reduction) as shown in Fig. 4.31. Also shown is the comparison of response of Hyper-3 to Hyper with H_2O_2 production in the wound of transgenic zebrafish. The pseudo first-order rate constants for the hyper response to H_2O_2 was found to be $k_s = 5 \times 10^5\ M^{-1}\ s^{-1}$ for HyPer, $1.2 \times 10^5\ M^{-1}\ s^{-1}$ for HyPer-2, and $2.5 \times 10^5\ M^{-1}\ s^{-1}$for HyPer-3.[92] However, the reduction rate was shown to be dependent on the glutathione system (i.e., GSH/Grx1), which can cause significant accumulation of the oxidized form in conditions that can compromise the gluthathione system such as during the loss of GSH or limited nicotinamide adenine dinucleotide phosphate (NADPH) levels.[89]

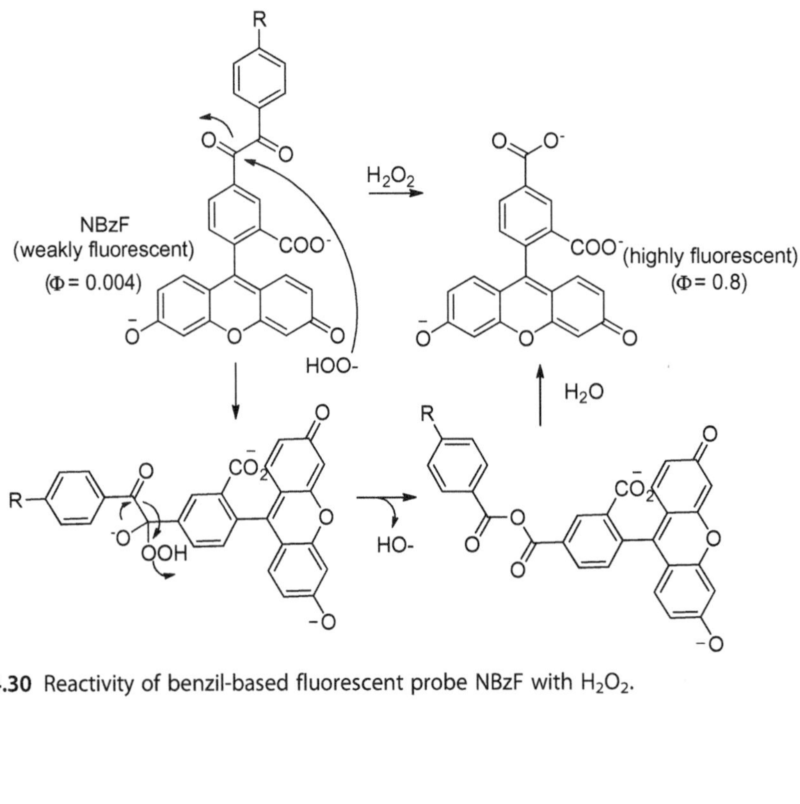

Figure 4.30 Reactivity of benzil-based fluorescent probe NBzF with H_2O_2.

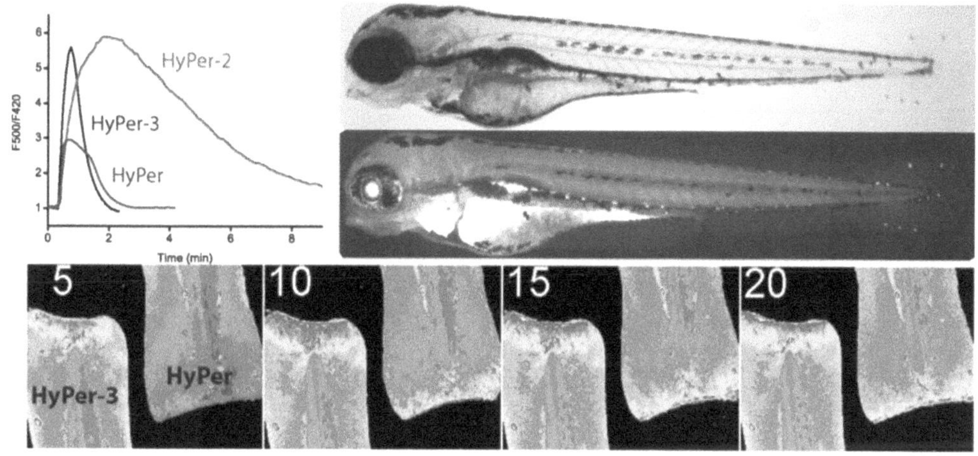

Figure 4.31 Plot of the timing of H_2O_2-induced ratiometric readings in individual cells expressing Hyper-3 (black line), Hyper (blue line), and Hyper-2 (red line). Images of HyPer-3 versus HyPer response to wound H_2O_2 production. (a) Brightfield (upper) and fluorescent (lower) images of β-actin:HyPer-3 transgenic zebrafish at 3.5 dpf stage. Source: *Reprinted with permission from Bilan DS, Pase L, Joosen L, Gorokhovatsky AY, Ermakova YG, Gadella TWJ, et al. ACS Chem Biol 2013;**8**:535.*[92] *Copyright © 2013 American Chemical Society.*

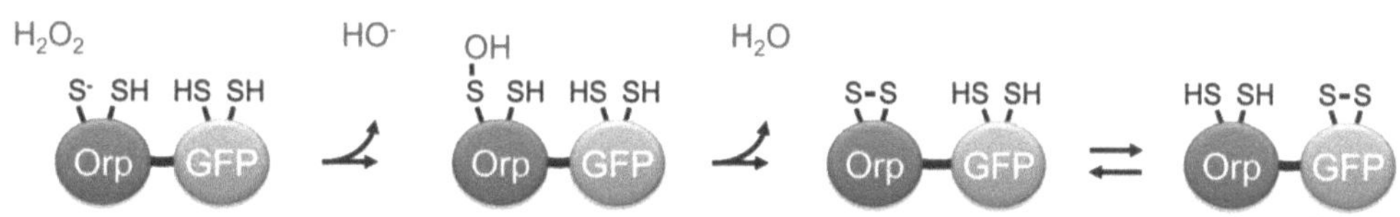

Figure 4.32 Mechanism of the roGFP2–Orp1sensing of H_2O_2. The Orp1 domain is oxidized by H_2O_2 to sulfenic acid residue and subsequently forms a disulfide bridge within the same domain. The roGFP2 disulfide bridge is formed via thiol–disulfide exchange between the Orp1 and roGFP2 domains. Source: *Adapted from Meyer et al.,* Antioxid Redox Signal *2010;**13**:621–50.*[89]

The high selectivity of peroxiredoxins and gluthathione peroxidases to H_2O_2 can also be exploited to detect H_2O_2 using roGFP fluorescence reporter. The oxidation of the transcription factor Yap1 by peroxidase Orp1 (also known as GPx3) via the Orp1-Yap1 redox relay serves as a basis for its application as an efficient redox relay for the oxidation of roGFP2 by H_2O_2 (Fig. 4.32). During H_2O_2-mediated oxidation of Orp1, sulfenic acid (Cys-SOH) is formed at the peroxidase cysteine (Cys36) site to form a mixed intermolecular disulfide bonds with redox-sensitive fluorescent protein (roGFP) when both are placed in close proximity, thus allowing fluorometric real-time measurement of redox state in both in vitro and in vivo systems and where the oxidized forms can be reversed by gluthathione or thioredoxin.[93]

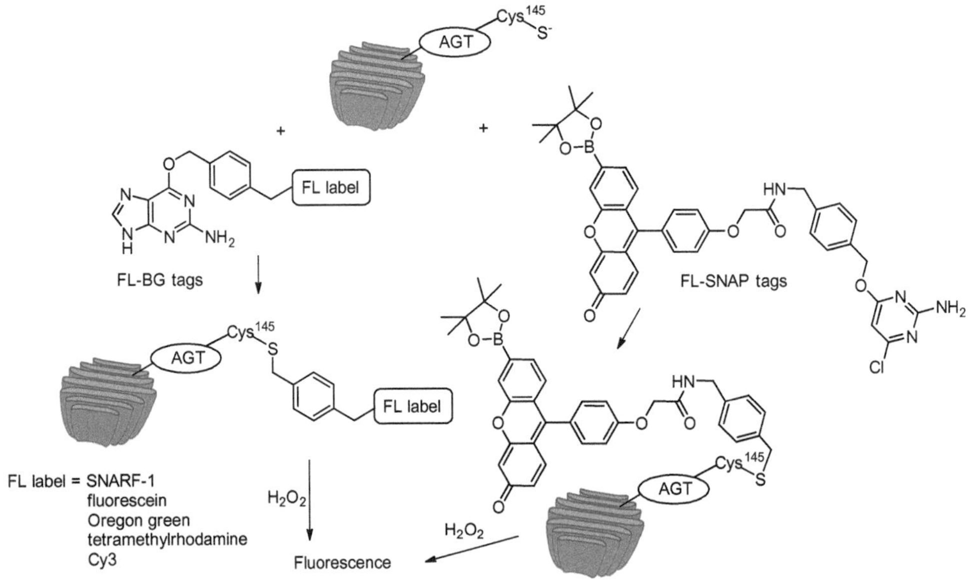

Figure 4.33 Subcellular target-specific fusion proteins AGT construct conjugated at the Cys145 site to FL-BG tags or FL–SNAP-tags bearing H_2O_2-sensitive fluorophores.

Fusion protein constructs of O6-alkylguanine-DNA alkyltransferase (AGT)[94] with different subcellular localizations to plasma membrane, nucleus, mitochondria, and endoplasmic reticulum can be labeled at Cys145 using O6-benzylguanine (BG) derivatives of various fluorescent probes[94] or with boronate-based fluorescent probes using SNAP-tag technology[95] (Fig. 4.33). Therefore, the resulting protein–fluorescent probe conjugates can give a fluorescence turn-on response with H_2O_2 treatment at a more specific cellular compartment.

4.4.2.3 Hypochlorous acid (HOCl)

Detection of hypochlorous acid in activated phagocytes is critical since they mediate in a variety of cellular processes such as signaling[96,97] and immune response[98,99] as well as oxidative damage of key biomolecular systems.[100] Several approaches had been developed to specifically detect HOCl in cellular system. Conventional approaches involve direct oxidation of the reduced probe to the fluorescent form such as in the case of the rhodamine-based R19-S bearing a thioester moiety (Fig. 4.34).[101] The probe showed high specificity toward HOCl as compared to other ROS, which only exhibited little induction of fluorescence. However, the selenoester analog gave poor selectivity to HOCl. This high specificity of R19-S to HOCl is due to the initial nucleophilic addition of ^{-}OCl to the thioester-carbon and the susceptibility of thioester-sulfur to oxidation by HOCl to form a sufenyl chloride (RSCl) that can

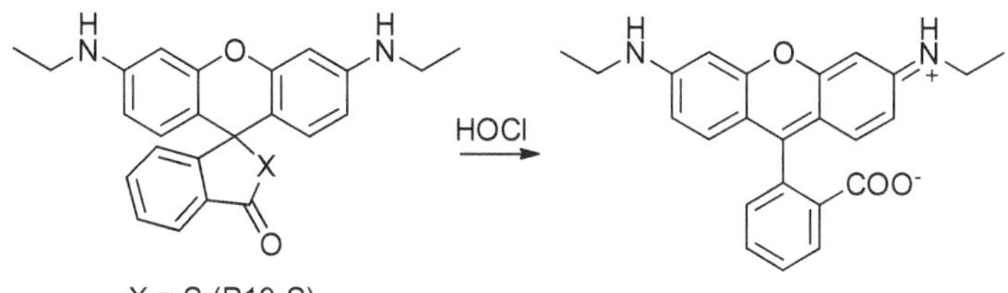

Figure 4.34 Oxidation of thioester by HOCl to yield a fluorescent signal using an off–on modality.

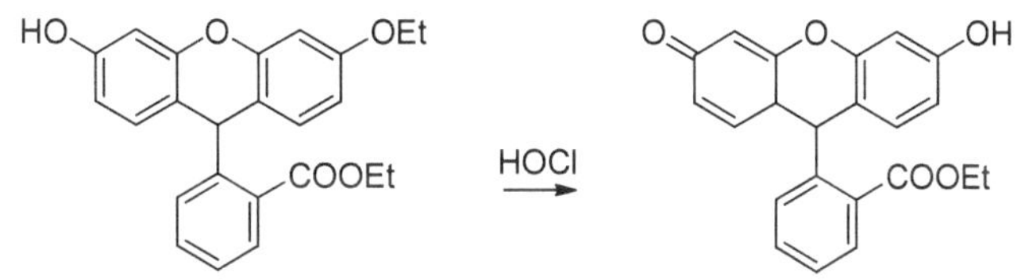

Figure 4.35 Oxidation of ether group by HOCl to yield a fluorescent signal using an off–on modality.

readily decompose thermally or photochemically or through metal-catalyzed oxidation.[102] Another similar approach was employed by exploiting the unique chemistry between hydrazide and HOCl (RH6G-hydrazide) where chlorination of the amino group is followed by elimination of HCl to form the acyl azo product, which can undergo hydrolysis to yield the rhodamine 19 (Fig. 4.34).[103]

Oxidation of ether group using the water-soluble probe FCN-2 was also exploited and showed high specificity to HOCl and negligible fluorescence enhancement with other reactive species (Fig. 4.35). Using confocal microscopy, FCN-2 was shown to detect HOCl formation in situ in NIH3T3 cells preincubated with NaCl, MPO, and H_2O_2 as well as in zebrafish larvae and adult ones incubated with HOCl.[104]

Another class of fluorescence probe is boron-dipyrromethene (BODIPY), whose *meso*-position can be modified to exploit the unique chemistry between the reaction of ClO^- and oxime (Fig. 4.36).[105] This BODIPY–oxime derivative, MitoClO, was successfully employed to detect HOCl in the mitochondria using exogenously introduced NaClO in MCF-7 cells or through lipopolysaccharide (LPS) or PMA stimulation of the cells, although inhibition of these fluorescence signals by SOD or catalase were not confirmed. Another BODIPY-based fluorometric probe for HOCl is HCS, which is based on the HOCl-mediated oxidation of methyl phenyl sulfide to sulfinyl ($R_2S=O$), allowing a 160-fold increase in quantum yield due to the suppression of the PET process from the sulfide group to the BODIPY by forming the sulfinyl on oxidation.[106] Formation of sulfinyl with lower HOMO than the BODIPY suppresses

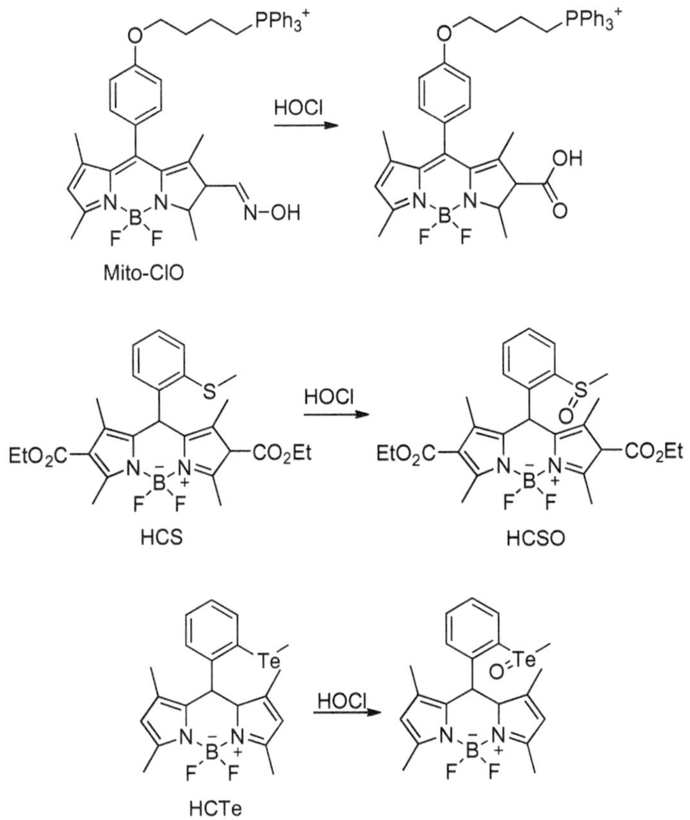

Figure 4.36 BODIPY-based HOCl probes that employ oxidation of oxime to carboxylic acid for Mito-ClO, oxidation of sulfide to sulfinyl for HCS, and oxidation of tellurium in HCTe.

the PET process, hence increasing the fluorescence intensity (Fig. 4.36). Another PET suppression approach was demonstrated using BODIPY as the fluorophore bearing a tellurium atom instead of the similar chalcogen, S, or Se as HCTe showing high specificity even against $ONOO^-$ (Fig. 4.36).[107]

Cyanine-based dyes were also employed for the HOCl detection. By incorporating selenomorpholine or thiamorpholine into the heptamethine cyanine dye structure, SeCy7 and SCy7 were reacted with HOCl and showed better fluorescence properties for in vivo application for the SeCy7 analog than the SCy7 (Fig. 4.37).[108] Except for the probe reactivity to $ONOO^-$ which was not reported, SeCy7 was selective toward HOCl. Using near-infrared (NIR) in vivo imaging, mouse that was injected with LPS and PMA showed increase fluorescence intensity through PET process suppression by the formed selenoxide moeity after 60 min of incubation.

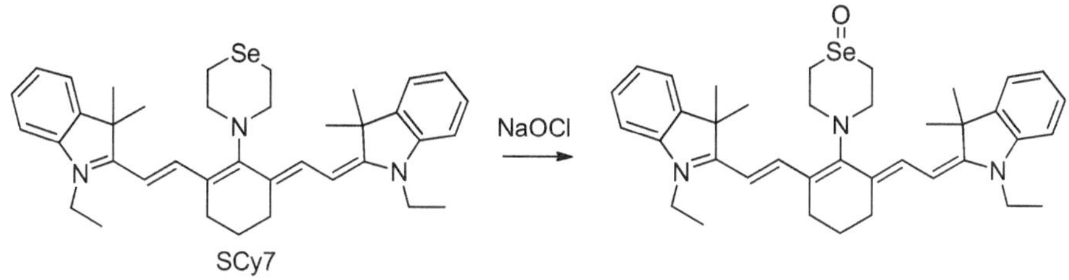

Figure 4.37 Cyanine-based HOCl probe showing suppression of PET process through oxidation of the selenide to selenoxide.

4.4.2.4 Nitric oxide (NO)

Nitric oxide (NO) generation in biological system is perhaps one of the most studied processes because its generation has been implicated in a variety of biological functions. Nitric oxide is a critical signaling molecule in regulating vascular smooth muscle relaxation,[109] inhibiting platelet aggregation,[110,111] and ensuring leukocyte adhesion to the endoltheliium.[112] NO is not only produced in the endothelium but also generated by phagocytes for immune and inflammatory responses.[113–115] In the neurons, NO is a critical cellular signaling molecule, but its overproduction could also lead to the pathogenesis of neurodegenerative diseases.[116–118] Nitric oxide is a relatively inactive molecule compared to other species due to the electron delocalization between the nitrogen and oxygen atoms, which makes them less susceptible to direct reaction with other organic functionalities. However, NO in biological system has been shown to form complex with the iron-heme center of hemoglobin or soluble guanylyl cyclase[119] and the [2Fe-2S] clusters[120,121] and directly react with O_2 or $O_2^{\bullet -}$ to form $NO_2^{\bullet}$ or $ONOO^-$, respectively.[122] Due to the selective nature of NO reactivity, the chemistry of its more reactive products such as $NO_2^{\bullet}$ and $ONOO^-$ were exploited to indirectly detect NO formation in aerobic or highly inflammatory conditions, respectively. In conditions in which $O_2^{\bullet -}$ levels are low or in aerobic conditions, NO reaction with O_2 is inevitable, forming $NO_2^{\bullet}$ and subsequently N_2O_3 as shown in Fig. 4.38. Although NO could react directly to amine to form diazeniumdiolate, this reaction is very slow.[123] Faster reactivity of NO to amine is typically catalyzed by O_2 to form an *N*-nitrosated products.[124] The *o*-phenylenediamine reaction to NO in the presence of O_2 yields an *N*-nitroso intermediate that can subsequently cyclically form the triazole via intramolecular nucleophilic displacement on the *N*-nitroso (or diazo hydroxide) as shown in Fig. 4.38.[125] With the formation of triazole, fluorescence can be "turned on" due to suppression of the PET process originating from the electron transfer between the amino group and fluorophore.

Diaminofluoresceins (DAFs) were perhaps one of the earliest probes employed for the detection of NO via *N*-nitrosation and subsequent formation of triazole to give

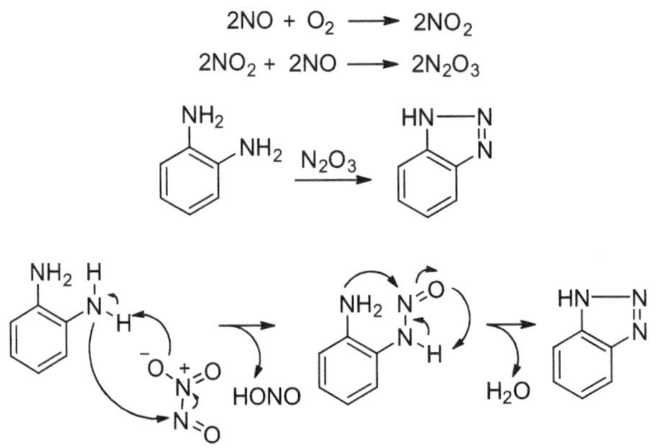

Figure 4.38 Indirect detection of NO via formation of $NO_2^{\bullet}/N_2O_3$ as the reactive intermediates and N_2O_3 reaction with *o*-phenylenediamine to form the triazole.

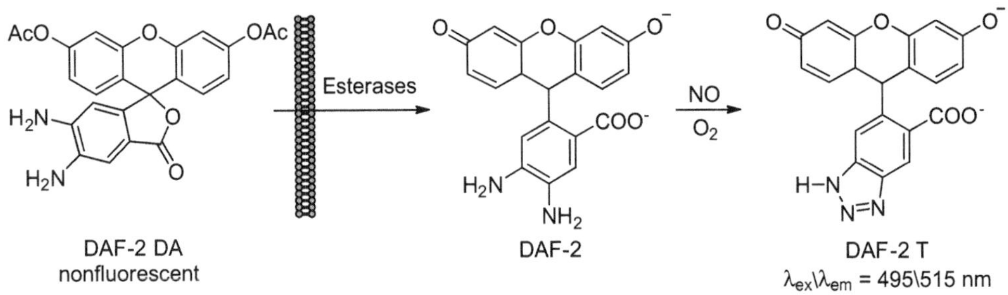

Figure 4.39 Internalization of DAF-2 DA into the cytosol and its subsequent oxidation by NO/O_2 to form the fluorescent triazole product.

green fluorescence with a very low detection limit of 5 nM. Conversion of diaminofluoroscein (DAF) to the triazole form imparts an increase in fluorescence efficiency by 100-fold, with high specificity against NO_2^-, NO_3^-, $O_2^{\bullet -}$, H_2O_2, and $ONOO^-$. The esterified membrane-permeable analog, DAF-2 DA (Fig. 4.39), was employed for NO detection in activated rat aortic smooth muscle cells.[126]

Following DAF-2 successes in NO detection, several analogs were synthesized that employ the same mode of chemistry using *o*-phenylenediamine as the trigger receptor. Since DAF analogs' main limitation is water solubility, several highly water-soluble NO probes were designed and synthesized. One example is the water-soluble, BODIPY-based fluorescent probe possessing two sulfonate groups as disodium 2,6-disulfonate-1,3,5,7-tetramethyl-8-(3′,4′-diaminophenyl) difluoroboradiaza-s-indance (TMDSDAB) (Fig. 4.40). As a turn-on fluorescent probe, TMDSDAB can react with NO efficiently in aqueous media with 540-fold enhancement in fluorescence. Using capillary electrophoresis strategy with laser-induced fluorescence detection,

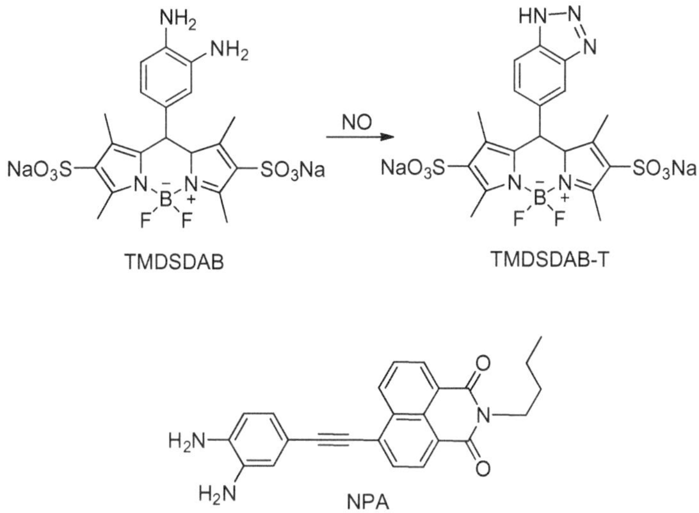

Figure 4.40 BODIPY- and napthalimide-based NO probes as TMDSDAB and NPA, respectively.

monitoring of NO released out of the cell and those that remained can be accomplished with limit of detection of 0.5 nM.[127] A napthalimide-based fluorescent probe (NPA) showed a fast rate of detection among *o*-phenylenediamine–based probes through inhibition of the internal charge transfer process (Fig. 4.40).[128]

Other than using the *o*-phenylenediamine receptor–based probes, fluorescence can be activated using metal complex such as in the case of Cu_2(FL2E), which was employed for the detection of NO in cell cultures and olfactory bulb brain slices. The detection of NO using Cu_2(FL2E) is considered a direct mode of NO detection because it is the NO adduct that is detected and not the indirect NO products.[129,130] The Cu(II)–fluorescein complex works by displacing Cu(II) through *N*-nitrosation of the fluorophore's amino moiety to form the displaced Cu(I) and FL1-NO (Fig. 4.41). Other metal-based NO probes were designed, including Co(II), Fe(II), Ru(II), and Rh(II), that work through a variety of mechanisms.[131]

Another mode of fluorescence activation is through the formation of a diazo product from NO/O_2 reaction with 2-amino-3′-dimethylaminobiphenyl-BODIPY (DMAB-BODIPY) as shown in Fig. 4.42. This approach is similar to those mentioned previously in which the PET is suppressed on conversion of the amine to azo where the absence of electron-donating amino groups allows BODIPY to fluoresce.[132] Oxidation of dihydropyrine (DHP) to pyridine could also lead to suppression of the PET process and lead to fluorescence formation (Fig. 4.42).[133]

The NO-mediated ring opening of spriolactam can also provide opportunity to "switch on" chromogenic and fluorogenic NO probes.[134] Under aerobic conditions, NO can release rhodamine from the spirolactam–rhodamine conjugate (SL-RhB)

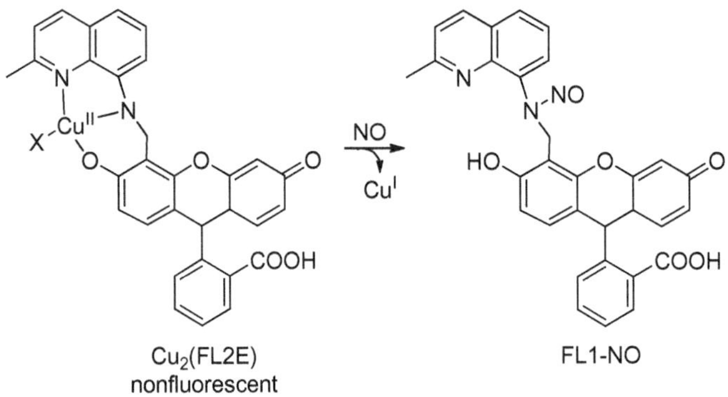

Figure 4.41 Metal complex–based NO probe showing the displacement of Cu(II) via *N*-nitrosation.

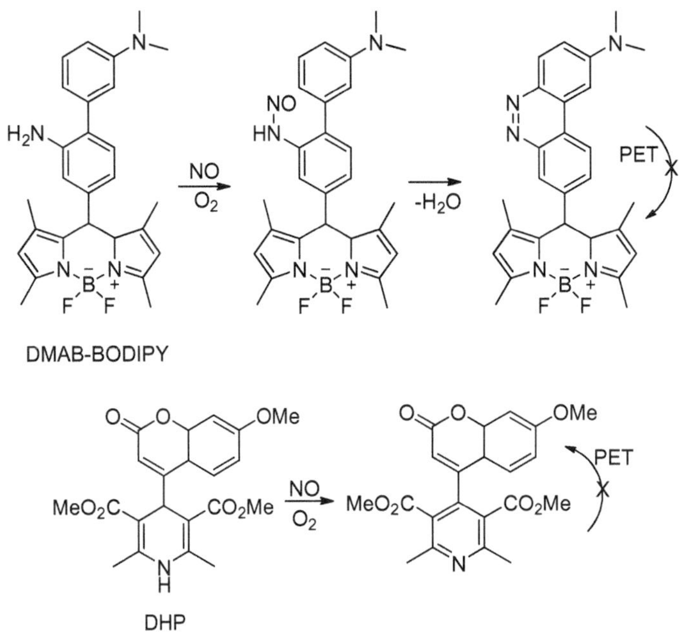

Figure 4.42 Fluorescence formation through suppression of the PET process from the reaction of NO/O_2 to DMAB-BODIPY and DHP to form the diazo and pyridine products, respectively.

according to Fig. 4.43. This strategy was further employed in the design of mitochondrial-targeted NO probe as *o*-phenylenediamine-locked rhodamine spirolactam (Mito-Rh-NO), which undergoes NO-mediated spirolactam ring opening to yield the fluorescent rhodamine Mito-Rh with high selectivity to NO (Fig. 4.43). Confocal microscopy imaging of NO in the MCF-7 cell mitochondria was demonstrated with Mito-Rh-NO using exogenously added NO or endogenously produced NO through addition of NO-stimulating agents.[135] Also, a lysosome-targeted

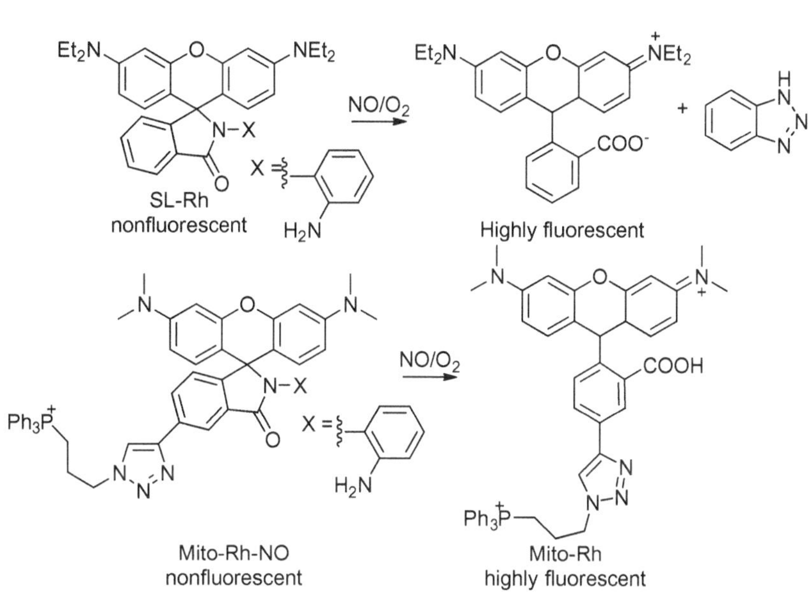

Figure 4.43 Fluorescence formation via NO-mediated ring opening of spriolactam to form the triazole from the nonfluorescent *o*-phenylenediamine-locked rhodamine.

ratiometric NO probe was designed by employing the *o*-phenylenediamine receptor approach.[136]

Deamination of primary aromatic monoamines was also exploited for fluorescence formation as in the case of FA-OMe by suppressing the PET process (see Fig. 4.44).[137] NIR-infrared application of NOC-13 allows for deep tissue monitoring of NO generation through the formation of triazole and the suppression of PET, according to Fig. 4.44. NOC-13 showed better reaction efficiency than DAF-2 as confirmed using isolated intact rat kidney.[138]

A fluorescence-quenching approach (on—off mechanism) was employed to assess NO formation in which the stable nitroxide, 2,2,6,6-tetramethylpiperidine-N-oxyl radical (TEMPO), was conjugated to the fluorescent acridine. Being an electron donor, TEMPO quenches the acridine fluorescence signal and, on complexation with Fe $(DTCS)_2$, allows for the suppression of PET from the nitroxide to the fluorophore. Using the NO-displacement approach, NO binds to $Fe(DTCS)_2$ and displaces the TEMPO—acridine conjugate hence activating PET within the molecule and suppressing the fluorescence signal. Although they offer a novel approach for monitoring NO production, probes that employ the fluorescence-quenching (on—off) approach are not as attractive as fluorescence formation (off—on).[139]

Diaminofluorescein was shown to react directly with ascorbic acid (AA) and may confound interpretation of NO production. Therefore, care should be taken when working with AA as a reagent.[140] The use of DAF-FM as a NO probe was found to

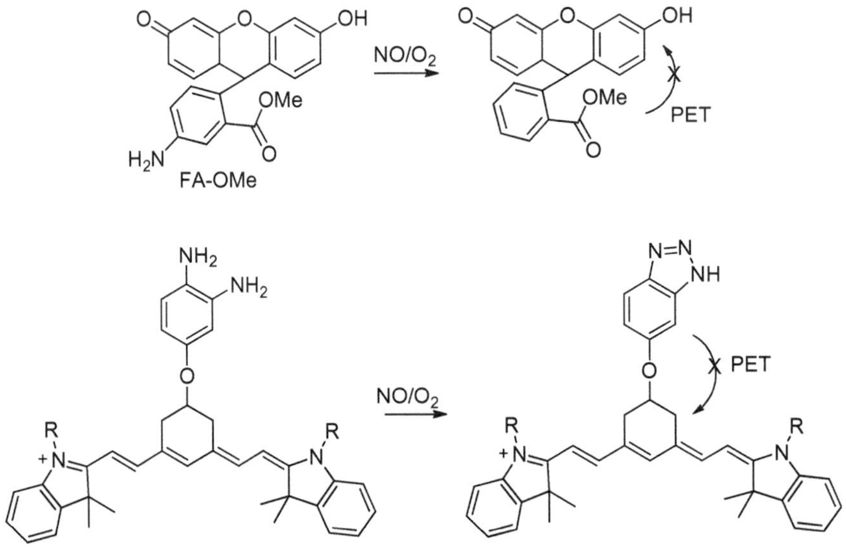

DAC-P (R = $CH_2CH_2CH_3$
DAC-S (R = $CH_2(CH_2)_3SO_3Na$, $CH_2(CH_2)_3SO_3$-

Figure 4.44 Deamination reaction by NO/O_2 of FA-OMe and triazole formation for DAC-P/S yielding a PET process—suppressed fluorescence.

be specific, but just like any other fluorescein-based probes they are inherently susceptible to photo-oxidation and oxidation by $O_2^{\bullet -}$, $ROO^{\bullet}$, and $ONOO^-$.[141] Similar photosensitivity was observed for DAF-2 DA[142]; it was also established that DAF-2's ability to detect NO is independent of the Ca^{2+} and Mg^{2+} ion concentrations in solution.[143] DAF-2 was also found to directly react with exogenously introduced reagents such as (+)-catechin and trans-resveratrol when studying induction of NO production in endothelial cells. It was shown that (+)-catechin can directly increase the triazolefluorescein, DAF-2T, and therefore proper control experiments are important when testing the stability of DAF-2 against exogenous reagents.[144]

4.4.2.5 Nitroxyl (HNO)

Nitroxyl (HNO or NO^-) detection in biological system is important due to the ability of NO^- to interact with cysteine thiols and thus affect physiological processes such as vasorelaxation[145] and has other pharmacological properties such as an antialcoholic drug, protection from reperfusion injury, and regulation of a *N*-methyl-D-aspartic acid receptor.[146] HNO dimerizes in solution with a rate constant of $8 \times 10^6\ M^{-1}\ s^{-1}$ to yield hyponitrous acid ($H_2N_2O_2$), which then undergoes dehydration to form N_2O. Therefore, in a biological system, the direct detection of HNO could be elusive. However, several chemistries have been exploited to detect HNO in both in vitro and

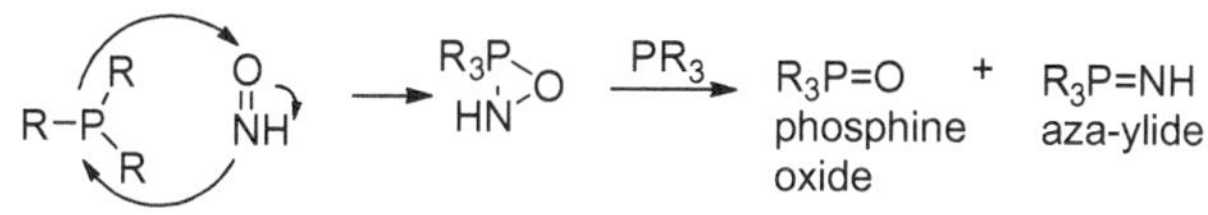

Figure 4.45 Reactivity of nitroxyl with phosphine to form phosphine oxide and aza-ylide.

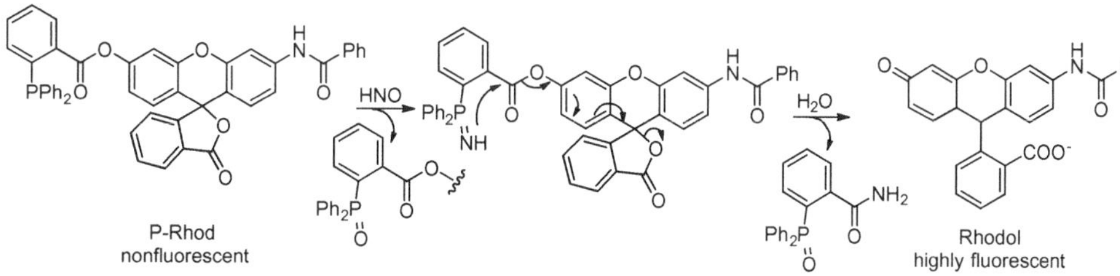

Figure 4.46 Mechanism of reaction of HNO with P-Rhod to yield the highly fluorescent Rhodol.

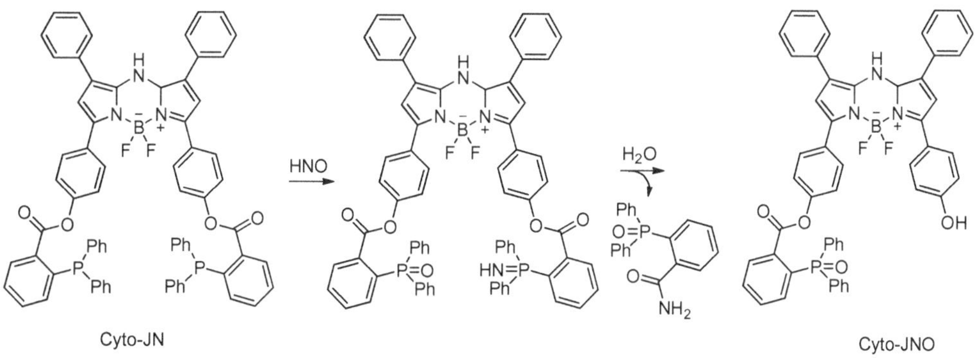

Figure 4.47 Reaction of HNO to Cyto-JN yielding the fluorescent Cyto-JNO.

in vivo conditions using Angeli's salt as the HNO source. The most widely exploited chemistry of HNO for its detection is its high reactivity toward phosphines. Fig. 4.45 shows the mechanism of phosphine oxide and aza-ylide formation from the reaction of two moles of phosphine with HNO via an intramolecular nucleophilic attack.

Through conjugation of the phosphine moiety to rhodamine as P-Rhod, HNO reaction with P-Rhod gives the phosphine oxide and aza-ylide products (Fig. 4.46). Subsequent hydrolysis of the aza-ylide yields the highly fluorescent Rhodol.[147] Conjugation of diphenylphosphines to BODIPY as Cyt-JN yields similar phosphine oxide and aza-ylide products, where the latter can further undergo a nucleophilic attack of the carbonyl-C releasing the fluorescent product Cyto-JNO according to Fig. 4.47.[148] Cyt-JN shows high specificity to HNO compared to other N- and O-derived reactive species such as GSNO, $ONOO^-$, NO, and $NO_2^{\bullet}$. Several other probes that exploit phosphine oxidation by HNO were synthesized,[149–151] with two-photon fluorescence properties for tissue imaging,[152] as well as using near-infrared fluorescent probe targeting the lysozyme in living cells and for in vivo HNO imaging.[153]

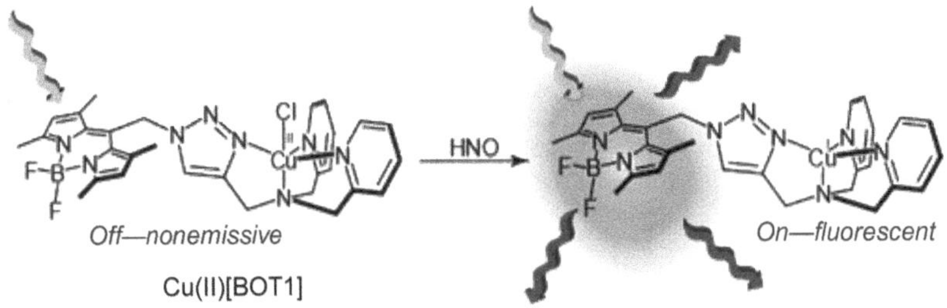

Figure 4.48 Metal complex–based HNO probe Cu(II)[BOT1] showing fluorescence formation on reduction of Cu(II) by HNO to Cu(I) complex. Source: *Reprinted with permission from Rosenthal J, Lippard SJ. J Am Chem Soc 2010;**132**:5536.*[154] *Copyright © 2010 American Chemical Society.*

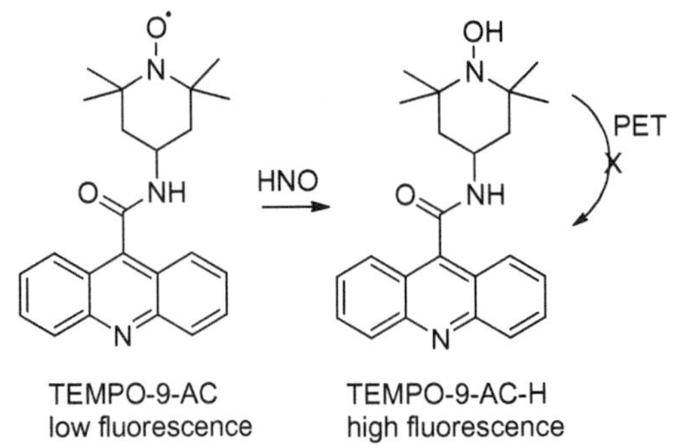

Figure 4.49 Fluorescence quenching by the nitroxyl group in TEMPO-9-AC by HNO to give the diamagnetic and fluorescent hydroxylamine TEMPO-9-AC-H.

Metal complex–based HNO probes were developed that exploit the redox activity of the Cu(II) center. In this strategy, BODIPY was used as a fluorophore; through its complexation with Cu(II) as Cu(II)[BOT1] (Fig. 4.48), the fluorescence was quenched. Reduction of Cu(II) to Cu(I) by HNO restores this fluorescence with high specificity against NO or other reactive nitrogen species (RNS) and ROS, and that Cu(II)[BOT1] was found to be membrane permeable with the capability of imaging HNO in live cells.[154]

By conjugating a stable nitroxide such as TEMPO to the fluorophore acridine to form TEMPO-9-Ac, the paramagnetic nitroxide suppresses the fluorescence via the PET process. Reduction of nitroxide to hydroxylamine by HNO via H-atom abstraction by the nitroxide from HNO gives the diamagnetic and highly fluorescent TEMPO-9-AC-H, according to Fig. 4.49.[155]

4.4.2.6 Peroxynitrite ($ONOO^-$)

The facile transformation of the lesser reactive NO and $O_2^{\bullet -}$ to much higher oxidizing forms $HO^\bullet$ and $NO_2^\bullet$ through the formation of $ONOO^-$ makes $ONOO^-$ one of the most cytotoxic reactive intermediates.[156] Peroxynitrite reacts with a variety of

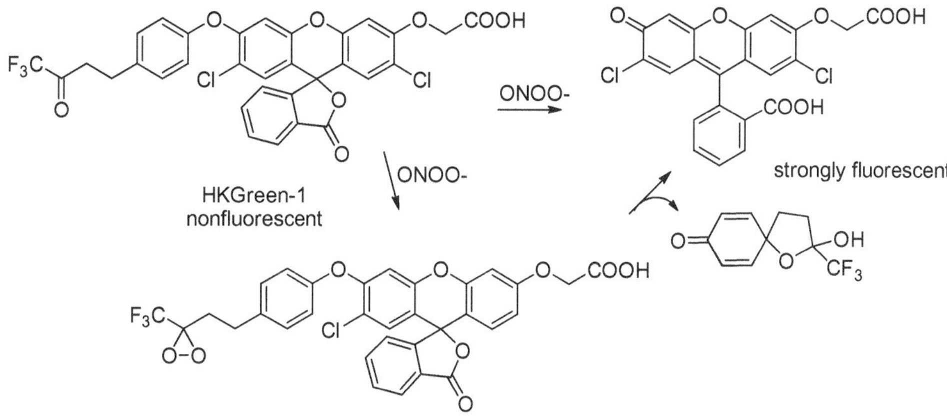

Figure 4.50 Fluorophore release mechanism as demonstrated by the reaction of HKGreen-1 to $ONOO^-$ showing oxidative cyclization liberating the fluorophore.

substrates such as sulfhydryl[157] and also causes lipid peroxidation[158] as well as tyrosine nitration.[159,160] Peroxynitrite was shown to be formed from macrophages,[161,162] endothelial cells[163] or neuronal cells[164] under pathophysiological conditions and is an important molecule in the regulation of transcription factors and signaling pathways that lead to apoptosis.[165]

Due to the high reactivity of $ONOO^-$, several traditional probes such as DCFH-DA, DCF, 5(and 6)-DH, and Rhodamine B hydrazide, as well as APF and HPF, were also employed for $ONOO^-$ detection.[166] One strategy to specifically detect $ONOO^-$ formation is through a release mechanism whereby the fluorophore is released via a reaction that is specific to $ONOO^-$. For example, as shown Fig. 4.50, the reaction of $ONOO^-$ to HKGreen-1 that has a trifluoromethylketone moiety that is conjugated to the para position of a phenoxy group forms a dioxirane intermediate that subsequently oxidizes the adjacent phenyl group to form the dienone, thereby releasing the fluorophore and showing high selectivity to $ONOO^-$.[9] A similar release strategy was employed using a rhodol-based probe, HKGreen-3[167]. Using the same reaction strategy of $ONOO^-$ with ketone, the formation of fluorescence could also be achieved using a nonreleasing mechanism by suppressing the PET process as shown for BODIPY-based HK-Green-2, according to Fig. 4.51.[168]

The high reactivity of boronate toward $ONOO^-$ could also be exploited since the rate constant for the $ONOO^-$ reaction with 4-acetylphenylboronic acid was determined to be $k = 1.6 \times 10^6\ M^{-1}\ s^{-1}$, which is faster than with H_2O_2 ($k = 2.2\ M^{-1}\ s^{-1}$) and HOCl ($k = 6.2 \times 10^3\ M^{-1}\ s^{-1}$) to yield the phenolic product and borate.[77] This chemistry could be applied to release the fluorophore on oxidation of a boronate–fluorescent conjugate such as in the case of coumarin-7-boronic acid (CBA) (see Fig. 4.52)[169] and that of benzothiazolyl iminocoumarin scaffold conjugated to boronate ester.[170]

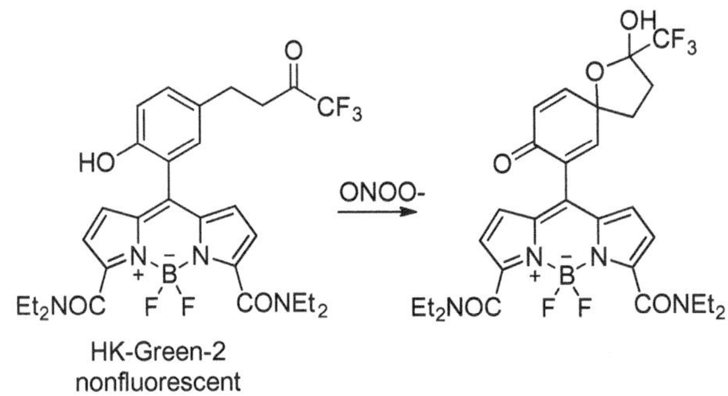

Figure 4.51 Fluorescence formation by HK-Green-2 through a nonreleasing mechanism by suppression of the PET process via oxidative cyclization on reaction with $ONOO^-$.

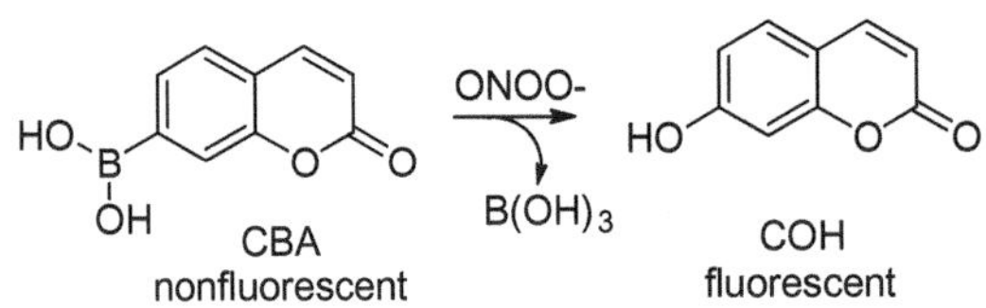

Figure 4.52 Peroxynitrite detection by CBA through boronate oxidation to yield borate and release of the fluorescent phenolic COH.

Similar to HOCl detection, chalcogen-based probes were also designed for NIR fluorescence detection of $ONOO^-$ in living cells in which the oxidation of the chalcogen atom suppresses PET-induced fluorescence quenching of cyanine moiety such as in the case of Cy-PSe, whose fluorescence due to $ONOO^-$ can be reversed by thiols but not ascobate.[171] However, a similarly designed probe, BzSe-Cy, the $ONOO^-$ oxidation can be reversed by reduced ascorbate[172] (Fig. 4.53). Tellurium-based probes 2Me TeR[173] (Fig. 4.53), and Cy-NTe[174] were also designed to exhibit thiol reversibility and near-infrared fluorescence properties. Furthermore, a release mechanism that employs oxidation of Se was demonstrated using BOD-Se in which the ICT process is modulated by the diaryl selenide; upon cleavage, fluorescence is formed (Fig. 4.53) as shown in the macrophage cell line RAW264.7.[175]

Since $ONOO^-$ probes can give false positive results in the presence of HOCl, a three-channel probe was designed that can distinguish between $ONOO^-$ from HOCl such as in the case of the coumarin-based fluorophore PN_{600}.[176] This detection relies on the oxidative coupling of phenol group with the amino group to form para-aminophenol by $ONOO^-$ or HOCl, which subsequently forms the final product iminoquinone on further oxidation by $ONOO^-$ but not by HOCl, hence achieving specificity for $ONOO^-$ (Fig. 4.54).

A mitochondrially targeted $ONOO^-$ probe was designed (MRhod123) using a release mechanism via an *N*-dearylation reaction by exploiting the $ONOO^-$

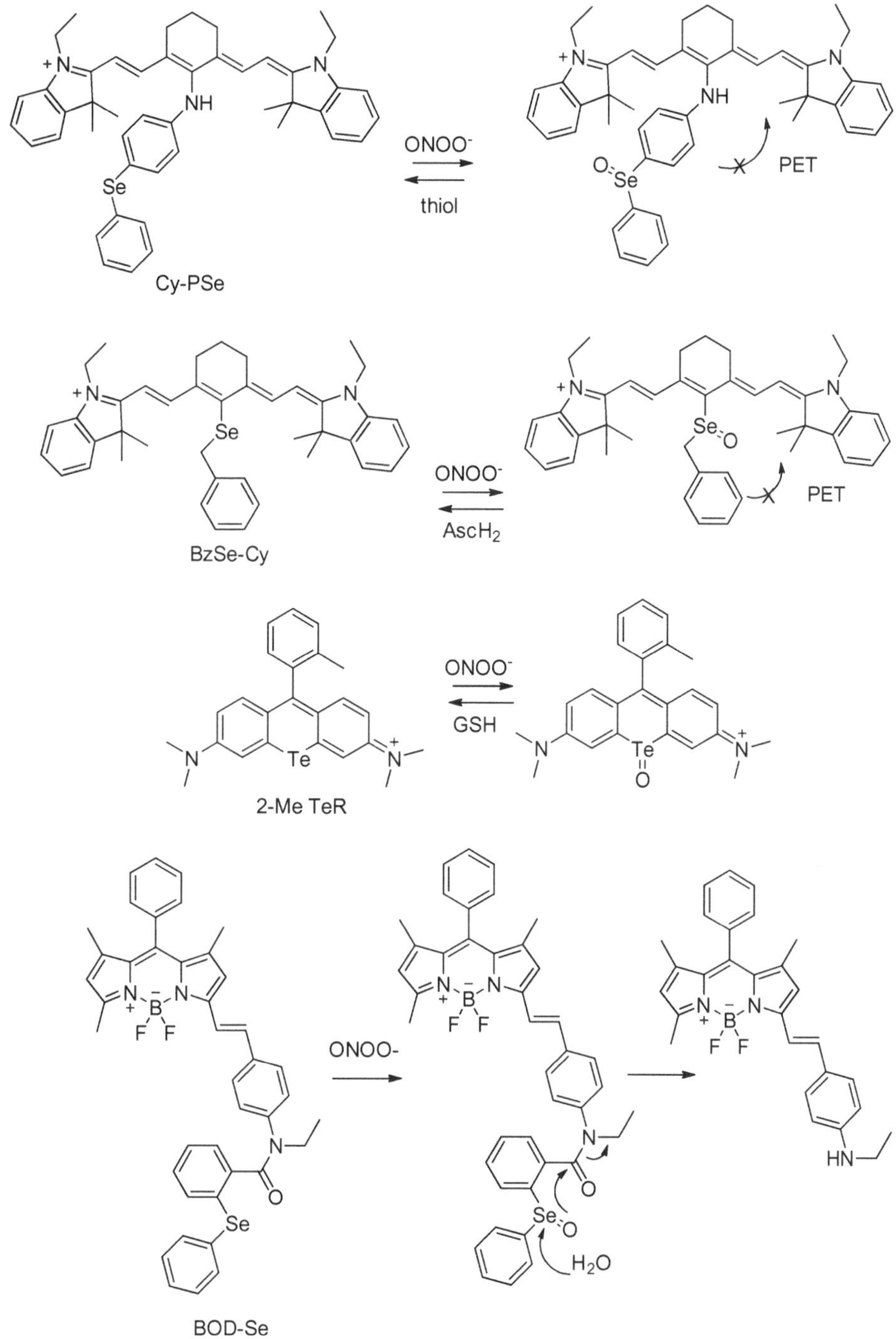

Figure 4.53 Chalcogen-based probes Cy-PSe, BzSe-Cy, 2-Me TeR, and BOD-Se for detecting $ONOO^-$ and showing various modes of fluorescence formation and reversibility in the presence of thiols or ascorbate.

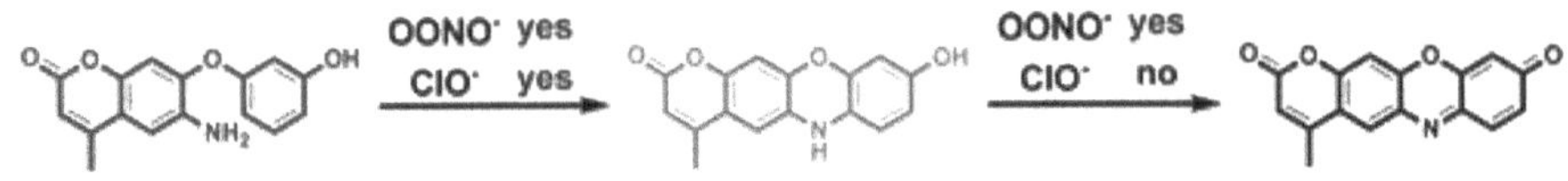

Figure 4.54 A three-channel fluorescent probe, PN_{600}, with the ability to distinguish between $ONOO^-$ and HOCl. Source: *Adapted from Zhang et al.,* J Am Chem Soc *2012;**134**:18479–82.*

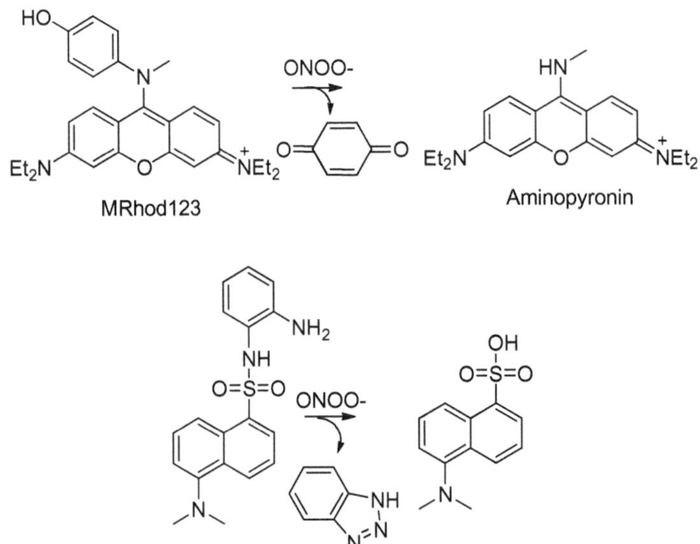

Figure 4.55 Fluorophore-release mechanism through $ONOO^-$ specific oxidation of phenol and *o*-phenylenediamine groups in MRhod123 and Ds-DAB, respectively.

reactivity to methyl(4-hydroxyphenyl)amino to form the quinone releasing the green emissive aminopyronin by eliminating the PET process, according to Fig. 4.55.[177] A similar release mechanism was employed using dansyl-fluorophore that was conjugated to *o*-phenylenediamaine via the aminosulfonyl linker group. Reaction of $ONOO^-$ to the secondary amine forms the nitrosamine-sulfonyl, which then cleaves to form the sulfonyl dansyl-acid and benzotriazol, according to Fig. 4.55.[178]

A novel genetically encoded unnatural amino acid, boronophenylalanine, into cpGFP was constructed as the mutant pnGFP derived from cpsGFP with high selectivity and sensitivity to $ONOO^-$ but not H_2O_2. Confocal imaging of $ONOO^-$ was achieved in live HEK 293T cells that were transiently transfected to express pnGFP.[179]

4.4.2.7 Singlet oxygen ($^1O_2$)

Singlet oxygen is endogenously generated through myeperoxidase catalysis in the presence of halides and H_2O_2 as an antimicrobial defense mechanism in polymorphonuclear leukocytes.[180,181] It is also formed from a variety of decomposition processes by H_2O_2, ROOH, $O_2^{\bullet -}$, and $ONOO^-$[182] and from in situ generation through exogenous stimulation by light using photosensitizing molecules in the presence of triplet O_2.[183,184] Although a variety of traditional fluorescence probe can detect $^1O_2$ such as Rhodamine 123 or APF, perhaps the most widely exploited chemistry for the detection of $^1O_2$ is the formation of endoperoxide from electron donors' moieties that suppress PET-induced fluorescence quenching such as in the case of DMAX[185] and the silicon-containing rhodamine moiety (Si-DMA) (Fig. 4.56). The latter allows for the visualization of $^1O_2$ generated in vitro through photo irradiation of live HeLA cells and RAW 264.7 macrophages in the presence of photosensitizers.[184] The metal chelate–based probe PATA-Tb^{3+} also allows for the detection of $^1O_2$ through formation of endoperoxide in the anthracene, which leads to the disappearance of the strong quenching effect between the triplet states of the anthracene and metal chelate.[186]

Ratiometric $^1O_2$ probe was designed by employing the ring opening of furan resulting from the reaction $^1O_2$ via an endoperoxide intermediate. Furan moiety was conjugated with fluorophores as DPBF, allowing suppression of fluorescence due to the presence of p-conjugation between the fluorophore and 1-phenylisobenzofuran. Upon reaction with $^1O_2$, a hypsochromatic shift can be observed for the di-ketone fluorescent product, allowing ratiometric measurement of $^1O_2$ as shown in Fig. 4.57.[187]

A different approach for $^1O_2$ detection was employed using His-Cy with NIR fluorescence capability. Here the 1,4-cycloaddition reaction of $^1O_2$ with imidazole ring occurs, and subsequent ring opening yields the hydroxyl and ketone groups, leading to the recovery of fluorescence through suppression of the PET process as shown in Fig. 4.58. Confocal imaging of $^1O_2$ was successful in PMA-stimulated living mice macrophages (RAW 264.7).[188]

4.4.2.8 Hydroperoxides (ROOH)

Perhaps the most ubiquitous form of hydroperoxide in cellular system is that of the lipid hydroperoxides (LOOH) whose formation has relevant physiological effects and are commonly considered biomarkers of oxidative stress.[189] Since LOOHs are relatively stable products of lipid peroxidation and compartmentalize in the lipid membrane, the design of fluorescence probes that can detect peroxide formation in the hydrophobic compartment of the cell is critical to discern reactive species formed in the cytosol and those in the membrane. Therefore, most of the LOOH probes that were designed incorporate both lipophilic and specificity to peroxides. The most widely exploited chemistry for the detection of ROOH is the ROOH-mediated oxidation of triphenylphosphine (TPP) to TPP oxide, according to Fig. 4.59. Examples of such probes are that of DPPP with several-fold greater sensitivity to methyl

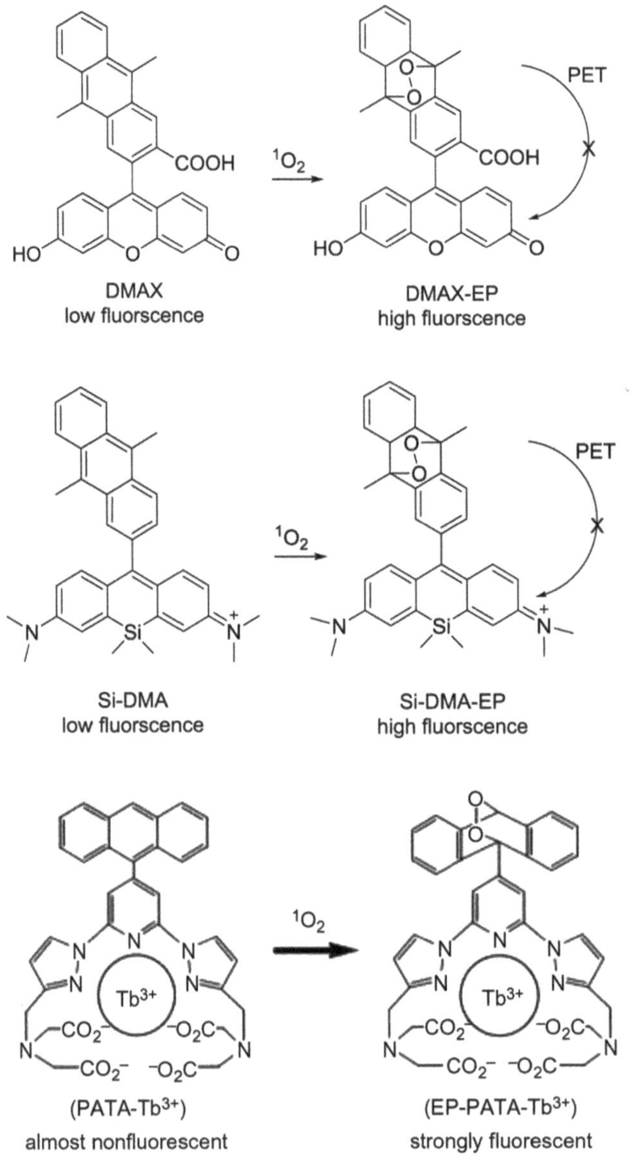

Figure 4.56 Singlet oxidation detection through formation of endoperoxide using DMAX, Si-DMA, or PATA-Tb^{3+}.

linoleate hydroperoxide than H_2O_2[190] and successful LOOH detection in both in vitro[191,192] and in situ detection in isolated intact rat lung.[193] Mitochondrially targeted DPPP (Mito-DPPP) was also used to detect LOOH formation in HepG2 cells on loading with LOOH.[194]

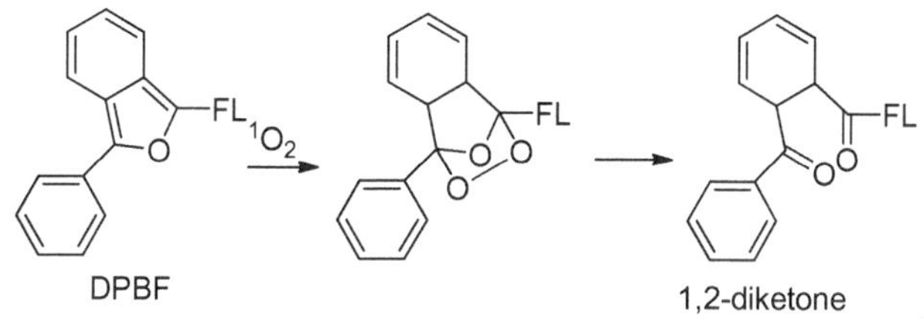

Figure 4.57 Diketone formation via endoperoxide formation from DPBF, resulting in hypsochromatic shift allowing ratiometric measurement of $^1O_2$.

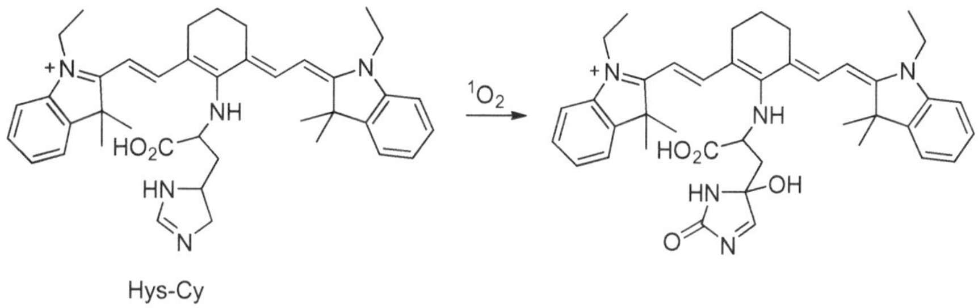

Figure 4.58 Cycloaddition reaction of $^1O_2$ to the histidine group, leading to the formation of ketone and hydroxyl groups and recovery of fluorescence.

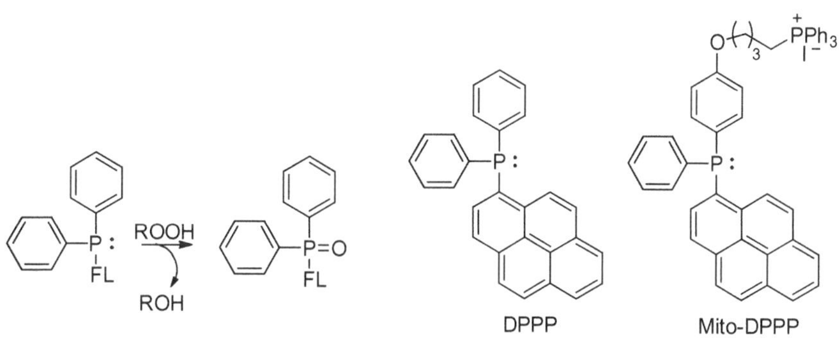

Figure 4.59 General chemistry of peroxide reaction with fluorophore–phosphine conjugate with DPPP and Mito-DPPP as examples.

In the use of fluorescent probe such as C11-BODIPY (Fig. 4.60), on oxidation of the unsaturated hydrocarbon chain the fluorescence could shift to the red range of the visible spectrum, demonstrating ratiometric imaging of hydroperoxides in fibroblasts.[195] While C11-BODIPY was used to evaluate lipid peroxidation, the suppression of fluorescence by antioxidants did not correlate with the suppression of lipid peroxidation, therefore caution should be taken in the interpretation of antioxidant capacity of certain molecules.[196] MitoPerOx combines both the triphenylphosphonium cation and C11-BODIPY in one molecular design (Fig. 4.60) such that lipid

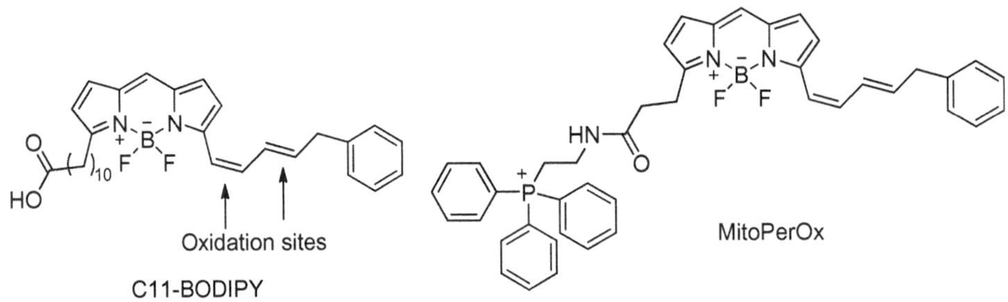

Figure 4.60 Peroxide probes C11-BODIPY and MitoPerOx, showing the sites of oxidation by LOOH.

peroxidation in the inner mitochondrial membrane in whole cells can be monitored in a ratiometric manner that gives a shift from ~590 to 520 nm on oxidation of the unsaturated hydrocarbon moiety.[197]

4.4.2.9 Hydrogen sulfide (H_2S)

It has become clear that hydrogen sulfide is an important small molecule mediator with diverse biological and pharmacological functions that are mostly dependent on O_2 such as inflammation, vasoactivity, angiogenesis, and respiratory inhibition as well as in ischemia and reperfusion.[198,199] Known to be generated in biological system in the cytosol and in the mitochondria by redox-sensitive enzymes using cysteine as substrate, H_2S can alter the cellular redox state and mediate in a variety of signaling pathways, hence its detection is of interest. The most commonly employed chemistry for the detection of H_2S is its reduction reaction with azide to form the amine. SF1 and SF2 are examples of conjugation rhodamine-based probes conjugated to the azide group (Fig. 4.61).[200] This approach allows for a turn-on response upon H_2S treatment with high specificity against ROS and other high-molecular-weight thiols such as GSH, cysteine, lipoic acid, sulfite, and thiosulfate. This H_2S-madiated azide-reduction approach was applied in the ratiometric measurement of H_2S, showing a shift in fluorescence emission as in the case of Cy-N3 (Fig. 4.61).[201]

Other reductive strategies involve reduction of nitro and hydroxylamine to yield the amino group as in the case for HSN1[202] and HAN1,[203] respectively, with high specificity toward H_2S compared to GSH and Cys (Fig. 4.62).

A turn-off mode of H_2S detection was applied using Rhod-CHO to distinguish H_2S from other higher-molecular-weight thiols. The turn-off mechanism exploits the Michael addition–mediated reductive cyclization of aldehyde–rhodamine conjugate as shown in Fig. 4.63. The availability of –SH after reaction to H_2S leads to intramolecular cyclization, which makes this a unique strategy to distinguish between H_2S and other primary thiols (RSHs).

High sensitivity to H_2S compared to Cys and GSH was observed with WSP-based probes, which use dual-electrophilic centers—i.e., a disulfide that reacts with the H_2S

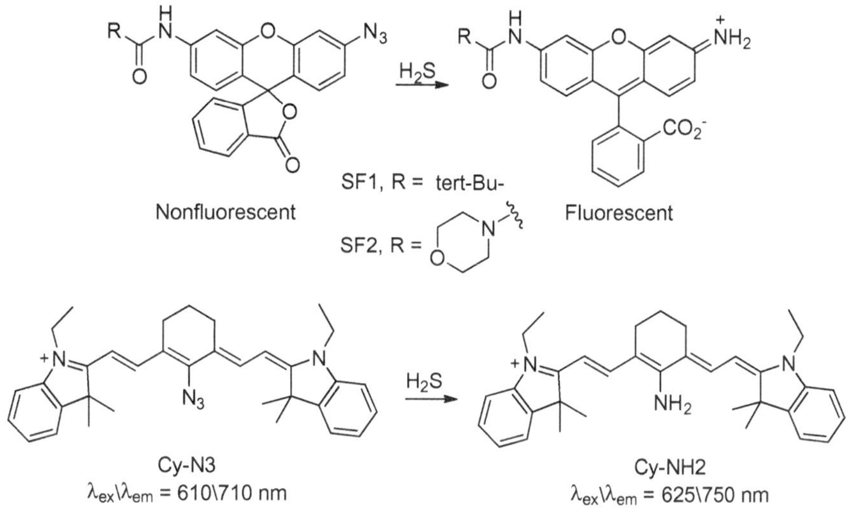

Figure 4.61 Examples of azide-reduction approach in the detection of H_2S showing both off–on modality for SF1 and SF2 probes and ratiometric approach for Cy-N3.

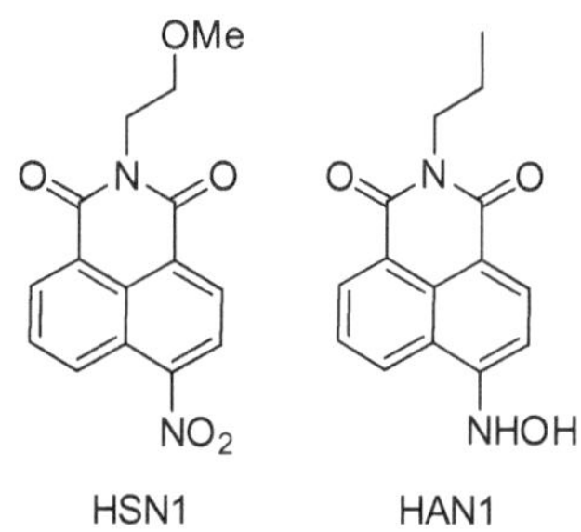

Figure 4.62 Examples of the nitro- and hydroxylamine-reduction approach in the detection of H_2S for HSN1 and HAN1, respectively.

and an ester linker group that can release the fluorophore on cyclization, according to Fig. 4.64.[204]

By linking 2,4-dinitrophenyl (DNP) to a fluorophore via an ether linkage as DNPPCy (Fig. 4.65) allows for a different strategy in H_2S detection in which the nucleophilic character of the thiol is exploited for thiolysis. Nucleophilic aromatic substitution reaction with H_2S leads to the liberation of the fluorophore as a phenolate that has a NIR emission property that allows for ratiometric measurement of H_2S. However, DNPPCy exhibits a slight fluorescence response to GSH and Cys.[205] Instead of an ether linker group, SBODIPY-DNP (Fig. 4.65) employs a sulfonyl linker group to release the fluorescent styryl-containing BODIPY from the DNP moiety on nucleophilic attack by H_2S; however, this is also specific to F^-.[206]

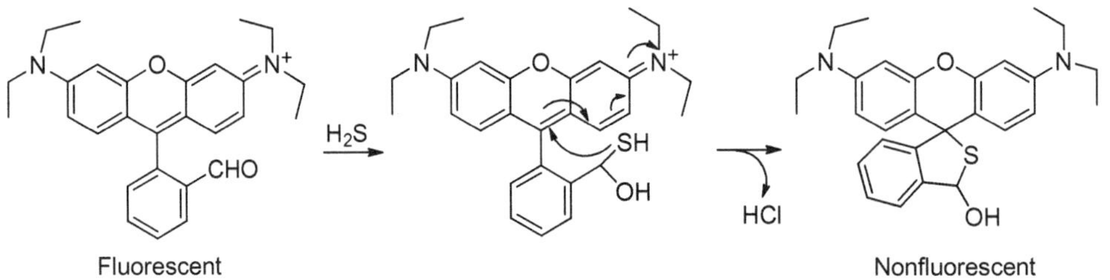

Figure 4.63 Reaction of H_2S to the aldehyde and reductive cyclization via Michael addition by the thiol.

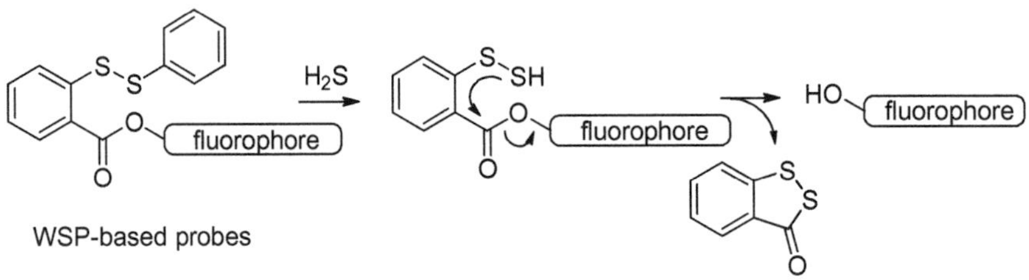

Figure 4.64 General mechanism of WSP-based probes' reaction to H_2S showing release of the fluorophore.

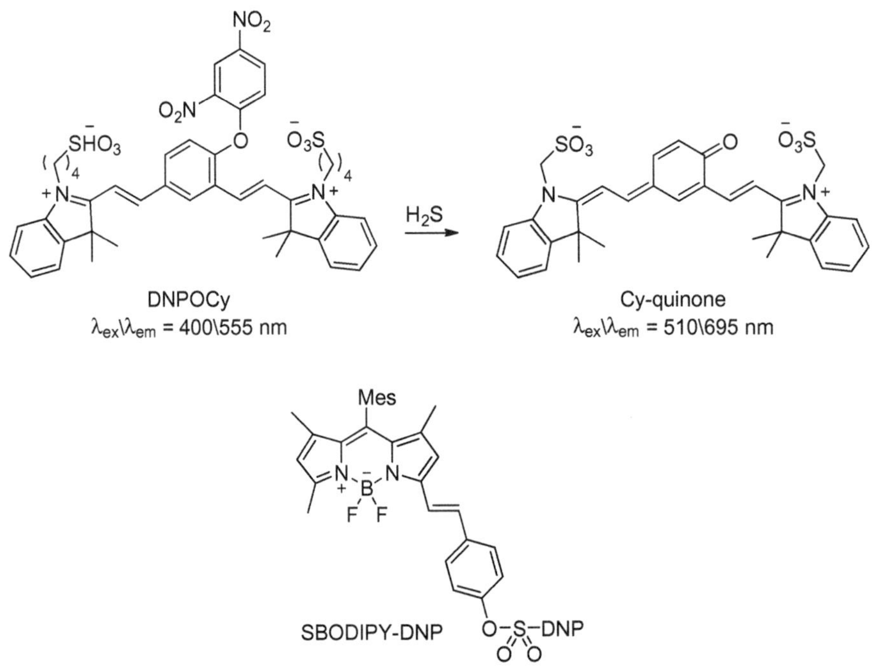

Figure 4.65 Fluorophore-release mechanism for H_2S detection via thiolysis of the ether linkage in DNPOCy and sulfonyl group in SBODIPY-DNP, both showing ratiometric measurement capability.

Finally, an azamacrocylic copper (II) complex conjugated to fluorescein was developed as a metal-based H_2S probe. Fluorescence is quenched in the presence of paramagnetic Cu^{2+}. On reaction of azamacrocylic copper (II) complex with H_2S, the Cu^{2+} is release, thus causing fluorescence recovery with high selectivity toward H_2S compared to higher-molecular-weight thiols. It was employed for the visualization of exogenously added Na_2S in HeLa cells.[207]

4.4.2.10 Thiols (RSH)

Shifts in redox equilibrium levels of sulfhydryl and disulfide (RSH/RSSR) indicate a response to oxidative insult and therefore that global thiols such as gluthathione, homocysteine, and cysteine protein residues are abundant extracellular and intracellular species and important in assessing oxidative stress. Using the fluorophore release modality, a rhodamine-based fluorescent probe with a Se−N bond was used for thiol detection. Nucleophilic attack by the sulfuhydryl of the selenium atom of the Se−N bond allows for the displacement of fluorophore by the thiol to form the selenylsulfide product and the freed rhodamine, according to Fig. 4.66.[208]

Another release mechanism was used with the sulfonamide (SO_2−N) as a linker group such as in the case of SDC (Fig. 4.67) to yield the SO_2, thioglycolic acid, and the fluorophore.[209] This thiol-mediated cleavage of the sulfonamide releases the aniline, thereby enhancing the push−pull character of the dye, and exhibits increased fluorescence and a bathochromic spectral shift. A similar release mechanism was employed for DNRh (Fig. 4.67) to yield rhodamine as the fluorophore and was successfully used to image thiol species in HeLA cells.[210] A cyanine-based NIR fluorescent probe was also employed using the same sulfonamide chemistry, which allows for the selective measurement of GSH in HeLA cell cultures and live mouse models. This selectivity of Cy-sulfonamide (Fig. 4.67) to GSH as compared to Cys and HCy was attributed to the conformation of piperazine ring and relevant H-bond interactions of the GSH with the probe.[211]

A turn-on probe that employs the facile nucleophilic aromatic substitution of phenyl-2,4-dinitrobenezensulfonate-fluorophore by thiols causes SO_2−O bond

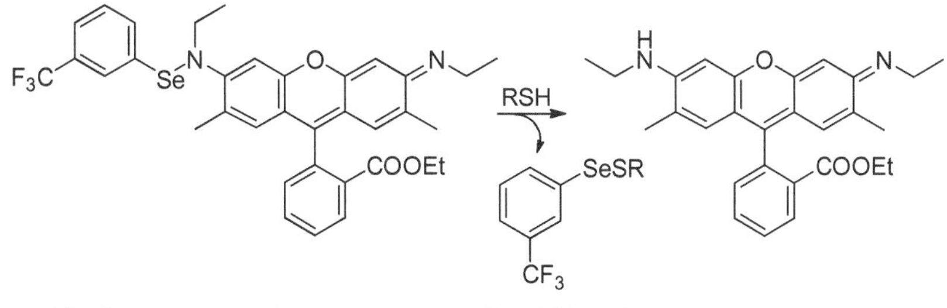

Figure 4.66 Thiol reaction to selenium−nitrogen (Se−N) bond.

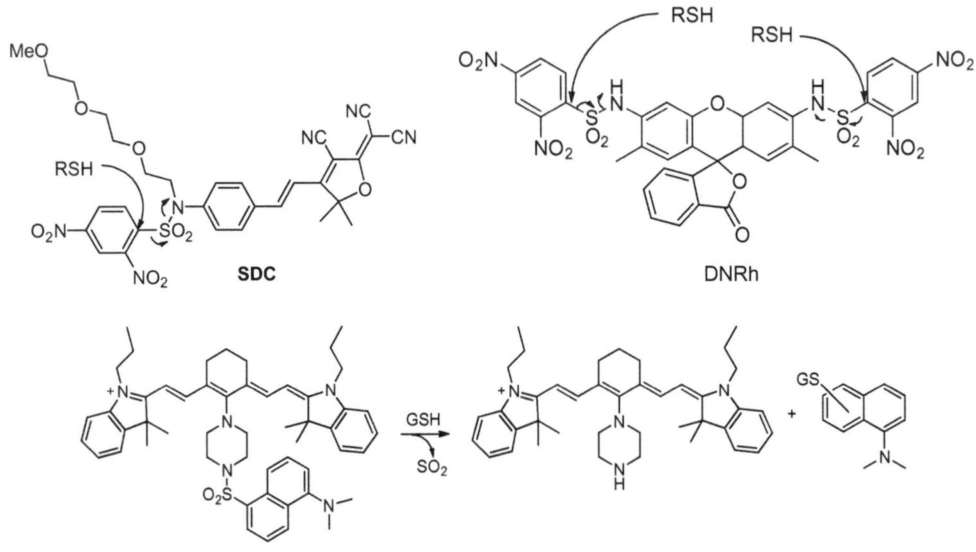

Figure 4.67 Reaction of thiol to sulfonamide (SO_2–N) bond.

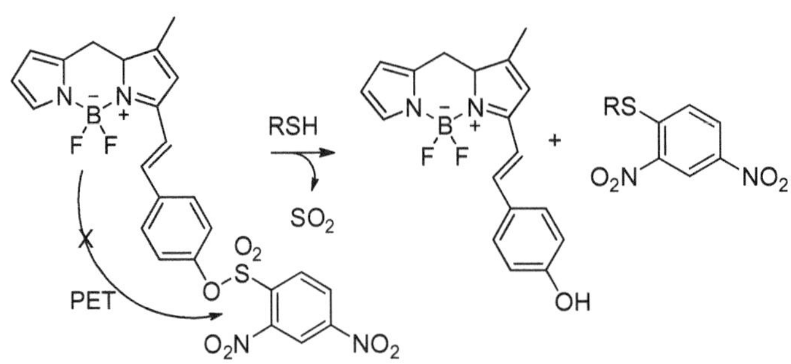

Figure 4.68 Reaction of thiol to sulfonate (SO_2–O) bond.

cleavage to give the (2,4- dinitrophenyl)(phenyl)sulfane, sulfur dioxide, and fluorescent phenolic fluorophore as shown in Fig. 4.68, where PET process is suppressed on release of the fluorophore to give strong fluorescence.[212]

A FRET-based probe for thiol measurement composed of BODIPY linked to a rhodamine via a thioester (NRFTP) (Fig. 4.69) employs a similar release mechanism but causes C(=O)–S bond cleavage via nucleophilic addition of the thiol to the carbonyl-C of the thioester. On separation of the two fluorophores, a decrease in the rhodamine emission can be observed and a new emission for the BODIPY appears at 510 nm, thus allowing ratiometric capability for the measurement of Cys, for example.[213]

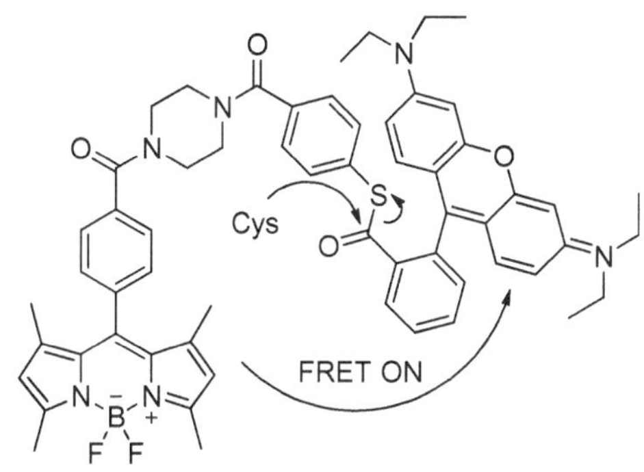

Figure 4.69 Nucleophilic thiol substitution reaction to thioester (C(O)–S) bond.

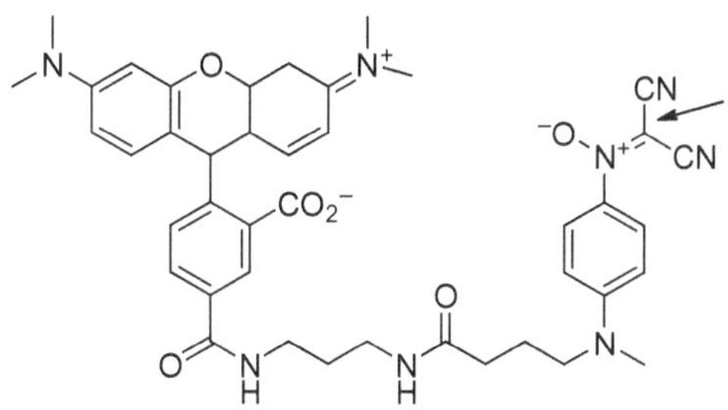

Figure 4.70 Nucleophilic addition of thiol to nitrone (C = N(O)).

A FRET-based probe for the formation of fluorescence was also employed using nucleophilic addition to the nitrone moiety as in the case of *N*-(dicyanomethylene)-aniline oxide that is conjugated to methylrhodamine (nitrone-TAMRA) (Fig. 4.70). Here, the addition of Cys, HCy, or GSH, e.g., to nitrone-TAMRA or through lipoic acid supplementation of 3T3 cells in the presence of nitrone-TAMRA results in a strong hypsochromic shift absorption, allowing fluorescence formation.[214]

A rapid turn-on HMBQ-Nap 1 probe (Fig. 4.71) exploits the reactivity of the carbamate linker group to thiol to release the 4-amino-9-(*n*-butyl)-1,8-naphthalimide fluorophore from the hydroxymethylbenzoquinone PET fluorescence quencher.[215]

A tandem approach was employed in which the initial disulfide bond cleavage by thiol occurs and subsequent intramolecular nucleophilic additions of the thiol to the carbamate-C form the cyclic thiolcarbonate. Here two naphthalimide-based fluorophore were linked via disulfide bond (Di-AN), and by exploiting the ability of thiols to reduce disulfide bonds to individual thiols could trigger subsequent cleavage of a disulfide-based carbamate to release the 4-aminonaphtalimide, imparting ratiometric fluorescence (Fig. 4.72).[216] A similar strategy was employed for FSeSeF but with a diselenide used instead of the disulfide linker for the fluorescent scaffolds.[217]

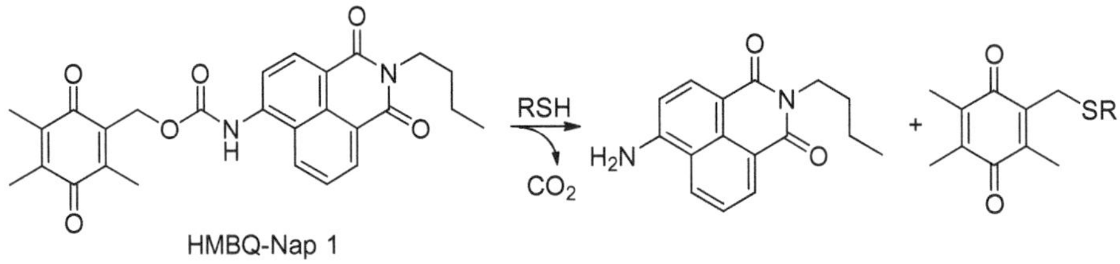

Figure 4.71 Thiol reaction to carbamate showing C–O bond cleavage with subsequent release of CO_2 and amino fluorophore.

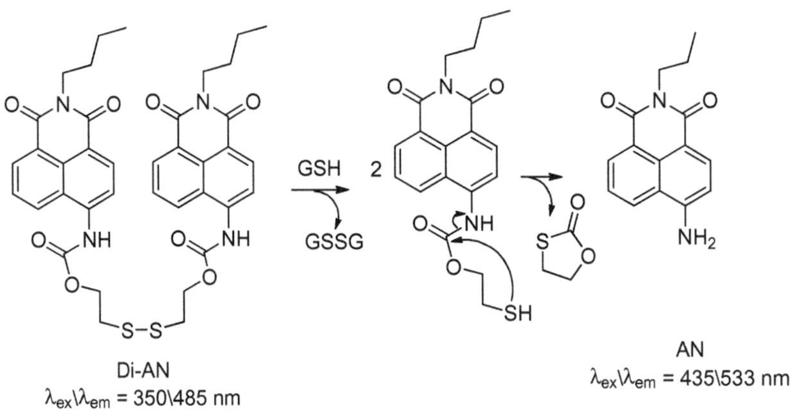

Figure 4.72 Tandem thiol reactions showing disulfide reduction followed by thiol nucleophilic addition to the carbamate-C.

Using the same tandem mechanism as for Di-AN, mitochondrial targeted thiol probe was developed for a two-photon microscopy application that uses lower energy excitation and thus allows increased penetration depth for tissue monitoring of thiols with ratiometric capability. Here SSH-Mito was designed with 6-(benzo[d]thiazol-20-yl)-2-(*N*,*N*-dimethylamino)naphthalene (BTDAN) as the fluorophore, a disulfide group as the thiol reaction site, and triphenylphosphonium salt (TPP) as the mitochondrial targeting site (Fig. 4.73).[218] An intracellular thiol probe to target integrin $\alpha_v\beta_3$, which is a biomarker of early tumor development, was also developed. Here the probe is composed of an intergrin-targeting cyclic arginylglycylaspartic acid peptide (RGD) (cRGD) peptide and a tetraphenylethene (TPE) fluorophore that are linked by a disulfide as TPE-SS-D5-cRGD (Fig. 4.73). Thiol reduction of the disulfide allows for the fluorescence formation.[219]

Thiol monitoring in the cell membrane was achieved with sialoglycoconjugate that was pretagged with azide, and the naphthalimide fluorophore containing a disulfide-carbamate moiety was introduced via a bioorthogonal ligation strategy using strain-promoted [3 + 2] azide–alkyne cycloaddition (Naph-T) (Fig. 4.74). Using the same tandem chemistry as noted previously that employs disulfide reduction and

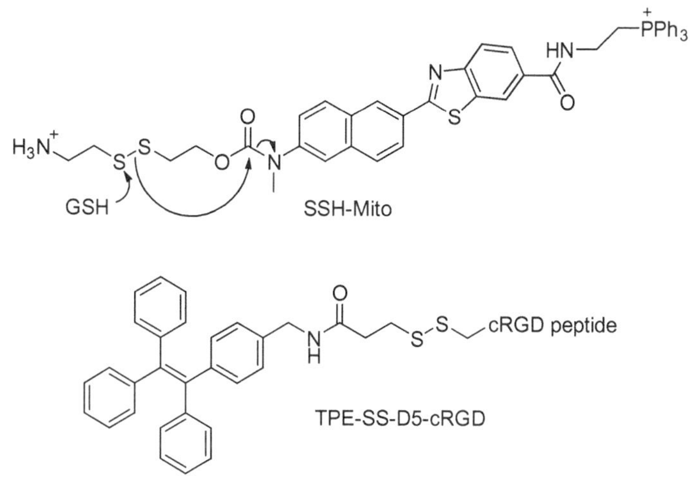

Figure 4.73 Subcellular-targeted thiol probes with tandem chemistry modality for fluorescence formation.

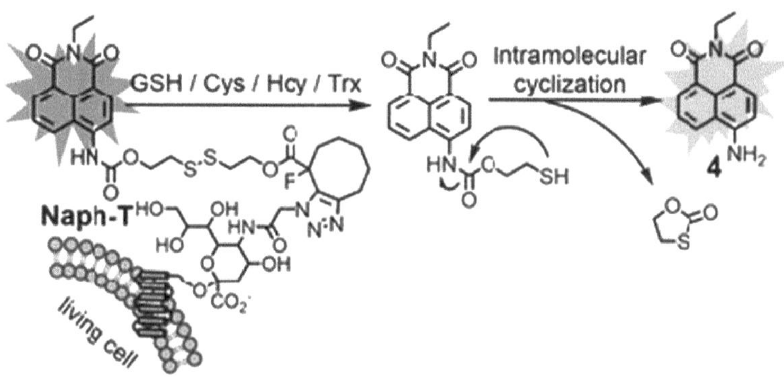

Figure 4.74 Cell membrane–targeted thiol probe (Naph-T) with sialoglycoconjugate bound to disulfide-linked naphthalimide via azide–octyne cycloaddition reaction. Source: *Reproduced from Rong L, Zhang C, Lei Q, Sun H-L, Qin S-Y, Feng J, et al. Chem Commun 2015;**51**:388.*[220] *with permission of The Royal Society of Chemistry.*

concomitant intramolecular cyclization to release the fluorophore allows for the long-term monitoring of thiols on the cell membrane for as long as 36 h.[220]

Thiol addition to the C = C bond of the malonitrile moieties via Michael addition reaction was employed for the ratiometric detection of thiols using a diketopyrroleopyrole-based probe (DPP) (Fig. 4.75).[221] Mass spectral analysis shows as many as four Cys can be added to the probe, including addition to the –CN group to form 4,5-dihydrothiazole exhibiting shifts in absorbance and emission and thus allowing for ratiometric imaging of live cells such as the MDA-231 cell line. A similar strategy was employed for biotin–disulfide–coumarin conjugate for cancer cell

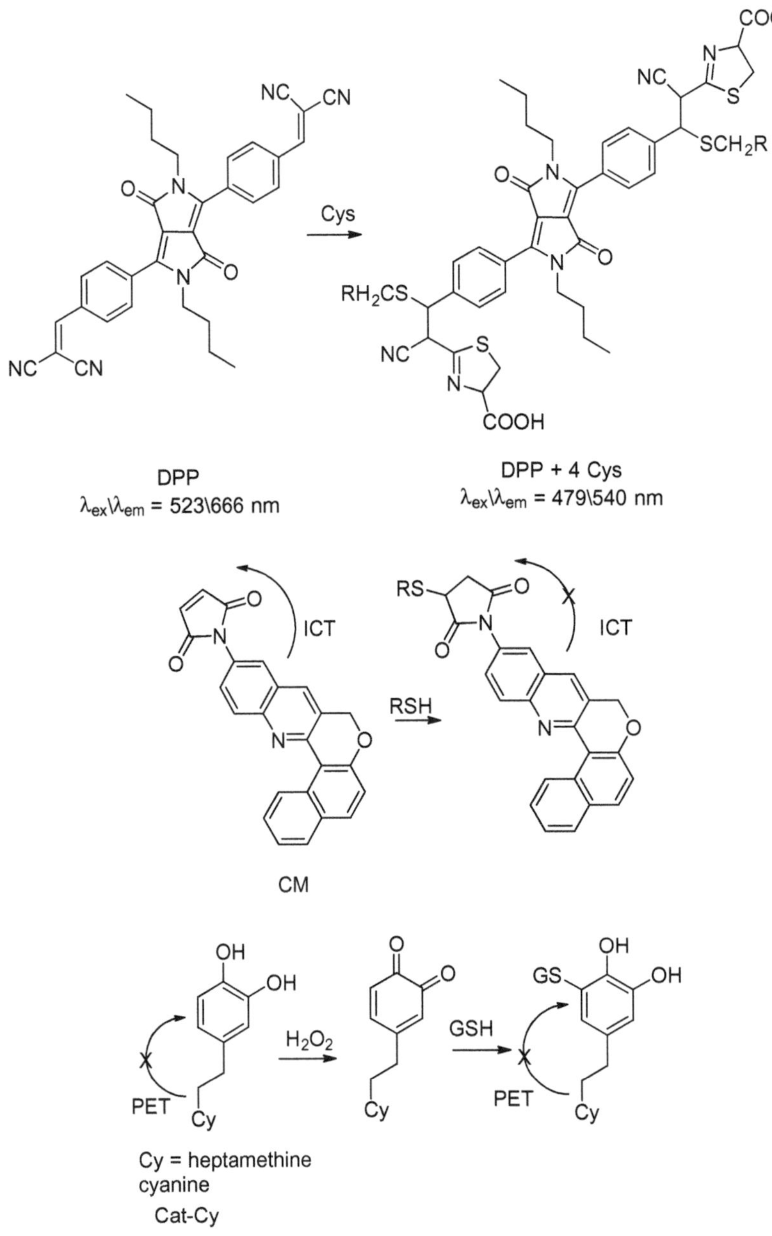

Figure 4.75 Michael addition reaction of thiol to malonitrile, maleimide, and o-quinone in DPP, CM, and Cat-Cy, respectively.

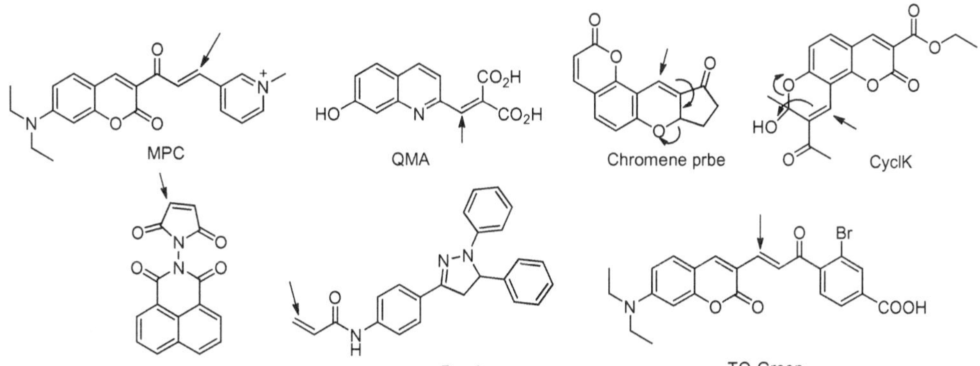

Figure 4.76 Other modes of Michael adduction processes are shown for various thiol probes and their sites of thiol addition.

targeting.[222] The Michael addition reaction of thiol to maleimide was also exploited for an off–on fluorescence probe using a chromenoquinoline-based fluorophore (CM) showing a 223-fold enhancement in fluorescence intensity on thiol addition (Fig. 4.75).[223] Tandem catechol oxidation by H_2O_2 and subsequent Michael addition of GSH to the formed ortho-quinone provided an opportunity to monitor H_2O_2-induced oxidative stress and thiol antioxidant property in living cells as exemplified by Cat-Cy (Fig. 4.75).[10]

Fig. 4.76 shows various structural motifs that exploit the Michael adduction process for thiol detection, showing the site of thiol addition such as in the case of coumarin conjugated to a methyl prydinium cation via unsaturated ketone (MPC)[224]; quinoline 2-methenyl malonic acid (QMA) that can react with thiols at acidic pH, whereby the malonic acid double bond undergoes addition reaction with thiols[225]; a chromene probe that forms a thiol adduct and can be reversed back to the probe by metal ions[226]; an addition reaction with subsequent ring opening to a nonfluorescent probe containing a cyclized α,β-unsaturated ketone (CyclK)[227]; in maleimide-based thiol probes[228]; or pyrazoline-based probe (Pyr-Ac), where the Michael adduction takes place at the acryloyl group of the probe to form the thioether.[229] Also, reversible Michael addition reaction using 7-amino coumarin fluorophore (TQGreen) that allows for the intracellular ratiometric measurement of GSH in living cells was demonstrated. TQGreen at 20 μM concentration can quantitatively measure large excesses of GSH.[230]

Michael adduction that undergoes a cascade of reactions involving an intramolecular ring closure with concomitant cleavage forms the phenoxy anion containing the fluorophore and β-benzoylacrylate moiety (FBA), according to Fig. 4.77.[231] Also, a turn-off approach using an imino-based fluorophore in which PET activation was observed on addition of thiol to the imino group is shown in Fig. 4.77.[232]

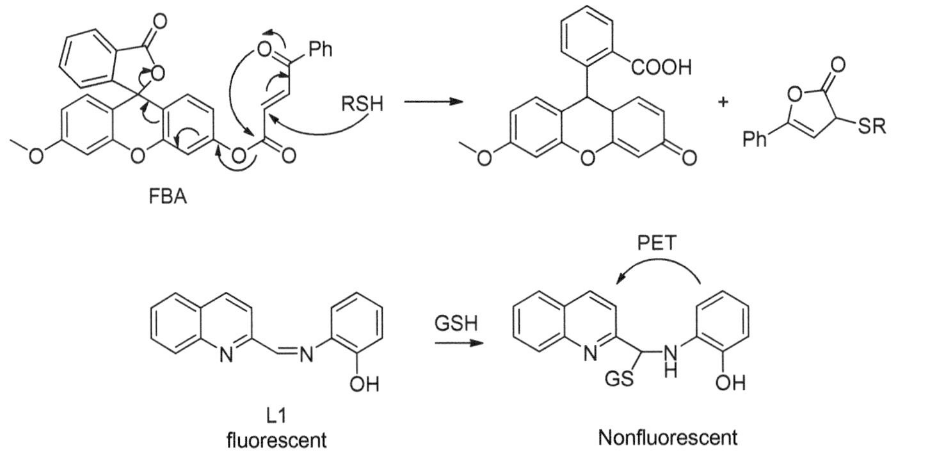

Figure 4.77 Uncommon Michael adduction processes showing intramolecular ring closure followed by fluorophore release with turn-on mechanism for thiol detection for FBA and a turn-off mechanism for L1 through PET activation.

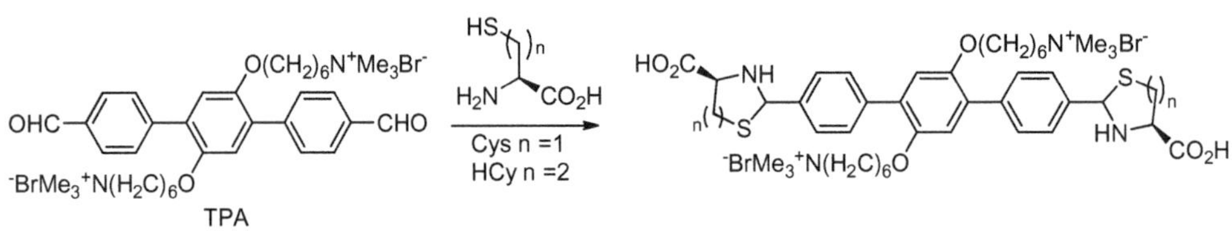

Figure 4.78 Nucleophilic addition of thiol to aldehyde group of TPA.

A water-soluble terphenyl derivative with aldehyde group (TPA) was developed. Direct addition to the aldehyde group was exploited to specifically detect Cys and Hcy and yield thiazolidine and thiazinane rings (Fig. 4.78), respectively, giving an enhanced fluorescence emission in live HeLa cells.[233]

Metal complex–based thiol sensors such as the iminifluroscein-Cu^{2+} (**1**-Cu^{2+})[234] were designed in which thiol complexation with the metal center displaces the fluorophore, leading to the fluorophore release as shown in Fig. 4.79. On release, the Schiff base is hydrolyzed by water to give the fluorescein–aldehyde product, which leads to increased fluorescence. A similar metal-based thiol fluorescent probe that exploits the displacement of metal from the metal–fluorophore complex was shown for the naphthalimide-Cu^{2+} (Fig. 4.79).[235]

Finally, redox-sensitive roGFP2 that was fused with human glutaredoxin-1 as Grx1-roGFP2 allowed for the dynamic real-time measurement of the GSH redox potential. This genetically encoded redox biosensor can measure nanomolar changes in oxidized GSH (GSSG) in a variety of cellular process such as mitochondrial depolarization, respiratory burst activity, or stimulation of immune response, to name a few.

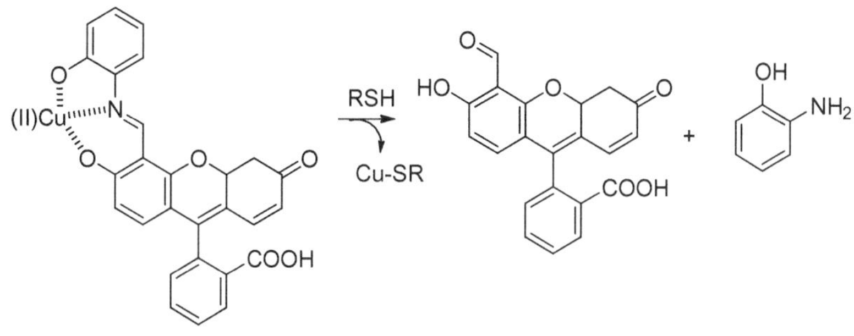

1—Cu^{2+}

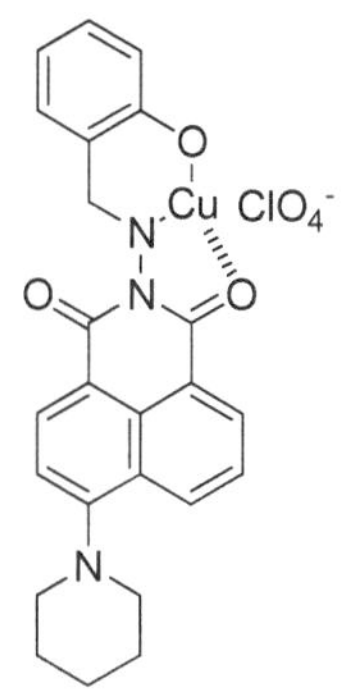

Naphthalimide-Cu^{2+}

Figure 4.79 Metal complex–based thiol probes 1-Cu^{2+} and naphthalimide-Cu^{2+} by fluorophore displacement by the thiol through thiol complexation to the metal center.

The H_2O_2 biosensor probe Hyper also responds to GSH but at a much lower maximal ratiometric change compared to Grx1-roGFP2, and the latter's redox state is mainly driven by the GSH–GSSG redox pair while Hyper involves H_2O_2-mediated oxidation alongside the GSH–GSSG system.[5]

4.4.2.11 Mixed ROS probes

A single-probe design that detects multiple reactive species is attractive since it can provide the opportunity to investigate the various species that may be independently produced and may have synergistic or interplaying roles in the regulation of various RS-mediated cellular processes. Studies of NO or H_2O_2 production or both was demonstrated by dual-color fluorescence imaging with the use of a FP-H_2O_2-NO probe to give distinctive fluorescence responses to NO or H_2O_2 alone or NO/H_2O_2 simultaneously. Although such measurements can be achieved using multiple probes, incorporation of such detection capability in one molecular design is desirable. With a FP-H_2O_2-NO probe, fluorescence patterns can be observed at various excitation and emission wavelengths—i.e., H_2O_2 alone ($\lambda_{ex440}/\lambda_{em460}$), NO alone ($\lambda_{ex550}/\lambda_{em580}$),

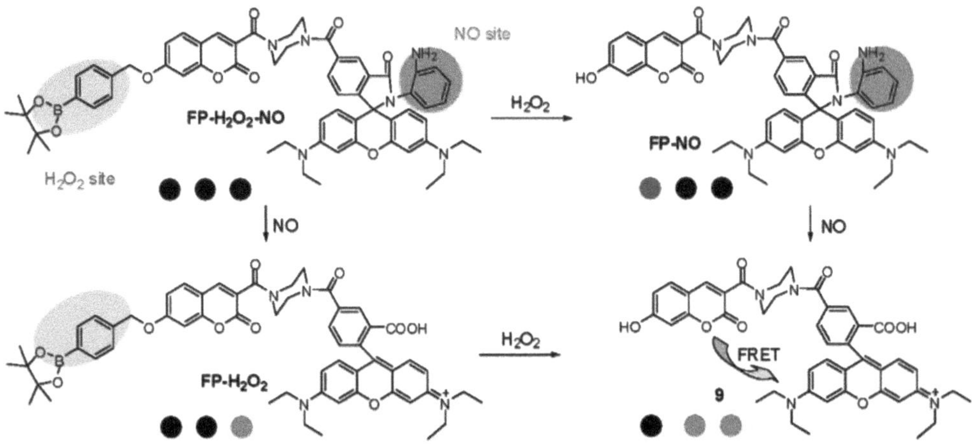

Figure 4.80 Single-probe design for the detection of H_2O_2, NO, or H_2O_2 and NO exhibiting three fluorescence patterns. Source: *Adapted from Yuan et al.,* J Am Chem Soc *2012;**134**:1305–1315.*[236]

and H_2O_2/NO ($\lambda_{ex400}/\lambda_{em580}$ and $\lambda_{ex550}/\lambda_{em580}$) as detected using two-color imaging according to Fig. 4.80.[236]

4.5 CONSIDERATIONS IN FLUORESCENCE PROBE APPLICATION

Several factors could have a profound effect on the intensity or lifetime of fluorescence such as the presence of metal ions, oxygen, exogenously added reagents, solution viscosity, ionic strength, pH, temperature, or light exposure to name a few, and therefore proper controls are necessary to eliminate these factors in interpreting fluorescence results. One serious issue in fluorescence measurements arises from the fact that continued or repeated irradiation could affect subsequent emission intensity, and this reduction of emission quantum yield (also known as *photobleaching*) is typically irreversible and occurs in the presence of O_2. On the other hand, the same reduction in emission intensity can be observed in the presence of the nonirradiative factors mentioned such as continued production of reactive species, metal ions, or oxygen alone and is usually referred to as *quenching*.

In choosing the type of probes for RS detection, the following must be considered.

1. *Type of biological system*. Is one of the most important consideration in the study of RS by fluorescence detection because it will determine what appropriate probe, technique and equipment/instumentation to use. Isolated proteins such as electrophoresis gels or blots are imaged using an instrument (fluorescence scanner) that is different from that used with dissolved proteins in solution; however, the former is seldom employed for RS detection due to the fact that in situ RS detection is the

most acceptable technique in accessing RS production. In vitro studies involve the use of isolated proteins, whole cells, lysed cells, or homogenized tissue while in vivo studies would involve anesthetized small animal models. Ex vivo studies involving blood draws, excised tissue, or whole organ using proper animal protocols.

2. *Type of instrumentation.* Several instruments are capable of measuring and imaging fluorescence, but the proper choice of instrumentation depends on the type of biological system being used. For example, the use of filter fluorometers or spectrofluorometers is appropriate for bulk fluorescence measurement of solutions of enzymes, plasma, or filtered-lysed cells with volumes ranging from μL to mL using single-cell cuvette or multiwell plate readers for high-throughput measurements. For whole cells (live or dead) or tissue slice, epifluorescence microscopy provides spectro-spatial (2-D or 3-D images), and through the use of confocal imaging technique can give high spatial resolution. Moreover, flow cytometry is employed for fluorescence measurement per single cell in a stream of fluid (or whole blood) containing suspended cells; this also allows for the measurement of multiple fluorescent labels at a single time and is more sensitive than confocal microscopy due to elimination of out-of-focus signals and faster signal collection for the former. Finally, an in vivo optical imaging system allows for noninvasive measurement of fluorescence in whole organs or living animals offering 3D tomography images and the use of multiple fluorescent probes due to large optical imaging windows ranging from 400 to 1200 nm[237] and gives high penetrability to tissue and less autofluorescence that would have been otherwise unachievable using probes absorbing at the visible wavelength.
3. *Type of expected radicals.* Knowing the type of expected radical to monitor would aid in choosing the right probe. For example, the relevant radicals to detect in the study of endothelial nitric oxide synthase (eNOS) uncoupling or S-nitrosylation are $O_2^{\bullet-}$, NO, or $ONOO^-$; or $O_2^{\bullet-}$, H_2O_2, or HOCl in the study of leukocyte activation or peroxidase activity; or $O_2^{\bullet-}$ to study initial oxidative bursts from mitochondria, xanthine oxidase, or NADPH oxidases. When not certain on the expected radicals to detect, detection of total thiols or total ROS in the study of global redox state in cells, tissues, or plasma is initially employed.
4. *Type of probe.* The degree of specificity of measurement as well as the type of imaging technique will highly determine what type of probe to use for a particular study. Probes that are highly specific, partially specific, or nonspecific to a particular reactive species can be purchased commercially, or the homemade ones can be obtained through collaboration and are usually available to investigators on request, especially if they were developed through government funding. The use of multiple probes could narrow down the number of possible ROS under investigation and allow for their identification. For example, probes to study global ROS production may be essential to roughly determine whether there was RS production;

Table 4.2 Summary of the various probes that are readily available commercially and their kinetics of reaction to various ROS

Probe	Reactive species	Rate constant (M^{-1} s^{-1})	Ref
Dichlorodihydrofluorescein	$ONOO^-$	5×10^4 (pH = 10)	36
	$CO_3^{\bullet-}$	2.6×10^8 (pH = 8.0–8.2)	29
	$NO_2^{\bullet}$	1.3×10^7 (pH = 7.4)	238
	$HO^{\bullet}$	1.3×10^{10} (pH = 7.4)	6
	$RS^{\bullet}$($GS^{\bullet}$or $CysS^{\bullet}$)	$2–4 \times 10^7$ (pH = 7.4)	
	$DCFH^{\bullet} + O_2$	5.3×10^8 (pH = 7.4)	
	$N_3^{\bullet}$	2×10^9 (pH = 7.4)	
Dihydrorhodamine	$CO_3^{\bullet-}$	6.7×10^8 (pH = 8.0–8.2)	29
	$HO^{\bullet}$	1.8×10^{10} (pH = 7.4)	30
	$NO_2^{\bullet}$	$4–7 \times 10^5$	
	$RhH^{\bullet\cdot} + O_2$	7×10^8	
4-acetylphenylboronic acid	$ONOO^-$	1.6×10^6	77
	H_2O_2	2.2	
	HOCl	6.2×10^3 (pH = 7.4)	
Coumarin-7-boronic acid	$ONOO^-$	1.1×10^6	169
	amino acid-OOH	7–23	239
Hydroethidine	HOCl	1.5×10^5 (pH = 7.4)	58
	$O_2^{\bullet-}$	2×10^6 (water–ethanol, 1:1)	54
Benzoate, coumarin, and phenoxazinone systems	$HO^{\bullet}$	$2–20 \times 10^9$	240

using a more specific probe right from the start could give a false positive result. Later, more specific probes could be employed to finally identify the RS of interest. In the study of tissues or whole animals, probes that absorb at the 700- to 900-nm range are preferred since this is outside the absorptive range of biological tissue, allowing deeper penetration of the NIR light compared to the visible wavelength.

5. *Possible interferences*. Table 4.2 shows the various kinetic rate constants for representative and commonly used probes that could be used to assess potential cross-reactivity with other radicals not under investigation, thus providing guidance for other confirmatory experiments, either nonoptical or optical, that should be carried out to identify a particular reactive species. For example, in the detection of $O_2^{\bullet-}$, although positive results may come with the use of MitoSox or HE, these data must be supplemented by other detection techniques such as HPLC analysis of the HE products or EPR spin trapping to confirm $O_2^{\bullet-}$-mediated fluorescence formation. The use of inhibitory experiments are also typically employed, but care must be taken in interpreting those data because, e.g., the use of SOD to quench $O_2^{\bullet-}$ may not get internalized by the cell due to its high molecular weight.

Lower-molecular-weight and cell-permeable SOD mimetics, however, may be more appropriate control reagents. Direct sequestration of the interfering RS or indirect sequestration through inhibition of sources are also typically employed to narrow down identification of reactive species.

REFERENCES

1. Valeur B, Berberan-Santos MN. A Brief History of Fluorescence and Phosphorescence before the Emergence of Quantum Theory. *J Chem Ed* 2011;**88**(6):731–8.
2. Stokes GG. On the change of refrangibility of light. *Philos Trans* 1852;**142**:463–562.
3. Bayer A. Ueber eine neue Klasse von Farbstoffen. *Ber Dtsch Chem Ges* 1871;**4**:555–8.
4. Heinrichs, A. Stains and fluorescent dyes Nature Milestone: Light Microscopy [Online], 2009.
5. Gutscher M, Pauleau A-L, Marty L, Brach T, Wabnitz GH, Samstag Y, et al. Real-time imaging of the intracellular glutathione redox potential. *Nat Methods* 2008;**5**(6):553–9.
6. Wrona M, Wardman P. Properties of the radical intermediate obtained on oxidation of 2',7'-dichlorodihydrofluorescein, a probe for oxidative stress. *Free Radic Biol Med* 2006;**41**(4):657–67.
7. Lin VS, Dickinson BC, Chang CJ. Boronate-based fluorescent probes: imaging hydrogen peroxide in living systems. *Methods Enzymol* 2013;**526**:19–43 (Hydrogen Peroxide and Cell Signaling, Part A)
8. Maeda H, Yamamoto K, Nomura Y, Kohno I, Hafsi L, Ueda N, et al. A design of fluorescent probes for superoxide based on a nonredox mechanism. *J Am Chem Soc* 2005;**127**(1):68–9.
9. Yang D, Wang H-L, Sun Z-N, Chung N-W, Shen J-G. A Highly Selective Fluorescent Probe for the Detection and Imaging of Peroxynitrite in Living Cells. *J Am Chem Soc* 2006;**128**(18):6004–5.
10. Yu F, Li P, Song P, Wang B, Zhao J, Han K. Facilitative functionalization of cyanine dye by an on-off-on fluorescent switch for imaging of H_2O_2 oxidative stress and thiols reducing repair in cells and tissues. *Chem Commun* 2012;**48**(41):4980–2.
11. Dong Z, Le X, Zhou P, Dong C, Ma J. An "off-on-off" fluorescent probe for the sequential detection of Zn2 + and hydrogen sulfide in aqueous solution. *New J Chem* 2014;**38**(4):1802–8.
12. Rota C, Chignell CF, Mason RP. Evidence for free radical formation during the oxidation of 2'-7'-dichlorofluorescin to the fluorescent dye 2'-7'-dichlorofluorescein by horseradish peroxidase: possible implications for oxidative stress measurements. *Free Radic Biol Med* 1999;**27**(7-8):873–81.
13. Marchesi E, Rota C, Fann YC, Chignell CF, Mason RP. Photoreduction of the fluorescent dye 2'-7'-dichlorofluorescein: a spin trapping and direct electron spin resonance study with implications for oxidative stress measurements. *Free Radic Biol Med* 1999;**26**(1-2):148–61.
14. Wardman P. Fluorescent and luminescent probes for measurement of oxidative and nitrosative species in cells and tissues: Progress, pitfalls, and prospects. *Free Radic Biol Med* 2007;**43**(7):995–1022.
15. Brubacher JL, Bols NC. Chemically de-acetylated 2',7'-dichlorodihydrofluorescein diacetate as a probe of respiratory burst activity in mononuclear phagocytes. *J Immunol Methods* 2001;**251**(1–2):81–91.
16. Tetz LM, Kamau PW, Cheng AA, Meeker JD, Loch-Caruso R. Troubleshooting the dichlorofluorescein assay to avoid artifacts in measurement of toxicant-stimulated cellular production of reactive oxidant species. *J Pharmacol Toxicol Methods* 2013;**67**(2):56–60.
17. O'Malley YQ, Reszka KJ, Britigan BE. Direct oxidation of 2',7'-dichlorodihydrofluorescein by pyocyanin and other redox-active compounds independent of reactive oxygen species production. *Free Radic Biol Med* 2004;**36**(1):90–100.
18. Tollefson KE, Kroczynski J, Cutaia MV. Time-Dependent Interactions of Oxidant-Sensitive Fluoroprobes with Inhibitors of Cellular Metabolism. *Lab Invest* 2003;**83**(3):367–75.
19. Hempel SL, Buettner GR, O'Malley YQ, Wessels DA, Flaherty DM. Dihydrofluorescein diacetate is superior for detecting intracellular oxidants: comparison with 2',7'-dichlorodihydrofluorescein diacetate, 5(and 6)-carboxy-2',7'-dichlorodihydrofluorescein diacetate, and dihydrorhodamine 123. *Free Radic Biol Med* 1999;**27**(1/2):146–59.

20. Karlsson M, Kurz T, Brunk UT, Nilsson SE, Frennesson CI. What does the commonly used DCF test for oxidative stress really show? *Biochem J* 2010;**428**(2):183–90.
21. Boulton S, Anderson A, Swalwell H, Henderson JR, Manning P, Birch-Machin MA. Implications of using the fluorescent probes, dihydrorhodamine 123 and 2',7'-dichlorodihydrofluorescein diacetate, for the detection of UVA-induced reactive oxygen species. *Free Radic Res* 2011;**45**(2):115–22.
22. Balcerczyk A, Kruszewski M, Bartosz G. Does the cellular labile iron pool participate in the oxidation of 2',7'-dichlorodihydrofluorescein? *Free Radic Res* 2007;**41**(5):563–70.
23. LeBel CP, Ischiropoulos H, Bondy SC. Evaluation of the probe 2',7'-dichlorofluorescin as an indicator of reactive oxygen species formation and oxidative stress. *Chem Res Toxicol* 1992;**5**(2):227–31.
24. Andoh Y, Mizutani A, Ohashi T, Kojo S, Ishii T, Adachi Y, et al. The antioxidant role of a reagent, 2',7'-dichlorodihydrofluorescin diacetate, detecting reactive-oxygen species and blocking the induction of heme oxygenase-1 and preventing cytotoxicity. *J Biochem* 2006;**140**(4):483–9.
25. Afri M, Frimer AA, Cohen Y. Active oxygen chemistry within the liposomal bilayer Part IV: Locating 2',7'-dichlorofluorescein (DCF), 2',7'-dichlorodihydrofluorescein (DCFH) and 2',7'-dichlorodihydrofluorescein diacetate (DCFH-DA) in the lipid bilayer. *Chem Phys Lipids* 2004;**131** (1):123–33.
26. Kuznetsov AV, Kehrer I, Kozlov AV, Haller M, Redl H, Hermann M, et al. Mitochondrial ROS production under cellular stress: comparison of different detection methods. *Anal Bioanal Chem* 2011;**400**(8):2383–90.
27. Diaz G, Liu S, Isola R, Diana A, Falchi AM. Mitochondrial localization of reactive oxygen species by dihydrofluorescein probes. *Histochem Cell Biol* 2003;**120**(4):319–25.
28. Baracca A, Sgarbi G, Solaini G, Lenaz G. Rhodamine 123 as a probe of mitochondrial membrane potential: evaluation of proton flux through F0 during ATP synthesis. *Biochim Biophys Acta (BBA)-Bioenergetics* 2003;**1606**(1–3):137–46.
29. Wrona M, Patel K, Wardman P. Reactivity of 2',7'-dichlorodihydrofluorescein and dihydrorhodamine 123 and their oxidized forms toward carbonate, nitrogen dioxide, and hydroxyl radicals. *Free Radic Biol Med* 2005;**38**(2):262–70.
30. Folkes LK, Patel KB, Wardman P, Wrona M. Kinetics of reaction of nitrogen dioxide with dihydrorhodamine and the reaction of the dihydrorhodamine radical with oxygen: Implications for quantifying peroxynitrite formation in cells. *Arch Biochem Biophys* 2009;**484**(2):122–6.
31. Royall JA, Ischiropoulos H. Evaluation of 2',7'-dichlorofluorescin and dihydrorhodamine 123 as fluorescent probes for intracellular hydrogen peroxide in cultured endothelial cells. *Arch Biochem Biophys* 1993;**302**(2):348–55.
32. Henderson LM, Chappell JB. Dihydrorhodamine 123: a fluorescent probe for superoxide generation? *Eur J Biochem* 1993;**217**(3):973–80.
33. Crow JP. Dichlorodihydrofluorescein and dihydrorhodamine 123 are sensitive indicators of peroxynitrite in vitro: implications for intracellular measurement of reactive nitrogen and oxygen species. *Nitric Oxide* 1997;**1**(2):145–57.
34. Esposito BP, Breuer W, Sirankapracha P, Pootrakul P, Hershko C, Cabantchik ZI. Labile plasma iron in iron overload: Redox activity and susceptibility to chelation. *Blood* 2003;**102**(7):2670–7.
35. Kooy NW, Royall JA, Ischiropoulos H, Beckman JS. Peroxynitrite-mediated oxidation of dihydrorhodamine 123. *Free Radic Biol Med* 1994;**16**(2):149–56.
36. Glebska J, Koppenol WH. Peroxynitrite-mediated oxidation of dichlorodihydrofluorescein and dihydrorhodamine. *Free Radic Biol Med* 2003;**35**(6):676–82.
37. Martin-Romero FJ, Gutierrez-Martin Y, Henao F, Gutierrez-Merino C. Fluorescence Measurements of Steady State Peroxynitrite Production Upon SIN-1 Decomposition: NADH versus Dihydrodichlorofluorescein and Dihydrorhodamine 123. *J Fluoresc* 2004;**14**(1):17–23.
38. Miranda KM, Espey MG, Yamada K, Krishna M, Ludwick N, Kim S, et al. Unique oxidative mechanisms for the reactive nitrogen oxide species, nitroxyl anion. *J Biol Chem* 2001;**276** (3):1720–7.
39. Sharpe MA, Cooper CE. Reactions of nitric oxide with mitochondrial cytochrome c: a novel mechanism for the formation of nitroxyl anion and peroxynitrite. *Biochem J* 1998;**332**(1):9–19.

40. Kelesidis T, Currier JS, Huynh D, Meriwether D, Charles-Schoeman C, Reddy ST, et al. A biochemical fluorometric method for assessing the oxidative properties of HDL. *J Lipid Res* 2011;**52**(12):2341–51.
41. Kelesidis T, Reddy ST, Huynh D, Meriwether D, Fogelman AM, Navab M, et al. Effects of lipid-probe interactions in biochemical fluorometric methods that assess HDL redox activity. *Lipids Health Dis* 2012;**11**:87.
42. Martin BD, Schoenhard JA, Sugden KD. Hypervalent Chromium Mimics Reactive Oxygen Species As Measured by the Oxidant-Sensitive Dyes 2',7'-Dichlorofluorescin and Dihydrorhodamine. *Chem Res Toxicol* 1998;**11**(12):1402–10.
43. Qin Y, Lu M, Gong X. Dihydrorhodamine 123 is superior to 2,7-dichlorodihydrofluorescein diacetate and dihydrorhodamine 6G in detecting intracellular hydrogen peroxide in tumor cells. *Cell Biol Int* 2008;**32**(2):224–8.
44. Walrand S, Valeix S, Rodriguez C, Ligot P, Chassagne J, Vasson M-P. Flow cytometry study of polymorphonuclear neutrophil oxidative burst: a comparison of three fluorescent probes. *Clin Chim Acta* 2003;**331**(1–2):103–10.
45. Imrich A, Ning YY, Kobzik L. Intracellular oxidant production and cytokine responses in lung macrophages: evaluation of fluorescent probes. *J Leukocyte Biol* 1999;**65**(4):499–507.
46. Setsukinai K, Urano Y, Kakinuma K, Majima HJ, Nagano T. Development of novel fluorescence probes that can reliably detect reactive oxygen species and distinguish specific species. *J Biol Chem* 2003;**278**(5):3170–5.
47. Flemmig J, Zschaler J, Remmler J, Arnhold J. The Fluorescein-derived Dye Aminophenyl Fluorescein Is a Suitable Tool to Detect Hypobromous Acid (HOBr)-producing Activity in Eosinophils. *J Biol Chem* 2012;**287**(33):27913–23.
48. Flemmig J, Schwarz P, Baecker I, Leichsenring A, Lange F, Arnhold J. Rapid and reliable determination of the halogenating peroxidase activity in blood samples. *J Immunol Methods* 2014;**415**:46–56.
49. Debordea M, von Gunten U. Reactions of chlorine with inorganic and organic compounds during water treatment—Kinetics and mechanisms: A critical review. *Water Res* 2008;**42**:13–51.
50. Keller A, Mohamed A, Droese S, Brandt U, Fleming I, Brandes RP. Analysis of Dichlorodihydrofluorescein and Dihydrocalcein as Probes for the Detection of Intracellular Reactive Oxygen Species. *Free Radic Res* 2004;**38**(12):1257–67.
51. Rohnstock A, Lehmann L. Evaluation of the probe dihydrocalcein acetoxymethylester as an indicator of reactive oxygen species formation and comparison with oxidative DNA base modification determined by modified alkaline elution technique. *Toxicol in Vitro* 2007;**21**(8):1552–62.
52. Jobsis PD, Rothstein EC, Balaban RS. Limited utility of acetoxymethyl (AM)-based intracellular delivery systems, in vivo: interference by extracellular esterases. *J Microsc* 2007;**226**:74–81 (Pt 1)
53. Zhao H, Kalivendi S, Zhang H, Joseph J, Nithipatikom K, Vasquez-Vivar J, et al. Superoxide reacts with hydroethidine but forms a fluorescent product that is distinctly different from ethidium: potential implications in intracellular fluorescence detection of superoxide. *Free Radic Biol Med* 2003;**34**(11):1359–68.
54. Zielonka J, Sarna T, Roberts JE, Wishart JF, Kalyanaraman B. Pulse radiolysis and steady-state analyses of the reaction between hydroethidine and superoxide and other oxidants. *Arch Biochem Biophys* 2006;**456**(1):39–47.
55. Zielonka J, Srinivasan S, Hardy M, Ouari O, Lopez M, Vasquez-Vivar J, et al. Cytochrome c-mediated oxidation of hydroethidine and mito-hydroethidine in mitochondria: Identification of homo- and heterodimers. *Free Radic Biol Med* 2008;**44**(5):835–46.
56. Zhao H, Joseph J, Fales HM, Sokoloski EA, Levine RL, Vasquez-Vivar J, et al. Detection and characterization of the product of hydroethidine and intracellular superoxide by HPLC and limitations of fluorescence. *Proc Natl Acad Sci U S A* 2005;**102**(16):5727–32.
57. Zielonka J, Vasquez-Vivar J, Kalyanaraman B. Detection of 2-hydroxyethidium in cellular systems: a unique marker product of superoxide and hydroethidine. *Nat Protoc* 2008;**3**(1):8–21.
58. Maghzal GJ, Cergol KM, Shengule SR, Suarna C, Newington D, Kettle AJ, et al. Assessment of Myeloperoxidase Activity by the Conversion of Hydroethidine to 2-Chloroethidium. *J Biol Chem* 2014;**289**(9):5580–95.

59. Zielonka J, Vasquez-Vivar J, Kalyanaraman B. The confounding effects of light, sonication, and Mn (III)TBAP on quantitation of superoxide using hydroethidine. *Free Radic Biol Med* 2006;**41**(7):1050–7.
60. Patsoukis N, Papapostolou I, Georgiou CD. Interference of non-specific peroxidases in the fluorescence detection of superoxide radical by hydroethidine oxidation: a new assay for H_2O_2. *Anal Bioanal Chem* 2005;**381**(5):1065–72.
61. Xu X, Arriaga EA. Qualitative determination of superoxide release at both sides of the mitochondrial inner membrane by capillary electrophoretic analysis of the oxidation products of triphenylphosphonium hydroethidine. *Free Radic Biol Med* 2009;**46**(7):905–13.
62. Zielonka J, Hardy M, Kalyanaraman B. HPLC study of oxidation products of hydroethidine in chemical and biological systems: ramifications in superoxide measurements. *Free Radic Biol Med* 2009;**46**(3):329–38.
63. Robinson KM, Janes MS, Pehar M, Monette JS, Ross MF, Hagen TM, et al. Selective fluorescent imaging of superoxide in vivo using ethidium-based probes. *Proc Natl Acad Sci U S A* 2006;**103**(41):15038–43.
64. Lyublinskaya OG, Zenin VV, Shatrova AN, Aksenov ND, Zemelko VI, Domnina AP, et al. Intracellular oxidation of hydroethidine: Compartmentalization and cytotoxicity of oxidation products. *Free Radic Biol Med* 2014;**75**:60–8.
65. Benov L, Sztejnberg L, Fridovich I. Critical evaluation of the use of hydroethidine as a measure of superoxide anion radical. *Free Radic Biol Med* 1998;**25**(7):826–31.
66. Wang W, Fang H, Groom L, Cheng A, Zhang W, Liu J, et al. Superoxide Flashes in Single Mitochondria. *Cell* 2008;**134**(2):279–90.
67. Schwarzlander M, Wagner S, Ermakova YG, Belousov VV, Radi R, Beckman JS, et al. The /'mitoflash/' probe cpYFP does not respond to superoxide. *Nature* 2014;**514**(7523):E12–14.
68. Cheng H, Wang W, Wang X, Sheu SS, Dirksen RT, Dong MQ. Cheng et al. reply. *Nature* 2014;**514**(7523):E14–15.
69. Ainley AD, Challenger F. Studies of the boron–carbon linkage. Part I. The oxidation and nitration of phenylboric acid. *J Chem Soc* 1930;2171–80.
70. Kuivila HG. Electrophilic displacement reactions. III. Kinetics of the reaction between hydrogen peroxide and benzeneboronic acid. *J Am Chem Soc* 1954;**76**:870–4.
71. Lo L-C, Chu C-Y. Development of highly selective and sensitive probes for hydrogen peroxide. *Chem Commun* 2003;**21**:2728–9.
72. Dickinson BC, Chang CJ, Targetable Fluorescent A. Probe for Imaging Hydrogen Peroxide in the Mitochondria of Living Cells. *J Am Chem Soc* 2008;**130**(30):9638–9.
73. Dickinson BC, Huynh C, Chang CJ. A Palette of Fluorescent Probes with Varying Emission Colors for Imaging Hydrogen Peroxide Signaling in Living Cells. *J Am Chem Soc* 2010;**132**(16):5906–15.
74. Chang MC, Pralle A, Isacoff EY, Chang CJ. A selective, cell-permeable optical probe for hydrogen peroxide in living cells. *J Am Chem Soc* 2004;**126**(47):15392–3.
75. Miller EW, Albers AE, Pralle A, Isacoff EY, Chang CJ. Boronate-based fluorescent probes for imaging cellular hydrogen peroxide. *J Am Chem Soc* 2005;**127**(47):16652–9.
76. Miller EW, Chang CJ. Fluorescent probes for nitric oxide and hydrogen peroxide in cell signaling. *Curr Opin Chem Biol* 2007;**11**(6):620–5.
77. Sikora A, Zielonka J, Lopez M, Joseph J, Kalyanaraman B. Direct oxidation of boronates by peroxynitrite: Mechanism and implications in fluorescence imaging of peroxynitrite. *Free Radic Biol Med* 2009;**47**(10):1401–7.
78. Purdey MS, Connaughton HS, Whiting S, Schartner EP, Monro TM, Thompson JG, et al. Boronate probes for the detection of hydrogen peroxide release from human spermatozoa. *Free Radic Biol Med* 2015;**81**:69–76.
79. Wen Y, Liu K, Yang H, Li Y, Lan H, Liu Y, et al. A Highly Sensitive Ratiometric Fluorescent Probe for the Detection of Cytoplasmic and Nuclear Hydrogen Peroxide. *Anal Chem* 2014;**86**(19):9970–6.

80. Srikun D, Miller EW, Domaille DW, Chang CJ. An ICT-Based Approach to Ratiometric Fluorescence Imaging of Hydrogen Peroxide Produced in Living Cells. *J Am Chem Soc* 2008;**130**(14):4596–7.
81. Chung C, Srikun D, Lim CS, Chang CJ, Cho BR. A two-photon fluorescent probe for ratiometric imaging of hydrogen peroxide in live tissue. *Chem Commun* 2011;**47**(34):9618–20.
82. Kumar M, Kumar N, Bhalla V, Sharma PR, Qurishi Y. A charge transfer assisted fluorescent probe for selective detection of hydrogen peroxide among different reactive oxygen species. *Chem Commun* 2012;**48**(39):4719–21.
83. Masanta G, Heo CH, Lim CS, Bae SK, Cho BR, Kim HM. A mitochondria-localized two-photon fluorescent probe for ratiometric imaging of hydrogen peroxide in live tissue. *Chem Commun* 2012;**48**(29):3518–20.
84. Hitomi Y, Takeyasu T, Funabiki T, Kodera M. Detection of Enzymatically Generated Hydrogen Peroxide by Metal-Based Fluorescent Probe. *Anal Chem* 2011;**83**(24):9213–16.
85. Maeda H, Fukuyasu Y, Yoshida S, Fukuda M, Saeki K, Matsuno H, et al. Fluorescent probes for hydrogen peroxide based on a non-oxidative mechanism. *Angew Chem Int Ed* 2004;**43**(18):2389–91.
86. Liao Y-X, Li K, Wu M-Y, Wu T, Yu X-Q. A selenium-contained aggregation-induced "turn-on" fluorescent probe for hydrogen peroxide. *Org Biomol Chem* 2014;**12**(19):3004–8.
87. Yu F, Li P, Song P, Wang B, Zhao J, Han K. Facilitative functionalization of cyanine dye by an on-off-on fluorescent switch for imaging of H_2O_2 oxidative stress and thiols reducing repair in cells and tissues. *Chem Commun (Cambridge, U. K)* 2012;**48**(41):4980–2.
88. Abo M, Urano Y, Hanaoka K, Terai T, Komatsu T, Nagano T. Development of a Highly Sensitive Fluorescence Probe for Hydrogen Peroxide. *J Am Chem Soc* 2011;**133**(27):10629–37.
89. Meyer AJ, Dick TP. Fluorescent Protein-Based Redox Probes. *Antioxid Redox Signaling* 2010;**13**(5):621–50.
90. Belousov VV, Fradkov AF, Lukyanov KA, Staroverov DB, Shakhbazov KS, Terskikh AV, et al. Genetically encoded fluorescent indicator for intracellular hydrogen peroxide. *Nat Methods* 2006;**3**(4):281–6.
91. Markvicheva KN, Bilan DS, Mishina NM, Gorokhovatsky AY, Vinokurov LM, Lukyanov S, et al. A genetically encoded sensor for H_2O_2 with expanded dynamic range. *Bioorg Med Chem* 2011;**19**(3):1079–84.
92. Bilan DS, Pase L, Joosen L, Gorokhovatsky AY, Ermakova YG, Gadella TWJ, et al. HyPer-3: A Genetically Encoded H_2O_2 Probe with Improved Performance for Ratiometric and Fluorescence Lifetime Imaging. *ACS Chem Biol* 2013;**8**(3):535–42.
93. Gutscher M, Sobotta MC, Wabnitz GH, Ballikaya S, Meyer AJ, Samstag Y, et al. Proximity-based Protein Thiol Oxidation by H_2O_2-scavenging Peroxidases. *J Biol Chem* 2009;**284**(46):31532–40.
94. Keppler A, Kindermann M, Gendreizig S, Pick H, Vogel H, Johnsson K. Labeling of fusion proteins of O6-alkylguanine-DNA alkyltransferase with small molecules in vivo and in vitro. *Methods* 2004;**32**(4):437–44.
95. Srikun D, Albers AE, Nam CI, Iavarone AT, Chang CJ. Organelle-targetable fluorescent probes for imaging hydrogen peroxide in living cells via SNAP-Tag protein labeling. *J Am Chem Soc* 2010;**132**(12):4455–65.
96. Bauer G. HOCl-dependent singlet oxygen and hydroxyl radical generation modulate and induce apoptosis of malignant cells. *Anticancer Res* 2013;**33**(9):3589–602.
97. Dupre-Crochet S, Erard M, Nubetae O. ROS production in phagocytes: why, when, and where? *J Leukoc Biol* 2013;**94**(4):657–70.
98. Prokopowicz ZM, Arce F, Biedron R, Chiang CL, Ciszek M, Katz DR, et al. Hypochlorous acid: a natural adjuvant that facilitates antigen processing, cross-priming, and the induction of adaptive immunity. *J Immunol* 2010;**184**(2):824–35.
99. Winterbourn CC, Kettle AJ. Redox reactions and microbial killing in the neutrophil phagosome. *Antioxid Redox Signaling* 2013;**18**(6):642–60.
100. Winterbourn CC. Biological reactivity and biomarkers of the neutrophil oxidant, hypochlorous acid. *Toxicology* 2002;**181–182**:223–7.

101. Chen X, Lee K-A, Ha E-M, Lee KM, Seo YY, Choi HK, et al. A specific and sensitive method for detection of hypochlorous acid for the imaging of microbe-induced HOCl production. *Chem Commun* 2011;**47**(15):4373–5.
102. Davies MJ, Hawkins CL. Hypochlorite-induced oxidation of thiols: formation of thiyl radicals and the role of sulfenyl chlorides as intermediates. *Free Radic Res* 2000;**33**(6):719–29.
103. Zhang Z, Zheng Y, Hang W, Yan X, Zhao Y. Sensitive and selective off–on rhodamine hydrazide fluorescent chemosensor for hypochlorous acid detection and bioimaging. *Talanta* 2011;**85**(1):779–86.
104. Zhou Y, Li J-Y, Chu K-H, Liu K, Yao C, Li J-Y. Fluorescence turn-on detection of hypochlorous acid via HOCl-promoted dihydrofluorescein-ether oxidation and its application in vivo. *Chem Commun* 2012;**48**(39):4677–9.
105. Cheng G, Fan J, Sun W, Sui K, Jin X, Wang J, et al. A highly specific BODIPY-based probe localized in mitochondria for HClO imaging. *Analyst* 2013;**138**(20):6091–6.
106. Liu SR, Vedamalai M, Wu SP. Hypochlorous acid turn-on boron dipyrromethene probe based on oxidation of methyl phenyl sulfide. *Anal Chim Acta* 2013;**800**:71–6.
107. Venkatesan P, Wu S-P. A turn-on fluorescent probe for hypochlorous acid based on the oxidation of diphenyl telluride. *Analyst* 2015;**140**(4):1349–55.
108. Cheng G, Fan J, Sun W, Cao J, Hu C, Peng X. A near-infrared fluorescent probe for selective detection of HClO based on Se-sensitized aggregation of heptamethine cyanine dye. *Chem Commun* 2014;**50**(8):1018–20.
109. Ignarro LJ. Nitric oxide: a unique endogenous signaling molecule in vascular biology (Nobel lecture). *Angew Chem Int Ed* 1999;**38**(13/14):1882–92.
110. Mellion BT, Ignarro LJ, Ohlstein EH, Pontecorvo EG, Hyman AL, Kadowicz PJ. Evidence for the inhibitory role of guanosine 3',5'-monophosphate in ADP-induced human platelet aggregation in the presence of nitric oxide and related vasodilators. *Blood* 1981;**57**(5):946–55.
111. Radomski MW, Palmer RMJ, Moncada S. The anti-aggregating properties of vascular endothelium: interactions between prostacyclin and nitric oxide. *Br J Pharmacol* 1987;**92**(3):639–46.
112. Kubes P, Kanwar S, Niu X-F, Gaboury JP. Nitric oxide synthesis inhibition induces leukocyte adhesion via superoxide and mast cells. *FASEB J* 1993;**7**(13):1293–9.
113. Bosca L, Zeini M, Traves PG, Hortelano S. Nitric oxide and cell viability in inflammatory cells: a role for NO in macrophage function and fate. *Toxicology* 2005;**208**(2):249–58.
114. O'Reilly P, Hickman-Davis JM, McArdle P, Young KR, Matalon S. The role of nitric oxide in lung innate immunity: Modulation by surfactant protein-A. *Mol Cell Biochem* 2002;**234/235**(1&2):39–48.
115. Grisham MB, Jourd'Heuil D, Wink DA, Nitric oxide I. Physiological chemistry of nitric oxide and its metabolites:implications in inflammation. *Am J Physiol* 1999;**276**:G315–21 (2 Pt 1)
116. Clementi E, Vecchio I, Corasaniti MT, Nistico G. Nitric oxide modulates agonist-evoked $Ca2+$ release and influx responses in PC12-64 cells. *Eur J Pharmacol* 1995;**289**(1):113–23.
117. Garthwaite J. Concepts of neural nitric oxide-mediated transmission. *Eur J Neurosci* 2008;**27**(11):2783–802.
118. Moncada S, Bolanos JP. Nitric oxide, cell bioenergetics and neurodegeneration. *J Neurochem* 2006;**97**(6):1676–89.
119. McCleverty JA. Chemistry of Nitric Oxide Relevant to Biology. *Chem Rev* 2004;**104**(2):403–18.
120. Foster MW, Cowan JA. Chemistry of Nitric Oxide with Protein-Bound Iron Sulfur Centers. Insights on Physiological Reactivity. *J Am Chem Soc* 1999;**121**(17):4093–100.
121. Tran CT, Williard PG, Kim E. Nitric Oxide Reactivity of [2Fe-2S] Clusters Leading to H_2S Generation. *J Am Chem Soc* 2014;**136**(34):11874–7.
122. Hughes MN. Chemistry of nitric oxide and related species. *Methods Enzymol* 2008;**436**:3–19 (Globins and Other Nitric Oxide-Reactive Proteins, Part A)
123. Bohle DS, Smith KN. Kinetics and Mechanism of Nucleophilic Addition to Nitric Oxide: Secondary Amine Diazeniumdiolation. *Inorg Chem* 2008;**47**(10):3925–7.

124. Lewis RS, Tannenbaum SR, Deen WM. Kinetics of N-Nitrosation in Oxygenated Nitric Oxide Solutions at Physiological pH: Role of Nitrous Anhydride and Effects of Phosphate and Chloride. *J Am Chem Soc* 1995;**117**(14):3933–9.
125. Uppu RM, Pryor WA. Nitrosation of 1,2-Phenylenediamine by Peroxynitrite/CO2: Evidence for a Free Radical Mechanism. *J Am Chem Soc* 1999;**121**(41):9738–9.
126. Kojima H, Nakatsubo N, Kikuchi K, Kawahara S, Kirino Y, Nagoshi H, et al. Detection and Imaging of Nitric Oxide with Novel Fluorescent Indicators: Diaminofluoresceins. *Anal Chem* 1998;**70**(13):2446–53.
127. Zhang Z-X, Guo X-F, Wang H, Zhang H-S. Capillary Electrophoresis Strategy to Monitor the Released and Remaining Nitric Oxide from the Same Single Cell Using a Specially Designed Water-Soluble Fluorescent Probe. *Anal Chem* 2015;**87**(7):3989–95.
128. Huang C-B, Huang J, Xu L. A highly selective fluorescent probe for fast detection of nitric oxide in aqueous solution. *RSC Adv* 2015;**5**(18):13307–10.
129. McQuade LE, Pluth MD, Lippard SJ. Mechanism of Nitric Oxide Reactivity and Fluorescence Enhancement of the NO-Specific Probe CuFL1. *Inorg Chem* 2010;**49**(17):8025–33.
130. Pluth MD, Chan MR, McQuade LE, Lippard SJ. Seminaphthofluorescein-Based Fluorescent Probes for Imaging Nitric Oxide in Live Cells. *Inorg Chem* 2011;**50**(19):9385–92.
131. Lim MH, Lippard SJ. Metal-Based Turn-On Fluorescent Probes for Sensing Nitric Oxide. *Acc Chem Res* 2007;**40**(1):41–51.
132. Lv X, Wang Y, Zhang S, Liu Y, Zhang J, Guo W. A specific fluorescent probe for NO based on a new NO-binding group. *Chem Commun* 2014;**50**(56):7499–502.
133. Ma S, Fang D-C, Ning B, Li M, He L, Gong B. The rational design of a highly sensitive and selective fluorogenic probe for detecting nitric oxide. *Chem Commun* 2014;**50**(49):6475–8.
134. Zheng H, Shang G-Q, Yang S-Y, Gao X, Xu J-G. Fluorogenic and Chromogenic Rhodamine Spirolactam Based Probe for Nitric Oxide by Spiro Ring Opening Reaction. *Org Lett* 2008;**10**(12):2357–60.
135. Yu H, Zhang X, Xiao Y, Zou W, Wang L, Jin L. Targetable Fluorescent Probe for Monitoring Exogenous and Endogenous NO in Mitochondria of Living Cells. *Anal Chem* 2013;**85**(15):7076–84.
136. Yu H, Xiao Y, Jin LA. Lysosome-Targetable and Two-Photon Fluorescent Probe for Monitoring Endogenous and Exogenous Nitric Oxide in Living Cells. *J Am Chem Soc* 2012;**134**(42):17486–9.
137. Shiue T-W, Chen Y-H, Wu C-M, Singh G, Chen H-Y, Hung C-H, et al. Nitric Oxide Turn-on Fluorescent Probe Based on Deamination of Aromatic Primary Monoamines. *Inorg Chem* 2012;**51**(9):5400–8.
138. Sasaki E, Kojima H, Nishimatsu H, Urano Y, Kikuchi K, Hirata Y, et al. Highly Sensitive Near-Infrared Fluorescent Probes for Nitric Oxide and Their Application to Isolated Organs. *J Am Chem Soc* 2005;**127**(11):3684–5.
139. Soh N, Katayama Y, Maeda M. A fluorescent probe for monitoring nitric oxide production using a novel detection concept. *Analyst* 2001;**126**(5):564–6.
140. Zhang X, Kim WS, Hatcher N, Potgieter K, Moroz LL, Gillette R, et al. Interfering with nitric oxide measurements. 4,5-diaminofluorescein reacts with dehydroascorbic acid and ascorbic acid. *J Biol Chem* 2002;**277**(50):48472–8.
141. Balcerczyk A, Soszynski M, Bartosz G. On the specificity of 4-amino-5-methylamino-2',7'-difluorofluorescein as a probe for nitric oxide. *Free Radic Biol Med* 2005;**39**(3):327–35.
142. Broillet M, Randin O, Chatton J. Photoactivation and calcium sensitivity of the fluorescent NO indicator 4,5-diaminofluorescein (DAF-2): implications for cellular NO imaging. *FEBS Lett* 2001;**491**(3):227–32.
143. Suzuki N, Kojima H, Urano Y, Kikuchi K, Hirata Y, Nagano T. Orthogonality of calcium concentration and ability of 4,5-diaminofluorescein to detect NO. *J Biol Chem* 2002;**277**(1):47–9.
144. Uhlenhut K, Hoegger P. Pitfalls and limitations in using 4,5-diaminofluorescein for evaluating the influence of polyphenols on nitric oxide release from endothelial cells. *Free Radic Biol Med* 2012;**52**(11–12):2266–75.

145. Fukuto JM, Bianco CL, Chavez TA. Nitroxyl (HNO) signaling. *Free Radic Biol Med* 2009;**47**(9):1318–24.
146. Fukuto JM, Bartberger MD, Dutton AS, Paolocci N, Wink DA, Houk KN. The Physiological Chemistry and Biological Activity of Nitroxyl (HNO): The Neglected, Misunderstood, and Enigmatic Nitrogen Oxide. *Chem Res Toxicol* 2005;**18**(5):790–801.
147. Kawai K, Ieda N, Aizawa K, Suzuki T, Miyata N, Nakagawa H, et al. A Reductant-Resistant and Metal-Free Fluorescent Probe for Nitroxyl Applicable to Living Cells. *J Am Chem Soc* 2013;**135**(34):12690–6.
148. Liu P, Jing X, Yu F, Lv C, Chen L. A near-infrared fluorescent probe for the selective detection of HNO in living cells and in vivo. *Analyst* 2015. Ahead of Print
149. Liu C, Wu H, Wang Z, Shao C, Zhu B, Zhang X. A fast-response, highly sensitive and selective fluorescent probe for the ratiometric imaging of nitroxyl in living cells. *Chem Commun* 2014;**50**(45):6013–16.
150. Mao G-J, Zhang X-B, Shi X-L, Liu H-W, Wu Y-X, Zhou L-Y, et al. A highly sensitive and reductant-resistant fluorescent probe for nitroxyl in aqueous solution and serum. *Chem Commun* 2014;**50**(43):5790–2.
151. Miao Z, Reisz JA, Mitroka SM, Pan J, Xian M, King SB. A selective phosphine-based fluorescent probe for nitroxyl in living cells. *Bioorg Med Chem Lett* 2015;**25**(1):16–19.
152. Zheng K, Lin W, Cheng D, Chen H, Liu Y, Liu K. A two-photon fluorescent turn-on probe for nitroxyl (HNO) and its bioimaging application in living tissues. *Chem Commun* 2015;**51**(26):5754–7.
153. Jing X, Yu F, Chen L. Visualization of nitroxyl (HNO) in vivo via a lysosome-targetable near-infrared fluorescent probe. *Chem Commun* 2014;**50**(91):14253–6.
154. Rosenthal J, Lippard SJ. Direct Detection of Nitroxyl in Aqueous Solution Using a Tripodal Copper(II) BODIPY Complex. *J Am Chem Soc* 2010;**132**(16):5536–7.
155. Cline MR, Toscano JP. Detection of nitroxyl (HNO) by a prefluorescent probe. *J Phys Org Chem* 2011;**24**(10):993–8.
156. Beckman JS, Beckman TW, Chen J, Marshall PA, Freeman BA. Apparent hydroxyl radical production by peroxynitrite: implications for endothelial injury from nitric oxide and superoxide. *Proc Natl Acad Sci U S A* 1990;**87**(4):1620–4.
157. Radi R, Beckman JS, Bush KM, Freeman BA. Peroxynitrite oxidation of sulfhydryls. The cytotoxic potential of superoxide and nitric oxide. *J Biol Chem* 1991;**266**(7):4244–50.
158. Radi R, Beckman JS, Bush KM, Freeman BA. Peroxynitrite-induced membrane lipid peroxidation: the cytotoxic potential of superoxide and nitric oxide. *Arch Biochem Biophys* 1991;**288**(2):481–7.
159. Ischiropoulos H, Zhu L, Chen J, Tsai M, Martin JC, Smith CD, et al. Peroxynitrite-mediated tyrosine nitration catalyzed by superoxide dismutase. *Arch Biochem Biophys* 1992;**298**(2):431–7.
160. Reiter CD, Teng R-J, Beckman JS. Superoxide reacts with nitric oxide to nitrate tyrosine at physiological pH via peroxynitrite. *J Biol Chem* 2000;**275**(42):32460–6.
161. Ischiropoulos H, Zhu L, Beckman JS. Peroxynitrite formation from macrophage-derived nitric oxide. *Arch Biochem Biophys* 1992;**298**(2):446–51.
162. Zhu L, Gunn C, Beckman JS. Bactericidal activity of peroxynitrite. *Arch Biochem Biophys* 1992;**298**(2):452–7.
163. Wattanapitayakul SK, Weinstein DM, Holycross BJ, Bauer JA. Endothelial dysfunction and peroxynitrite formation are early events in angiotensin-induced cardiovascular disorders. *FASEB J* 2000;**14**(2):271–8.
164. Drechsel DA, Estevez AG, Barbeito L, Beckman JS. Nitric Oxide-Mediated Oxidative Damage and the Progressive Demise of Motor Neurons in ALS. *Neurotoxic Res* 2012;**22**(4):251–64.
165. Spear N, Estévez AG, Radi R, Beckman JS. Peroxynitrite and Cell Signaling. In: Forman HJ, Cadenas E, editors. *Oxidative Stress and Signal Transduction*. Boston, MA: Springer US; 1997. p. 32–51.
166. Gomes A, Fernandes E, Lima JLFC. Use of fluorescence probes for detection of reactive nitrogen species: A review. *J Fluoresc* 2006;**16**(1):119–39.

167. Peng T, Yang D. HKGreen-3: A Rhodol-Based Fluorescent Probe for Peroxynitrite. *Org Lett* 2010;**12**(21):4932–5.
168. Sun Z-N, Wang H-L, Liu F-Q, Chen Y, Tam PKH, Yang D. BODIPY-Based Fluorescent Probe for Peroxynitrite Detection and Imaging in Living Cells. *Org Lett* 2009;**11**(9):1887–90.
169. Zielonka J, Sikora A, Joseph J, Kalyanaraman B. Peroxynitrite Is the Major Species Formed from Different Flux Ratios of Co-generated Nitric Oxide and Superoxide: Direct Reaction with Boronate-based Fluorescent Probe. *J Biol Chem* 2010;**285**(19):14210–16.
170. Kim J, Park J, Lee H, Choi Y, Kim Y. A boronate-based fluorescent probe for the selective detection of cellular peroxynitrite. *Chem Commun* 2014;**50**(66):9353–6.
171. Yu F, Li P, Li G, Zhao G, Chu T, Han K, et al. Fluorescent Probe Modulated by Selenium for Monitoring Peroxynitrite and Imaging in Living Cells. *J Am Chem Soc* 2011;**133**(29):11030–3.
172. Xu K, Chen H, Tian J, Ding B, Xie Y, Qiang M, et al. A near-infrared reversible fluorescent probe for peroxynitrite and imaging of redox cycles in living cells. *Chem Commun* 2011;**47**(33):9468–70.
173. Koide Y, Kawaguchi M, Urano Y, Hanaoka K, Komatsu T, Abo M, et al. A reversible near-infrared fluorescence probe for reactive oxygen species based on Te-rhodamine. *Chem Commun* 2012;**48**(25):3091–3.
174. Yu F, Li P, Wang B, Han K. Reversible Near-Infrared Fluorescent Probe Introducing Tellurium to Mimetic Glutathione Peroxidase for Monitoring the Redox Cycles between Peroxynitrite and Glutathione in Vivo. *J Am Chem Soc* 2013;**135**(20):7674–80.
175. Wang B, Yu F, Li P, Sun X, Han KA. BODIPY fluorescence probe modulated by selenoxide spirocyclization reaction for peroxynitrite detection and imaging in living cells. *Dyes Pigm* 2013;**96**(2):383–90.
176. Zhang Q, Zhu Z, Zheng Y, Cheng J, Zhang N, Long Y-T, et al. A three-channel fluorescent probe that distinguishes peroxynitrite from hypochlorite. *J Am Chem Soc* 2012;**134**(45):18479–82.
177. Zhang H, Liu J, Sun Y-Q, Huo Y, Li Y, Liu W, et al. A mitochondria-targetable fluorescent probe for peroxynitrite: fast response and high selectivity. *Chem Commun* 2015;**51**(13):2721–4.
178. Lin K-K, Wu S-C, Hsu K-M, Hung C-H, Liaw W-F, Wang Y-M. A N-(2-Aminophenyl)-5-(dimethylamino)-1-naphthalenesulfonic Amide (Ds-DAB) Based Fluorescent Chemosensor for Peroxynitrite. *Org Lett* 2013;**15**(16):4242–5.
179. Chen Z-j, Ren W, Wright QE, Ai H-w. Genetically Encoded Fluorescent Probe for the Selective Detection of Peroxynitrite. *J Am Chem Soc* 2013;**135**(40):14940–3.
180. Krinsky NI. Singlet excited oxygen as a mediator of the antibacterial action of leukocytes. *Science* 1974;**186**(4161):363–5.
181. Rosen H, Klebanoff SJ. Formation of singlet oxygen by the myeloperoxidase-mediated antimicrobial system. *J Biol Chem* 1977;**252**(14):4803–10.
182. Miyamoto S, Martinez GR, Medeiros MHG, Di Mascio P. Singlet molecular oxygen generated by biological hydroperoxides. *Journal of Photochemistry and Photobiology B: Biology* 2014;**139**(0):24–33.
183. Kessel D, Price M. Evaluation of diethyl-3-3'-(9,10-anthracenediyl)bis acrylate as a probe for singlet oxygen formation during photodynamic therapy. *Photochem Photobiol* 2012;**88**(3):717–20.
184. Kim S, Tachikawa T, Fujitsuka M, Majima T. Far-Red Fluorescence Probe for Monitoring Singlet Oxygen during Photodynamic Therapy. *J Am Chem Soc* 2014;**136**(33):11707–15.
185. Tanaka K, Miura T, Umezawa N, Urano Y, Kikuchi K, Higuchi T, et al. Rational Design of Fluorescein-Based Fluorescence Probes. Mechanism-Based Design of a Maximum Fluorescence Probe for Singlet Oxygen. *J Am Chem Soc* 2001;**123**(11):2530–6.
186. Tan M, Song B, Wang G, Yuan J. A new terbium(III) chelate as an efficient singlet oxygen fluorescence probe. *Free Radic Biol Med* 2006;**40**(9):1644–53.
187. Song D, Cho S, Han Y, You Y, Nam W. Ratiometric Fluorescent Probes for Detection of Intracellular Singlet Oxygen. *Org Lett* 2013;**15**(14):3582–5.
188. Xu K, Wang L, Qiang M, Wang L, Li P, Tang B. A selective near-infrared fluorescent probe for singlet oxygen in living cells. *Chem Commun* 2011;**47**(26):7386–8.
189. Niki E. Lipid peroxidation: Physiological levels and dual biological effects. *Free Radic Biol Med* 2009;**47**(5):469–84.

190. Okimoto Y, Watanabe A, Niki E, Yamashita T, Noguchi N. A novel fluorescent probe diphenyl-1-pyrenylphosphine to follow lipid peroxidation in cell membranes. *FEBS Lett* 2000;**474**(2,3):137–40.
191. Okimoto Y, Warabi E, Wada Y, Niki E, Kodama T, Noguchi N. A novel method of following oxidation of low-density lipoprotein using a sensitive fluorescent probe, diphenyl-1-pyrenylphosphine. *Free Radic Biol Med* 2003;**35**(6):576–85.
192. Takahashi M, Shibata M, Niki E. Estimation of lipid peroxidation of live cells using a fluorescent probe, Diphenyl-1-pyrenylphosphine. *Free Radic Biol Med* 2001;**31**(2):164–74.
193. Matot I, Manevich Y, Al-Mehdi A-B, Song C, Fisher AB. Fluorescence imaging of lipid peroxidation in isolated rat lungs during nonhypoxic lung ischemia. *Free Radic Biol Med* 2003;**34**(6):785–90.
194. Shioji K, Oyama Y, Okuma K, Nakagawa H. Synthesis and properties of fluorescence probe for detection of peroxides in mitochondria. *Bioorg Med Chem Lett* 2010;**20**(13):3911–15.
195. Drummen GPC, van Liebergen LCM, Op den Kamp JAF, Post JA. C11-BODIPY581/591, an oxidation-sensitive fluorescent lipid peroxidation probe: (micro)spectroscopic characterization and validation of methodology. *Free Radic Biol Med* 2002;**33**(4):473–90.
196. Itoh N, Cao J, Chen Z-H, Yoshida Y, Niki E. Advantages and limitation of BODIPY as a probe for the evaluation of lipid peroxidation and its inhibition by antioxidants in plasma. *Bioorg Med Chem Lett* 2007;**17**(7):2059–63.
197. Prime TA, Forkink M, Logan A, Finichiu PG, McLachlan J, Pun PBL, et al. A ratiometric fluorescent probe for assessing mitochondrial phospholipid peroxidation within living cells. *Free Radic Biol Med* 2012;**53**(3):544–53.
198. Stein A, Bailey SM. Redox biology of hydrogen sulfide: Implications for physiology, pathophysiology, and pharmacology. *Redox Biol* 2013;**1**(1):32–9.
199. Kolluru GK, Shen X, Bir SC, Kevil CG. Hydrogen sulfide chemical biology: Pathophysiological roles and detection. *Nitric Oxide* 2013;**35**(0):5–20.
200. Lippert AR. Designing reaction-based fluorescent probes for selective hydrogen sulfide detection. *J Inorg Biochem* 2014;**133**:136–42.
201. Yu F, Li P, Song P, Wang B, Zhao J, Han K. An ICT-based strategy to a colorimetric and ratiometric fluorescence probe for hydrogen sulfide in living cells. *Chem Commun* 2012;**48**(23):2852–4.
202. Montoya LA, Pluth MD. Selective turn-on fluorescent probes for imaging hydrogen sulfide in living cells. *Chem Commun* 2012;**48**(39):4767–9.
203. Xuan W, Pan R, Cao Y, Liu K, Wang W. A fluorescent probe capable of detecting H_2S at submicromolar concentrations in cells. *Chem Commun* 2012;**48**(86):10669–71.
204. Peng B, Chen W, Liu C, Rosser EW, Pacheco A, Zhao Y, et al. Fluorescent Probes Based on Nucleophilic Substitution-Cyclization for Hydrogen Sulfide Detection and Bioimaging. *Chem -Eur J* 2014;**20**(4):1010–16.
205. Maity D, Raj A, Samanta PK, Karthigeyan D, Kundu TK, Pati SK, et al. A probe for ratiometric near-infrared fluorescence and colorimetric hydrogen sulfide detection and imaging in live cells. *RSC Adv* 2014;**4**(22):11147–51.
206. Fu L, Tian F-F, Lai L, Liu Y, Harvey PD, Jiang F-L. A ratiometric "two-in-one" fluorescent chemodosimeter for fluoride and hydrogen sulfide. *Sens Actuators, B* 2014;**193**:701–7.
207. Sasakura K, Hanaoka K, Shibuya N, Mikami Y, Kimura Y, Komatsu T, et al. Development of a Highly Selective Fluorescence Probe for Hydrogen Sulfide. *J Am Chem Soc* 2011;**133**(45):18003–5.
208. Tang B, Xing Y, Li P, Zhang N, Yu F, Yang G, et al. Probe Containing a Se-N Bond for Detecting Thiols and Its Application in Living Cells. *J Am Chem Soc* 2007;**129**(38):11666–7.
209. Bouffard J, Kim Y, Swager TM, Weissleder R, Hilderbrand SA. A Highly Selective Fluorescent Probe for Thiol Bioimaging. *Org Lett* 2008;**10**(1):37–40.
210. Shibata A, Furukawa K, Abe H, Tsuneda S, Ito Y. Rhodamine-based fluorogenic probe for imaging biological thiol. *Bioorg Med Chem Lett* 2008;**18**(7):2246–9.

211. Yin J, Kwon Y, Kim D, Lee D, Kim G, Hu Y, et al. Cyanine-Based Fluorescent Probe for Highly Selective Detection of Glutathione in Cell Cultures and Live Mouse Tissues. *J Am Chem Soc* 2014;**136**(14):5351–8.
212. Li X, Qian S, He Q, Yang B, Li J, Hu Y. Design and synthesis of a highly selective fluorescent turn-on probe for thiol bioimaging in living cells. *Org Biomol Chem* 2010;**8**(16):3627–30.
213. Long L, Lin W, Chen B, Gao W, Yuan L. Construction of FRET-based ratiometric fluorescent thiol probe. *Chem Commun* 2011;**47**(3):893–5.
214. Egloff C, Jacques SA, Nothisen M, Weltin D, Calligaro C, Mosser M, et al. Bio-specific and bio-orthogonal chemistries to switch-off the quencher of a FRET-based fluorescent probe: application to living-cell biothiol imaging. *Chem Commun* 2014;**50**(70):10049–51.
215. Nawimanage RR, Prasai B, Hettiarachchi SU, McCarley RL. Rapid, Photoinduced Electron Transfer-Modulated, Turn-on Fluorescent Probe for Detection and Cellular Imaging of Biologically Significant Thiols. *Anal Chem* 2014;**86**(24):12266–71.
216. Zhu B, Zhang X, Li Y, Wang P, Zhang H, Zhuang X. A colorimetric and ratiometric fluorescent probe for thiols and its bioimaging applications. *Chem Commun* 2010;**46**(31):5710–12.
217. Lou Z, Li P, Sun X, Yang S, Wang B, Han K. A fluorescent probe for rapid detection of thiols and imaging of thiols reducing repair and H_2O_2 oxidative stress cycles in living cells. *Chem Commun* 2013;**49**(4):391–3.
218. Lim CS, Masanta G, Kim HJ, Han JH, Kim HM, Cho BR. Ratiometric Detection of Mitochondrial Thiols with a Two-Photon Fluorescent Probe. *J Am Chem Soc* 2011;**133**(29):11132–5.
219. Yuan Y, Kwok RTK, Feng G, Liang J, Geng J, Tang BZ, et al. Rational design of fluorescent light-up probes based on an AIE luminogen for targeted intracellular thiol imaging. *Chem Commun* 2014;**50**(3):295–7.
220. Rong L, Zhang C, Lei Q, Sun H-L, Qin S-Y, Feng J, et al. Long-term thiol monitoring in living cells using bioorthogonal chemistry. *Chem Commun* 2015;**51**(2):388–90.
221. Deng L, Wu W, Guo H, Zhao J, Ji S, Zhang X, et al. Colorimetric and Ratiometric Fluorescent Chemosensor Based on Diketopyrrolopyrrole for Selective Detection of Thiols: An Experimental and Theoretical Study. *J Org Chem* 2011;**76**(22):9294–304.
222. Jung D, Maiti S, Lee JH, Lee JH, Kim JS. Rational design of biotin-disulfide-coumarin conjugates: a cancer targeted thiol probe and bioimaging. *Chem Commun* 2014;**50**(23):3044–7.
223. Kand D, Kalle AM, Varma SJ, Talukdar P. A chromenoquinoline-based fluorescent off-on thiol probe for bioimaging. *Chem Commun* 2012;**48**(21):2722–4.
224. Zhou X, Jin X, Sun G, Li D, Wu X. A cysteine probe with high selectivity and sensitivity promoted by response-assisted electrostatic attraction. *Chem Commun* 2012;**48**(70):8793–5.
225. Song Q-H, Wu Q-Q, Liu C-H, Du X-J, Guo Q-X. A novel fluorescent probe for selective detection of thiols in acidic solutions and labeling of acidic organelles in live cells. *J Mater Chem B* 2013;**1**(4):438–42.
226. Yang Y, Huo F, Yin C, Zheng A, Chao J, Li Y, et al. Thiol-chromene click chemistry: A coumarin-based derivative and its use as regenerable thiol probe and in bioimaging applications. *Biosens Bioelectron* 2013;**47**:300–6.
227. Long L, Zhou L, Wang L, Meng S, Gong A, Du F, et al. A coumarin-based fluorescent probe for biological thiols and its application for living cell imaging. *Org Biomol Chem* 2013;**11**(47):8214–20.
228. Qu L, Yin C, Huo F, Li J, Chao J, Zhang Y. A maleimide-based thiol fluorescent probe and its application for bioimaging. *Sens Actuators, B* 2014;**195**:246–51.
229. Zhang R-R, Zhang J-F, Wang S-Q, Cheng Y-L, Miao J-Y, Zhao B-X. Novel pyrazoline-based fluorescent probe for detecting thiols and its application in cells. *Spectrochim Acta, Part A* 2015;**137**:450–5.
230. Jiang X, Yu Y, Chen J, Zhao M, Chen H, Song X, et al. Quantitative Imaging of Glutathione in Live Cells Using a Reversible Reaction-Based Ratiometric Fluorescent Probe. *ACS Chem Biol* 2015;**10**(3):864–74.
231. Liu J, Sun Y-Q, Lv X, Guo W. Fluorescent detection of biothiols based on a novel cascade reaction. *Anal Methods* 2013;**5**(15):3642–5.

232. Das P, Mandal AK, Reddy GU, Baidya M, Ghosh SK, Das A. Designing a thiol specific fluorescent probe for possible use as a reagent for intracellular detection and estimation in blood serum: kinetic analysis to probe the role of intramolecular hydrogen bonding. *Org Biomol Chem* 2013;**11**(38):6604–14.
233. Son S-H, Kim Y, Heo MB, Lim YT, Lee TS. A fluorescence turn-on probe for the detection of thiol-containing amino acids in aqueous solution and bioimaging in cells. *Tetrahedron* 2014;**70**(11):2034–9.
234. Wang H, Zhou G, Chen X. An iminofluorescein-Cu^{2+} ensemble probe for selective detection of thiols. *Sens Actuators, B* 2013;**176**:698–703.
235. Shi YG, Yao JH, Duan YL, Mi QL, Chen JH, Xu QQ, et al. 1,8-Naphthalimide-Cu(II) ensemble based turn-on fluorescent probe for the detection of thiols in organic aqueous media. *Bioorg Med Chem Lett* 2013;**23**(9):2538–42.
236. Yuan L, Lin W, Xie Y, Chen B, Zhu S. Single Fluorescent Probe Responds to H_2O_2, NO, and H_2O_2/NO with Three Different Sets of Fluorescence Signals. *J Am Chem Soc* 2012;**134**(2):1305–15.
237. Zhang X, Bloch S, Akers W, Achilefu S. Near-infrared Molecular Probes for In Vivo Imaging. *Current Protocols in Cytometry* 2012.
238. Wrona M, Patel KB, Wardman P. The roles of thiol-derived radicals in the use of 2',7'-dichlorodihydrofluorescein as a probe for oxidative stress. *Free Radic Biol Med* 2008;**44**(1):56–62.
239. Michalski R, Zielonka J, Gapys E, Marcinek A, Joseph J, Kalyanaraman B. Real-time Measurements of Amino Acid and Protein Hydroperoxides Using Coumarin Boronic Acid. *J Biol Chem* 2014;**289**(32):22536–53.
240. Newton GL, Milligan JR. Fluorescence detection of hydroxyl radicals. *Radiat Phys Chem* 2006;**75**(4):473–8.

CHAPTER 5

EPR Spin Trapping

5.1 INTRODUCTION

There has never been a more desirable way to detect radicals than by direct detection, and this can be achieved through the use of electron paramagnetic resonance (EPR) spectroscopy. Although it is possible to directly detect free radicals alone, the single most important limitation for detecting them is their extremely short half-life in aqueous solution. For example, although $O_2^{\bullet -}$ is a weak base (pK_a of its conjugate acid = 4.69), it can abstract protons from water, according to the Eq. (5.1), with an equilibrium constant of $K_{eq} = 0.9 \times 10^9$, thus favoring its almost instantaneous decay to the diamagnetic products.[1]

$$2O_2^{\bullet -} + H_2O \rightleftharpoons HO_2^- + O_2 + HO^- \quad (5.1)$$

Direct detection of $HO^\bullet$ itself in solution is even more difficult due to its high bimolecular reaction with $k_2 = 0.5 \times 10^{10}\ M^{-1}\,s^{-1}$, according to Eq. (5.2).[2]

$$2HO^\bullet \rightarrow H_2O_2 \quad (5.2)$$

The favorability of radicals to decompose in aqueous solution translates to short lifetimes; in the presence of biological milieu, their lifetimes are expected to be even shorter.

In the case of $O_2^{\bullet -}$, its direct detection by EPR spectroscopy was achieved in glassy protic and aprotic solvents through rapid freezing techniques at temperatures in the range of 4–280K.[3–6] Direct $O_2^{\bullet -}$ detection was also accomplished in the presence of colloidal TiO_2 at room temperature[7] as well in the presence of Ti^{3+} and ionomeric polymers in the temperature range of 220–320K.[8]

Fig. 5.1 shows the EPR spectra of $O_2^{\bullet -}$ from the electrochemical reduction of oxygen in butyronitrile at various temperatures.[3] While it is possible to detect free radicals alone in solution, it requires special solvents and stabilizing agents such as metal ions, so this means limiting the application of radical detection in biological systems at ambient conditions. Since cellular organization, cell viability, energy utilization, chemical and biomechanical activities, enzymatic function, metabolic regulation, and cell structural integrity all depend on temperature and microenvironment conditions, it is desirable to detect radicals at the site of their formation—i.e., from their native source and in normal conditions. Direct radical detection in biology faces difficult technical challenges

Reactive Species Detection in Biology
DOI: http://dx.doi.org/10.1016/B978-0-12-420017-3.00004-9

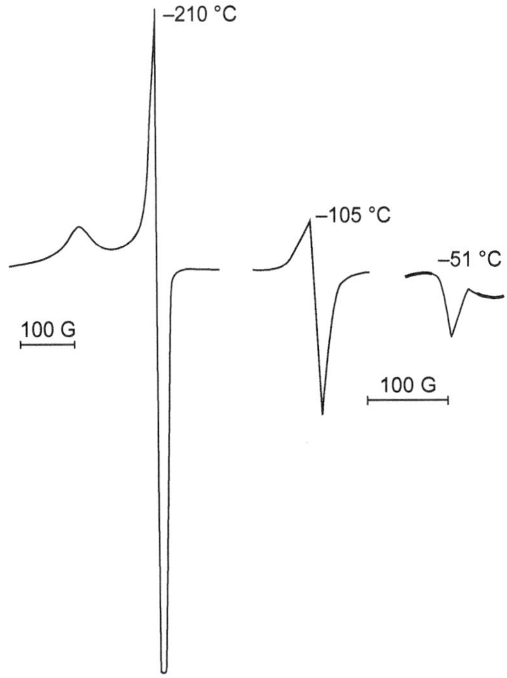

Figure 5.1 Anisotropic EPR spectra of $O_2^{\bullet-}$ in the liquid phase from electrochemical reduction of molecular oxygen with $g_{\parallel} = 2.072$ and $g_{\perp} = 2.009$.[3] *Reprinted with permission from Bagchi et al. Characterization of the ESR spectrum of the superoxide anion in the liquid phase. J Am Chem Soc 1989;**111**:8270–1. Copyright © 1989 American Chemical Society.*

since radicals are mostly produced at ambient conditions; freezing the cells or tissue will inhibit radical formation. Although there is a basal radical generation in cells in the absence of external stimuli, their concentrations are too low to be detected by EPR spectroscopy, which brings us to another technical aspect in radical detection in biological systems because most radicals are produced from the oxidative burst generated from the stimulation of enzymes or cells by external stimulants. Radicals are generated almost instantaneously with the introduction of exogenous agents, so timing is critical when it is desirable to detect radicals in situ immediately after stimulation. When radical initiators are added, users only have a few seconds to place the sample in the EPR cavity, and a few more precious seconds are lost during the initial acquisitions.

The mode by which radical initiators are added into the solution has been a challenge, and stopped-flow EPR techniques have been developed at time resolutions of seconds to minutes,[9] but those with millisecond time resolution require a much more specialized loop-gap resonator technology[10] and were found to be useful for investigating the kinetics of radical production in various biological systems. Although stopped-flow EPR seems useful for the direct detection of transient radicals, their

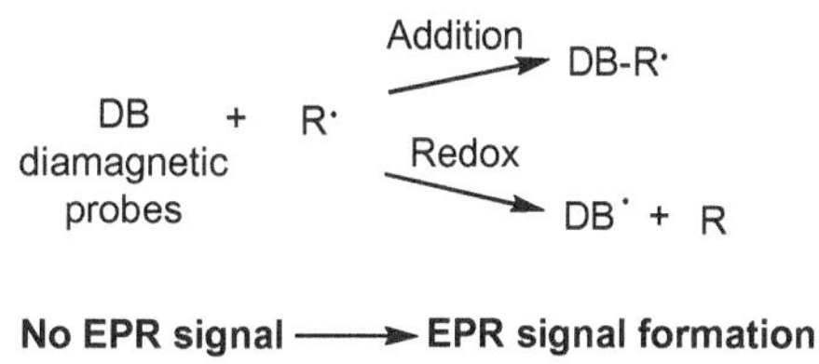

Scheme 5.1 Routes of diamagnetic probe to paramagnetic probe formation.

time resolution is still not short enough to detect short-lived radicals such as $O_2^{\bullet-}$ and $HO^{\bullet}$ in solution due to their very short half-lives, which range from a nanosecond to a microsecond. Hence, the use of diamagnetic or paramagnetic probes to "stabilize" or "probe" these radicals has become essential for the detection of short-lived radicals in solution. Since probes are reagents used to detect radicals, technically one could not claim such a technique to be direct radical detection anymore since the reaction between the probe and the radical is the one that is being monitored by EPR and not the radicals directly themselves.

Probes could be broadly defined as reagents that are employed to qualitatively and quantitatively measure radical species. Probes could be categorized into two main classes: (1) diamagnetic and (2) paramagnetic. Two commonly used diamagnetic probes are nitrone and hydroxylamine, and both impart EPR signals by spin formation after addition or redox reaction with the radical species, respectively, as shown in Scheme 5.1:

The development of small molecule probes, both diamagnetic and paramagnetic, paved the way for a more convenient measurement of free radical formation under normal conditions. The most commonly used diamagnetic probe to date are the nitrone based (commonly referred to as *spin traps*) and the hydroxylamines, while the paramagnetic ones are stable radicals such as nitroxides and trityl radicals. Due to the ability of the EPR spectrum of the probes to be formed, quenched, or transformed in the presence of radicals, the EPR technique is a powerful tool that can provide key information on the following and even more: (1) concentration of reactive species formed, (2) nature of the radical formed, (3) kinetics of radical formation, and (4) oxygen concentration. Therefore, the biomedical values of spin trapping include understanding the origins of radical production, the toxicology of xenobiotics, and the antioxidant activities of some compounds as well as improving the accuracy of radical identification from the cellular level to whole animals using EPR spectroscopy.

5.2 ELECTRON PARAMAGNETIC RESONANCE SPECTROSCOPY

Electron paramagnetic resonance (EPR) or electron spin resonance (ESR) spectroscopy is a magnetic resonance technique commonly used in the detection of paramagnetic species.[11] This involves absorption of microwave radiation by the unpaired electron in the presence of an external magnetic field, resulting in the transition between two spin states. The energy absorbed corresponds to the energy of splitting between the two spin states and is directly proportional to the applied magnetic field strength commonly called the *Zeeman effect* and is described by the fundamental Eq. (5.3):

$$\Delta E = hv = g\Delta H \tag{5.3}$$

where ΔE corresponds to the energy of absorbed microwave energy, h is Planck's constant, v is the frequency of applied electromagnetic radiation, g is a factor equal to 2.0023 for a free electron, and H is the applied field strength. Any species bearing an unpaired electron such as a free radical, transition metals, or biradicals with triplet ground states are typically detectable by EPR using conventional x-band spectrometer at ~9.8 GHz microwave frequency, which is generated by a solid-state Gunn diode. The sample is placed in the cavity between the poles of two magnets and in the presence of microwave radiation. Typically, microwave absorption by the sample is measured by increasing the magnetic-field strength at constant frequency, and the spectrum is presented as a first-derivative spectrum for improved sensitivity. However, detection of radicals in biological systems is often limited by their short half-lives and, therefore EPR techniques that use spin traps and redox probes that yield (or quench) persistent or stable radicals are desirable.

5.3 CHEMISTRY OF SPIN TRAPPING

Spin trapping is an analytical technique employed to detect and identify transient free radicals, although this is not limited to radical species because nitrones are susceptible to nucleophilic addition reactions as well, which will be discussed in the succeeding section. Nitrone is a functionality that is chemically similar in reactivity to carbonyl compounds such as ketones and aldehydes. The carbon doubly bonded to the nitrogen called nitronyl-carbon has significant positivity that it is susceptible to nucleophilic addition reaction, which is analogous to the carbonyl-carbons.

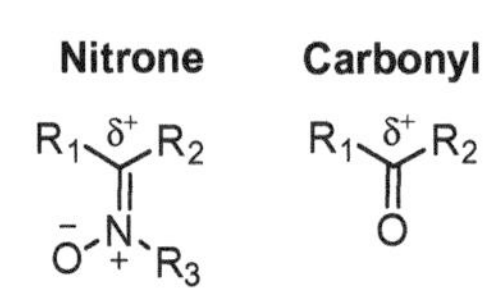

$R_1 = R_2 =$ alkyl	Ketonitrone	Ketone
$R_1 =$ alkyl; $R_2 =$ H	Aldonitrone	Aldehyde

But what makes nitrones uniquely different from carbonyl compounds is their ability to undergo addition reaction with radical species to form a more persistent

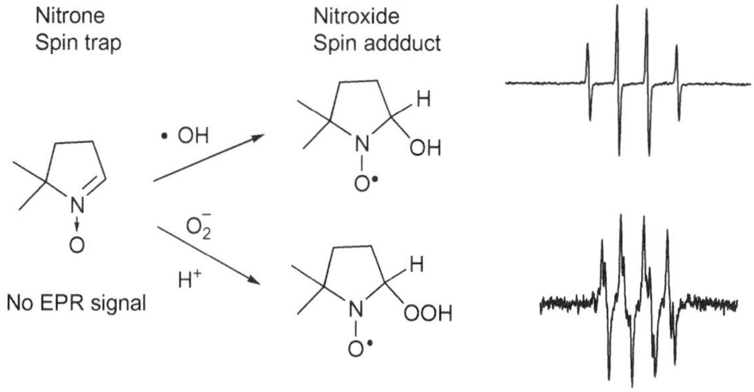

Figure 5.2 EPR signals from spin or radical adduct formation.

nitroxide radical (Eq. 5.4). Since the radical species being trapped contain an unpaired electron with an intrinsic spin, nitrones are usually referred to as *spin trap* and the nitroxide as a *spin adduct.*

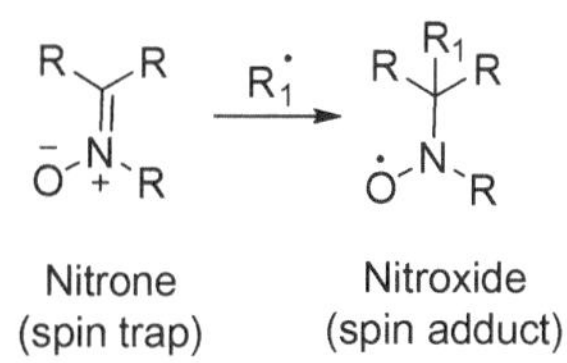

(5.4)

Spin trapping is one of the very few analytical techniques that can identify the nature or structure of a radical intermediate. The EPR signal that is produced by the nitroxide spin adduct provides a wealth of information due to their ability to not only *quantify* the amount of radicals produced but also to identify radicals rather than just detect them, which is one unique feature of spin trapping over the use of other techniques. The most attractive feature of the spin-trapping technique is its ability to discern various radicals formed from each other based on their spectral profiles (e.g., hyperfine splitting constants, line width, or symmetry) of their spin adduct. For example, in Fig. 5.2, two different EPR spectral profiles can be seen from the addition of $O_2^{\bullet -}$ or $HO^{\bullet}$ to the spin trap, 5,5-dimethylpyrroline-*N*-oxide (DMPO). Therefore, with spin trapping, one could differentiate O-centered radicals such as $O_2^{\bullet -}$, $HO^{\bullet}$, and $RO^{\bullet}$ as well as C, N, and S radicals from each other and is typically referred to as *spectral fingerprinting*, which is similar in concept to ^{1}H—nuclear magnetic resonance and infrared spectroscopic techniques.

5.4 CLASSIFICATION OF SPIN TRAPS

Two major classes of spin traps are commonly employed in spin trapping: (1) cyclic and (2) linear nitrones as shown in Fig. 5.3. The most common cyclic nitrone is the 5,5-

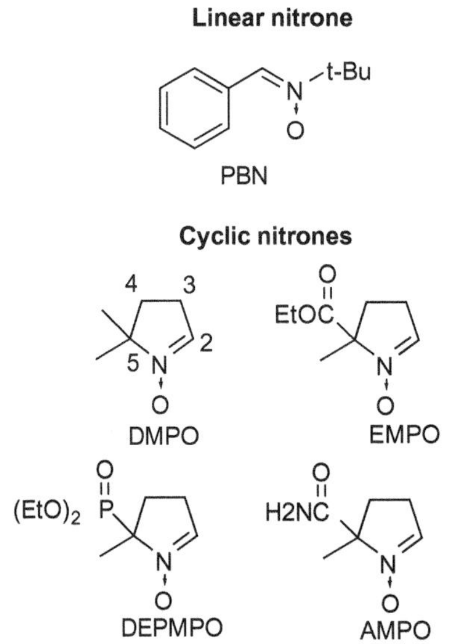

Figure 5.3 Common types of nitrone spin traps.

dimethylpyrroline-*N*-oxide (DMPO), while for the linear one it is *N*-tert-butyl-α-phenyl nitrone (PBN). A number of cyclic spin traps have been synthesized over the past years (e.g., imidazole,[12,13] isoindole,[14] indole-one[15] nitrones), but they do not trap $O_2^{\bullet -}$. In the case of trifluoromethyl-pyrroline,[16] isoquinoline, and benzazepine[17] nitrones, the resulting $O_2^{\bullet -}$ adducts have short half-lives. Moreover, bicyclic nitrones exhibited a very complex spectrum.[18,19] Most PBN-type nitrones are mostly limited by their inability to trap $O_2^{\bullet -}$, but even if they do the adducts are too unstable to be detectable by EPR.[17] Although the possibility for designing new nitrones seems endless, hurdles are typically encountered in the application of new nitrones in biological systems.

To date, only DMPO-type nitrones exhibit the most promising spin-trapping properties compared to any nitrone-based spin traps. This is due to their low cytoxicity, faster reactivity to $O_2^{\bullet -}$ compared to PBN-type nitrones,[20] longer half-lives of the respective $O_2^{\bullet -}$ adducts, and their ability to give distinguishable EPR spectrum on reaction with radicals, specifically those of O-centered radicals due to the conformational rigidity of the pyrroline-ring system, which is perhaps the most important feature of such class of compounds. To improve DMPO spin-trapping properties, its derivatization at the C-5 position with alkoxyphosphoryl, alkoxycarbonyl, or carbamoyl (Fig. 5.3) as exemplified by 5-(diethoxyphosphoryl)-5-methyl-1-pyrroline *N*-oxide (DEPMPO),[21] 5-ethoxycarbonyl-5-methyl-1-pyrroline-*N*-oxide (EMPO),[22] and 5-carbamoyl-5-methyl-1-pyrroline-*N*-oxide (AMPO),[23] respectively, gave longer half-lives for their respective $O_2^{\bullet -}$ adducts or improved reactivities to $O_2^{\bullet -}$ or both. However, 5-substituted nitrones still suffer certain limitations as will be discussed in the succeeding section.

5.5 KINETICS AND THERMODYNAMICS OF SPIN TRAPPING AND pH EFFECT

Two important kinetic properties have to be considered in employing spin traps for radical detection—i.e., the kinetics of spin trapping and adduct decay. While nitric oxide (NO) is unreactive toward DMPO, the reported experimental apparent rate constant for the reaction of DMPO with $O_2^{\bullet-}$ is $1-170\ M^{-1}s^{-1}$,[24,25] $600\ M^{-1}s^{-1}$ with $HO_2^{\bullet}$,[26] $1.1\times10^6\ M^{-1}s^{-1}$ with $CH_3{}^{\bullet}CHOH$,[27] $1.1\times10^6\ M^{-1}s^{-1}$ with glutathiyl radical $GS^{\bullet}$,[28] and $1.93\times10^9\ M^{-1}s^{-1}$ with $HO^{\bullet}$.[29] This shows that the order of increasing rate of addition of radicals to DMPO is $NO < O_2^{\bullet-} < HO_2^{\bullet} < CH_3{}^{\bullet}CHOH < GS^{\bullet} < HO^{\bullet}$. These kinetics of radical addition to DMPO correlate well with the reduction potentials of the respective radicals showing the same order of increasing oxidizing capability (reported as $E^{o\prime}$ values, V): $NO < (-0.80) < HS^{\bullet}$ or $GS^{\bullet}$ (0.92)[31] $< O_2^{\bullet-}$ (0.94) $< HO_2^{\bullet}$ (1.06) $< {}^{\bullet}CH_3$ (1.90) $< HO^{\bullet}$ (2.31),[30,31] with the exception of the thiyl radicals. However, the calculated free energies of reaction (ΔG_{rxn}, kcal/mol) in aqueous phase of various radicals to DMPO at the PCM/B3LYP/6-31 + G(d,p)//B3LYP/6-31G(d) level gave the following order of increasing favorability: $NO < (15.1) < O_2^{\bullet-}$ (3.4) $< HO_2^{\bullet}$ (−12.8) $< HS^{\bullet}$ (−15.5) $< {}^{\bullet}CH_3$ (−30.2) $< HO^{\bullet}$ (−42.1).[32] The calculated average free energies of reaction (ΔG_{rxn}, kcal/mol) from 10 different spin traps with various radicals according to increasing favorability is $NO < (14.6) < O_2^{\bullet-}$ (−7.5) $< HO_2^{\bullet}$ (−13.9) $< HS^{\bullet}$ (−16.6) $< {}^{\bullet}CH_3$ (−32.2) $< HO^{\bullet}$ (−43.7).[32] Regardless of the spin trap being used, thiyl, carbon-centered radical and hydroxyl radicals always exhibit the highest favorability and rates of reaction to nitrones and therefore would be the easiest to detect in solution using the minimal concentrations typically used for spin trapping (~0.1−1 mM). However, with $O_2^{\bullet-}$ and $HO_2^{\bullet}$—whose reactivities are orders of magnitude less favorable compared to thiyl—carbon-centered radical and hydroxyl radicals would require higher concentrations of the spin traps in the range of 10−100 mM.

Based on the kinetic and thermodynamic data of spin trapping, the detection of $O_2^{\bullet-}$ or $HO_2^{\bullet}$ or both are the two most challenging due to their low reactivities compared to other radicals. However, in spin trapping, it is the $HO_2^{\bullet}$ adduct that is detected, even if the $O_2^{\bullet-}$ was initially trapped. There are three conditions by which $HO_2^{\bullet}$ adduct could be formed, depending on the pH of the solution: (1) at pH above the pK_a of the conjugate acid (i.e., $HO_2^{\bullet}$) of $O_2^{\bullet-}$; (2) at pH below the pK_a of the nitrone−H^+; and (3) at pH below the pK_a of $HO_2^{\bullet}$. The pK_a of $HO_2^{\bullet}$ was experimentally determined to be 4.8[33] (or 4.4)[34] as shown in Eq. (5.5)

$$HO_2^{\bullet} \overset{pK_a = 4.8}{\rightleftharpoons} O_2^{\bullet-} + H^+ \quad (5.5)$$

And therefore at pH > 4.8, the spin trapping of $O_2^{\bullet -}$ predominates, according to Eq. (5.6), which shows the calculated free energies:

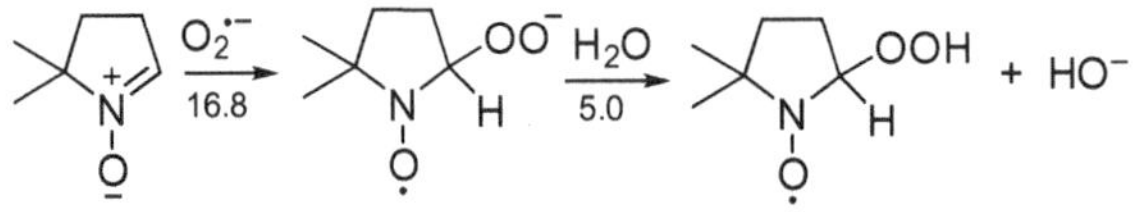

(5.6)

The high positive $\Delta G_{rxn,aq}$ of 16.8 kcal/mol for the addition of $O_2^{\bullet -}$ to DMPO is reflected in its low rate of reaction with $k = 2\ M^{-1}\,s^{-1}$.[35]

A second scenario is spin trapping at pH below the pK_a of the nitrone–H^+. Because the pK_a of the conjugate acid of DMPO was determined to be 6.0,[36] the protonation of DMPO would be more favored than the protonation of $O_2^{\bullet -}$ at slightly acidic pH, according to Eq. (5.7):

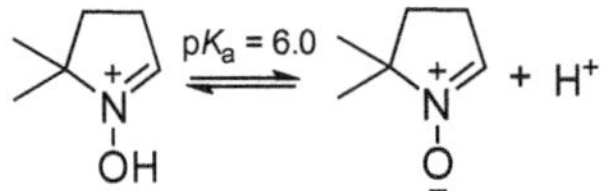

(5.7)

This could lead to higher favorability of $O_2^{\bullet -}$ addition to DMPO–H^+ due to a higher

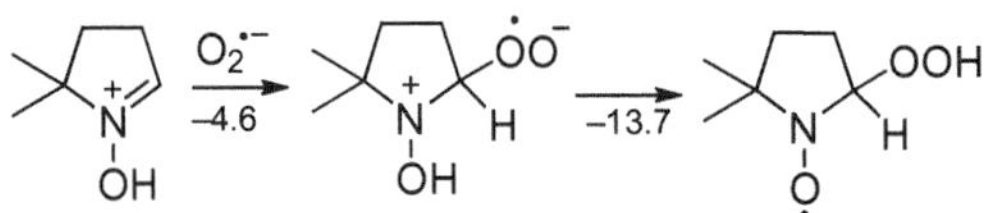

(5.8)

partial positive character on the nitronyl-C, making it more susceptible to nucleophilic addition reactions, similar to those observed for the enhanced nucleophilic addition reaction of carbonyl compounds under acidic condition. This is evidenced by the exoergic free energy of reaction of $O_2^{\bullet -}$ to DMPO–H^+ of −4.6 kcal/mol (Eq. 5.8) and subsequent intramolecular proton transfer with free energy of reaction of −13.7 kcal/mol.[36] This translates to a higher experimental rate constant of $k \sim 27\ M^{-1}\,s^{-1}$ for the spin trapping of DMPO at pH 6.2 which was found to be higher compared to that in pH 7.2 with $k \sim 2\ M^{-1}\,s^{-1}$ or pH 9.3 with $k \sim 0.35\ M^{-1}\,s^{-1}$.[26]

Spin trapping below the pK_a of $HO_2^{\bullet}$ of 4.8 offers another possibility of adduct formation via direct addition of $HO_2^{\bullet}$ to the nitrone. Since $HO_2^{\bullet}$ is a stronger oxidizer than $O_2^{\bullet -}$, it is expected that this reaction would impart a higher rate of reaction as well as favorability, according to Eq. (5.9),

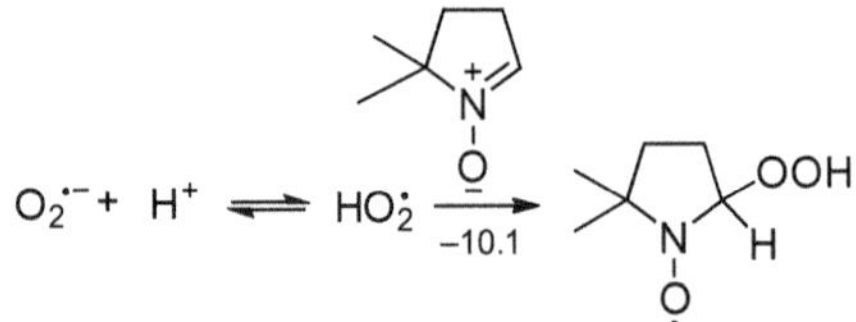

(5.9)

with an exoergic free energy of reaction of $\Delta G_{rxn,aq} = -10.1$ kcal/mol. The experimental rate constants for $O_2^{\bullet-}$ addition to DMPO were reported to be 10 $M^{-1}s^{-1}$ at pH 7 and 6600 $M^{-1}s^{-1}$ at pH 5.[37] In summary, solution pH affects the rate of $O_2^{\bullet-}$ trapping. As shown in Fig. 5.4, the pH dependence of apparent rate constant of spin trapping of $O_2^{\bullet-}$ by DMPO follows an inverse correlation in which the spin trapping is faster in acidic conditions compared to basic solutions.[37]

Aside from the solution pH, the rate of $O_2^{\bullet-}$ trapping is also affected by several factors, depending on the type of (1) radical generating system, (2) spin trap, (3) solvent system, and (4) kinetic model used in the calculation. Table 5.1 is a compilation of the various apparent rate constants for the $O_2^{\bullet-}$ trapping by DMPO, DEPMPO, 5-butoxycarbonyl-5-methyl-1-pyrroline-*N*-oxide (BMPO), EMPO, and 5,5-diethoxycarbonyl-1-pyrroline-*N*-oxide (DEPO). For example, using DMPO, rate constants can range from 2 to 170 $M^{-1}s^{-1}$, and no general trend in rate constant can be seen across all the spin-trapping agents used, which indicates that it is very important to use DMPO or DEPMPO as the gold standard for rate measurements when determining rate constants for newly developed spin traps.

What is clear, however, is that DFT studies reveal the direct proportionality of the charge density on the nitronyl-carbon to the favorability and rate of $O_2^{\bullet-}$ reaction to nitrones as shown in Fig. 5.5. Computational study shows that electron-withdrawing substituents such as *N*-monoalkylsubstituted amide or an ester as substituent at the C-5 position of the nitrone gave improved favorability and rate of spin-trapping properties, further demonstrating the nucleophilic nature of $O_2^{\bullet-}$. The synthesis of AMPO provided two important insights into the factors that govern high reactivity of $O_2^{\bullet-}$ to nitrones (Fig. 5.6): (1) increased electrophilicity of C-2 (the site of $O_2^{\bullet-}$ addition)

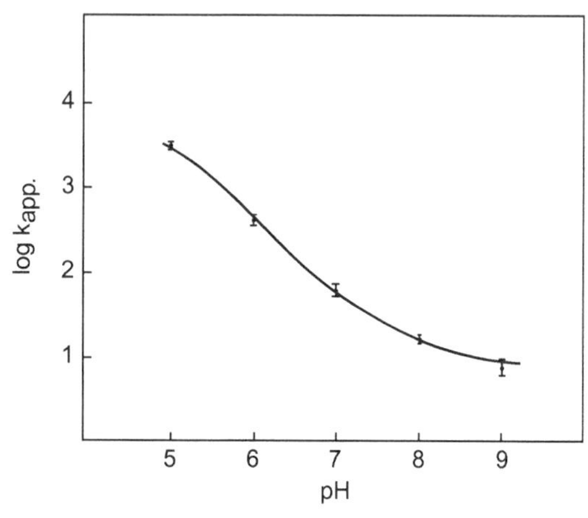

Figure 5.4 Effect of pH on the apparent rate constant for the trapping of $O_2^{\bullet-}/HO_2^{\bullet}$ by DMPO using a light-riboflavin superoxide generating system. The solid line yields rate constants of $k_{HO2} = 6.6 \times 10^3\ M^{-1}s^{-1}$ and $k_{O2} = 10\ M^{-1}s^{-1}$. *Reprinted with permission from Finkelstein et al. Spin trapping. Kinetics of the reaction of superoxide and hydroxyl radicals with nitrones. JACS 1980;**102**:4994–9. Copyright © 1980 American Chemical Society.*

Table 5.1 Survey of the reported apparent rate constants k_{app} ($M^{-1}\,s^{-1}$) for the spin trapping of superoxide radical by DMPO-type nitrones using various methodologies (technique used, competitor, radical generating system and pH)[24]

Method	Rate constants k_{app} ($M^{-1}\,s^{-1}$)				
	DMPO	DEPMPO	BMPO[a]	EMPO	DEPO
I.[38] EPR, superoxide dismutase (SOD), light-riboflavin, pH 7.0	N/A[b]	58	7	N/A	N/A
II.[37] EPR, SOD, light-riboflavin, pH 8.0	15.7	N/A	N/A	N/A	N/A
III.[21,39] EPR, light-riboflavin, pH 7.0	N/A	90c	N/A	N/A	N/A
IV.[40–42] EPR, $O_2^{\bullet-}$ dismutation X/XO, 7.2	2.0	3.95	3.45	10.9	31.1
V.[43] EPR, ferricytochrome-c, X/XO, 7.0	78.5 ± 2.1	N/A	77.0 ± 5.0	74.5 ± 6.4	N/A
VI.[44] EPR, $O_2^{\bullet-}$ dismutation, X/XO, 7.4	2.4	0.53	0.24	N/A	N/A
VII.[45] Pulse radiolysis, γ-irradiation of N_2O	170 ± 40	N/A	<3	N/A	N/A
VIII[24] UV visible (UV vis), stopped-flow, phenol red, KO_2, pH 10.5	1.72 ± 0.01	0.65 ± 0.01	N/A	104.6 ± 4.6	N/A
IX[25] EPR, $O_2^{\bullet-}$ dismutation, X/XO, 7.4	30	N/A	N/A	N/A	N/A
X[46] EPR, nicotinamide adenine dinucleotide phosphate (NADPH)−cytochrome P-450 reductase-paraquat, 7.4	1.2	N/A	N/A	N/A	N/A

[a]5-tert-Butoxylcarbonyl-5 methyl-pyrroline *N*-oxide, also known as BMPO.
[b]Not available.
[c]Apparent rate constant is relative to $k_{2,DMPO} = 60\ M^{-1}\,s^{-1}$.
Reprinted with permission from F.A. Villamena, et al. Reactivity of Superoxide Radical Anion with Cyclic Nitrones: Role of Intramolecular H-Bond and Electrostatic Effects. J Am Chem Soc 129, 2007, 8177.[24]

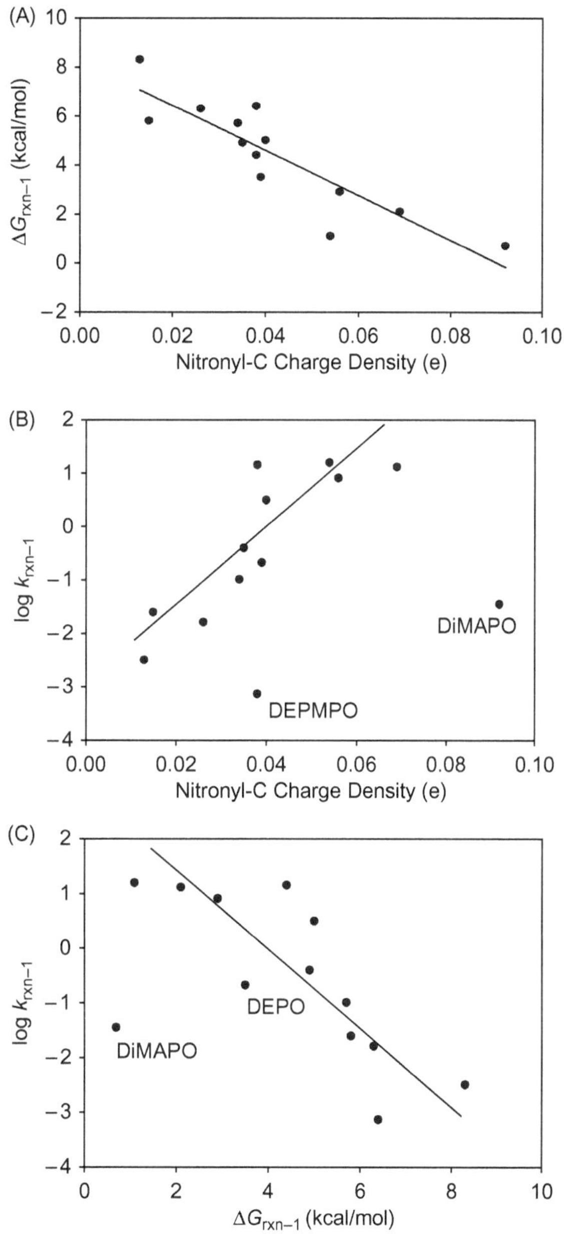

Figure 5.5 Correlation of the nitronyl-C (C-2) charge densities with that of the (A) free energies ($\Delta G_{rxn-1,298K}$, kcal/mol) ($r^2 = 0.78$); (B) rate constants (log k_{rxn--1}) of $O_2^{\bullet -}$ addition to nitrones in the aqueous phase at the PCM(water)/mPW1K/6-31 + G(d,p) level of theory at 298K ($r^2 = 0.76$, excluding the outliers DEPMPO and DIMAPO); and (C) shows the degree of correlation between log k_{rxn-1} and ΔG_{rxn-1} at the same level of theory with $r^2 = 0.73$, excluding the outliers DIMAPO and DEPO). *Reprinted with permission from F.A. Villamena, et al. Reactivity of Superoxide Radical Anion with Cyclic Nitrones: Role of Intramolecular H-Bond and Electrostatic Effects. J Am Chem Soc* ***129****, 2007, 8177.*[24] *Copyright © 2007 American Chemical Society.*

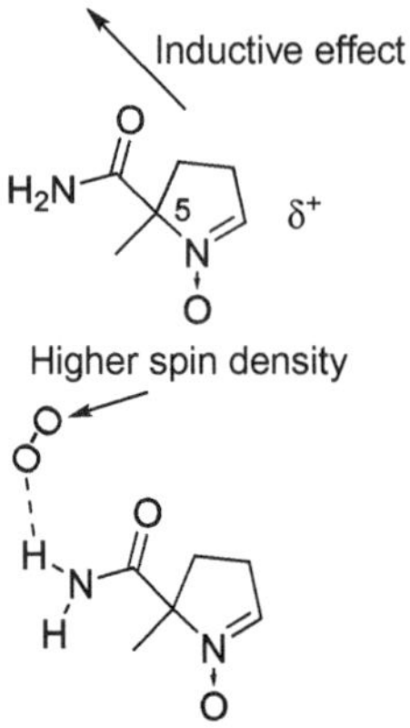

Figure 5.6 Two major effects of amide substituents on $O_2^{\bullet-}$ reactivity to nitrones.

due to the inductive effect by C-5 substitution with primary and secondary amide groups and (2) polarization of the electron distribution on $O_2^{\bullet-}$ by the amide N–H group via α-effect, which causes an increase in electron density of the terminal O and therefore increases $O_2^{\bullet-}$ reactivity to nitrones. This leads to an improved rate constant of 25 $M^{-1}s^{-1}$ compared to the absence of an amide group—i.e., 2 $M^{-1}s^{-1}$ for DMPO using the same spin-trapping conditions.[24]

More innovative spin traps that exploit these two important factors for increased $O_2^{\bullet-}$ reactivity had been synthesized. As shown in Fig. 5.7 e.g., having several hydroxyl groups cyclodextrin can increase H-bonding interaction with $O_2^{\bullet-}$ and was tethered to DMPO via an amide bond for increase positivity on the nitronyl-C, which led to increased rate of $O_2^{\bullet-}$ trapping to 60 $M^{-1}s^{-1}$ in comparison to DMPO of 2 $M^{-1}s^{-1}$.[47]

Since $O_2^{\bullet-}$ carries a negative charge, the use of an anion receptor that carries H-bond donors such as the pyrrole–NH calix[4]pyrrole may exhibit selectivity toward $O_2^{\bullet-}$. It was hypothesized that calix[4]pyrrole conjugation with spin traps can offer improved spin-trapping properties in terms of increased reactivity to $O_2^{\bullet-}$ and longer $O_2^{\bullet-}$ adduct half-life. Computational studies at the PCM/B3LYP/6-31 + G(d,p)//B3LYP/6-31G(d) suggest that a pendant-type linkage between the calix [4]pyrrole and the nitrone would be the most efficient design for spin trapping of $O_2^{\bullet-}$, giving exoergic reaction enthalpies ($\Delta H_{298K,aq}$) and free energies ($\Delta G_{298K,aq}$) of −16.9 and −2.1 kcal/mol, respectively, which are the first exoergic reaction energies ever observed for any nitrones studied thus far.[48] Using UV vis competitive stopped-flow kinetics in DMF/H_2O using KO_2 as $O_2^{\bullet-}$ source, the rate constant for the formation of the CalixMPO–$O_2^{\bullet-}$ adduct was approximated to be $k_2 = 680\ M^{-1}s^{-1}$ compared to AMPO of 130 $M^{-1}s^{-1}$, EMPO 105 $M^{-1}s^{-1}$, DMPO of 2 $M^{-1}s^{-1}$, and DEPMPO of 1 $M^{-1}s^{-1}$ (Fig. 5.8). Although the solubility of CalixMPO is limited in water, this is the highest rate constant observed so far for a

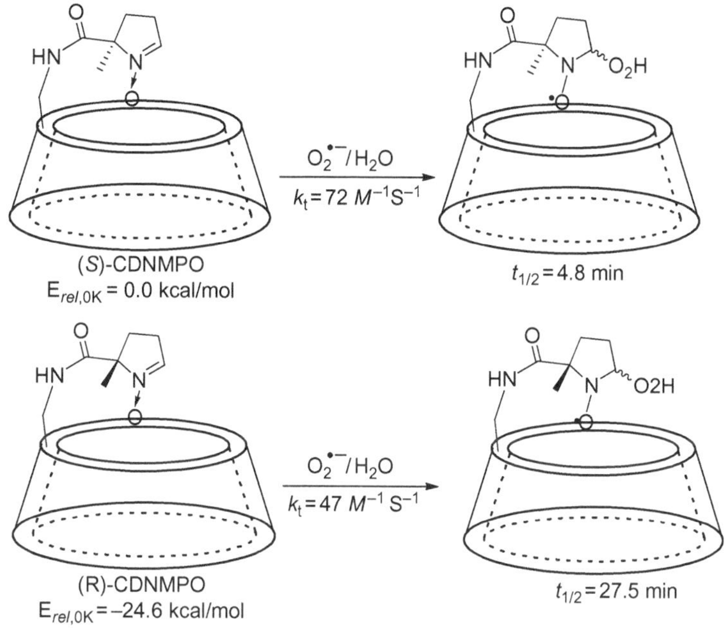

Figure 5.7 Spin-adduct formation from the two stereoisomeric forms of CD-MPO showing their rate constants of formation and half-lives. *Reprinted with permission from Han, Y. et al. Improved Spin Trapping Properties by beta-Cyclodextrin-Cyclic Nitrone Conjugate. J Org Chem* ***73****, 2008,7108. Copyright © 2008 American Chemical Society.*

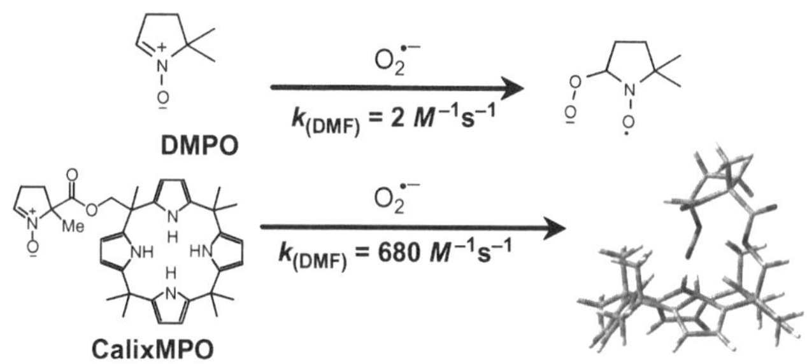

Figure 5.8 Comparison of the rate constants of formation of the superoxide adduct of DMPO versus CalixMPO. *Reprinted with permission from Kim S-U, Liu Y, Nash KM, Zweier JL, Rockenbauer A, Villamena FA. Fast Reactivity of Cyclic Nitrone-Calix[4]pyrrole Conjugate with Superoxide Radical Anion: Theoretical and Experimental Studies. J Am Chem Soc 2010;132:17157. Copyright © 2001 American Chemical Society.*

nitrone at the experimental conditions used in the study. The unusually high reactivity of CalixMPO to $O_2^{\bullet-}$ was rationalized to be due to the synergy between α- and electrostatic effects by the calix[4]pyrrole moiety on the $O_2^{\bullet-}$ and nitrone, respectively.

5.6 KINETICS AND THERMODYNAMICS OF ADDUCT DECAY

The half-life of nitrone radical adducts varies, depending on several factors, including (1) the nature of the radical trapped, (2) the nature of the spin trap, (3) solution pH,

(4) geometric isomerism, and (5) the presence of oxidoreductants. At neutral pH, the DMPO adducts of various radicals gave adduct half-lives ($t_{1/2}$) of 55 min[29] for $HO^{\bullet}$ and >1 h for C-centered ones, making these two spin adducts the easiest to detect in solution. However, the first-order decay half-life of DMPO adduct with $HO_2^{\bullet}$ is only 50 s.[25]

As a function of the spin trap, adducts such as DMPO–OH, EMPO–OH, BMPO–OH, DEPMPO–OH, and 5-diisopropoxyphosphoryl-5-methyl-1-pyrroline *N*-oxide–OH (DIPPMPO–OH) gave $t_{1/2}$ (min) = 55, 127, 37, 132, 158, respectively.[29] These reported experimental half-lives of $HO^{\bullet}$ adducts with various nitrones are relatively longer compared to their respective $HO_2^{\bullet}$ adducts—i.e., for DMPO, EMPO, BMPO, DEPMPO, and DIPPMPO, the $t_{1/2}$ (min) are ~1, 10, 9, 15, and 21, respectively.[21,38,49] The $HO_2^{\bullet}$ half-lives are therefore shorter compared to their $HO^{\bullet}$ adduct counterparts. Even when the $HO_2^{\bullet}$ adducts of 5-substituted spin traps impart longer half-lives compared to DMPO, these $HO_2^{\bullet}$ adducts are inherently less persistent and unstable in aqueous solution.[50] The $HO_2^{\bullet}$ adduct is known to decompose to the more persistent $HO^{\bullet}$ adducts. Several factors have been suggested to affect $HO_2^{\bullet}$ adduct stability such as presence of trace iron ions in solution, pH, and $O_2^{\bullet -}$ itself. A study by Buettner[50] suggests the dependence of $HO_2^{\bullet}$ adduct half-life on iron concentration where at concentrations of 1–100 μM of iron decomposition to the $HO^{\bullet}$ adduct occurs and at a faster rate at higher iron concentrations. The mechanism proposed for the generation of $HO^{\bullet}$ adduct was through the iron-catalyzed Haber-Weiss reaction where $HO^{\bullet}$ is generated from H_2O_2 and $O_2^{\bullet -}$, according to Eqs. (5.10)–(5.12):

$$Fe^{2+} + O_2^{\bullet -} \xrightarrow{H^+} Fe^{3+} + H_2O_2 \tag{5.10}$$

$$Fe^{2+} + H_2O_2 \rightarrow Fe^{3+} + HO^{-} + HO^{\bullet} \tag{5.11}$$

$$Fe^{3+} + O_2^{\bullet -} \rightarrow Fe^{2+} + O_2 \tag{5.12}$$

However, the presence of chelators such as ethylenediaminetetraacetate, diethylene triamine pentaacetic acid (DTPA), and deferoxamine did not prevent the decomposition of $HO_2^{\bullet}$ adducts or the formation of $HO^{\bullet}$ adducts.[50,51] Nevertheless, in spin-trapping experiments, buffers containing small amount of a chelator such as DTPA are recommended due to the metal-mediated nucleophilic addition reaction of water to nitrone, and its subsequent oxidation could produce the artifactual $HO^{\bullet}$ adduct, which will be discussed in detail in the succeeding section.

Solution pH has been shown to affect the $HO_2^{\bullet}$ adduct's half-life. According to Fig. 5.9, the half-life of DMPO-O_2H decreases in basic pH and is longest in acidic pH.[50]

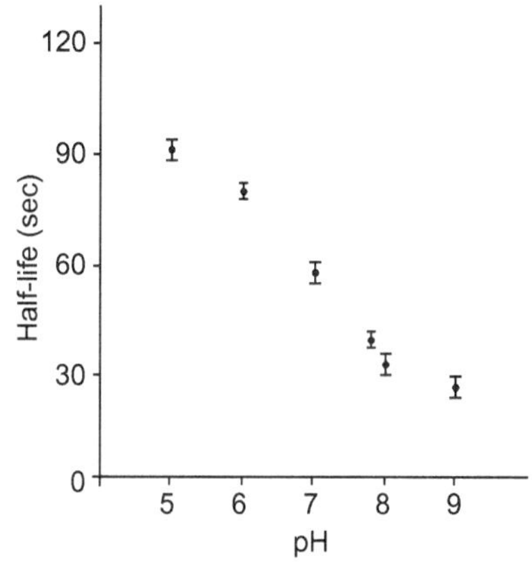

Figure 5.9 The half-life of the $HO_2^{\bullet}$ spin-adduct signal of DMPO in aqueous solution as a function of pH. (Adapted from G.R. Buettner and L.W. Oberley, *Biochem Biophys Res Commun* ***83***, *1978,69).*[50]

A threefold increase in the half-life of the $HO_2^{\bullet}$ adduct from pH 9 ($t_{1/2} = 27$ s) to pH 5 ($t_{1/2} = 91$ s) can be observed, and this trend is consistent with the pH dependence of $HO_2^{\bullet}$ adduct half-life of BMPO as well.[38]

Computational studies show that proton abstraction of the β-H of the $HO_2^{\bullet}$ adduct is highly exoergic with free energy of reaction of −78 kcal/mol (Eq. 5.13). However, EPR evidence for the formation of the keto-nitroxide is lacking, and the mechanism does not explain the short half-life regardless of the pH.[32]

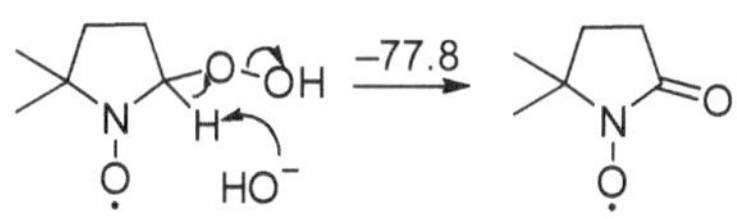

(5.13)

Thermodynamic instability of the adduct was then explained through its unimolecular decomposition via homolytic cleavage of the O−O peroxyl moiety and intramolecular ring opening to produce the nitrosoaldehyde and $HO^{\bullet}$. This reaction was found to be only slightly exoergic by 2.7 kcal/mol as shown in Eq. (5.14).[32] Using a different level of theory, the calculated unimolecular $\Delta G_{rxn,aq}$ of decomposition for DMPO−O_2H was found to be the most exoergic (−13.4 kcal/mol) compared to other $HO_2^{\bullet}$ adduct of various 5-substituted spin traps (−6.8 to 0.6 kcal/mol), but these values do not explain the trends in the relative stability between the adducts of these 5-substituted spin traps since DEPMPO−O_2H gave the most exoergic decomposition free energy of $\Delta G_{rxn,aq} = -4.7$ and −6.8 in this group when it is the most long-lived adduct among the spin traps studied.[52]

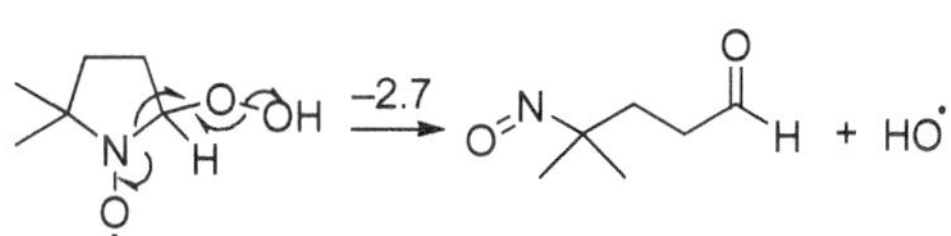

(5.14)

The redox properties of various nitrone$-O_2H$ were then theoretically investigated,[53] and results show that *cis*- and *trans*-DEPMPO$-O_2H$ gave the lowest electron affinity but the highest ionization potential, indicating that DEPMPO$-O_2H$ is hardest to reduce and oxidize compared to the $HO_2^{\bullet}$ adducts of DMPO, EMPO, and AMPO, thus offering an explanation for the long half-life of DEMPO$-O_2H$ in solution. In general, the IPs of $HO_2^{\bullet}$ adducts are higher compared to the more stable nitroxides such as 2,2,6,6-tetramethylpiperidine-*N*-oxyl (TEMPO), PROXYL, DMPO$-$OH, and DMPO$-CH_3$ adducts, which indicates a more facile oxidation for these latter class of nitroxides. The order of preference for nitroxide decomposition is via reduction by $O_2^{\bullet-}$ followed by oxidation by O_2 and then oxidation by $O_2^{\bullet-}$. Fig. 5.10 shows the thermodynamically preferred decomposition pathway for the one-electron reduction of nitrone$-O_2H$ by $O_2^{\bullet-}$ via formation of the hydroxylamine and subsequent ring opening to form the aldehyde and concomitant elimination of HNO_2. For the one-electron oxidation of nitrone$-O_2H$ by $O_2^{\bullet-}$, the preferred decomposition is the formation of the hydroxamic acid. This theoretical study predicts the redox properties of the nitrones and of the $O_2^{\bullet-}$ adducts as well as their subsequent decomposition to form biological active species such as NO, H_2O_2, and hydroxamic acid (Fig. 5.10).

Most of the adduct half-lives were obtained using chemical and enzymatic radical generating systems. While the half-lives of nitrone$-O_2H$ could be enhanced by 5-substitution with electron-withdrawing groups, their half-lives in the presence of biological milieu are often times too short to be observable due to the susceptibility of the aminoxyl (nitroxyl) group of the spin adduct to be degraded by oxidoreductants and even by the radicals themselves. Application of spin trapping in the detection of $O_2^{\bullet-}$has been more successful at the enzymatic level using purified enzymes than using non-leukocyte cells. Stability of various spin adducts using different cyclic nitrones in the presence of Chinese hamster ovary (CHO) and 9L tumor cells was evaluated and, in the absence of cells, showed $t_{1/2}$ ranges of 19$-$90 min, 3$-$22 min, and 3$-$8 min for $SO_3^{\bullet-}$, $HO^{\bullet}$, and $^{\bullet}CH_3$ adducts, respectively. These half-lives were greatly reduced in cell suspension with $t_{1/2}$ ranges of 6$-$30 min, 1.5$-$5 min, and 1.5$-$5 min for $SO_3^{\bullet-}$, $HO^{\bullet}$, and $^{\bullet}CH_3$ adducts, respectively.[54]

The presence of methylated-cyclodextrin could increase the half-life of DMPO$-O_2H$ and DEPMPO$-O_2H$ in solution as previously demonstrated with the

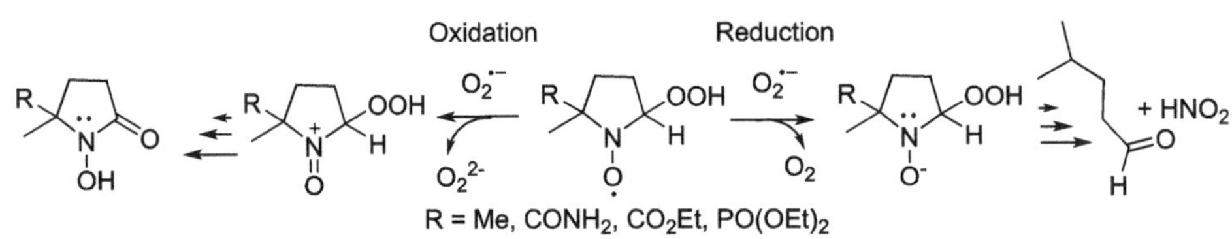

Figure 5.10 Mechanisms of nitrone$-O_2^{\bullet-}$ adduct decomposition after reduction and oxidation.

sevenfold increases in their half-lives.[55] The stabilizing effect of cyclodextrin was extended to esterified-nitrones with enhanced half-life in the presence of ascorbate.[56] And in some in vitro and most in vivo systems, $O_2^{\bullet -}$ production could only be deduced through the formation of HO-adduct and through additional control experiments involving SOD and catalase addition to show the origin of HO-adduct—i.e., whether from $O_2^{\bullet -}$ or H_2O_2, respectively.[57,58] Several methods, however, have been employed to increase nitrone$-O_2H$ half-life in solutions containing biological milieu—e.g., using EMPO and methylated-cylcodextrin or 2-hydroxypropyl-β-cyclodextrin, robust EMPO$-O_2H$ adduct formation was observed compared to without the aid of cyclodextrin.[59] Radical detection from endothelial cells is often times very difficult due to the low levels of radical produced on cell stimulation. Using the same EMPO−cyclodextrin system, the formation of C-centered $HO_2^{\bullet}$ and $HO^{\bullet}$ was observed from oxidatively challenged bovine aortic endothelial cells.[60]

Conjugation of DEPMPO with permethylated-β-cyclodextrin proved to increase the half-life of the HO_2-adduct by 250% (i.e., from 17 min for DEPMPO$-O_2H$ to 40 min for CD−DEPMPO$-O_2H$) in the absence of biological milieu.[61] Also, conjugation of nonmethylated-β-cyclodextrin with DMPO (CDNMPO) via amide bond showed enhanced HO_2-adduct half-life of approximately 30-fold from 1 min for DMPO$-O_2H$ to 27 min for CDNMPO$-O_2H$ in aqueous solutions.[47] Whereas, conjugation of DMPO with calix[4]pyrole (CalixMPO) gave an improved HO_2-adduct half-life in dimethyl sulfoxide (DMSO) of ~25 min compared to DMPO$-O_2H$ of 6 min.[48] In the presence of rat liver microsomes, the of HO_2-adduct of nitrones−cyclodextrin conjugates—in particular, that of CD−DEPMPO and CD−DIPPMPO—showed half-lives of ~80 and 110 min, respectively, similar in stability to that in the absence of microsomes. For the unconjugated-cyclodextrin spin traps, half-lives were significantly shortened in the presence of microsomes. All nitrones, however, showed diminished HO_2-adduct stability in the presence of rat liver cytosol.[62]

5.7 BIOSTABILITY AND CYTOTOXICITY OF SPIN TRAPS

Computational studies show that in general, electron-withdrawing group substituted nitrones at the C-5 position results in higher calculated electron affinity and ionization potential making them more susceptible to reduction but are more difficult to oxidize.[53] Fig. 5.11 shows the resonance structures of oxidized and reduced nitrones.

Cyclic voltammetric studies in aqueous medium reveal cathodic peak potential (V vs SCE) of −1.92 V for DMPO compared to −1.88 V for PBN,[63] indicating a more facile reduction for the latter. DMPO also exhibits high oxidation potential (vs SCE) of 1.63 and 1.47 V for PBN,[63] giving DMPO a wide potential window for application in biological systems. Therefore, in a biological system and given the range

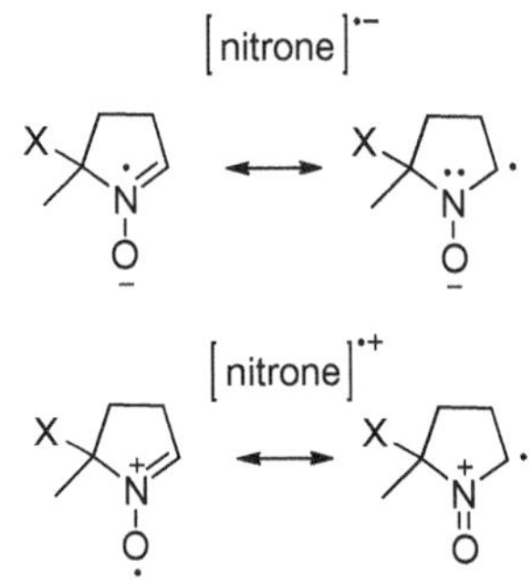

Figure 5.11 Resonance structures for the radical anion (top) and cation (bottom) showing the location of the unpaired electron.

of reduction potential for various oxidoreductants of 2.3 to −1.5 V (vs SCE),[30] it is expected that DMPO would generally be stable and that only those oxidants such as $HO^{\bullet}$ (2.31 V) or $CO_3^{\bullet -}$ (2.1 V)[64] with high reduction potentials above −1.92 V would be capable of oxidizing DMPO.

Nitrones have been shown to be tolerated in aortic endothelial cells with DMPO exhibiting the lowest toxicity with an IC_{10} of 86.8 mM compared to PBN of 5.4 mM after 24 h of incubation with highest toxicity for the nitroso compounds.[65] In a separate study using CHO cells, after 6 h of incubation no significant decrease in cell viability was observed using DMPO, BMPO, CMPO, EMPO, and DEPMPO in up to 50 mM concentrations except for BMPO, which gave significant toxicity at 50 mM and was only nontoxic up to 25 mM.[54] The effect of the spin traps on the number of colony formations of CHO and 9L-tumor cells was found to depend on cell type.[54]

The cytotoxicity of various derivatives of EMPO on different carcinoma cell lines also show BMPO to be the most toxic ($IC_{50}\sim$5−6 mM) for all cell lines,[66] with DEPMPO and *i*-5-n-propyl-5-methyl-1-pyrroline-*N*-oxide (PrMPO) as the least toxic ($IC_{50}\sim$100−300 mM) but dependent on cell line. The cell-type dependence of spin-trap toxicity, e.g., with DMPO was found to be only toxic to bovine pulmonary artery endothelial cells up to 1 mM but not toxic to lung microvascular endothelial cells up to 100 mM (unpublished data). Conjugation of an amphiphilic carrier to PBN and DMPO to form fluorinated amphiphilic PBN conjugate (FAPBN) and e fluorinated amphiphilic DMPO conjugate (FAMPO), respectively, only affects FAMPO toxicity but not FAPBN. FAMPO showed 20% and 50% decrease in cell viability at 0.5 and 1 mM after 24 h of incubation, respectively, while the carrier alone gave cell toxicity of 5−25%.[67] These results indicate that cellular permeability plays an important role on nitrone toxicity. Therefore, during spin trapping where cells are freshly harvested and the EPR spectrum is acquired immediately, toxicity would not be an issue.

In rats, DMPO was found to be not toxic up to 120 mg/100 g (w/w) compared to PBN with a lethal dose of 100 mg/100 g.[68] Evidence from gross pathology and histopathological examinations showed that spin traps injected intraperitoneally gave no cellular damage—neither DMPO (232 mg/100 g by wt.) nor PBN (100 mg/100 g by wt)—while 4-PyOBN was found to be lethal at 100 and 200 mg/100 g by weight.[69] Spin trapping in vivo using DMPO was employed and found to have relative stability of 90 min after intraperitoneal injection (15 mmol/kg) in rats with concentrations high enough to trap C-centered radicals. After 2 h, 40% of the injected DMPO was still available. However, only the more stable adduct—i.e., the C-centered ones—could be detected since the half-life of DMPO-OH was too short to be observed, disappearing after 1 min of its generation.[70] In spite of the in vivo application of spin trapping using DMPO, $O_2^{\bullet -}$ production has also been indirectly observed as an $HO^{\bullet}$ adduct (see Table 5.3).

5.8 SYNTHESIS OF SPIN TRAPS

The general synthesis of nitrones is shown in Scheme 5.2. Several review articles, books, and papers have appeared over the years describing the general synthesis of nitrones.[71–74] The most efficient routes to synthesize nitrone functionality are by condensing carbonyl compounds with hydroxylamine and oxidizing amine, imine, or hydroxylamine moieties.

Reaction I. The reaction proceeds via condensation of the aldehyde or ketone with hydroxylamine to give aldonitrone or ketonitrone, respectively, but often yields crude products that are difficult to separate.[75]

Reaction II. For cyclic or DMPO-type nitrones, this procedure will typically involve the synthesis of aldo–nitro or keto–nitro compounds via a Michael addition reaction of a nitroalkane (i.e., 2-nitropropane) to activated olefins such as α,β-unsaturated carbonyl compounds. Subsequent reductive cyclization using

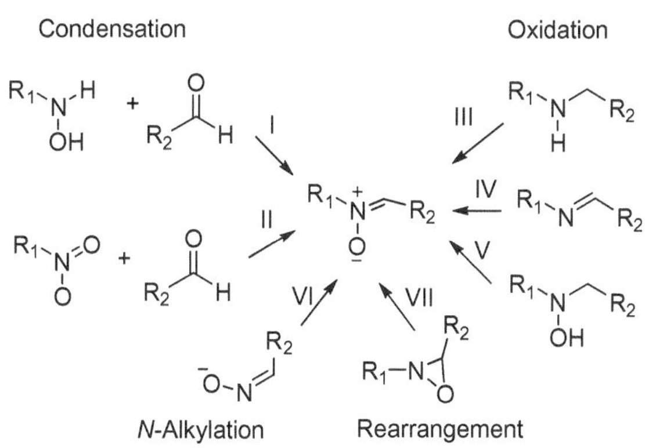

Scheme 5.2 Various routes for nitrone synthesis.

reducing agents such as Zn metal in acidic conditions forms the nitrone.[76] In some cases, it is necessary to protect the aldehyde as an ethylene–glycol acetal[38,43,77,78] to minimize side reactions. Selective reductive coupling of nitro compounds with aldehydes to nitrones could also be accomplished in H_2 using carbon-supported Pt nanoparticles.[79]

Reaction III. Oxidation of secondary amines is usually carried out with H_2O_2 using molecular sieves;[80] Na_2WO_4[81] and SeO_2,[82] as catalysts; alkyl hydroperoxides and titanium alkoxides as catalyst;[83] or with mCPBA.[84]

Reaction IV. Imines can be oxidized to nitrones by dioxiranes such as dimethyldioxirane[85] and methyl(trifluoromethyl)dioxirane[86,87] or oxidation by peroxyacid.[85] Imines can be reduced to amine with $NaBH_4$ followed by oxidation with H_2O_2/Na_2WO_4 to nitrones.[88]

Reaction V. Oxidation of hydroxylamines to nitrones can be carried out using bleach,[89] MnO_2,[90] H_2O_2 catalyzed by methylrhenium trioxide $MeReO_3$ [91], or $Cu(OAc)_2/O_2$.[92]

Reaction VI. Benzaldoxime is *N*-alkylated by various alkyl halides to form the nitrone, however, consideration of geometric isomerism and steric factors is important due to the formation of O-alkylated products.[93]

Reaction VII. Oxaziridines prepared from the oxidation of imines by *m*-chloroperoxybenzoic acid can undergo thermal rearrangement to form the corresponding nitrones and can be catalyzed by Lewis acids.[94]

5.9 INTERPRETATION OF EPR SPECTRA

Spin-trapping EPR spectra could provide a myriad of information about the identity, molecular size, and nature of the radical species being trapped as well as their concentration. Three important pieces of information can be obtained from the *x*-axis.

First, the g value provides information into the nature of the radical species (e.g., organic versus metal-based). For the spin adducts, the solution g value is typical of aminoxyl compounds ($g = 2.006$)[95] with g values ranging from 2.005 to 2.007.[28,96–98]

Second, hyperfine splitting (or coupling) constant is the most informative spectral parameter in EPR spin trapping such that it provides information about the identity of the radical being trapped. The interaction of the unpaired electron with the nuclei of the trapped radical that is bound to the C-2 gives rise to additional splitting patterns for the magnetic nuclei or shows that it could also result in variations in the hyperfine splitting constants of the nitronyl-N and β-hydrogen as shown in Table 5.2. Noteworthy is the $a_N/a_H = 1$ observed for the $HO^•$ adducts generated from various radical generating system and $a_N/a_H = 1.21–1.27$, which is unique for $HO_2^•$, depending on the $O_2^{•-}$-generating system used. The smallest a_N/a_H ratio of 0.65–0.71 can be observed for the C-centered adducts for phenyl, phenyl-substituted, and alkyl radicals.

Table 5.2 Survey of the various hyperfine splitting constants observed for various spin adducts of DMPO in aqueous solution[a]

Trapped radical	a_N (G)	$a_{\beta\text{-H}}$ (G)	Additional a_x (G)	a_N/a_H	Ref
$^{\bullet}$H	16.0–16.6	21.5–23.2 (2 H's)		0.70–0.75	
$^{\bullet}C_6H_5$-X	15.0–15.4	21.2–25.0		0.65–0.70	
$^{\bullet}$R	14.9–16.5	19.6–24.1		0.68–0.71	
$^{\bullet}CONH_2$	15.5	20.5		0.75	
$^{\bullet}CO_2^-$	15.7	18.7		0.84	99
$^{\bullet}OCO_2^-$	14.3	10.7	1.37 (γ-H)	1.33	100
$^{\bullet}$OH	14.3–15.0	14.3–15.0		1.0	
$^{\bullet}NH_2$, $^{\bullet}$NHR	15.6–15.9	18–19.3	1.6–2.5 (N)	0.82–0.86	
$^{\bullet}$NHCOR	15.4	20.3	2.4 (N)	0.76	
$^{\bullet}N_3$	14.5–16.9	14.5–16.9	3.0–3.2 (N)	1.00–1.05	101
$^{\bullet}O_2R$	14.25–14.5	10.5–11.5	1.1–1.25 (γ-H)	1.28–1.38	
$^{\bullet}O_2H$	13.8–14.4	11.1–11.7	1.22–1.37(γ-H)	1.21–1.27	
$^{\bullet}$OR	14.5–15.0	15.7–16.7		0.89–0.93	
$^{\bullet}$SH, SR	15.1–15.4	16.2–18.0		0.89–0.94	
$^{\bullet}SO_3^-$	14.5–14.7	15.9–16.5		0.89–0.91	102
$^{\bullet}OSO_3^-$	13.7–13.9	10.1–10.5	1.3–1.5 (γ-H)	1.36–1.37	102
DMPO-X	7.0–7.2	4.0–4.2		1.69–1.71	

[a]All the values were taken from the Spin Trap Database by the National Institute of Environmental Health Sciences and their respective references can be obtained therein. Additional references are provided.

Moreover, hyperfine splitting constants are strongly influenced by the solvent polarity where protic solvents results in higher hyperfine splitting constants (HFSCs) compared to aprotic solvents.[103]

Third, the line width and shape of the spectrum are affected by the molecular size of the trapped radical, intermolecular attractive forces such as H-bonding between adducts, or viscosity of the solution, giving information on the local molecular tumbling or rotational dynamics. Moreover, the presence of conformational isomerism and, in the case of 5-substituted nitrones, geometric isomerism also contributes to line width and shape of the spectrum.

In general, EPR spectra resulting from spin trapping are all isotropic because of the fast tumbling motion of the molecules, which allows for the averaging of the rotational motion and hence gives more symmetrical line patterns. However, small differences in the spectral profile can be observed such as the ones shown in Fig. 5.12 in which the type of trapped radical has a drastic effect on the hyperfine splitting pattern as well as line width of the spectrum of the adduct formed. Although the molecular size of each trapped radical is approximately the same, their conformational and inductive effects as well as their ability to exhibit intermolecular interaction could

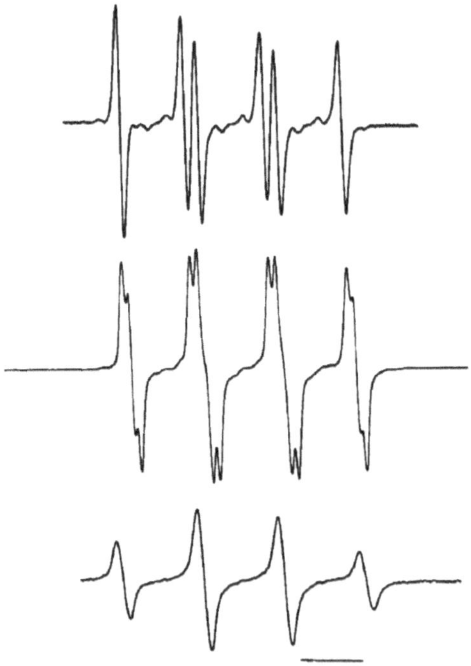

Figure 5.12 EPR spectra of DMPO thiyl adducts from (top) 2-methyl-2-propanethiol, (middle) α-toluenethiol, and (bottom) 2-hydroxyethanethiol.[104] *(Adapted from Josephy et al. Tet Lett 1984;**25**:1685).*

affect spectral profile. The line width broadening for 2-hydroxyethanethiol-adduct could be the result of a strong H-bond interaction between the adducts, and that can hinder molecular tumbling motion (which is more evident at the highest field peak), so the hyperfine splitting pattern resulting from the β-H as seen from 2-methyl-2-propanethiol- and α-toluenethio-adducts was obscured.

The effect of molecular size on the EPR spectrum of thiyl adducts is shown in Fig. 5.13. Comparison of DMPO−glutathione disulfide (DMPO-SG) and DMPO-SHb (hemoglobin thiyl adduct) shows a much broader line width for the latter, although still in the fast motion regime, which indicates a slower rotational motion (τ_c) compared to the former.

Because a spectral profile is influenced by several of the factors just mentioned, it is possible to rationalize the origins of hyperfine splitting patterns using a computational approach[106] or by isotopic labeling.[107] Fig. 5.14 shows the characteristic spectral profiles exhibited by various DMPO spin adducts. Of particular interest is the distinctive spectrum exhibited by DMPO−OH compared to that of DMPO−O_2H in spite of the very small difference in their chemical compositions—i.e., the latter has only one more oxygen atom than the former. Aside from the difference in the $a_N/a_{\beta\text{-H}}$ ratio of DMPO−OH and that of DMPO−O_2H, the presence of additional splitting

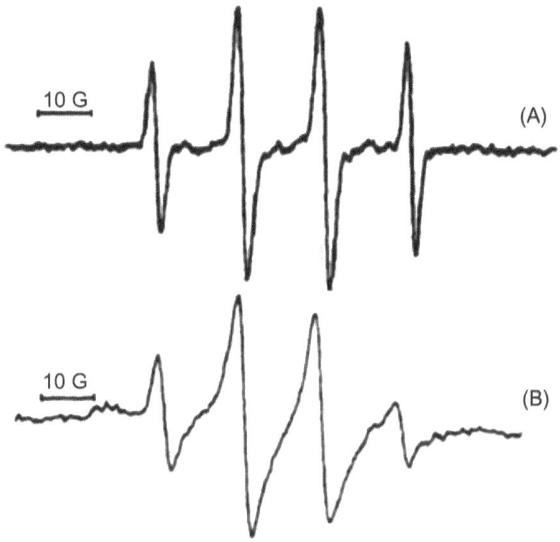

Figure 5.13 EPR spectra of DMPO-SG adduct generated after 1 minute of Hg/Xe light irradiation of (A) 10 mM DMPO and 50 mM GSSG and (B) DMPO-SHb generated 30 min after addition of 175 μM phenylhydroxylamine in 20% intact rat red cell suspension containing 100 mM DMPO.[105] *(Adapted from Bradshaw et al. Free Radic Biol Med 1995;**18**:279).*

due to γ-H (i.e., hydrogen attached at the 3- or 4-carbon position) of ~1.2–1.4 G (see Table 5.2) gives rise to a 12-line spectrum that is distinctive of the DMPO–O_2H. However, extensive spectral simulation and low-level quantum mechanical calculations led to the proposition that the additional splitting pattern in the spectrum of DMPO–O_2H was due to the overlapping spectra from two different conformers of DMPO–O_2H and not from γ-H splitting.[108] Deuterium isotopic labeling of the γ-H's at C-3 only gave a six-line spectrum, confirming ~1.2 G contribution of γ-H's to the hyperfine structure.[107] Density functional theory approach showed that only the $a_{\beta\text{-H}}$ is sensitive to conformational changes of the –OOH moiety and that both γ-H's that are *syn* and *anti* to –OOH gave the highest HFSC (1.1–1.3 G) compared to other γ-H's attached at C-4,[106] further confirming the contribution of these H's to the hyperfine structure. In addition to the conformational effects to the hyperfine structure, the presence of diastereoisomers[29,109] in the case of 5-substituted DMPO analogs such as DEPMPO can give rise to line asymmetry and line-width broadening as shown in Fig. 5.14 for the various spin adducts of DEPMPO, further complicating spectral interpretation.[110]

Qualitative spectral analysis focuses mainly on the identification of the radical trapped. Hyperfine splitting patterns as well as the line shape and width of the spin-adduct spectrum serves as a spectral fingerprint that can provide information into the

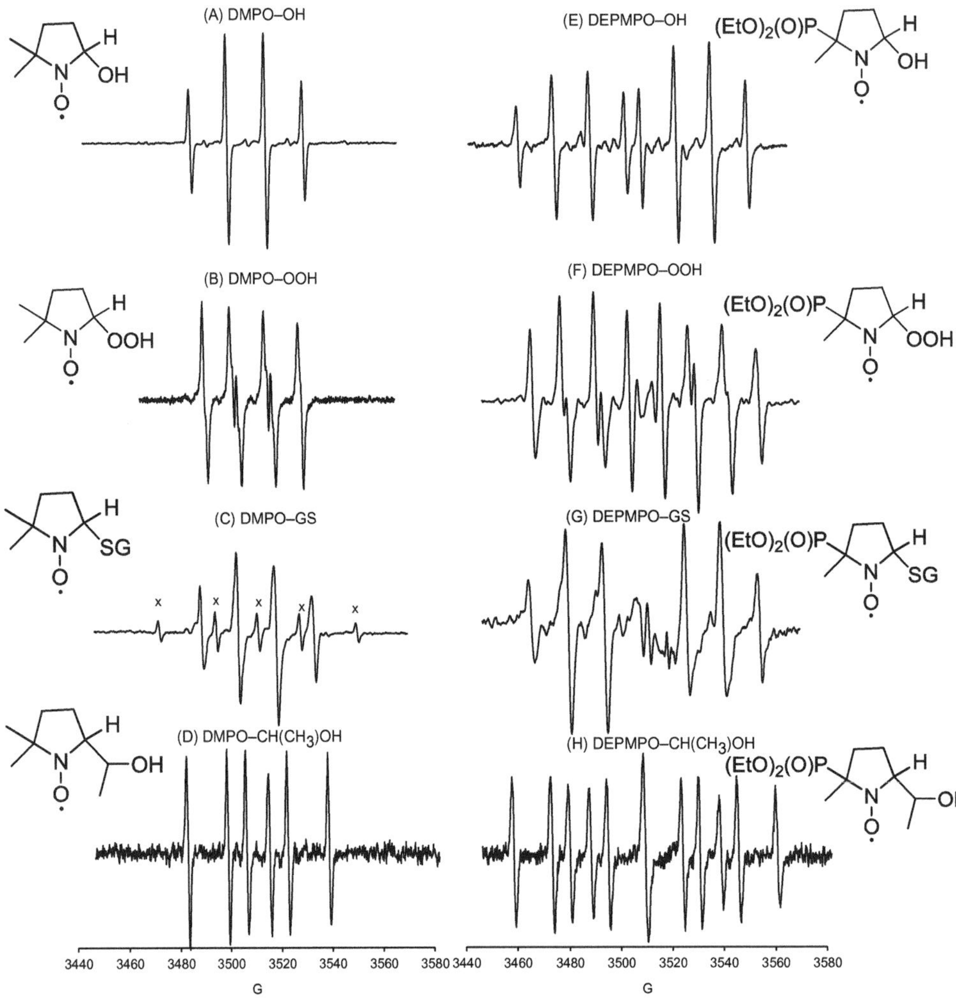

Figure 5.14 EPR spectra of various adducts of DMPO and DEPMPO (Adapted from Villamena, F. A. *Antioxid Redox Signal* **6**, *2004, 619)* .[111]

nature, type, and identity of the radical. Typically, the isotropic solution spectrum is simulated using the following software:

1. WinSim—This public domain software is available from the National Institute of Environmental Health Sciences, Public Electron Paramagnetic Resonance Software Tools. It is intended for EPR simulations of experimental isotropic spectra—either the continuous wave (CW) or the fourier transform (FT) spectrum. The program can simulate as many as 10 species at varying concentrations, with each species containing 16 magnetic nuclei.[112]
2. EasySpin—This MATLAB-based application developed by Dr. Stephan Stoll.[113]

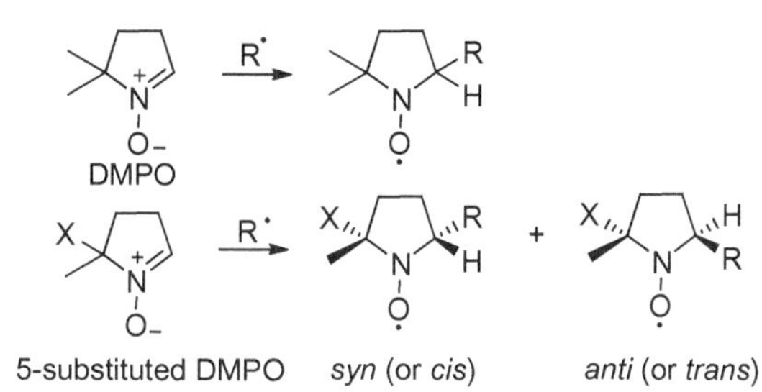

Figure 5.15 Comparison of formation of various adducts from DMPO and 5-subtituted nitrone

3. ROKI—This DOS-based application was developed by Professor Antal Rockenbauer.[114]

The type of spin traps and the relative concentration of spin adducts are considered during simulation. The use of DMPO simplifies spectral analysis such that it avoids consideration of diasterioisomer formation, so there are fewer species to consider in the simulation. Typically with DMPO, only one species is considered, but with 5-substituted nitrones such as EMPO, DEPMPO, BMPO, and AMPO, because radical addition is not regiospecific, the presence of diastereoisomers are inevitable as shown in Fig. 5.15.

The relative concentrations of the spin adducts generated from two or more type of radicals as well as formation of diastereomeric adducts can vary and could primarily affect the line shape and width of the spectrum. Slight differences in the *g*-factor of the various adducts formed can also affect overall spectral profile. Simulation of spectral parameters should give optimized parameters for each species such as hyperfine splitting constants for relevant atoms, *g*-factor, line width, and fractional amount of the species.

Quantitative analysis of spin adducts can provide important information about the kinetics and favorability of radical production. The rate of radical production can be measured by competitive kinetic techniques using enzymatic or chemical systems as mentioned elsewhere.[25,37,38,41] While these techniques mostly assess the rate of $O_2^{\bullet -}$ trapping of newly developed spin traps, in principle they can be employed to study kinetics of radical production from cells, but none of those have been carried out yet. In determining the rate of radical reaction by an antioxidant (new spin traps or non–spin traps alike), competitive kinetic technique uses a known radical scavenger (e.g., for $O_2^{\bullet -}$, SOD, ferricytochrome c, or DMPO) whose rate constants for $O_2^{\bullet -}$ trapping are known and the relative rates are then calculated. Radical production can be initiated, depending on the nature of radical generation. Photolysis and stopped-flow technique allow for the acquisition of critical initial rates of radical formation since the sample is already positioned in the cavity before radical production commences. In cases when it is impractical to initiate radical production in the EPR cavity, radical initiator is added first into the tube and the sample holder is immediately placed inside the cavity. However, by doing so, critical time is lost during sample placement in the cavity and microwave tuning, therefore, this technique is not recommended for fast reactions such as studying the rate of $HO^•$ production, although this technique is reasonable for studying the rate of $O_2^{•-}$ reaction due to the slower production of $O_2^{•-}$ and its reaction to spin traps.

Data acquisition can be carried out in two ways: (1) by using time scan at constant field where the growth of either the first or second low-field peak is monitored over a period of time provided a stable magnetic field and (2) by an incremental field sweep in which the whole or part of the spectrum is repetitively recorded over time. With time-scan acquisition, the initial rate of reaction is determined from the slope of peak amplitude (or intensity) as plotted against time, while with incremental field sweep, peak intensity, or area of a selected peak or of the whole spectrum is determined and plotted against time. Quantitation of radical production employs a standard curve of stable nitroxide such as TEMPO or proxyl nitroxides whose double integrated area of a peak or whole spectrum or peak amplitude of the low-field peak is plotted against a range of concentrations ($\sim$0.1–10 μM in buffer).[115]

5.10 APPLICATIONS OF SPIN TRAPPING

There are three major scenarios in radical detection using EPR spin trapping: (1) no signal, (2) weak signal, and (3) strong signal. The major limitation in radical detection using spin trapping is the quantity of radicals being produced by the system. The presence of biological milieu, which drastically shortens the adduct half-life, is perhaps the most significant setback in the application of spin traps for biological radical detection. Furthermore, the detection of $O_2^{•-}$ is even more difficult due to its slow reactivity to spin traps and its inherently short half-life. Therefore, the absence of EPR signal during spin trapping may not necessarily indicate that radicals were not generated but could be due to any of the factors previously mentioned. Conversely, the presence of a signal does not necessarily indicate radical production. However, several experimental "tricks" and procedures have been developed to overcome these shortcomings.

1. *Avoid basic pH conditions.* Spin-adduct half-life is shortest in basic conditions as mentioned previously and, therefore pH conditions between normal physiological pH and below would be appropriate for spin trapping.
2. *Sequester trace amounts of transition metal ions.* Artifacts are commonly encountered in spin-trapping experiments. The metal-catalyzed Forrester-Hepburn mechanism shown in Eq. (5.15) has been proposed as a nonradical pathway for adduct formation and could be misleading.[116] Formation of DMPO–OH adduct due to the nucleophilic addition of H_2O in the presence of Fe^{III} or Cu^{II} was confirmed using ^{17}O-enriched water. While catalase does not affect this adduct formation, the presence of metal chelators exerts inhibition. Because metal impurities are only present at submicromolar levels, nonradical metal-mediated reactions can be

significantly minimized through the use of chelating agents showing both DTPA and nitrilotriacetic acid to effectively prevent DMPO−OH formation in the presence of high-valent transition metal ions.[117] Typically, buffer solutions containing 100 μM of DTPA is a reasonable precautionary measure for carrying out spin trapping under biological conditions.

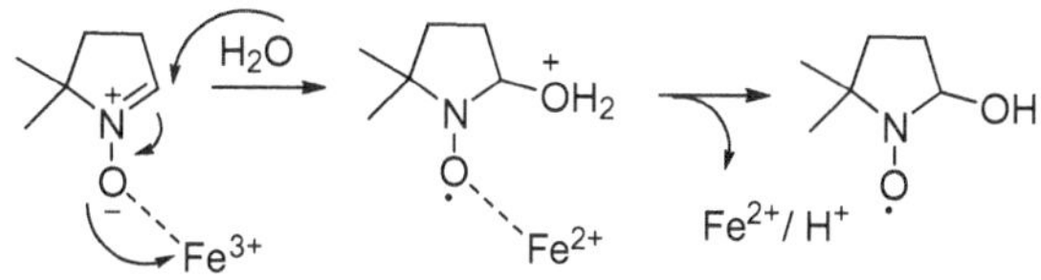

(5.15)

3. *Proper choice of media.* Sometimes mixed solvents are necessary to solubilize reagents such as drugs or newly developed spin traps used for radical generation. When the formation of O-centered radicals is desired, it is desirable to use solvents that do not form C-centered radicals. Secondary radical adduct formation from H-atom abstraction from solvents by $HO^{\bullet}$ can complicate spectral interpretation. Heats of reaction (ΔH) of H-atom abstraction from various solvents by $HO^{\bullet}$ in water at 300 K shows the following order of increasing exoergicity: CH_3CN (3.7) < CH_3CH_2OH (−2.8) < CH_3OH (−3.5) < $(CH_2OH)_2$ (−8.1) < $(CH_3)_2CHOH$ (−8.5).[118] In general, the use of acetonitrile (CH_3CN) could minimize C-centered adduct formation. Also recommended is the use of phenol red free cell culture media (preferably just plane buffer) because phenol red can also compete with spin traps for radicals, especially that of $O_2^{\bullet-}$.[20]
4. *Sequence of reagent addition matters.* Knowing the molecular basis of radical production would aid in the design of successful protocols for radical detection. Due to the very short half-lives of radicals, having the spin trap in solution before initiating radical production increases the chance of detecting spin adducts. Examples of radical initiators are visible light for a riboflavin system to form $O_2^{\bullet-}$; UV for the homolytic cleavage of disulfides and peroxides to formthiyl and alkoxyl radicals, respectively; electric current for electrochemical generation of radicals; chemicals such as KO_2 or 3-morpholino-sydnonimine (SIN-1) to generate $O_2^{\bullet-}$ or peroxynitrite, respectively; substrates such as xanthine for xanthine oxidase; cell activators such as lipopolysaccharide (LPS), menadione, phorbol 12-myristate 13-acetate (PMA), or calcium ionophore; enzyme could also be an initiator such as xanthine oxidase or horseradish peroxidase; and O_2 in hypoxia and reoxygenation models.

The most obvious source of free radicals such as O_2 is often overlooked in spin trapping where its presence is necessary for radical production. Bubble air or pure oxygen in solution can increase radical production particularly in radical generating systems such as light/riboflavin, NADPH oxidases, xanthine/xanthine oxidase. Use of appropriate light source is necessary to generate free radicals such that one would not use visible light to generate $HO^{\bullet}$ from H_2O_2.

5. *Toxicity of spin traps.* As mentioned previously, although spin traps are mostly non-toxic, their toxicity depends on the spin trap and cell type. When working on new cell lines that have not been investigated for spin trapping, it is advisable to do quick colorimetric cell viability studies using tetrazolium dye, 3-(4,5-dimethylthiazol-2-yl)-2,5-diphenyltetrazolium bromide (MTT assay) , or lactate dehydrogenase cytotoxicity assays.
6. *Choice of spin trap.* For the detection of $O_2^{\bullet -}$, because the rates of trapping by commercially available spin traps are pretty much the same, the choice will greatly depend on the spin trap that will impart the longest $O_2^{\bullet -}$ adduct half-life. Typically, EMPO, BMPO, or DEPMPO are the nitrones of choice. For more reactive radicals such as OH, C-centered or thiyl radicals, the choice of spin trap is probably not that important for the most part because the rate constants are of the same orders of magnitude with all of the commonly used nitrones, and their relative spin-adduct half-lives are long enough to be detected by EPR. The choice between PBN and cyclic nitrones will depend on whether one wants to perform in vivo spin trapping of free radicals utilizing ex vivo detection since the use of PBN is more established and tested than DMPO for this type of application.[119]
7. *Type of cell lines.* All cell types have distinctive biological roles and, therefore their ability to generate radicals could vary from one type to another. Although spin trapping has been employed on various cells such as endothelial cells[60,120] and epithelial cells[121] and bacteria,[122] radical generation from leukocytes had been shown to be the most robust such for neutrophils[123] or monocytes and macrophages.[124,125]
8. *Integrity of samples and spin traps used.* Enzyme activity should be carried out to assess enzyme viability. Cells must be freshly cultured, detached, pelletized, and resuspended in an appropriate medium (preferably just buffer) before any spin-trapping experiment. A cytotoxicity assay to test spin-trap toxicity on cells must also be carried out. Pure spin traps or aliquot solutions should be stored at least at −20°C or at −80°C for long-term storage. Spin traps could be degraded with long-term exposure to light, heat, air, or metals ions. Although spin traps are sold and advertised as 99.9% pure, some of the commercially available spin traps sometimes have paramagnetic impurities and therefore the EPR spectrum of the solution of the spin trap alone should first be tested for the presence of paramagnetic impurities before starting actual studies. One should note that the typically sensitivity of EPR for nitroxides is ~ 1-10 μM and therefore a 50 mM spin trap with a 1% paramagnetic impurity would give a signal that could greatly interfere with the rest of the spectrum. Spin traps could also contain hydroxylamine as impurity and could produce nitroxide on exposure to O_2 or other oxidizing agents in the solution. Aside from running an EPR spectrum of

the spin-trap solution alone, one should also run a positive control experiment of the nitrone using a well-established radical-generating system such as the Fenton system composed of 1–100 μM Fe^{2+} and 0.1–10 mM H_2O_2.

9. *Miscellaneous issues.* Before planning any experiment, instrument should be tested to ensure that it is working properly by running a spectrum of a known solution of stable nitroxides such as that of TEMPO. Critically couple the cavity by properly positioning the sample, locating the microwave frequency, adjusting the iris screw for better coupling, or increasing the microwave energy. Depending on the radical-generating system being used, optimize the EPR experimental parameters such as receiver gain, modulation amplitude, time constant, and scan rate. Choose the appropriate sample cell to be used. For example, standard 50-μL glass microhematocrit capillary tube is typically used for spin-trapping experiments in aqueous solutions where one end of the tube is sealed by synthetic wax sealant (white synthetic wax is preferred to the brownish clay sealant because the latter contains paramagnetic manganese ions that interfere with the spectrum), but they are not preferred for UV irradiation where quartz capillary tube is used instead and is commercially available. Quartz sample tube (135–250 mm long × 4 mm outside diameter, where the longer one is used with finger dewar) is preferred for solid samples and organic liquids with lower dielectric constants ($\varepsilon = 2-4$). Flat cell (10 mm) is used for solvents with high dielectric constants and is preferable for the study of radical production from cells because this allows for the use of higher volume (at least 300 μL) and cell density needed to achieve upper limits of radical production.
10. *Use of control experiments.* To exclude nonradical pathway for spin-adduct formation, adding SOD and catalase before initiating radical production should inhibit $O_2^{\bullet -}$ and $HO^{\bullet}$ (via H_2O_2) adduct formation, respectively. Ethanol, methanol, or DMSO can also be added to confirm the generation of $HO^{\bullet}$ via formation of alcohol-derived secondary C-centered radicals. Since DMPO-OOH has a short half-life and is known to decompose to DMPO–OH, performing competitive experiments with alcohols, DMSO, and catalase could verify the origin of DMPO–OH—i.e., from DMPO–OOH decomposition or direct addition of $HO^{\bullet}$ to DMPO. Also, due to the instability of the primary peroxyl radical, conversion to its corresponding alkoxyl radical adduct could occur. For example, in the case of polyunsaturated fatty acid (PUFA) peroxyl radicals could gave rise to PUFA-derived alkoxyl radical adducts rather than the expected primary PUFA peroxyl radical adducts, thus complicating spectral interpretation.[126]

Since $O_2^{\bullet -}$ is the major precursor of the more highly oxidizing (or reducing) reactive species, its detection is highly desirable but also the most challenging. Shown in Table 5.3 are the various systems in which $O_2^{\bullet -}$ has been directly (or indirectly) detected using spin trapping, which includes chemicals, enzymes, organelles, cells, organs (tissues), and whole animals. The ease of $O_2^{\bullet -}$ detection is more evident from

Table 5.3 Survey of the various chemical, enzymatic, organelle, cellular, and in vivo systems in which radical generation has been detected using EPR spin trapping

System	Method	Spin trap	Radical detected	Ref
Chemical				
Fullerenes	Ultraviolet A (UVA) irradiation with or without nicotinamide adenine dinucleotide (NADH) addition	DMPO	$O_2^{\bullet-}$	127
Phenanthraquinone	P450 reductase, NADPH	DMPO	$O_2^{\bullet-}$, $HO^{\bullet}$	128
Dichlorofluorescein	Oxidation by horseradish Peroxidase, NADH	DMPO	$O_2^{\bullet-}$	129
Cyanine dyes	Irradiation at 350-nm light	DMPO	$O_2^{\bullet-}$, $HO^{\bullet}$	130
Advanced glycation end products	UVA irradiation	DMPO	$O_2^{\bullet-}$	131
Aqueous cigarette extract	Incubation in air saturated DMSO or water	DMPO	$O_2^{\bullet-}$, $HO^{\bullet}$	132
Curcumin	420-nm irradiation in benzene	DMPO	$O_2^{\bullet-}$, $HO^{\bullet}$	133
Enzymatic				
eNOS	NADPH, Ca^{2+}, calmodulin	DMPO	$O_2^{\bullet-}$	134
nNOS	NADPH, Ca^{2+}, calmodulin	DMPO	$O_2^{\bullet-}$	135–137
iNOS	NADPH, Ca^{2+}, calmodulin	DMPO	$O_2^{\bullet-}$	58,138
Soybean lypoxygenase	Linoleic acid	DMPO	$HO^{\bullet}$, $O_2^{\bullet-}$, $RO^{\bullet}$, $R^{\bullet}$	139
Cell wall–bound peroxidase	Incubation	DEPMPO	$O_2^{\bullet-}$, $HO^{\bullet}$	140
Xanthine oxidase	Xanthine	DMPO	$O_2^{\bullet-}$, $HO^{\bullet}$	50
Aldehyde oxidase	NADH	DEPMPO	$O_2^{\bullet-}$	141
Fe^{3+}cyt c	NADH, H_2O_2	DMPO, Me-β-CD	$O_2^{\bullet-}$	142
Complex I (NQR)	NADH	DEPMPO	$O_2^{\bullet-}$	143
Succinate-cytochrome c reductase	Succinate	DEPMPO	$O_2^{\bullet-}$	144
NADH dehydrogenase	NADH	DEPMPO	$O_2^{\bullet-}$	145

Organelle				
Intact mitochondria	Succinate	DEPMPO, mito-DEPMPO	$O_2^{\bullet -}$	146
Microsomes	Uroporphyrin, NADPH	DMPO	$O_2^{\bullet -}$, $HO^{\bullet}$	147
Chloroplast	Light photolysis in the presence of O_2	DMPO	$O_2^{\bullet -}$	148
Thylakoid membranes and PS II core particles	White light illumination	EMPO, Me-β-CD or HP-β-CD	$O_2^{\bullet -}$	59
Cellular				
Neutrophils	PMA	DMPO	$O_2^{\bullet -}$	149
	Chemotactic peptide, formylmethionylleucylphenylalanine	DMPO	$O_2^{\bullet -}$	147
Macrophages (RAW 264.7 cells)	LPS, PMA	DMPO	$HO^{\bullet}$	58
Human mammary epithelial cells (MCF-10A cells)	1,6-Benzo[a]pyrenequinone	DMPO, BMPO, DEPMPO, DEPPEPO	$O_2^{\bullet -}$	121
Microglial cells	Opsonified zymosan	DEPMPO	$O_2^{\bullet -}$, $HO^{\bullet}$	150
Osteoclasts	NADPH	DMPO	$O_2^{\bullet -}$ as $HO^{\bullet}$	151
Endothelial cells	O_2 purge	DMPO	$O_2^{\bullet -}$, $HO^{\bullet}$	152
	Anoxia or reoxygenation	DMPO	$O_2^{\bullet -}$, $HO^{\bullet}$	57,153
	SIN-1 treatment, then Ca I	EMPO, Me-β-CD	$O_2^{\bullet -}$, $HO^{\bullet}$, $R^{\bullet}$	60
Organ (perfusates)				
Rat liver	I/R	DMPO	$O_2^{\bullet -}$, $HO^{\bullet}$	154
Rat heart	I/R	DMPO	$O_2^{\bullet -}$, $HO^{\bullet}$, $R^{\bullet}$	155,156

(*Continued*)

Table 5.3 (Continued)

System	Method	Spin trap	Radical detected	Ref
Animal (ex-vivo)				
Humans	Excised skins; UVA, UVB, or UVC irradiation	DMPO	$R^{\bullet}$, $RO^{\bullet}$	157
Rabbits	Injury-induced excised aorta, NADH, NADPH	DMPO	$O_2^{\bullet -}$ as $HO^{\bullet}$	158
Rats	Blood, excessive resistive loading of the respiratory muscles	PBN	$R^{\bullet}$	159
	Bile and urine samples, ethylene glycol poisoning	POBN	$^{\bullet}[^{13}C]$Et-Glycol radical	160
	Lung extracts of intratracheally instilled asbestos	POBN	$R^{\bullet}$	161
Swine	Blood, ischemia, and reperfusion of the heart	PBN	$RO^{\bullet}$	162
Mice	Brain, spleen, and liver lipid extracts after gamma irradiation	PBN	$R^{\bullet}$	163
	Skin, topical treatment with phenol	PBN	$R^{\bullet}$	164

chemical to leukocytic cellular systems. Superoxide detection in cells other than leukocytes can be aided through the use of adduct stabilizing agent such as the permethylated-β-cyclodextrin (Me-β-CD). However, with tissues, the robustness of $O_2^{\bullet -}$ adduct formation diminishes and is mostly detected as secondary C-centered and alkoxyl-adducts in ex vivo measurements from animals. From cellular systems to in vivo, the source of radical production is not clear due to the nontarget specificity of DMPO and its analogs, so inhibition experiments of key enzymatic system may realistically work using cells. However, with tissues and in vivo, enzyme inhibition studies could be a challenge because inhibitors could become systemic especially in the latter.

REFERENCES

1. Sawyer DT, Valentine JS. How super is superoxide? *Acc Chem Res* 1981;**14**:393.
2. Buxton GV, Greenstock CL, Helman WP, Ross AB. Critical review of rate constants for reactions of hydrated electrons, hydrogen atoms and hydroxyl radicals (OH/O) in aqueous aolution. *J Phys Chem Ref* 1988;**17**:513.
3. Bagchi RN, Bond AM, Scholz F, Stoesser R. Characterization of the ESR spectrum of the superoxide anion in the liquid phase. *J Am Chem Soc* 1989;**111**:8270.
4. Nakanishi I, Ohkubo K, Fujita S, Fukuzumi S, Konishi T, Fujitsuka M, et al. Direct detection of superoxide anion generated in C60-photosensitized oxidation of NADH and an analogue by molecular oxygen. *J Chem Soc Perkin Trans* 2002;**2**:1829.
5. Petr A, Kataev V, Buechner B. First Direct In Situ EPR Spectroelectrochemical Evidence of the Superoxide Anion Radical. *J Phys Chem B* 2011;**115**:12036.
6. Symons MCR, Eastland GW, Denny LR. Effect of solvation on the electron spin resonance spectrum of the superoxide ion. *J Chem Soc Faraday Trans 1* 1980;**76**:1868.
7. Yu J, Chen J, Li C, Wang X, Zhang B, Ding H. ESR Signal of Superoxide Radical Anion Adsorbed on TiO2 Generated at Room Temperature. *J Phys Chem B* 2004;**108**:2781.
8. Bosnjakovic A, Schlick S. Nafion perfluorinated membranes treated in Fenton media: radical species detected by ESR spectroscopy. *J Phys Chem B* 2004;**108**:4332.
9. Mendiara SN, Ghibaudi E, Perissinotti LJ, Colussi AJ. Free radicals and diradicals in the reaction between nitrous acid and bisulfite in acid aqueous media. *J Phys Chem* 1992;**96**:8089.
10. Jiang J, Bank JF, Scholes CP. Subsecond time-resolved spin trapping followed by stopped-flow EPR of Fenton reaction products. *J Am Chem Soc* 1993;**115**:4742.
11. Weil JA, Bolton JR, Wertz JE. *Electron Paramagnetic Resonance: Elementary Theory and Practical Applications*. New York: John Wiley and Sons; 1994.
12. Klauschenz E, Haseloff RF, Volodarskii LB, Blasig IE. Spin Trapping Using 2,2-Dimethyl-2H-Imidazole-1-Oxides. *Free Radical Res* 1994;**20**:103.
13. Dikalov S, Kirilyuk I, Grigor'ev IA. Spin trapping of O-, C-, and S-centered radicals and peroxynitrite by 2H-imidazole-1-oxides. *Biochem Biophys Res Commun* 1996;**218**:616.
14. Bottle SE, Micallef AS. Synthesis and EPR spin trapping properties of a new isoindole-based nitrone: 1,1,3-trimethylisoindole N-oxide (TMINO). *Org Biomol Chem* 2003;**1**:2581.
15. Rosen GM, Tsai P, Barth ED, Dorey G, Casara P, Spedding M, et al. A one-step synthesis of 2-(2-Pyridyl)-3H-indol-3-one N-oxide: is it an efficient spin trap for hydroxyl radical? *J Org Chem* 2000;**65**:4460.
16. Khramtsov VV, Reznikov VA, Berliner LJ, Litkin AK, Grigor'ev IA, Clanton TL. NMR spin trapping: detection of free radical reactions with a new fluorinated DMPO analog. *Free Radical Biol Med* 2001;**30**:1099.
17. Thomas CE, Ohlweiler DF, Carr AA, Nieduzak TR, Hay DA, Adams G, et al. Characterization of the radical trapping activity of a novel series of cyclic nitrone spin traps. *J Biol Chem* 1996;**271**:3097.

18. Stolze K, Udilova N, Nohl H. ESR analysis of spin adducts of alkoxyl and lipid-derived radicals with the spin trap Trazon. *Biochem Pharmacol* 2002;**63**:1465.
19. Sankuratri N, Janzen EG. Synthesis and spin trapping chemistry of a novel bicyclic nitrone: 1,3,3-trimethyl-6-azabicyclo[3.2.1]oct-6-ene-N-oxide (Trazon). *Tetrahedron Lett* 1996;**37**:5313.
20. Durand G, Choteau F, Pucci B, Villamena FA. Reactivity of superoxide radical anion and hydroperoxyl radical with alpha-phenyl-N-tert-butylnitrone (PBN) derivatives. *J Phys Chem A* 2008;**112**:12498.
21. Frejaville C, Karoui H, Tuccio B, Le Moigne F, Culcasi M, Pietri S, et al. 5-(Diethoxyphosphoryl)-5-methyl-1-pyrroline N-oxide: a new efficient phosphorylated nitrone for the *in vitro* and *in vivo* spin trapping of oxygen-centered radicals. *J Med Chem* 1995;**38**:258.
22. Olive G, Mercier A, Moigne FL, Rockenbauer A, Tordo P. 2-ethoxycarbonyl-2-methyl-3,4-dihydro-2H-pyrrole-1-oxide: evaluation of the spin trapping properties. *Free Radical Biol Med* 2000;**28**:403.
23. Villamena FA, Rockenbauer A, Gallucci J, Velayutham M, Hadad CM, Zweier JL. Spin trapping by 5-carbamoyl-5-methyl-1-pyrroline N-oxide (AMPO): theoretical and experimental studies. *J Org Chem* 2004;**69**:7994.
24. Villamena FA, Xia S, Merle JK, Lauricella R, Tuccio B, Hadad CM, et al. Reactivity of Superoxide Radical Anion with Cyclic Nitrones: Role of Intramolecular H-Bond and Electrostatic Effects. *J Am Chem Soc* 2007;**129**:8177.
25. Buettner GR. On the reaction of superoxide with DMPO/.OOH. *Free Radic Res Commun* 1990;**10**:11.
26. Allouch A, Lauricella RP, Tuccio BN. Effect of pH on superoxide/hydroperoxyl radical trapping by nitrones: an EPR/kinetic study. *Mol Phys* 2007;**105**:2017.
27. Pou S, Ramos CL, Gladwell T, Renks E, Centra M, Young D, et al. A Kinetic Approach to the Selection of a Sensitive Spin Trapping System for the Detection of Hydroxyl Radical. *Analytical Biochemistry* 1994;**217**:76.
28. Davies MJ, Forni LG, Shuter SL. Electron spin resonance and pulse radiolysis studies on the spin trapping of sulphur-centered radicals. *Chem-Biol Interact* 1987;**61**:177.
29. Villamena FA, Hadad CM, Zweier JL. Kinetic Study and Theoretical Analysis of Hydroxyl Radical Trapping and Spin Adduct Decay of Alkoxycarbonyl and Dialkoxyphosphoryl Nitrones in Aqueous Media. *J Phys Chem A* 2003;**107**:4407.
30. Buettner GR. The pecking order of free radicals and antioxidants: lipid peroxidation, alpha-tocopherol, and ascorbate. *Arch Biochem Biophys* 1993;**300**:535.
31. Madej E, Wardman P. The oxidizing power of the glutathione thiyl radical as measured by its electrode potential at physiological pH. *Arch Biochem Biophys* 2007;**462**:94.
32. Villamena FA, Merle JK, Hadad CM, Zweier JL. Superoxide Radical Anion Adduct of 5,5-Dimethyl-1-pyrroline N-Oxide (DMPO). 2. The Thermodynamics of Decay and EPR Spectral Properties. *J Phys Chem A* 2005;**109**:6089.
33. Behar D, Czapski G, Rabani J, Dorfman LM, Schwarz HA. Acid dissociation constant and decay kinetics of the perhydroxyl radical. *J Phys Chem* 1970;**74**:3209.
34. Czapski G, Bielski BHJ. The formation and decay of H_2O_3 and HO_2 in electron-irradiated aqueous solutions. *J Phys Chem* 1963;**67**:2180.
35. Villamena FA, Merle JK, Hadad CM, Zweier JL. Superoxide radical anion adduct of 5,5-dimethyl-1-pyrroline n-oxide (DMPO). 1. The thermodynamics of formation and its acidity. *J Phys Chem A* 2005;**109**:6083.
36. Burgett RA, Bao X, Villamena FA. Superoxide radical anion adduct of 5,5-dimethyl-1-pyrroline N-oxide (DMPO). 3. Effect of mildly acidic pH on the thermodynamics and kinetics of adduct formation. *J Phys Chem A* 2008;**112**:2447.
37. Finkelstein E, Rosen GM, Rauckman EJ. Spin trapping. Kinetics of the reaction of superoxide and hydroxyl radicals with nitrones. *J Am Chem Soc* 1980;**102**:4994.
38. Villamena FA, Zweier JL. Superoxide radical trapping and spin adduct decay of 5-tert-butoxycarbonyl-5-methyl-1-pyrroline N-oxide (BocMPO): kinetics and theoretical analysis. *J Chem Soc Perkin Trans* 2002;**2**:1340.

39. Frejaville C, Karoui H, Tuccio B, Le Moigne F, Culcasi M, Pietri S, et al. 5-(Diethoxyphosphoryl)-5-methyl-1-pyrroline N-oxide: a new phosphorylated nitrone for the efficient in vitro and in vivo spin trapping of oxygen-centered radicals. *J Chem Soc Chem Commun* 1994;1793.
40. Allouch A, Roubaud V, Lauricella R, Bouteiller J-C, Tuccio B. Spin trapping of superoxide by diester-nitrones. *Org Biomol Chem* 2005;**3**:2458.
41. Lauricella R, Allouch A, Roubaud V, Bouteiller J-C, Tuccio B. A new kinetic approach to the evaluation of rate constants for the spin trapping of superoxide/hydroperoxyl radical by nitrones in aqueous media. *Org Biomol Chem* 2004;**2**:1304.
42. Lauricella RP, Bouteiller J-CH, Tuccio BN. Evidence of overestimation of rate constants for the superoxide trapping by nitrones in aqueous media. *Phys Chem Chem Phys* 2005;**7**:399.
43. Tsai P, Ichikawa K, Mailer C, Pou S, Halpern HJ, Robinson BH, et al. Esters of 5-carboxyl-5-methyl-1-pyrroline N-oxide: A family of spin traps for superoxide. *J Org Chem* 2003;**68**:7811.
44. Kezler A, Kalyanaraman B, Hogg N. Comparative investigation of superoxide radical trapping by cyclic nitrone spin traps. *Free Radical Biol Med* 2003;**35**:1149.
45. Goldstein S, Rosen GM, Russo A, Samuni A. Kinetics of spin trapping superoxide, hydroxyl, and aliphatic radicals by cyclic nitrones. *J Phys Chem A* 2004;**108**:6679.
46. Yamazaki I, Piette LH, Grover TA. Kinetic studies on spin trapping of superoxide and hydroxyl radicals generated in NADPH-cytochrome P-450 reductase-paraquat systems. Effect of iron chelates. *J Biol Chem* 1990;**265**:652.
47. Han Y, Tuccio B, Lauricella R, Villamena FA. Improved Spin Trapping Properties by beta-Cyclodextrin-Cyclic Nitrone Conjugate. *J Org Chem* 2008;**73**:7108.
48. Kim S-U, Liu Y, Nash KM, Zweier JL, Rockenbauer A, Villamena FA. Fast Reactivity of Cyclic Nitrone-Calix[4]pyrrole Conjugate with Superoxide Radical Anion: Theoretical and Experimental Studies. *J Am Chem Soc* 2010;**132**:17157.
49. Chalier F, Tordo P. 5-Diisopropoxyphosphoryl-5-methyl-1-pyrroline N-oxide, DIPPMPO, a crystalline analog of the nitrone DEPMPO: synthesis and spin trapping properties. *J Chem Soc Perkin Trans* 2002;**2**:2110.
50. Buettner GR, Oberley LW. Considerations in the spin trapping of superoxide and hydroxyl radical in aqueous systems using 5,5-dimethyl-1-pyrroline-1-oxide. *Biochem Biophys Res Commun* 1978;**83**:69.
51. Burkitt MJ, Tsang SY, Tam SC, Bremner I. Generation of 5,5-dimethyl-1-pyrroline N-oxide hydroxyl and scavenger radical adducts from copper/H2O2 mixtures: effects of metal ion chelation and the search for high-valent metal-oxygen intermediates. *Arch Biochem Biophys* 1995;**323**:63.
52. Villamena FA. Superoxide Radical Anion Adduct of 5,5-Dimethyl-1-pyrroline N-Oxide. 5. Thermodynamics and Kinetics of Unimolecular Decomposition. *J Phys Chem A* 2009;**113**:6398.
53. Villamena FA. Superoxide Radical Anion Adduct of 5,5-Dimethyl-1-pyrroline N-Oxide. 6. Redox Properties. *J Phys Chem A* 2009;**114**:1153.
54. Khan N, Wilmot CM, Rosen GM, Demidenko E, Sun J, Joseph J, et al. Spin traps: in vitro toxicity and stability of radical adducts. *Free Radic Biol Med* 2003;**34**:1473.
55. Karoui H, Rockenbauer A, Pietri S, Tordo P. Spin trapping of superoxide in the presence of beta-cyclodextrins. *Chem Commun (Cambridge, UK)* 2002;3030.
56. Bardelang D, Rockenbauer A, Karoui H, Finet J-P, Biskupska I, Banaszak K, et al. Inclusion complexes of EMPO derivatives with 2,6-di-O-methyl-beta-cyclodextrin: synthesis, NMR and EPR investigations for enhanced superoxide detection. *Org Biomol Chem* 2006;**4**:2874.
57. Zweier JL, Broderick R, Kuppusamy P, Thompson-Gorman S, Lutty GA. Determination of the mechanism of free radical generation in human aortic endothelial cells exposed to anoxia and reoxygenation. *J Biol Chem* 1994;**269**:24156.
58. Xia Y, Zweier JL. Superoxide and peroxynitrite generation from inducible nitric oxide synthase in macrophages. *Proc Natl Acad Sci USA* 1997;**94**:6954.
59. Snyrychova I. Improvement of the sensitivity of EPR spin trapping in biological systems by cyclodextrins: A model study with thylakoids and photosystem II particles. *Free Radic Biol Med* 2010;**48**:264.
60. Das A, Gopalakrishnan B, Druhan LJ, Wang TY, De Pascali F, Rockenbauer A, et al. Reversal of SIN-1-induced eNOS dysfunction by the spin trap, DMPO, in bovine aortic endothelial cells via eNOS phosphorylation. *Br J Pharmacol* 2014;**171**:2321.

61. Hardy M, Bardelang D, Karoui H, Rockenbauer A, Finet J-P, Jicsinszky L, et al. Improving the Trapping of Superoxide Radical with a Beta-Cyclodextrin- 5-Diethoxyphosphoryl-5-methyl-1-pyrroline-N-oxide (DEPMPO) Conjugate. *Chem - Eur J* 2009;**15**:11114.
62. Beziere N, Hardy M, Poulhes F, Karoui H, Tordo P, Ouari O, et al. Metabolic stability of superoxide adducts derived from newly developed cyclic nitrone spin traps. *Free Radical Biol Med* 2014;**67**:150.
63. McIntire GL, Blount HN, Stronks HJ, Shetty RV, Janzen EG. Spin trapping in electrochemistry. 2. Aqueous and nonaqueous electrochemical characterizations of spin traps. *J Phys Chem* 1980;**84**:916.
64. Koppenol WH. Thermodynamics of reactions involving nitrogen-oxygen compounds. *Methods Enzymol* 1996;**268**:7.
65. Haseloff RF, Mertsch K, Rohde E, Baeger I, Grigor'ev IA, Blasig IE. Cytotoxicity of spin trapping compounds. *FEBS Lett* 1997;**418**:73.
66. Rohr-Udilova N, Stolze K, Marian B, Nohl H. Cytotoxicity of novel derivatives of the spin trap EMPO. *Bioorg Med Chem Lett* 2006;**16**:541.
67. Durand G, Prosak RA, Han Y, Ortial S, Rockenbauer A, Pucci B, et al. Spin trapping and cytoprotective properties of fluorinated amphiphilic carrier conjugates of cyclic versus linear nitrones. *Chem Res Toxicol* 2009;**22**:1570.
68. Janzen EG, Poyer JL, Schaefer CF, Downs PE, DuBose CM. Biological spin trapping. II. Toxicity of nitrone spin traps: dose-ranging in the rat. *J Biochem Biophys Methods* 1995;**30**:239.
69. Schaefer CF, Janzen EG, West MS, Poyer JL, Kosanke SD. Blood chemistry changes in the rat induced by high doses of nitronyl free radical spin traps. *Free Radic Biol Med* 1996;**21**:427.
70. Liu K, Jiang J, Ji L, Shi X, Swartz H. An HPLC and EPR investigation on the stability of DMPO and DMPO spin adducts in vivo. *Res Chem Intermed* 1996;**22**:499.
71. Breuer E, Aurich HG, Nielsen A. *Nitrones, Nitronates and Nitroxides*. New York: John Wiley and Sons; 1989.
72. Torsell KBG. *Nitrile Oxides, Nitrones, and Nitronates in Organic Synthesis*. VCH Verlagsgesellschaft mbH: Weinheim; 1988.
73. Murahashi SI. Synthetic apects of metal-catalyzed oxidations of amines and related reactions. *Angew Chem Int Ed Engl* 1995;**34**:2443.
74. Rosen GM, Britigan BE, Halpern HJ, Pou S. *Free Radicals: Biology and Detection by Spin Trapping*. New York: Oxford University Press; 1999.
75. Villamena FA, Dickman MH, Crist DR. Nitrones as Ligands in Complexes of Cu(II), Mn(II), Co (II), Ni(II), Fe(II), and Fe(III) with N-tert-Butyl–(2-pyridyl)nitrone and 2,5,5-Trimethyl-1-pyrroline-N-oxide. *Inorg Chem* 1998;**37**:1446.
76. Zhang Y-K, Lu D-H, Xu GZ. Synthesis and radical additional sterochemistry of two trimethyl-1-pyrroline 1-oxides as studied by EPR spectroscopy. *J Chem Soc Perkin Trans* 1991;**2**:1855—60.
77. Turner MJ, Rosen GM. Spin trapping of superoxide and hydroxyl radicals with substituted pyrroline 1-oxides. *J Med Chem* 1986;**29**:2439.
78. Villamena F, Zweier J. Superoxide radical trapping and spin adduct decay of 5-tert-butoxycarbonyl-5-methyl-1-pyrroline N-oxide (BocMPO): kinetics and theoretical analysis. *J Chem Soc Perkin Trans* 2002;**2**:1340.
79. Pou S, Rosen GM, Wu Y, Rosen GM. Synthesis of deuterium and ^{15}N-containing pyrroline 1-oxides: Spin trapping study. *J Org Chem* 1990;**55**:4438.
80. Cisneros L, Serna P, Corma A. Selective reductive coupling of nitro compounds with aldehydes to nitrones in H2 using carbon-supported and -decorated platinum nanoparticles. *Angew Chem Int Ed* 2014;**53**:9306.
81. Joseph R, Sudalai A, Ravindranathan T. Catalytic oxidation of secondary amines with H_2O_2 over molecular sieves: A high yield and single step preparation. *Synlett* 1995;1177.
82. Murahashi SI, Mitsui H, Shiota T, Tsuda T, Watanabe S. Tungstate-catalyzed oxidation of secondary amines to nitrones. a-Substitution of secondary amines via nitrones. *J Org Chem* 1990;**55**:1736.
83. Murahashi SI, Shiota T. Selenium dioxide catalyzed oxidation of secondary amines with hydrogen peroxide. Simple synthesis of nitrones from secondary amines. *Tetrahedron Lett* 1987;**28**:2383.
84. Forcato M, Nugent WA, Licini G. A "waterproof" catalyst for the oxidation of secondary amines to nitrones with alkyl hydroperoxides. *Tetrahedron Lett* 2003;**44**:49.
85. Tokuyama H, Kuboyama T, Amano A, Yamashita T, Fukuyama T. A novel transformation of primary amines to N-monoalkylhydroxylamines. *Synthesis* 2000;1299.

86. Boyd DR, Coulter PB, McGuckin MR, Sharma ND, Jennings WB, Wilson VE. Imines and derivatives. Part 24. Nitrone synthesis by imine oxidation using either a peroxyacid or dimethyldioxirane. *J Chem Soc Perkin Trans* 1990;**1**:301.
87. Ferrer M, Sanchez-Baeza F, Casas J, Messeguer A. Decomposition of dioxiranes induced by dialkyl ethers. *Tetrahedron Lett* 1994;**35**:2981.
88. Busque F, de March P, Figueredo M, Font J, Gallagher T, Milan S. Efficient synthesis of (S)-3,4-dihydro-2-[(pivaloyloxy)methyl]-2H-pyrrole 1-oxide. *Tetrahedron: Asymmetry* 2002;**13**:437.
89. Dehnel A, Griller D, Kanabus-Kaminska JM. Designer spin traps with a cyclic nitrone structure. *J Org Chem* 1988;**53**:1566.
90. Cicchi S, Corsi M, Goti A. Inexpensive and environmentally friendly oxidation of hydroxylamine to nitrones with bleach. *J Org Chem* 1999;**64**:7243.
91. Cicchi S, Marradi M, Goti A, Brandi A. Manganese dioxide oxidation of hydroxylamines to nitrones. *Tetrahedron Lett* 2001;**42**:6503.
92. Zauche TH, Espenson JH. Kinetics and mechanism of the oxidation of secondary hydroxylamines to nitrones with hydrogen peroxide, catalyzed by methylrhenium trioxide. *Inorg Chem* 1997;**36**:5257.
93. Bonnett R, Brown RFC, Clark VM, Sutherland IO, Todd A. Experiments towards the synthesis of corrins. Part II. The preparation and reactions of D^1-pyrroline 1-oxides. *J Chem Soc* 1959;2094.
94. Buehler E. Alkylation of syn- and anti-benzaldoximes. *J Org Chem* 1967;**32**:261.
95. Christensen D, Jorgensen KA, Hazell RG. Rearrangement reactions of oxaziridines to nitrones. X-Ray crystal and molecular structure of N-t-butyl-[small alpha](o-hydroxyphenyl)nitrone. *J Chem Soc Perkin Trans* 1990;**1**:2391.
96. Griffith OH, Cornell DW, McConnell HM. Nitrogen hyperfine tensor and g tensor of nitroxide radicals. *J Chem Phys* 1965;**43**:2909.
97. Taniguchi H, Madden KP. DMPO-alkyl radical spin trapping: an in situ radiolysis steady-state ESR study. *Radiat Res* 2000;**153**:447.
98. Taniguchi H, Madden KP. In Situ Radiolysis Steady-State ESR Study of Carboxyalkyl Radical Trapping by 5,5-Dimethyl-1-pyrroline-N-oxide: Spin Adduct Structure and Stability. *J Phys Chem A* 1998;**102**:6753.
99. Cholvad V, Szaboova K, Stasko A, Nuyken O, Voit B. ESR parameters of 5,5-dimethylpyrrolidine 1-oxide (DMPO) spin adducts in the photochemical decomposition of azo compounds. *Magn Reson Chem* 1991;**29**:402.
100. Davies MJ, Forni LG, Shuter SL. Electron spin resonance and pulse radiolysis studies on the spin trapping of sulfur-centered radicals. *Chem-Biol Interact* 1987;**61**:177.
101. Villamena FA, Locigno EJ, Rockenbauer A, Hadad CM, Zweier JL. Theoretical and experimental studies of the spin trapping of inorganic radicals by 5,5-dimethyl-1-pyrroline N-oxide (DMPO). 1. Carbon dioxide radical anion. *J Phys Chem A* 2006;**110**:13253.
102. Villamena FA, Locigno EJ, Rockenbauer A, Hadad CM, Zweier JL. Theoretical and experimental studies of the spin trapping of inorganic radicals by 5,5-dimethyl-1-pyrroline N-oxide (DMPO). 2. Carbonate radical anion. *J Phys Chem A* 2007;**111**:384.
103. Nash KM, Rockenbauer A, Villamena FA. Reactive nitrogen species reactivities with nitrones: theoretical and experimental studies. *Chem Res Toxicol* 2012;**25**:1581.
104. Zamora PL, Villamena FA. Theoretical and Experimental Studies of the Spin Trapping of Inorganic Radicals by 5,5-Dimethyl-1-pyrroline N-Oxide (DMPO). 3. Sulfur Dioxide, Sulfite, and Sulfate Radical Anions. *J Phys Chem A* 2012;**116**:7210.
105. Reszka K, Bilski P, Chignell CF. EPR spectra of DMPO spin adducts of superoxide and hydroxyl radicals in pyridine. *Free Rad Res Commun* 1992;**17**:377.
106. Josephy PD, Rehorek D, Janzen EG. Electron spin resonance spin trapping of thiyl, radicals from the decomposition of thionitrites. *Tetrahedron Lett* 1984;**25**:1685.
107. Bradshaw TP, McMillan DC, Crouch RK, Jollow DJ. Identification of free radicals produced in rat erythrocytes exposed to hemolytic concentrations of phenylhydroxylamine. *Free Radic Biol Med* 1995;**18**:279.

108. Villamena FA, Liu Y, Zweier JL. Superoxide radical anion adduct of 5,5-dimethyl-1-pyrroline N-oxide. 4. Conformational effects on the EPR hyperfine splitting constants. *J Phys Chem A* 2008;**112**:12607.
109. Clement JL, Ferre N, Siri D, Karoui H, Rockenbauer A, Tordo P. Assignment of the EPR spectrum of 5,5-dimethyl-1-pyrroline N-oxide (DMPO) superoxide spin adduct. *J Org Chem* 2005;**70**:1198.
110. Rosen GM, Beselman A, Tsai P, Pou S, Mailer C, Ichikawa K, et al. Influence of conformation on the EPR spectrum of 5,5-dimethyl-1-hydroperoxy-1-pyrrolidinyloxyl: a spin trapped adduct of superoxide. *J Org Chem* 2004;**69**:1321.
111. Villamena FA, Hadad CM, Zweier JL. Theoretical study of the spin trapping of hydroxyl radical by cyclic nitrones: a density functional theory approach. *J Am Chem Soc* 2004;**126**:1816.
112. Karoui H, Chalier F, Finet J-P, Tordo P. DEPMPO: an efficient tool for the coupled ESR-spin trapping of alkylperoxyl radicals in water. *Org Biomol Chem* 2011;**9**:2473.
113. Villamena FA, Zweier JL. Detection of reactive oxygen and nitrogen species by EPR spin trapping. *Antioxid Redox Signal* 2004;**6**:619.
114. Duling DR. Simulation of Multiple Isotropic Spin-Trap EPR Spectra. *Journal of Magnetic Resonance Series B* 1994;**104**:105.
115. Stoll S, Schweiger A. EasySpin, a comprehensive software package for spectral simulation and analysis in EPR. *J Magn Reson* 2006;**178**:42.
116. Rockenbauer A, Korecz L. Automatic Computer Simulations of ESR Spectra. *Appl Magn Reson* 1996;**10**:29.
117. Samouilov A, Roubaud V, Kuppusamy P, Zweier JL. Kinetic analysis-based quantitation of free radical generation in EPR spin trapping. *Analytical Biochemistry* 2004;**334**:145.
118. Hanna PM, Chamulitrat W, Mason RP. When are metal ion-dependent hydroxyl and alkoxyl radical adducts of 5,5-dimethyl-1-pyrroline N-oxide artifacts? *Arch Biochem Biophys* 1992;**296**:640.
119. Eberson L. *Toxicology of the Human Environment: The Critical Role of Free Radicals*. New York, NY: Taylor and Francis, Inc; 2000.
120. Kanabus-Kaminska JM, Gilbert BC, Griller D. Solvent effects on the thermochemistry of free-radical reactions. *J Am Chem Soc* 1989;**111**:3311.
121. Mason R, Kadiiska M. In: Eaton S, Eaton G, Berliner L, editors. *Biomedical EPR, Part A: Free Radicals, Metals, Medicine, and Physiology*, Vol. 23. US: Springer; 2005. p. 93.
122. Zweier JL, Kuppusamy P, Thompson-Gorman S, Klunk D, Lutty GA. Measurement and characterization of free radical generation in reoxygenated human endothelial cells. *Am J Physiol* 1994;**266**: C700.
123. Shi H, Timmins G, Monske M, Burdick A, Kalyanaraman B, Liu Y, et al. Evaluation of spin trapping agents and trapping conditions for detection of cell-generated reactive oxygen species. *Arch Biochem Biophys* 2005;**437**:59.
124. Hougaard AB, Arneborg N, Andersen ML, Skibsted LH. ESR spin trapping for characterization of radical formation in Lactobacillus acidophilus NCFM and Listeria innocua. *J Microbiol Methods* 2013;**94**:205.
125. Ranguelova K, Rice AB, Khajo A, Triquigneaux M, Garantziotis S, Magliozzo RS, et al. Formation of reactive sulfite-derived free radicals by the activation of human neutrophils: an ESR study. *Free Radic Biol Med* 2012;**52**:1264.
126. Deschacht M, Horemans T, Martinet W, Bult H, Maes L, Cos P. Comparative EPR study of different macrophage types stimulated for superoxide and nitric oxide production. *Free Radic Res* 2010;**44**:763.
127. Bannister JV, Bannister WH. Production of oxygen-centered radicals by neutrophils and macrophages as studied by electron spin resonance (ESR). *Environ Health Perspect* 1985;**64**:37.
128. Dikalov SI, Mason RP. Spin trapping of polyunsaturated fatty acid-derived peroxyl radicals: reassignment to alkoxyl radical adducts. *Free Radic Biol Med* 2001;**30**:187.
129. Zhao B, He Y-Y, Bilski PJ, Chignell CF. Pristine (C60) and Hydroxylated [C60(OH)24] Fullerene Phototoxicity towards HaCaT Keratinocytes: Type I vs Type II Mechanisms. *Chem Res Toxicol* 2008;**21**:1056.

130. Sugimoto R, Kumagai Y, Nakai Y, Ishii T. 9,10-Phenanthraquinone in diesel exhaust particles downregulates Cu,Zn-SOD and HO-1 in human pulmonary epithelial cells: Intracellular iron scavenger 1,10-phenanthroline affords protection against apoptosis. *Free Radical Biol Med* 2005;**38**:388.
131. Rota C, Fann YC, Mason RP. Phenoxyl free radical formation during the oxidation of the fluorescent dye 2',7'-dichlorofluorescein by horseradish peroxidase. Possible consequences for oxidative stress measurements. *J Biol Chem* 1999;**274**:28161.
132. Chen C, Zhou B, Lu D, Xu G. Electron transfer events in solutions of cyanine dyes. *J Photochem Photobiol A* 1995;**89**:25.
133. Masaki H, Okano Y, Sakurai H. Generation of active oxygen species from advanced glycation end-products (AGEs) during ultraviolet light A (UVA) irradiation and a possible mechanism for cell damaging. *Biochim Biophys Acta Gen Subj* 1999;**1428**:45.
134. Zang L-y, Stone K, Pryor WA. Detection of free radicals in aqueous extracts of cigarette tar by electron spin resonance. *Free Radical Biol Med* 1995;**19**:161.
135. Chignell CF, Bilski P, Reszka KJ, Motten AG, Sik RH, Dahl TA. Spectral and photochemical properties of curcumin. *Photochem Photobiol* 1994;**59**:295.
136. Xia Y, Tsai A-L, Berka V, Zweier JL. Superoxide generation from endothelial nitric-oxide synthase. A Ca2+/calmodulin-dependent and tetrahydrobiopterin regulatory process. *J Biol Chem* 1998;**273**:25804.
137. Vasquez-Vivar J, Martasek P, Hogg N, Karoui H, Masters BSS, Pritchard KA, et al. Electron spin resonance spin-trapping detection of superoxide generated by neuronal nitric oxide synthase. *Methods Enzymol* 1999;**301**:169.
138. Xia Y, Berlowitz CO, Zweier JL. PIN inhibits nitric oxide and superoxide production from purified neuronal nitric oxide synthase. *Biochim Biophys Acta, Gen Subj* 2006;**1760**:1445.
139. Cardounel AJ, Xia Y, Zweier JL. Endogenous Methylarginines Modulate Superoxide as Well as Nitric Oxide Generation from Neuronal Nitric-oxide Synthase: Differences in the Effects of Monomethyl- and Dimethylarginines in the Presence and Absence of Tetrahydrobiopterin. *J Biol Chem* 2005;**280**:7540.
140. Xia Y, Roman LJ, Masters BSS, Zweier JL. Inducible nitric-oxide synthase generates superoxide from the reductase domain. *J Biol Chem* 1998;**273**:22635.
141. Xia Y, Zweier JL. Superoxide and peroxynitrite generation from inducible nitric oxide synthase in macrophages. *Proc Natl Acad Sci USA* 1997;**94**:6954.
142. Chamulitrat W, Hughes MF, Eling TE, Mason RP. Superoxide and peroxyl radical generation from the reduction of polyunsaturated fatty acid hydroperoxides by soybean lipoxygenase. *Arch Biochem Biophys* 1991;**290**:153.
143. Kukavica B, Mojovic M, Vuccinic Z, Maksimovic V, Takahama U, Jovanovic SV. Generation of hydroxyl radical in isolated pea root cell wall, and the role of cell wall-bound peroxidase, Mn-SOD and phenolics in their production. *Plant Cell Physiol* 2009;**50**:304.
144. Kundu TK, Velayutham M, Zweier JL. Aldehyde Oxidase Functions as a Superoxide Generating NADH Oxidase: An Important Redox Regulated Pathway of Cellular Oxygen Radical Formation. *Biochemistry* 2012;**51**:2930.
145. Velayutham M, Hemann C, Zweier JL. Removal of H2O2 and generation of superoxide radical: Role of cytochrome c and NADH. *Free Radical Biol Med* 2011;**51**:160.
146. Chen J, Chen C-L, Rawale S, Chen C-A, Zweier JL, Kaumaya PTP, et al. Peptide-based Antibodies against Glutathione-binding Domains Suppress Superoxide Production Mediated by Mitochondrial Complex I. *J Biol Chem* 2010;**285**:3168.
147. Chen Y-R, Chen C-L, Yeh A, Liu X, Zweier JL. Direct and Indirect Roles of Cytochrome b in the Mediation of Superoxide Generation and NO Catabolism by Mitochondrial Succinate-Cytochrome c Reductase. *J Biol Chem* 2006;**281**:13159.
148. Chen Y-R, Chen C-L, Zhang L, Green-Church KB, Zweier JL. Superoxide Generation from Mitochondrial NADH Dehydrogenase Induces Self-inactivation with Specific Protein Radical Formation. *J Biol Chem* 2005;**280**:37339.

149. Hardy M, Chalier F, Ouari O, Finet J-P, Rockenbauer A, Kalyanaraman B, et al. Mito-DEPMPO synthesized from a novel NH2-reactive DEPMPO spin trap: a new and improved trap for the detection of superoxide. *Chem Commun (Cambridge, UK)* 2007;1083.
150. Morehouse KM, Moreno SNJ, Mason RP. The one-electron reduction of uroporphyrin I by rat hepatic microsomes. *Arch Biochem Biophys* 1987;**257**:276.
151. Harbour JR, Bolton JR. Superoxide formation in spinach chloroplasts: Electron spin resonance detection by spin trapping. *Biochem Biophys Res Commun* 1975;**64**:803.
152. Gopalakrishnan B, Nash KM, Velayutham M, Villamena FA. Detection of nitric oxide and superoxide radical anion by electron paramagnetic resonance spectroscopy from cells using spin traps. *J Visualized Exp* 2012. E2810/1.
153. Shi H, Timmins G, Monske M, Burdick A, Kalyanaraman B, Liu Y, et al. Evaluation of spin trapping agents and trapping conditions for detection of cell-generated reactive oxygen species. *Arch Biochem Biophys* 2005;**437**:59.
154. Sankarapandi S, Zweier JL, Mukherjee G, Quinn MT, Huso DL. Measurement and characterization of superoxide generation in microglial cells: evidence for an NADPH oxidase-dependent pathway. *Arch Biochem Biophys* 1998;**353**:312.
155. Silverton SF, Mesaros S, Markham GD, Malinski T. Osteoclast radical interactions: NADPH causes pulsatile release of NO and stimulates superoxide production. *Endocrinology* 1995;**136**:5244.
156. Zweier JL, Duke SS, Kuppusamy P, Sylvester JT, Gabrielson EW. Electron paramagnetic resonance evidence that cellular oxygen toxicity is caused by the generation of superoxide and hydroxyl free radicals. *FEBS Lett* 1989;**252**:12.
157. Zweier JL, Kuppusamy P, Lutty GA. Measurement of endothelial cell free radical generation: evidence for a central mechanism of free radical injury in postischemic tissues. *Proc Natl Acad Sci USA* 1988;**85**:4046.
158. Togashi H, Shinzawa H, Matsuo T, Takeda Y, Takahashi T, Aoyama M, et al. Analysis of hepatic oxidative stress status by electron spin resonance spectroscopy and imaging. *Free Radical Biol Med* 2000;**28**:846.
159. Zweier JL. Measurement of superoxide-derived free radicals in the reperfused heart. Evidence for a free radical mechanism of reperfusion injury. *J Biol Chem* 1988;**263**:1353.
160. Zweier JL, Kuppusamy P, Williams R, Rayburn BK, Smith D, Weisfeldt ML, et al. Measurement and characterization of postischemic free radical generation in the isolated perfused heart. *J Biol Chem* 1989;**264**:18890.
161. Haywood R, Andrady C, Kassouf N, Sheppard N. Intensity-dependent direct solar radiation- and UVA-induced radical damage to human skin and DNA, lipids and proteins. *Photochem Photobiol* 2011;**87**:117.
162. Souza HP, Souza LC, Anastacio VM, Pereira AC, Junqueira ML, Krieger JE, et al. Vascular oxidant stress early after balloon injury: evidence for increased NAD(P)H oxidoreductase activity. *Free Radic Biol Med* 2000;**28**:1232.
163. Hartell MG, Borzone G, Clanton TL, Berliner LJ. Detection of free radicals in blood by electron spin resonance in a model of respiratory failure in the rat. *Free Radic Biol Med* 1994;**17**:467.
164. Kadiiska MB, Mason RP. Ethylene glycol generates free radical metabolites in rats: an ESR in vivo spin trapping investigation. *Chem Res Toxicol* 2000;**13**:1187.
165. Ghio AJ, Kadiiska MB, Xiang QH, Mason RP. In vivo evidence of free radical formation after asbestos instillation: an ESR spin trapping investigation. *Free Radic Biol Med* 1998;**24**:11.
166. Mergner GW, Weglicki WB, Kramer JH. Postischemic free radical production in the venous blood of the regionally ischemic swine heart. Effect of deferoxamine. *Circulation* 1991;**84**:2079.
167. Lai EK, Crossley C, Sridhar R, Misra HP, Janzen EG, McCay PB. In vivo spin trapping of free radicals generated in brain, spleen, and liver during gamma radiation of mice. *Arch Biochem Biophys* 1986;**244**:156.
168. Murray AR, Kisin E, Castranova V, Kommineni C, Gunther MR, Shvedova AA. Phenol-induced in vivo oxidative stress in skin: evidence for enhanced free radical generation, thiol oxidation, and antioxidant depletion. *Chem Res Toxicol* 2007;**20**:1769.

CHAPTER 6

UV–Vis Absorption and Chemiluminescence Techniques

6.1 INTRODUCTION

Spectrophotometric techniques involving ultraviolet and colorimetric detection offer a convenient way of detecting reactive species (RS) formation due to the prevalence of UV–Vis spectrophotometer in research laboratories. Conventional spectrophotometric measurement of RS exploits their unique chemical reactivity with a small organic molecule and/or enzyme that has specificity to these RS where formation (or disappearance) of absorption peak/s at a particular wavelength is observed at the UV to the visible region of the electromagnetic spectrum. Direct detection of $O_2^{\bullet -}$ from xanthine oxidase in the past involves rapid-freeze technique using electron paramagnetic resonance (EPR) spectroscopy.[1] At the same time, spectrophotometric detection of $O_2^{\bullet -}$ generated from xanthine oxidase/xanthine system was also employed for the investigation of the enzymatic property of superoxide dismutase (SOD) where the $O_2^{\bullet -}$ levels were measured through reduction of ferricytochrome *c*, tetranitromethane, or oxidation of epinephrine to adenochrome.[2] Several approaches have been developed since then to improve sensitivity with the use of submicromolar probe concentration thus allowing minimal interferences of the probe on the biological process being investigated. Specificity has also been improved to increase the reaction rate of probes to certain RS through synthesis of new and innovative analogues that exploit the unique chemistry between the probe and RS. Chromophore stability was also achieved through improved molecular design, optimized experimental conditions, or addition of supramolecular reagents since chromophores impart inherent thermodynamic stability as a function of its chemical structure, solvent polarity, pH, temperature, or due to presence of oxido-reductants or other reactive substances. Improved sample preparation and high-throughput analysis were also developed in order to maximize efficiency in the measurement markers of oxidative stress and determination of antioxidant capacity (AOC) of known molecules, food, biological fluid, or tissue. Therefore, spectrophotometric techniques for RS measurement have found broad application in the fields of biomedical research, clinical chemistry, plant biology, food chemistry, environmental chemistry, radiation chemistry, pharmaceuticals, toxicology, or material science to name a few, or just simply for the investigation of RS production in simple chemical systems. However, unlike the fluorescence probes,

Reactive Species Detection in Biology
DOI: http://dx.doi.org/10.1016/B978-0-12-420017-3.00006-2

spectrophotometric as well as most applications of chemiluminescence probes do not provide spectrospatial image of the RS localization in cells, hence, one cannot deduce the site of radical production unless multiple probes and/or inhibitors are used with varying compartmentalization property (i.e., extracellular or intracellular). For example, in the investigation of radical production in cellular NADPH oxidase, several methods had been suggested such as measurement of O_2 consumption, use of SOD-inhibitable probe such as ferricytochrome *c* and horseradish peroxidase (HRP)/Amplex Red for extracellular $O_2^{\bullet-}$ and H_2O_2, respectively, and the HPLC analysis of the 2-OH-E^+ marker for the quantification of intracellular $O_2^{\bullet-}$.[3]

6.2 SUPEROXIDE RADICAL PROBES

6.2.1 Nitro blue tetrazolium

One of the earliest application of nitro blue tetrazolium (NBT or NBT^{2+}) was in the study of $O_2^{\bullet-}$ −dependent formation of oxygen-free radicals by aldehyde oxidase as evidenced by its reduction to form the water-insoluble diformazan (Fig. 6.1). However, the formation of the water-insoluble diformazan reduction product requires dissolution in organic solvent prior to colorimetric measurement which makes this method cumbersome. The solubilized reduction product of NBT^{2+} can be spectrophotometrically monitored at 535 nm and was used for the detection of $O_2^{\bullet-}$ involving assay for SOD activity in human red blood cells,[4] leukocytes,[5] or various photosynthetic organisms.[6] Rate constant of $O_2^{\bullet-}$ reduction of NBT^{2+} using pulse radiolytic technique at pH 7−11 was found to be $k = 5.88 \times 10^4\ M^{-1}\ s^{-1}$ while no reduction was observed between the pH of 2 and 3 indicating that $HO_2^{\bullet}$ does not react with NBT^{2+}.[7] Kinetic evidence showed that $O_2^{\bullet-}$ reaction with NBT^{2+} yields $NBT^{\bullet+}$ and O_2 where $NBT^{\bullet+}$ ($\lambda_{max} = 405$ nm) undergoes second-order disproportionation reaction to produce the monoformazan ($\lambda_{max} = 530$ nm) and the original reactant NBT^{2+}.

Under anaerobic conditions, NBT^{2+} could also act as direct electron acceptor to form the $NBT^{\bullet+}$ in the presence of aldehyde oxidase[8] or with deoxygenated or oxygenated hemoglobin (which can reduce O_2 to $O_2^{\bullet-}$ therefore regenerating back the NBT^{2+}).[9] NBT was shown to inhibit NAD(P)H-dependent oxidation such as in the case of glyceryl trinitrate metabolism to NO by bovine aortic smooth muscles but

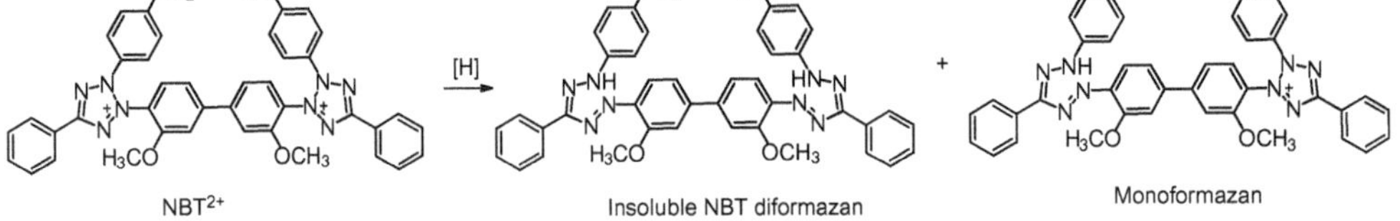

Figure 6.1 Reduction of NBT^{2+} to form the water-insoluble NBT diformazan.

NBT does not affect NO formation from L-arginine.[10] Therefore, the unsuitability of NBT^{2+} as $O_2^{\bullet -}$ probe was suggested for NBT^{2+} for certain systems and calls for other means of $O_2^{\bullet -}$ to be performed to complement NBT^{2+} assay.

Nevertheless, NBT^{2+} assay is widely employed to assess free radical scavenging properties of various compounds in in vitro systems,[11] or enzyme activity involving $O_2^{\bullet -}$ production such as in the case of NADH-cytochrome b5 reductase[12] and NADPH oxidase 4.[13] NBT^{2+} is most commonly used for assay of SOD and SOD-mimic activity using xanthine–xanthine oxidase $O_2^{\bullet -}$-generating system[14,15] from blood,[16] cultured cells,[17] and tissue[18] as percent inhibition of NBT^{2+} reduction. Microtiter analysis using water-soluble tetrazolium was also employed for SOD activity assay of red cell extracts and tissue homogenates that is simple and reproducible.[19] Also, NBT^{2+} is used as chromogenic substrate to evaluate phagocytosis and oxidative metabolism in phagocytes under electron microscopy[20] although these phagocytes could produce $O_2^{\bullet -}$, direct and superoxide-independent reduction of NBT^{2+} could give false-positive results for $O_2^{\bullet -}$ production and must be verified by other techniques.

6.2.2 Cytochrome c

The reactivity of $O_2^{\bullet -}$ to cytochrome *c* (cyt *c*) is a redox reaction that involves the reduction of the iron-heme center of ferricytochrome *c* (Fe^{3+}) to ferrocytochrome *c* (Fe^{2+}) (Eq. (6.1)) that can be monitored through formation of an absorption peak at λ_{max} of 550 nm.

$$\text{ferricytochrome } c\ (Fe^{3+}) + O_2^{\bullet -} \rightarrow \text{ferrocytochrome } c(Fe^{2+}) + O_2 \qquad (6.1)$$

This redox reaction has a rate constant of $k = 1.4 \times 10^6\ M^{-1}\ s^{-1}$ between pH 4.7 and 6.7 and $k = 2\text{–}2.5 \times 10^5\ M^{-1}\ s^{-1}$ above pH of 7.45.[21,22] Although $O_2^{\bullet -}$ does not react with ferrocytochrome *c*, $HO_2^{\bullet}$ can oxidize ferrocytochrome *c* with a rate constant of $\sim k = 5 \times 10^5\text{–}5 \times 10^6\ M^{-1}\ s^{-1}$,[23] which can compete with the spontaneous dismutation of $O_2^{\bullet -}$ with $HO_2^{\bullet}$ ($k = 8.8 \times 10^7\ M^{-1}\ s^{-1}$).[24] Effect of SDS detergents on the reactivity of cyt *c* toward $O_2^{\bullet -}$ showed increased favorability of $O_2^{\bullet -}$ reaction with cyt *c* as evidenced by negative free energy rate ($\Delta G_{H2O} = 1.9$ kcal/mol vs $\Delta G_{SDS} = -4.8$ kcal/mol).[25] Reoxidation of ferrocytochrome *c* to ferricytochrome *c* by the formed H_2O_2 from $O_2^{\bullet -}$ dismutation was suggested to only occur at the mM levels and is highly improbable due to low amounts of H_2O_2 formed in solution.[26] This high reactivity of cyt *c* with $O_2^{\bullet -}$ makes it attractive probe for assaying $O_2^{\bullet -}$ production. Early studies involving reduction of O_2 by reduced flavins and quinones,[27] or $O_2^{\bullet -}$ generation from heart mitochondrial particles[28] were measured using cyt *c* reduction assay method. Also, cyt *c* assay was used to assess $O_2^{\bullet -}$ from NADH-cytochrome *b5* reductase,[12] from neutrophils in myocardial ischemia–reperfusion (I–R) injury mice models,[29] or from rat spinal cord using cyt *c* microperfusate.[30] In spite of the high reactivity of cyt *c* with $O_2^{\bullet-}$, there are limitations encountered with its application. EPR spin trapping with DEPMPO showed higher sensitivity than cyt *c* assay by 40-fold.[31] Comparison of the EPR technique using the hydroxylamine probe (CMH) vs spectrophotometric assay using cyt *c* showed that the former is more sensitive than the latter for $O_2^{\bullet-}$ detection in phorbol myristate acetate (PMA)-stimulated human lymphoblasts by 4.3 times while cyt *c* only detects extracellular $O_2^{\bullet-}$.[32] The high molecular weight of the enzyme limits its intracellular compartmentalization and, hence, could only extracellularly detect $O_2^{\bullet-}$ in the case of endothelial cells.[33]

In simple system, cytochrome *c* binds with imidazole and decreases rate of cyt *c* reduction by ascorbate at high imidazole concentration but increases it with guanidine hydrochloride.[34] However, cyt *c* stability in plasma showed that intravenously injected cyt *c* in rat disappeared in the circulation with a half-life of only 4 min which was accounted to the susceptibility of cyt *c* to be reduced by low molecular weight reductants, mostly at physiological concentrations of ascorbic acid[35] but human red blood cells were also able to reduce cyt *c*.[36] Due to the positive and higher reduction potential of cyt *c* of E_p (cyt *c* Fe^{3+}/cyt *c* Fe^{2+}) = +0.254 V,[37] it is susceptible to reduction by other species such as sodium dithionate,[38] DOPA intermediate during oxidation of by polyphenoloxidase,[39] pyridine coenzymes as mediated by thyroid hormones and other 1,4-dihydronicotinamides,[40] ferrocytochrome *b5*,[41] or by ferrocytochrome *c* peroxidase.[42] Pulse radiolytic studies showed reduction of cyt *c* by radicals of benzoate, formate, NAD^+, or FMN to name a few with rate constants ranging from 10^7 to 10^9 M^{-1} s^{-1}.[43] Cytochrome *c* is also rapidly reduced by hydrated electrons and H-atoms, moderately by malate, lactate, and ethyl alcohol radicals,[44] and also by hydroxyl radical with 40% yield.[45] It was also demonstrated that cyt *c* can be reduced by thiols and, therefore, the rate of $O_2^{\bullet-}$ production in the presence of thiols could be underestimated.[46] Cytochrome *c* reduction was found to be not primarily due to $O_2^{\bullet-}$ in the presence of the diabetogenic drug, alloxan, which requires more SOD to inhibit cyt *c* reduction than in the absence of alloxan. This was due to the formation of alloxan radical that could indirectly inhibit cyt *c* reduction and could have implication in the interpretation of SOD-inhibitable effects of some drugs.[47]

Therefore, changes in the absorbance of ferricytochrome *c* may not be due to $O_2^{\bullet-}$ alone, hence, inhibition studies must be implemented. For example, the use of SOD to inhibit $O_2^{\bullet-}$ reaction and addition of catalase to remove excess H_2O_2 that can potentially reduce ferricytochrome c via $HO^{\bullet}$ generation since it has been demonstrated that ferrocytochrome *c* can be reoxidized back to ferricytochrome *c* by H_2O_2 thereby affecting the estimated absorbance reading leading to ambiguous data interpretation.[48] Given the nonselectivity of cyt *c*, detection of $O_2^{\bullet-}$ at low concentrations may result in artifactual signal responses.[49] In summary, cyt *c* assay is more appropriate for in vitro extracellular studies of $O_2^{\bullet-}$ generation and in systems where the $O_2^{\bullet -}$ flux could be higher as in the case of oxidative burst from phagocytic cells as compared to vessels, vascular smooth muscle and endothelial cells where $O_2^{\bullet -}$ levels are much lower.

6.2.3 Epinephrine (adrenaline)

A simple SOD activity assay based on the oxidation of epinephrine to a colored adrenochrome product ($\lambda_{max} = 480$ nm, $\varepsilon = 4020\ M^{-1}\ cm^{-1}$) was employed by Misra et al.[50] While epinephrine oxidation was found to be the simplest methodology for multiple assay with high sensitivity, the extent of epinephrine oxidation by $O_2^{\bullet -}$ increases with increasing pH (7.8–10.2) which involves complex mechanism characterized by free radical-mediated chain-propagating autoxidation of epinephrine via initial oxidation to a semiquinone. This initial oxidation to semiquinone leads to further oxidation to the *o*-quinone. the *o*-quinone undergoes intramolecular Michael addition reaction with the amino group to form bicyclic leucochrome that is further oxidized by $O_2^{\bullet -}$ to give the adrenochrome semiquinone and, ultimately, to the adrenochrome form according to Fig. 6.2.[51] From Fig. 6.2, one would expect further production of $O_2^{\bullet -}$ from the semiquinone forms in the presence of O_2 and, therefore, could amplify the reduction of epinephrine and leucochrome intermediate thereby increasing the yield of the adrenochrome.

Early applications of epinephrine to assess $O_2^{\bullet -}$ production were performed in systems such as oxy-Hb,[50] or NADPH-cytochrome *c* reductase-catalyzed oxidation of NADPH,[52] Ehlich ascites tumor cells in the presence of tetradecanoylphorbol myristate acetate (TPA) tumor promoter,[53] or hypoxia–reoxygenation injury in isolated hepatocytes.[54] The fact that there are two absorption bands for adrenochrome

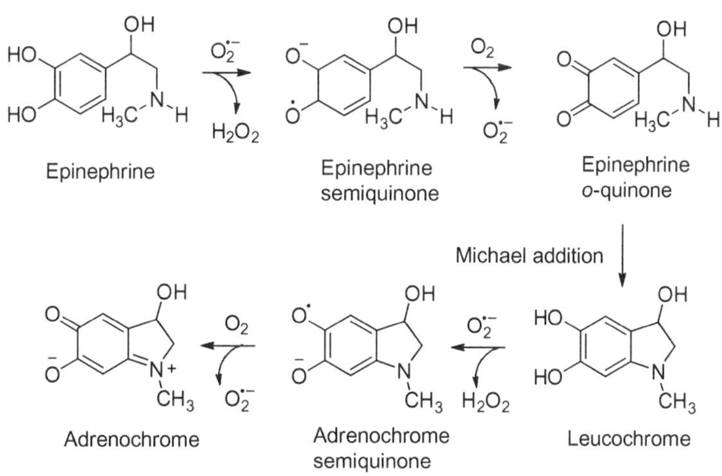

Figure 6.2 Free radical reaction of epinephrine with $O_2^{\bullet -}$ yielding the adrenochrome. Shown are further productions of RS during epinephrine oxidation.

(i.e., 480 and 320 nm) offered alternative for monitoring epinephrine oxidation at 320 nm since the 480 nm band is weak, broader, and unstable at high pH.[55] Monitoring of absorption at 310–320 nm was employed for SOD activity assay with significant improvement in stability, sensitivity, and consistency as well as improved linearity during time course measurements as compared to measurements at 480 nm. One has to note that the adrenochrome reacts with thiols which can lead to overestimation of $O_2^{\bullet-}$ production in thiol-rich systems[46] and epinephrine causes amplification of $O_2^{\bullet-}$ production by 102% from incubation of epinephrine with human coronary artery endothelial cells (HCAECs) for 1 hour where the increase in $O_2^{\bullet-}$ production was found to be due to β_1-adrenoceptor activation which upregulates SOD mRNA and protein expression.[56] Therefore, independent technique for $O_2^{\bullet-}$ measurement must supplement epinephrine assay.

6.2.4 Free radical-based probes

Early work on the reactivity of trityl radical such as TAM OX063 to $O_2^{\bullet-}$ shows loss of EPR signal due to the conversion of the paramagnetic trityl to a diamagnetic species and was demonstrated in systems such as light/riboflavin, xanthine/xanthine oxidase, and stimulated neutrophils.[57] This reaction by TAM OX063 trityl with $O_2^{\bullet-}$ could also be spectrophotometrically monitored through formation of an absorption peak at 546 nm that is inhibitable by SOD.[58] Other analogues such as PTM-TC was synthesized and the peak disappearance at 380 nm can be monitored as a function of $O_2^{\bullet-}$ formation[59] or peak formation at 540 nm using CT-02[60] (Fig. 6.3) with high rate constants of $1.7 \times 10^4\ M^{-1}\ s^{-1}$ and $8.6 \times 10^8\ M^{-1}\ s^{-1}$, respectively. The PTM-TC probe was employed for the detection of $O_2^{\bullet-}$ in actual biological systems such as plant root using EPR imaging,[61] but in spite of its high specificity to $O_2^{\bullet-}$, trityl compounds have yet to find more biological applications.

6.3 HYDROXYL RADICAL PROBES

6.3.1 Salicylic acid[62]

Hydroxyl radical undergoes electrophilic addition reaction (aromatic hydroxylation) to salicylic acid (SA) with high reaction rate constant of $k = 5 \times 10^9\ M^{-1}\ s^{-1}$.[63] The most common route for the generation of $HO^{\bullet}$ in biological system is via Fenton reaction of low-valent transition metal ions such as Fe^{2+} or Cu^{+} with H_2O_2 (formed from $O_2^{\bullet-}$ dismutation). Hydroxyl radical could also be formed via Haber–Weiss reaction involving $O_2^{\bullet-}$ and H_2O_2. In xanthine–xanthine oxidase system, hydroxylation, could only occur in the presence of both $O_2^{\bullet-}$ and H_2O_2, and low concentrations of $FeSO_4$ or $FeCl_3$ that is inhibitable by excess iron chelator such as DTPA indicating that H_2O_2 reacts with $O_2^{\bullet-}$ via an Fe-catalyzed Haber–Weiss reaction to

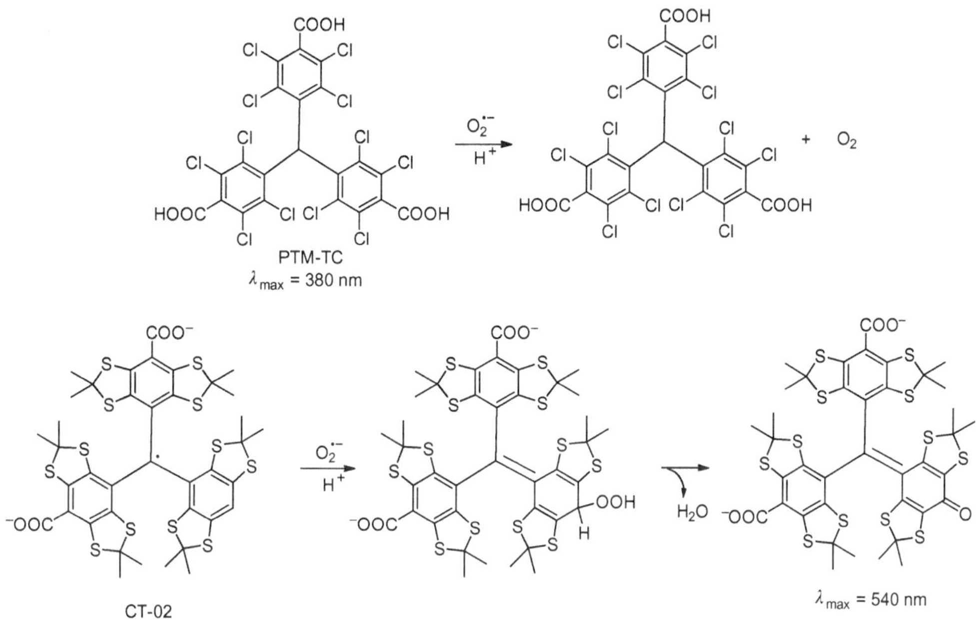

Figure 6.3 Reaction of trityl, PTM-TC, and CT-02 with superoxide forming diamagnetic products that can be monitored spectrophotometrically.

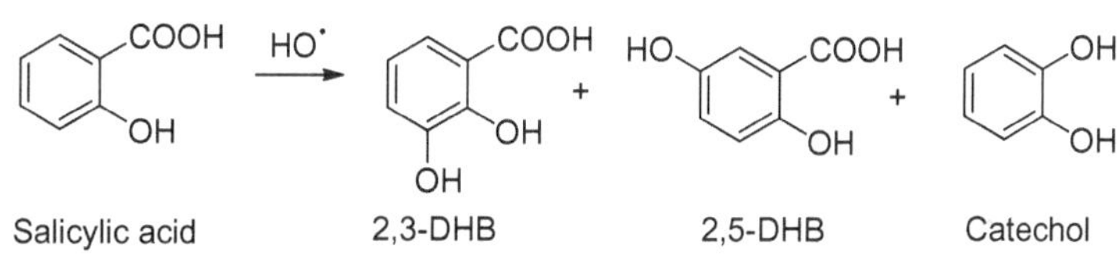

Figure 6.4 Aromatic hydroxyl of SA producing DHB products and catechol.

produce $HO^{\bullet}$.[64] In the presence of ascorbate, $HO^{\bullet}$ is produced when ascorbate is oxidized by H_2O_2 in the presence of catalytic amounts of Cu^+ or Fe^{2+}.[65] Furthermore, peroxynitrite can also cause hydroxylation of SA along with nitration via homolytic fission of peroxynitrous to NO_2 and $HO^{\bullet}$.[66]

Detection of SA hydroxylation is important since it is a metabolic product of acetylsalicyclic acid or aspirin after esterase hydrolysis in the cytosol. Aspirin being the most widely used analgesic, its hydroxylation in vivo is relevant.[67] Nonenzyme-mediated $HO^{\bullet}$ addition to SA yields 2,3-DHB, 2,5-DHB, and small amount of catechol decarboxylated products (Fig. 6.4).

In vivo detection of 2,3-/2,5-DHB ratio is more important than the detection of the absolute levels of each of the hydroxylated species where their ratio can vary between 1.2 and 5 or to as low as 0.15 in plasma[68] and is dependent on the type of system being investigated. In vivo studies, however, suggest that under oxidative stress conditions, salicylate hydroxylation could occur via NADPH-dependent metabolism of salicylate by cytochrome P450 system in the liver microsomes with increased 2,5-DHB but no change in 2,3-DHB levels, therefore, caution should be observed when only 2,5-DHB is measured.[69] Consideration of individual cytochrome P450 shows that the recombinant human P450s, 2E1, is the major human P450 involved in the hydroxylation at the 3- and 5-positions while 2A6 and 2B6 only contributed to 5-hydroxylation.[67] Electrolytically generated $HO^•$ in the presence of SA showed 2,3-DHB being the major product and that the highest yield achievable for both 2,3- and 2,5-DHB was ~22%.[70] The use of SA to investigate in vitro or ex vivo production of $HO^•$ is a viable alternative to the use of spin traps using EPR instrument. Improvements in the detection of hydroxylated SA have been achieved through the use of HPLC with UV, fluorescence, and electrochemical detection. For example, HPLC analysis of 2,5-dihydroxybenzoic acid (2,5-DHBA) was found to be the most sensitive analyte for the determination of $HO^•$ production in rat polymorphonuclear leukocytes using fluorescence detection.[71] The 2,3-DHB/SA ratios were determined to assess $HO^•$ generation in human patients with coronary artery disease and in patients with myocardial infarction complicated by heart failure.[72] In perfusion of isolated rat brain following hemorrhagic stroke,[73] or in ischemic or postischemic isolated rat heart,[74,75] 2,5-DHBA levels as a function of ROS formation were monitored using HPLC technique. Moreover, under anaerobic medium, negligible DHBA production was observed indicating O_2-derived $HO^•$ formation.[75] The production of 2,3-DHBA was monitored to investigate $HO^•$ production from hippocampus after carbon monoxide poisoning or hypoxia in the rat.[76] Both 2,3-DHBA and 2,5-DHBA were measured using HPLC-ED from various tissues such as the lung, heart, muscles, brain, and blood in adriamacin-treated rats.[77] Therefore, analysis of various hydroxylated products using HPLC technique for the assessment of $HO^•$ in cells and tissue could lead to a better correlation with RS production.

6.3.2 Phenylalanine

Similar to the chemistry of SA, the hydroxylation of dietary L-phenylalanine (L-Phe) could occur nonenzymatically through direct hydroxylation by $HO^•$ as well as enzymatically by L-Phe hydroxylase. Addition of $HO^•$ to L-Phe and D-Phe both yield isomeric mixtures of tyrosine and ortho-, para-, and meta-hydroxylated products as shown in Fig. 6.5 which can be separated using HPLC with UV and/or electrochemical detection.

Pulsed-radiolytic studies revealed 50% ortho-, 30% para-, and 14% meta-tyrosine isomers are formed.[78] In biological system, L-Phe was used to measure $HO^•$ production from leghemoglobin–H_2O_2 mixture or activated human neutrophils.[79] Stimulation of PMNs causes hydroxylation of Phe and SA and is accelerated in the

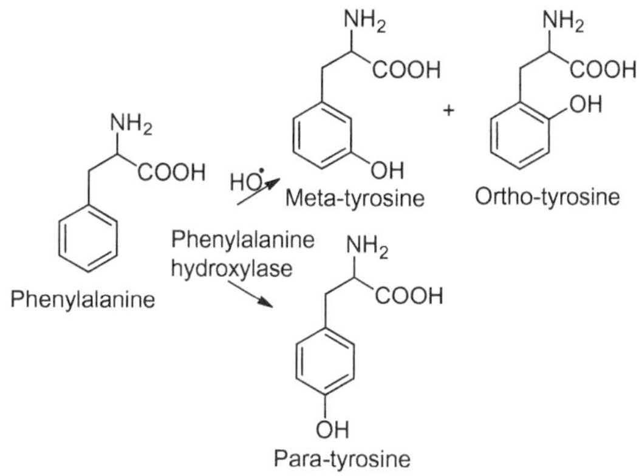

Figure 6.5 Nonenzymatic and enzymatic hydroxylation products of phenylalanine.

presence of glutathione.[80] Hydroxylation along with nitration can also occur with $ONOO^-$ reaction with Phe and Tyr.[81] To distinguish between enzymatic and nonenzymatic hydroxylation, the specificity of D-Phe was explored for a more reliable method of $HO^•$ detection in vivo since the D-isomer cannot be hydroxylated by Phe hydroxylase. Using resting human neutrophils, D-Phe hydroxylation did not occur but L-Phe was converted to *p*-Tyr. Stimulation of PMN in the presence of redox cycling iron results in the production of *p*-Tyr from D-Phe by 9.4-fold compared to L-Phe of only 3.6-fold. Therefore, using D-Phe instead of L-Phe could eliminate the possibility of Phe hydroxylation by Phe hydroxylase.[82] Hydroxylation of D-Phe was therefore exploited as an alternative probe for the detection of $HO^•$ in intact isolated rat heart undergoing I–R eliminating endogenous enzyme-mediated hydroxylation of Phe.[83]

Comparison of SA and Phe hydroxylation in in vitro using Fenton system did not show stark difference in sensitivity while in vivo experiments using rats implanted with microdialysis probes and striatal perfusion with a neurotoxin, 6-hydroxydopamine, showed increased production of *o*- and *m*-tyrosine for Phe, or 2,3-DBHD for SA at the same time course and peak levels.[84] This is supported by the high reactivity to $HO^•$ of both Phe and SA (i.e., $k_{Phe} = 3.5 \times 10^9\ M^{-1}\ s^{-1}$, pH = 6, and $k_{SA} = 5.6 \times 10^9\ M^{-1}\ s^{-1}$, pH = 9).[63] For biological studies, Phe could be a more appropriate probe since Phe is less toxic than SA. In fact, aromatic hydroxylation in animal models of diabetes mellitus was demonstrated using Phe where the degree of oxidation was correlated with levels of lipid peroxidation and glycooxidation.[85] Phenylalanine hydroxylation was shown to be a direct evidence of $HO^•$-mediated myocardial stunning but can act as $HO^•$ scavenger thereby attenuating postischemic dysfunction of intact heart.[86] In simple systems, Phe hydroxylation was proven effective in the measurement of $HO^•$ production in myeloperoxidase–H_2O_2 system,[87] riboflavin illumination,[88] hypoxanthine–xanthine oxidase (HX–XO),[89] aerobic solution of 6,7-dimethyl-5,6,7,8-tetrahydropteridine,[90] and ascorbic acid–H_2O_2 system.[91]

6.4 ANTIOXIDANT CAPACITY ASSAYS

These assays involve detection of RS as well as end-products of RS reaction with biochemical system. These methods are commonly employed for the determination of AOC in foods and dietary supplements, cells, body fluids, and tissue. AOC assays employ two major mechanisms, one that involves hydrogen atom transfer (HAT), and the other a single electron transfer (SET), both from the antioxidant (AOH) to the RS or probe ($R^{\bullet}$) (Fig. 6.6). HAT mechanism is typically found in radical reactions of AOH possessing an H–X (X = O or S) bond such as in the case of polyphenols, thiols, or vicinal C–H to the amino group undergoing homolytic bond-breaking process. In SET mechanism, AOH for example can transfer an electron to RS where $HO^{\bullet}$ is converted to HO^{-} or $HOO^{\bullet}$ to HOO^{-} along with formation of $AOH^{\bullet+}$. Proton abstraction from $AOH^{\bullet+}$ by the reduced RS yields the neutral species, i.e., HO^{-} to H_2O, or HOO^{-} to H_2O_2. Typical structural feature for antioxidants involved in SET reactions are those that bear electron-rich systems with pi-orbitals, highly ionizable lone pair of electrons such as in the case of conjugated aromatic or olefins systems, and those with amino, carbonyl, or sulfonyl moieties. Therefore, the AOH's ionization potential and deprotonation are important factors in SET reaction, hence, pH plays an important role in facilitating this type of reaction. Hydrogen atom abstraction by radicals may not apply to those antioxidants that do not contain reactive H-atoms and, therefore, in systems with known composition, careful examination of the structure of the antioxidants is necessary for the accurate interpretation of AOC data. However, when dealing with complex systems such as food and nutraceuticals, antioxidant property must be careful interpreted and should reflect both SET and HAT mechanisms and, therefore, it is essential to perform more than one AOC assay. Whether in simple or complex system, HAT and SET could simultaneously occur but one has to note that the latter reaction is slower than the former. In the succeeding section, HAT mechanism will be represented by oxygen radical absorbance capacity (ORAC) assay, while SET with DPPH and ABTS assays. Efforts had been made to

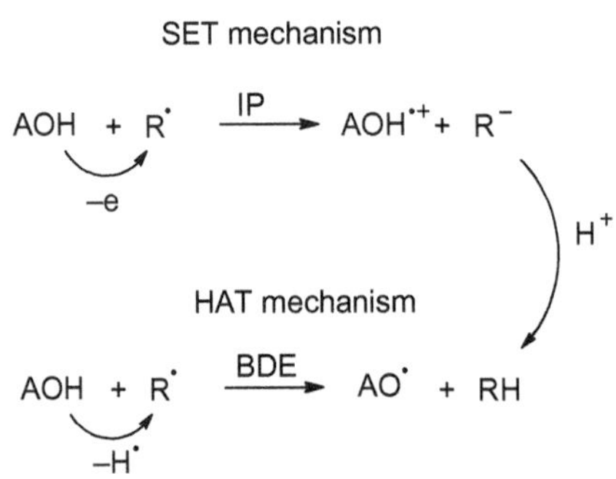

Figure 6.6 Major mechanisms of AOC assays involving hydrogen and SET.

standardize AOC for the purpose of addressing issues as far as analytical range, recovery, repeatability, reproducibility, and recognition of interfering substance are concerned.[92] This section will not cover all the AOC assays that are currently being employed but it will only describe the three major assays commonly employed for AOC determination, that is, DPPH, TAEC and ORAC assays.

6.4.1 2,2-Diphenyl-1-picrylhydrazyl (DPPH) assay

It should be emphasized that this reagent is not commonly used to detect RS but rather used to assess antioxidant activity (AUC) of certain molecules with radical scavenging properties. In this assay, instead of assessing reactivity of antioxidants with transient radicals, $X^•$, such as $O_2^{•-}$, $HO^•$, $LOO^•$, $LO^•$, or $NO^•$, AUC is assessed against a longer-lived radical, DPPH, as shown in Fig. 6.7. The assay uses the DPPH free radical probe which contains a tri-nitro hydrazyl moiety where the unpaired electron is localized between the *N*–*N* hydrazyl group constrained by the two adjacent phenyl rings. The DPPH has a deep violet coloration with an absorption at 517 nm, where upon reduction by an antioxidant, decolorizes to the yellow product and, therefore, the assay simply involves monitoring the decrease of the absorption band at 517 nm.

DPPH assay was employed for the determination of AOC of certain compounds such as thiols, ascorbic acid, tocopherol, polyphenols, isoflavones, and its metabolites[93] or aromatic amines.[94] Also, AOC determination of more complex matrices such plant extracts,[95] and tomato seed oil.[96] DPPH assay was coupled with HPLC for screening of antioxidants from complex matrix,[97,98] and in clinical studies where DPPH assay was employed to measure antioxidant property of human serum.[99] Automated methods involving multisyringe flow injection analysis was applied for AOC determination from several food products with excellent reproducibility and detection limit.[100]

AOC is typically reported as IC_{50} (i.e., concentration of AOH needed to reduce 50% of the DPPH, therefore, the initial DPPH concentration is typically reported).

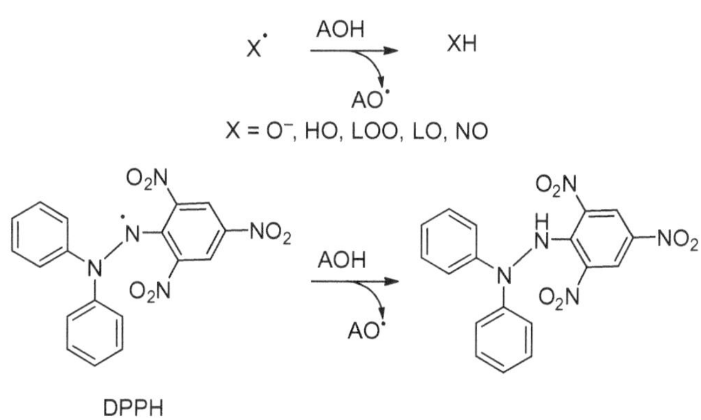

Figure 6.7 Reaction of DPPH with antioxidant showing competition reaction with other RS.

Due to the poor solubility of DPPH in water, studies are usually performed in pure alcohol or alcoholic solution in ethyl acetate, chloroform, or buffers at pH of 5−6.5. The extent of DPPH reduction could be affected by the presence of transition metal ions, incubation time, concentration, and molecular size of the antioxidant.[101] Due to dependence of IC_{50} to various experimental conditions, IC_{50} value for a particular AOH could vary from one study to another and, therefore, is very qualitative and is only presented in terms of AOC relative to standards such as Trolox or ascorbic acid. Several recommendations were suggested for a more meaningful interpretation of AOC using DPPH assay such as: (1) measurement of initial rate of reaction (0−2 min) preferably using stopped-flow mixing rather than the final absorbance after long incubation that ignores the antioxidant concentration effects, reaction saturation, and effect of the antioxidant molecular size; and (2) use of range of antioxidant concentration rather than just one in order to identify valid linear response range, saturation levels, and differentiates reactivity patterns.[102] Unless working with pure compounds, interpretation of AOC using complex mixtures such as in extracts should only reflect the overall AOC of the mixture itself without referencing specific chemistry.

6.4.2 2,2′-Azino-bis(3-ethylbenzothiazoline-6-sulphonic acid) radical cation (ABTS) or TAEC assay

The 2,2′-azino-bis(3-ethylbenzothiazoline-6-sulphonic acid) radical cation ($ABTS^{\bullet+}$) has a maxima at 405 or 734 nm and is generated from the reduction of 2,2′-azino-bis (3-ethylbenzothiazoline-6-sulphonate) by ferrylmyoglobin radical species formed from H_2O_2−metmyoglobin,[103] HRP/H_2O_2,[104] or potassium persulfate[105] (Fig. 6.8). This ferrylmyoglobin/ABTS and persulfate/ABTS assay are collectively known as Trolox equivalent antioxidant capacity (TAEC) assay since the antioxidants (AOH) tested are compared to that of the water-soluble vitamin E analogue, Trolox, and that AOC is calculated as Trolox equivalents.

The persulfate/ABTS assay is an improvement over the ferrylmyoglobin/ABTS since it avoids the formation of an intermediate radical (i.e., ferrylmyoglobin radical species) that can potentially oxidize the antioxidant and not the $ABTS^{\bullet+}$ (the species that is being monitored). Also, the persulfate/ABTS assay allows for the measurement of AOC both in aqueous and lipophilic systems, therefore, eliminating the need for solubilization of the samples as in the case of ferrylmyoglobin/ABTS assay.[105] The stability of $ABTS^{\bullet+}$, hence, the AOC values, is pH dependent with loss of sensitivity at acidic pH where iron reducing agent power (FRAP) assay is instead employed using a different iron salt, but this behavior of $ABTS^{\bullet+}$ at various pH is also dependent on the chemical structure of the antioxidant wherein structurally related antioxidants give the same trends in the duration and pH dependency of their oxidation.[106] Typical TAEC assay involves measurement of absorbance at fixed time point usually between 4 and 6 minutes but with certain antioxidants (e.g., resveratrol), the reaction with $ABTS^{\bullet+}$ is not complete at these time points

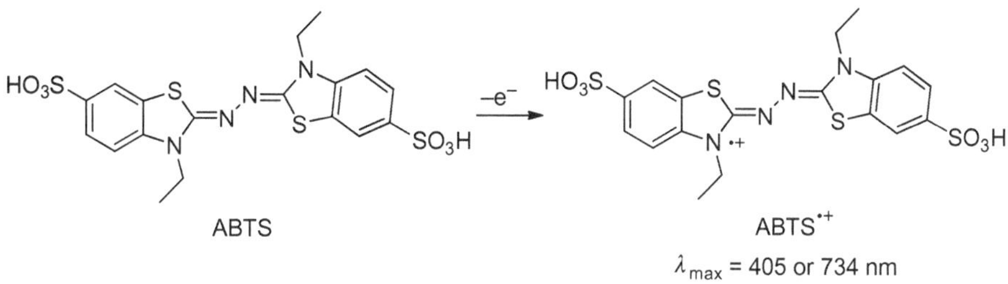

Ferrylmyoglobin/ABTS assay

$$\text{Metmyoglobin} + H_2O_2 \longrightarrow \text{Ferrylmyoglobin}^{\bullet} + H_2O$$

$$\text{ABTS} + \text{Ferrylmyoglobin}^{\bullet} \longrightarrow \text{ABTS}^{\bullet+} + \text{Metmyoglobin}$$

$$\text{ABTS} + \text{AOH} \not\longrightarrow \text{ABTS}^{\bullet+} + \text{Metmyoglobin}$$

Persulfate/ABTS assay

$$\text{ABTS} + K_2S_2O_8 \longrightarrow \text{ABTS}^{\bullet+}$$

$$\text{ABTS}^{\bullet+} + \text{AOH} \longrightarrow \text{ABTSH}^{+} + \text{AO}^{\bullet}$$

Figure 6.8 Formation of ABTS$^{\bullet+}$ from reduction by ferrylmyoglobin radical or potassium persulfate and its redox reaction with antioxidant (AOH).

and, therefore, the final absorbance reading does not reflect the kinetics of ABTS$^{\bullet+}$ reaction with antioxidants. Consideration of ABTS$^{\bullet+}$ oxidation of antioxidants as a function of time provides better insight into the relative AOC of various antioxidants.[107] To account for the varying rates of reaction of antioxidants with ABTS$^{\bullet+}$ and variation in the TAEC values with time (as in the case of ferrylmyoglobin/ABTS assay), relative AUC could be measured using the persulfate/ABTS assay and calculated from the antioxidant concentration derived from the area under the curve (vs Trolox) as a function of time.[105] Studies involving AOC of peptides show that dipeptides that were highly reactive to ABTS$^{\bullet+}$ and peroxyl radicals were inert toward DPPH, suggesting that DPPH have limited application.[108] Optimization of ABTS assay for high-throughput application and improved resolution of phenolics analysis from natural products were accomplished using online two-dimensional liquid chromatography which is a combination of hydrophilic interaction chromatography separation and reversed-phase liquid chromatography separation that address analyte polarity and hydrophobicity, respectively.[109]

6.4.3 Oxygen radical absorbance capacity (ORAC) assay

This assay is based on the inhibition of peroxyl radical reaction by antioxidants and, therefore, mimics radical-mediated oxidation in biological systems much closer than

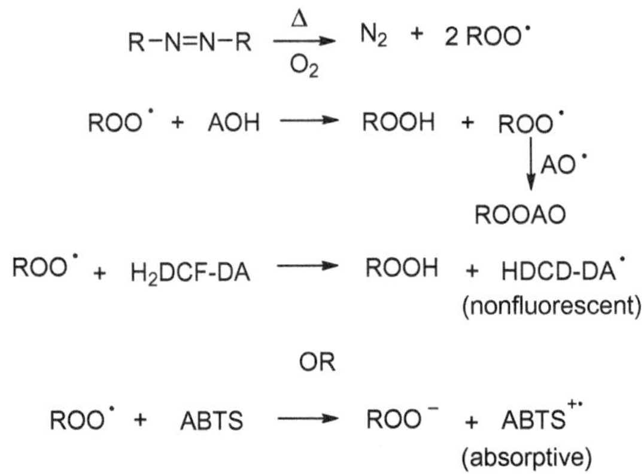

Figure 6.9 Generation of peroxyl radical and its competition reactions with AOH and $DCFH_2$-DA or ABTS.

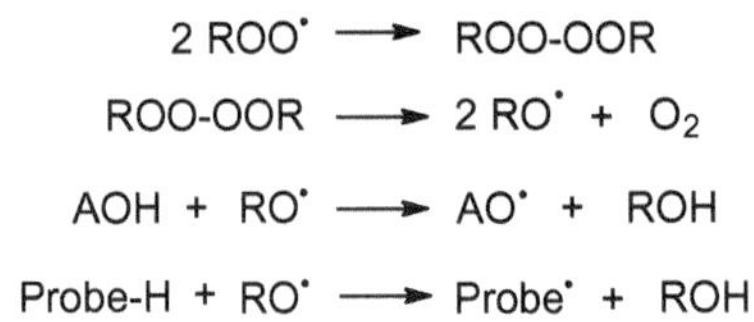

Figure 6.10 Formation of alkoxyl radicals from the disproportionation of reaction of peroxyl radical.

that of DPPH and $ABTS^{\bullet +}$. As shown in Fig. 6.9, thermolysis of an azo-bis compound (e.g., 2,2′-azo-bis(2-amidinopropane) dihydrochloride) (AAPH) leads to its decomposition to N_2 gas and alkyl radicals where the latter further reacts with dissolved O_2 to form the peroxyl radical. AOC is monitored through competition reaction of the antioxidant of interest with a dye that is either fluorescent (e.g., $DCFH_2$-DA) or absorptive (e.g., ABTS) when reacted with peroxyl radical using Trolox as standard.[110]

Others had employed EPR technique to investigate the nature of radicals formed from aerobic azo-bis thermolysis using competition reaction between spin traps and antioxidants with the generated radicals. The spin adduct EPR spectral profile of the generated radical indicates an alkoxyl radical ($RO^{\bullet}$) but not $ROO^{\bullet}$.[111] This is further confirmed by kinetic studies which revealed that the dominant radical formed from the ORAC protocol was $RO^{\bullet}$ and not $ROO^{\bullet}$ which can be explained in terms of disproportionation of the ROO−OOR to alkoxyl radicals and oxygen according to Fig. 6.10.[112]

Also, the type of radical generated is radical source-dependent. For example, when the azo-radical initiator used is AAPH, $RO^{\bullet}$ is mainly detected by EPR spin trapping but $ROO^{\bullet}$ is predominantly formed when the lipid-soluble initiator, azo-bis(isobutyronitrile), is used in nonaqueous system.[113] However, caution should be observed in

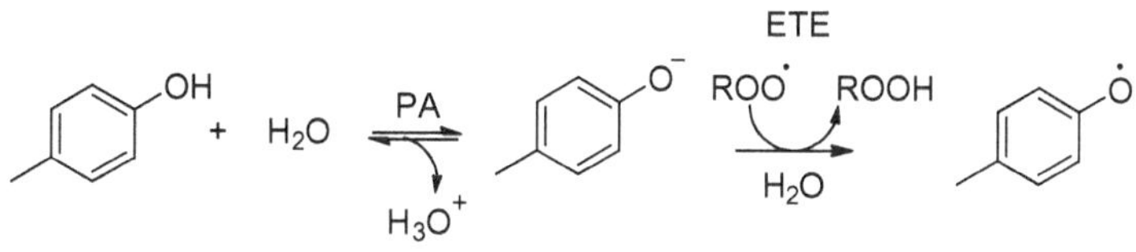

Figure 6.11 ORAC reaction as described by PA and ETE.

interpreting ROO• vs RO• production by ESR spin trapping since the half-life of peroxyl adduct vs alkoxyl must be taken into consideration where previous studies indicate that *tert*-ROO-adduct decay rates are several thousand-fold faster than RO-adducts.[114]

Further investigation of the mechanism of ORAC reaction was carried out using computational approach. In the study, the relationship between flavonoid structure and AOC value were computationally rationalized and showed that neither O–H bond dissociation enthalpy nor ionization potential values of flavonoids correlate with ORAC values. Instead, proton affinity (PA) and electron transfer enthalpy (ETE) values, via the sequential proton-loss electron transfer (Fig. 6.11) were more consistent with the trends in ORAC values obtained from using pyrogallol Red and fluorescein probes. Hence, ORAC value is highly dependent on the pH and the stability of the deprotonated flavonoids is predicted by ETE values.[115]

Even the type of competing probe could also affect the interpretation of AOC. For example, in the case alizarin red, ORAC value for certain antioxidants was several hundred times higher than other antioxidants while there is not much difference in their ORAC values when using fluorescein.[116] Comparison of various probes such as fluorescein, alizarin red, pyrogallol red, and pyranine show that trends in ORAC indices from using pure or complex samples vary from one probe to another.[117] Other competitive probes such as *p*-aminobenzoic acid was employed to increase the assays's specificity to ROO•,[118] or the use of pyragallol red allowed for the determination of AOC in near-gastric conditions at $pH = 2$,[119] or by using other detection techniques such as EPR[120] were employed to address the common limitations of ORAC assay such as irreproducibility, photobleaching, or probe–antioxidant interaction with the aim to expand their applications to a wide range of biological systems. Comparison of the AOC in human serum using the three assay, i.e., ORAC, TAEC, and ferric reducing ability assay showed weak or no correlation between the three methods which was accounted to the differences in their ability to measure SH-containing antioxidants, breath of their reactivity, and time required to complete the reactions.[121] Moreover, pure compounds and food juices treated with ethylenediaminetetraacetate (EDTA) showed markedly higher AOC readings than those without the metal chelator.[122] Therefore, the presence of metal ions affects AOC reading due to rapid autoxidation of the antioxidant by the metal.

6.5 NITRIC OXIDE AND METABOLITES PROBES

6.5.1 Hemoglobin

The presence of non-bonding electron pair donors in O_2 and NO is reflected on the similarity of their binding properties with the iron-heme. For example, the rate constant of NO or O_2 binding to deoxygenated hemoglobin, Hb(II), is in the order of $k \sim 10^7\ M^{-1}\ s^{-1}$ while it is lower with hemoglobin, Hb(III) with $k = 10^3 - 10^4\ M^{-1}\ s^{-1}$.[123] With oxyhemoglobin, the reaction of NO proceeds according to the Eq. (6.2):

$$HbO_2 + NO \rightarrow MetHb + NO_3^- \qquad (6.2)$$

The conversion of oxyhemoglobin Fe(II) to methemoglobin Fe(III) can be spectrophotometrically monitored through formation of absorbance at 401 nm with an isosbestic point at 411 nm.[124] This method was employed in the kinetic investigation of NO and $O_2^{\bullet -}$ reaction[125] and in vitro NO production from blood PMN leukocytes.[126] In vivo studies involving intracerebral microdialysis in rats allows for a detection limit of 7 nM for exitotoxin-induced NO production, although serious challenges in establishing NO basal levels in extracellular fluid and in vivo calibration were encountered.[127] However, hemoglobin assay is highly dependent on temperature, autoxidation of HbO_2, presence of other Fe-heme proteins, pH, and reducing agents, and therefore, its sensitivity to interferences is its major drawback.[124]

6.5.2 Triiodide method

Biologically bound NO in the form of Hb[FeNO], SNO-HB, and SNO-albumin can be measured by chemiluminescence technique through light emission from the reaction of the released NO with excess amount of O_3 according to the Eqs. (6.3)−(6.4):

$$NO + O_3 \rightarrow NO_2^* + O_2 \qquad (6.3)$$

$$NO_2^* \rightarrow NO_2 + h\nu(\lambda \sim 1200\ \text{nm}) \qquad (6.4)$$

About 8% of the excited NO_2^* formed decays to NO_2 with the emission of light that is measured by a photomultiplier at 600−875 nm. This technique uses an NO analyzer with a configuration that allows removal of NO under vacuum from the sample, typically via reductive denitrosylation using acidic iodide according to Fig. 6.12 [128].

Therefore, NO must have to be extracted into the gas phase before it can be detected and is the major technical challenge in employing this technique. For example, since NO_2^* can be quenched by the carrier gas, gas pressure is one of the main sources of error.[129] Along with the manner by which ozone is delivered, the reaction of ozone with ethylenic hydrocarbons, sulfur compounds, and carbonyls are potential

$$4HI + O_2 \longrightarrow 2I_2 + 2H_2O$$
$$I_2 + I^- \longrightarrow I_3^-$$
$$I_3^- + 2\,RSNO \longrightarrow 3\,I^- + 2\,RS^{\bullet} + 2\,NO^+$$
$$2\,RS^{\bullet} \longrightarrow RS\text{-}SR$$
$$2\,NO^+ + 2\,I^- \longrightarrow 2\,NO + I_2$$

Figure 6.12 Generation of triiodide and its reductive denitrosylation of RSNO showing formation of nitrosonium cation and oxidation of thiol.

sources of chemiluminescence other than NO and is concerning due to the high sensitivity of the photomultiplier which is capable of measuring NO in as little 10 pmol.[129]

The triiodide chemistry is complicated and not well-understood where changes in pH and altered ionic composition could affect its overall reaction. Triiodide method could also reduce nitrite to NO but this is remedied by adding hydroquinone to establish that RSNO as the NO source and not nitrite under neutral condition. However, in acidic pH, pretreatment with ammonia sulfamate prevents NO release from nitrite, while mercury pretreatments prevents the release of NO from RSNO.[128] The iodine produced from acidified KI was also known to be consumed by thiols thereby reducing the yield of I_3^- which translates to fewer release of NO from RSNO.[128] The formation of NO^+ from the triiodide assay yields NOI according to Eq. (6.5):

$$NO^+ + I^- \rightarrow NOI \tag{6.5}$$

whose potent nitrosylating properties could affect the yield of NO. Therefore, this assay is highly dependent on sample composition where those with high thiol content could have increased nitrosylation affecting final unbound NO levels. This raises concern over the accuracy of triiodide method compared to photolysis chemiluminescence technique in determining actual levels of NO from SNOs, metal nitrosyls, or nitrite due to low and variable yields in the former.[130] Triiodide assay in general give lower metabolite values compared to other methods that assess bound NO levels. This difference was suggested to be due to the autocapture of nitric oxide by deoxygenated cell-free heme in the reaction chamber which could be remedied by the addition of potassium ferricyanide to the triiodide reagent allowing measurement of NO levels ranging from 10^{-4} to 10^{-3} mol NO/mol Hb.[131] A comparison of the triiodide, copper/CO/cysteine, and modified copper/cysteine methods showed that all are sensitive, and could be effectively used for the measurement of RSNO provided proper experimental controls, and complementary techniques for Fe-NO formation such as low-temperature EPR must be performed.[132] In spite of the few limitations, triiodide method was successfully employed for the determination of basal circulatory levels of Hb-SNO, albumin-SNO, and nitrite in human circulation and could provide insights into their role in maintaining basal vascular tone.[133,134]

6.5.3 Griess assay

Level of NO_2^-/NO_3^- in blood, tissue, and even urine is mostly associated with degree of NO availability as a function of nitric oxide synthase (NOS) activity. Diminished NOS activity has been correlated with endothelial dysfunction that is manifested in various disease etiologies such as atherosclerosis,[135] hypertension,[136,137] I–R injury,[138] inflammation,[139] and diabetes.[140] Autoxidation of nitric oxide in the presence of O_2 yields nitrite, while NO reaction to $O_2^{\bullet -}$ yields $ONOO^-$ which isomerizes to nitrate with concomitant release of NO_2 and $HO^{\bullet}$.[141] Nitric oxide conversion to nitrate can also be mediated by oxyhemoglobin to form methemoglobin, while nitrite formation from NO can be mediated by cytochrome *c* oxidase or ceruplasmin.[142] Nitrite can be oxidized to nitrate by oxygenated-Hb while nitrate can be reduced back to nitrite by bacterial reductase.[143] These thermodynamically stable nitrite and nitrate products can be measured using Griess assay.[144,145] Griess assay involves diazotization reaction via the initial reaction of acidified nitrite (as nitrous acid) with sulfanilic acid or sulfanilamide to form the diazonium ion. Azo-coupling to an amino group-containing dye (e.g., *N*-alpha-naphthyl-ethylenediamine) subsequently occurs to produce a red-violet water-soluble azo dye with a $\lambda_{max} = 540$ nm (Fig. 6.13). The diamine is preferred over the use of 1-aminonaphthalene due to the carcinogenicity of the latter and higher solubility in acidic aqueous solution of the former.

Nitrate does not undergo diazotization reaction and has to be reduced first to nitrite before the assay could be performed. Commonly used reducing agents were zinc dust,[146] vanadium (III) chloride,[147] cadmium-activated column,[148] copper-plated cadmium fillings,[149] or using the nitrate reductase enzyme.[150] Griess assay has a limit of detection of ~0.5 μM which is less sensitive than the chemiluminescence technique which is in the 1–10 pmol range. However, at low nitrite concentration, significant discrepancy between the measured and actual concentrations is evident due to the absorbance interference from heme and flavoproteins present in the medium. For example, myoglobin and iNOS can interfere in the absorbance reading at low

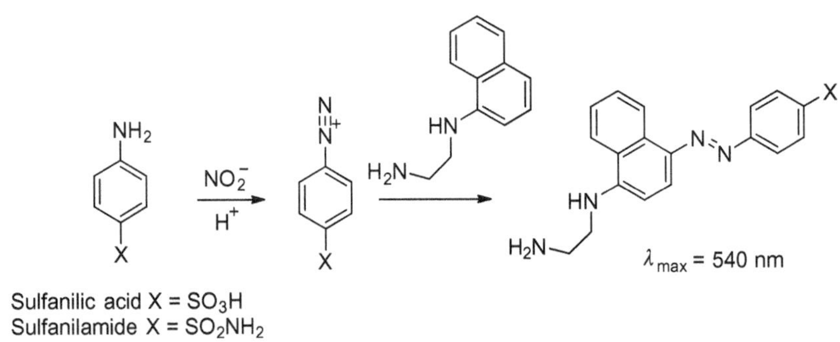

Figure 6.13 Griess assay reaction mechanism showing diazotization reaction of acidified nitrite and subsequent formation of the azo dye via azo-coupling reaction.

nitrite concentration. This can be eliminated using H-point standard addition method providing improved nitrite measurements with high precision and accuracy.[151] The type of medium could also affect nitrite measurements. Using an NO donor, it was found that Griess assay was the most inaccurate method for measuring NO (as nitrite) in various media such as plasma, urine, saliva, and PBS in contrast to chemiluminescence or electrochemical techniques.[152] The type of acid used for the assay (i.e., HCl, phosphoric acid, acetic acid, or TCA) at pH range of 0.6–3.3 gave varying results with lower absorbance reading at lower pH. It was suggested that 6% phosphoric acid to be the most suitable reagent for nitrite measurements in complex biological matrix such as plasma which contains NADPH that can potentially interfere with nitrite measurements at high pH.[153] In vivo and in vitro studies where L-arginine-derived NOS inhibitors such as L-NAME, NG-nitro-L-arginine, or high doses of aminoguanidine can interfere with the NO_2^-/NO_3^- assay as well as from using chemiluminescence technique. Therefore, advising against the use of such inhibitors when measuring NO_2^-/NO_3^- in in vivo and in vitro systems.[154] L-NAME and L-NNA (but not serotonin) interfere with the NO_2^-/NO_3^- analysis, therefore, elimination of these compounds by SPE cation exchanger is recommended prior to analysis.[155] *N*-L-arginine can also interfere with cadmium reduction of nitrate where the arginine can directly increase formation of nitrite independent of nitrate.[156] Furthermore, enzymatic or chemical oxidation of 6-nitrochatecholamines such as 6-nitrodopamine and 6-nitronorepinephrine results in formation of Griess-positive pigments which were identified as indoles and whose formation are independent of the azo-compound.[138]

For optimal measurement of nitrite in plasma/serum, several recommendations were suggested which involve ideal λ_{max} for monitoring absorbance, correct dilution, and addition of ethanol to avoid nonspecific binding of the azo with plasma proteins.[153] HPLC equipped with postcolumn reactions and detection at 540 nm using Griess reagents gave high sensitivity of up to 30 pmol, a stark improvement over the colorimetric technique.[157] Rapid and sensitive measurement of NO_2^-/NO_3^- in clinical setting was demonstrated using enzymatic batch Griess assay with improved correlation of results with ion-pairing HPLC and gas chromatography-negative ion chemical ionization (GC-NICI-MS) techniques, however, this requires deproteinization of the biological samples prior to the analysis.[158]

6.6 THIOLS PROBES

Ellman's reagent is the most widely used reagent for the spectrophotometric analysis of free thiols in biological system.[159] The reagent is composed of 5,5′-dithiobis (2-nitrobenzoic acid) (DTNB) which reacts with free thiolates in a simple SN2 mechanism to form the mixed disulfide with liberation of the dianion, 2-nitro-5-thiobenzoic acid (TNB) (Fig. 6.14).[160] Since TNB has a pK_a of 4.53, it exists as 99.8%

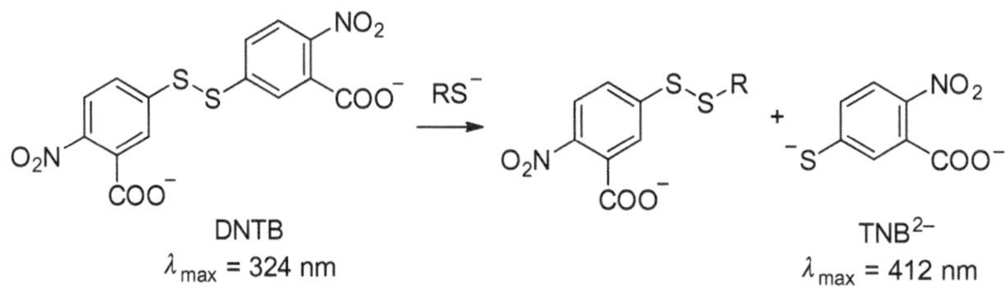

Figure 6.14 Mechanism of DTNB reaction with thiolate showing nucleophilic substitution reaction to form mixed disulfide and the chromophoric TNB^{2-}.

dianionic (as TNB^{2-}) at neutral pH which is intensely colored with a $\lambda_{max} = 412$ nm (ε ~14,000 M^{-1} cm^{-1}).[161]

DTNB can decompose in aqueous solution at 25°C and pH = 7.0 at a rate of 0.02% per hour and 10 times faster at basic pH of 8.0, therefore, it can be assumed that even at neutral pH, some degree of DTNB decomposition could occur but is very minimal.[161] Several factor could affect the kinetics of DNTB's reaction with thiols such presence of low concentrations of Cu(II) and air. Dithiols gave higher rate constant ($k = 10^5$ M^{-1} s^{-1}) than the monothiols ($k = 10^2$–10^4 M^{-1} s^{-1}) with DTNB, and follow Brϕnsted relation where the reactivity of the thiolate is directly proportional to the strength of their basicity, i.e., the higher the rate constant, the higher the pK_a of the thiol and/or the pH of the solution.[162] In protein systems, Brϕnsted coefficients were determined for the rates of thiol–disulfide interchange between low molecular weight thiols and the disulfide moieties of DNase, lysozyme, papain, or adenylate kinase, which were found similar to the coefficients derived from the reduction of GSSG and Ellman's reagent by the same low molecular weight thiols used in the study.[163] Since protein systems possess unique electronic properties, the accessibility of the thiol group to the negatively charged DTNB is crucial. Hence, modified Ellman's reagents were synthesized that possess different charge types such as in the case of 5,5′-dithio bis-(2-nitro-*N*-trimethylbenzyl ammonium iodide) and 5,5′-dithiobis-(2-nitro-*N*-2′-hydroxyethyl benzamide) as positive and neutral analogues of DTNB, respectively, and showed marked difference in their reactivity to thiols with different electrostatic environment.[164]

Several other modified DTNB were employed such as 4,4′-dithiopyridine. 4,4′-Dithiopyridine has the ability to measure thiols under acidic conditions (pH = 1.9),[165] nonpolar thiols in organic solvent,[166] and poorly accessible protein thiols due to its small molecular size, amphiphilic character, and neutral charge.[167] A selenium analogue of DTNB was also used but although its respective chromophore has lower extinction coefficient (10,000 M^{-1} cm^{-1}) than TNB^{2-}, it is stable at pH > 9.[168] The use of papain-$SSCH_3$ in place of DTNB could amplify spectrophotometric signals through hydrolysis of the chromophoric substrate, *N*-benzoyl-L-arginine-*p*-nitroanilide (L-BAPNA), by the reactivated papain with 100-fold higher

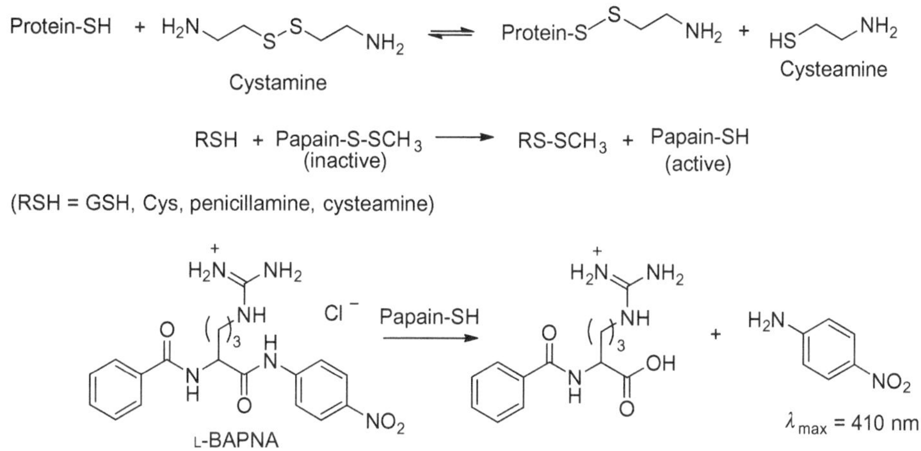

Figure 6.15 Use of papain-$SSCH_3$ for the amplification of thiol detection sensitivity.

sensitivity than DTNB (Fig. 6.15).[169] As shown in Fig. 6.15, low molecular weight thiols or inorganic sulfides can react with papain-$SSCH_3$ to form the activated papain-SH. Furthermore, protein-SH reaction with cystamine yields stoichiometric amount of cysteamine which can further react with papapin-S-SCH_3. Therefore, the total active papain-SH formed represents all the free thiols, both low and high molecular weights. Finally, papain-SH then cleaves L-BAPNA to release the chromophoric nitroaniline with a maximum absorption at 410 nm. This technique has also been employed in the determination of maleimide groups in proteins.[170]

Total thiol determination involves reduction of the disuflide to thiol for improved sensitivity and reproducibility since DTNB only detects thiolates but not disulfides. A combination of sodium borohydride reduction and using 4,4′-dithiopyridine reagent were employed thus allowing for the complete reduction of disulfides to sulfide as analyzed by reversed-phased HPLC with detection limit range in the picomolar at low pH.[171] Borohydride reduction coupled with DTNB also allowed for the determination of thiol and disulfide levels using HPLC technique.[172] It should be noted, however, that interference in thiol measurement can be caused by outdoor stray light. Outside daylight ($\lambda \sim 325$ nm) could increase the rate of DTNB degradation but has no effect on TNB^{2-}, therefore, outdoor light coming through window glass but not artificial indoor lighting must be avoided when transporting or performing this type of assay.[173]

Ellman's reagent was employed using a variety of biological samples. In urine samples, combined HPLC and postcolumn reaction with DTNB gave detection limit of 60 µg/mL penicillin for those patients taking 250 and 750 ng penicillin/day.[174] With liposomal formulations, DTNB can undergo both exo- and endo-transvesicular reactions with lipophilic thiols such as thiocholesterol, thiophenol, 2-thionaphthol, DL-cysteine, glutathione, 1-butanethiol, and 1-octanethiol.[175] Detection of exofacial (surface) thiols in CHO cells was demonstrated[176] as well as improved detection sensitivity was accomplished by precipitation of human-serum albumin—thiol (HSA-SH) from serum using ammonium sulfate with high yield allowing for a more precise and reliable measurement of HSA-SH using DTNB in type 2 diabetic patients.[177] Liquid chromatography-mass spectrometry (LC/MS) analysis of DTNB-derivatized thiols such as GSH, GSSG, CysSH, HCysSH, and homocystine from various tissue samples showed high selectivity and sensitivity with a detection limit of 3—15 pmol.[178]

6.7 PEROXIDES PROBES

6.7.1 Thiobarbituric acid reactive substance-malondialdehyde assay

Although not a direct technique for the detection of peroxyl radicals, thiobarbituric acid reactive substance (TBARS) is widely employed for the spectrophotometric determination of biomarkers of lipid peroxidation, in particular, malonaldialdehyde (MDA). MDA along with 4-hydroxy-2-nonenal are end-products of lipid peroxidation and prostaglandin synthesis, and is known to be mutagenic and carcinogenic. Other nonlipid biomolecular sources of MDA such as nucleic acid, bile pigments, proteins, and carbohydrates had been reported.[179] Also, observations of MDA formation from unsaturated fatty acids in malignant and normal tissue,[180] from NADPH-dependent lipid peroxidation in rat liver microsomes,[181] oxidation of 2-deoxyribose by HIO_4,[182] or CCl_4-induced hepatotoxicity in rats had been reported.[183] MDA accumulations was reported in disease conditions such as diabetes[184,185] atherosclerosis,[186–188] hypertension,[189] cancer,[190] and neurodegeneration.[191]

The nucleophilic reaction of thiobarbituric acid (TBA) with the dicarbonyl, MDA, forms the colored adduct according to Fig. 6.16 with a maximum absorbance at ~530 nm. However, correlation between TBA-MDA adduct formation, MDA production, and other lipid peroxidation indices sometimes do not correlate with each other, and this inconsistency was suggested to arise from the degree of unsaturation of the PUFA being oxidized, the nonexclusivity of MDA as the end-aldehydic product of lipid peroxidation, microenvironment, degradation of MDA, and nonlipid-

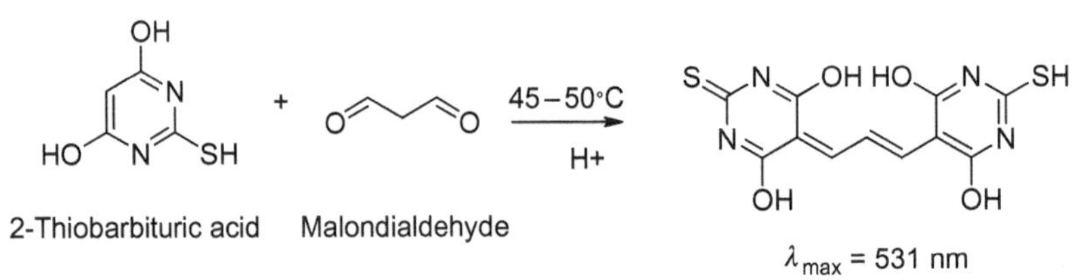

Figure 6.16 Reaction of thiobarbituric acid with malondialdehyde to form the chromophoric adduct.

based formation of MDA.[179] Accuracy in analysis also may vary from one medium to another. For example, TBARS concentration in dairy cattle serum, heparinized plasma, and plasma-EDTA do not show correlation with each other in which TBARS concentration in heparinized plasma vary considerably and showed plasma-EDTA to be the most suitable medium for TBARS determination.[192] When working with NO sources, one has to note that organic nitrates and nitrites can inhibit lipid peroxidation and could affect TBARS assay.[193]

Improved detection of total plasma MDA with high correlation to plasma TBARS was accomplished through HPLC using diode array detector (DAD), ultraviolet-fluorescence (UV-FL), and LC/MS-DAD(UV-2)-MS, the last being the most accurate detection compared to the spectrophotometric technique.[194] However, although HPLC methods was shown to lower detection limit and improve specificity for MDA in biological fluids such as serum and plasma, HPLC analysis can be affected by the type of column used, type of acid for sample hydrolysis and type of anticoagulant.[195] Fluorometric determination of TABRS in serum showed optimal condition at highly acidic pH of 2.4–2.6 with a detection limit of 0.1 μmol/L but following careful collection, preservation, and storage protocols.[196] Using fractionation technique, protein-bound MDA along with free MDA can be simultaneously and spectrofluorimetrically measured from plasma at $\lambda_{ex/em} = 535/550$ nm as TBA adducts along with hydroperoxides.[197]

6.7.2 Ferrous oxidation with xylenol orange assay

Another convenient method for determining lipid peroxidation in plasma and serum is through the use of Fe(II)/xylenol orange (FOX) assay. Unlike TBARS where the by-product of lipid peroxidation is measured, FOX directly detects the formed lipid peroxide (LOOH).

Iron (II) reduces peroxides (or hydrogen peroxide) to alkoxide (or hydroxide) and respective $LO^{\bullet}$ (or $HO^{\bullet}$) via Fenton-type reaction (Fig. 6.17). The Fe^{3+} then coordinates with xylenol orange through its amino carboxylate moieties forming a blue-purple complex that can be detected at 560 nm. To prevent the autoxidation of Fe^{2+}, this assay is typically performed in acidic pH and sorbitol is added to further oxidize the Fe^{2+} to Fe^{3+} according to Fig. 6.18, hence, enhancing the color formation.[198,199]

Inclusion of 100 mM sorbitol into the assay solution can increase the molar absorptivity from 4.46×10^4 to 2.24×10^5 M^{-1} cm^{-1} using H_2O_2 but only doubled with organic and blood serum hydroperoxides.[198] Two versions of FOX assay, i.e., FOX1 and FOX2, were designed to determine levels of water-soluble and lipid-soluble peroxides, respectively. FOX2 was developed to account for the high background signal arising from the nonperoxidized lipids and uses a chain-breaking antioxidant, butylated-hydroxytoluene, to prevent reaction of alkoxyl radical with native lipids to produce more hydroperoxides. Here, sorbitol is omitted due to the presence of already

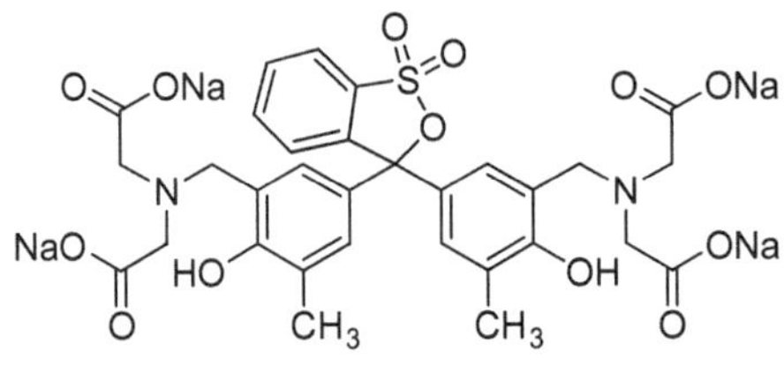

Xylenol-orange

$$Fe^{2+} + LOOH\ (HOOH) \longrightarrow Fe^{3+} + (LO^-)\ HO^- + LO^{\bullet}(HO^{\bullet})$$

$$Fe^{3+} + \text{xylenol orange} \longrightarrow \text{Xylenol-}Fe^{3+}\text{ complex } (\lambda_{max} = 560\text{ nm})$$

Figure 6.17 FOX assay mechanism of peroxide detection by xylenol orange as mediated by Fe^{2+} oxidation.

$$RO^{\bullet} + \text{sorbitol-CH} \longrightarrow \text{sorbitol-C}^{\bullet} + ROH$$

$$O_2 + \text{sorbitol-C}^{\bullet} \longrightarrow \text{sorbitol-CO}_2^{\bullet} \longrightarrow \text{sorbitol-C=O} + HO_2^{\bullet}$$

$$Fe^{2+} + HO_2^{\bullet} \longrightarrow Fe^{3+} + H_2O_2$$

Figure 6.18 Color enhancement by sorbitol via further oxidation of Fe(II) to Fe(III) by hydroperoxyl radical.

high concentration of methanol for solubilization where the methanol could also act as scavenger of $RO^{\bullet}$.[199] FOX2 assay is also used in conjunction with triphenylphosphine as hydroperoxide reductant which could be used to distinguish hydroperoxide generated in plasma from the background signal generated by ferric ions present in plasma.[200] The use of sucrose instead of sorbitol could also amplify the reaction by 60% for FOX1 and 7–8 times for FOX2 method allowing detection of up to 0.1 μM concentration of peroxide.[201]

Perhaps, one major limitation of FOX assay is that in order to prevent Fe^{2+} autoxidation, assay is performed in highly acidic pH (1.7–1.8) using sulfuric acid. Substitution of sulfuric acid with perchloric acid at pH 1.1 gave higher sensitivity through increased molar absorptivity but uses higher perchloric acid concentration of 110 mM compared to the conventional 25 mM sulfuric acid.[202] Other solvents that were shown to be appropriate were 50% acetic acid and acidified 90% MeOH.[203] Although FOX assay showed to be not specific to hydroperoxides of highly oxidized lipids especially that of arachidonyl-containing lipids,[204] it found numerous applications in the assessment of hydroperoxide formation from chemical, enzymatic, and cellular systems such as in the quantification of tyrosine/tyramine peroxide from peptides and amino acids,[205,206] in the assessment of cell viability by monitoring H_2O_2 levels in cell culture,[207] in the quantification of enzyme activity such as L-amino acid oxidase by detecting H_2O_2 production,[208] as well as myeloperoxidase-mediated generation of tyrosine peroxide by neutrophils.[209]

Methanolic extract of animal tissue showed less interference from peroxidizable lipids.[210] In biological fluid such as human blood serum, both lipid and protein peroxidation (as BSA-OOH) can be assayed using FOX reagents. Protein hydroperoxide are formed immediately after $HO^{\bullet}$ radical generation and that neither serum antioxidants nor ascorbate were able to prevent protein peroxidation, however, the degree of lipid peroxidation is dependent on the presence of ascorbate.[211] In the rabbit aqueous humor containing high levels of ascorbate, H_2O_2 can still be detected with 5 μM detection limit under nitrogen atmosphere.[212] From urine samples of diabetes patients, H_2O_2 concentrations were approximated using FOX1 assay showing higher H_2O_2 excretion in diabetic patients compared to healthy controls and gave good correlation showing increased plasma hydroperoxides and decreased red blood cells GSH levels.[213] Also, H_2O_2 formation was shown to correlate with O_2 consumption when human urine is exposed to 21% O_2.[214]

FOX assay can also be used in conjunction with other assays in order to obtain a more generalized assessment of the redox state of a system. For example, TBARS and FOX assays were carried out on fractionated human plasma for the simultaneous analysis of free MDA/peroxides as well as protein-bound MDA/peroxides with sensitivity of 20 and100 pmol for MDA and peroxide samples, respectively.[197] FOX and iodometric assays were shown to correlate with each other in LOOH determination from plant tissue under UVB light exposure. Moreover, FOX showed to have the ability to measure initial stages of fatty acid oxidation rather than the advance stages and is, therefore, appropriate for studying early membrane-associated stresses in plant extracts.[215] Study of lipid peroxidation in liposomes showed that FOX assay gave comparable results to that of other techniques such as iodometric method and conjugated diene measurement, but TBARS underestimated the levels of peroxidation.[216] Comparing all five analytical methods for the determination of peroxide (i.e., three versions of iodometric titration and two colorimetric techniques, ferric thiocyanate and FOX assay) in oxidized glee samples revealed that FOX method gave the highest correlation coefficient for peroxide value or flavor score with time of storage.[217] Efforts to modify FOX assay for improved throughput, precision, and sensitivity led to the use of multiwell plate reader and HPLC. A 96-well plate FOX assay in lipid emulsion was shown to increase reagents stability for at least 7 days compared to <6 day with older reagent formulations, ease of use and precision,[218] and that reversed-phase HPLC was used to monitor lipid hydroperoxide in photosensitized oxidation of rat plasma with high sensitivity of 40 pmol to 2 nmol.[219]

6.7.3 Iodometric assay

Direct reaction of lipid hydroperoxides using excess iodide leads to the reduction of the peroxides to alcohol as shown in Fig. 6.19. Formation of iodine forms an

$$ROOH + 2H^{+} + 2I^{-} \longrightarrow ROH + H_2O + I_2$$

$$I^{-} + I_2 \rightleftharpoons I_3^{-}$$

Figure 6.19 Mechanism of triiodide method for the detection of peroxides showing formation of iodine and its reaction to excess iodide to form the chromophoric triiodide.

equilibrium species with iodide to yield the chromophoric triiodide that can be measure colorimetrically at 290 or 360 nm. Common interference is O_2 and, therefore, must be performed anaerobically. Metal ions have also been reported to be a major interference which can be addressed by the addition of metal chelator. Other interference can be corrected by organic extraction of enzymatically hydrolyzed lipids from plasma followed by treatment of the organic extract with potassium iodide in acetic acid and methylene chloride to yield the triiodide. Absorbance reading arising from nonoxidizing chromophores and nonperoxide oxidants can be corrected using back titration of the triiodide to iodide using thiosulfate and resistance to reduction by gluthathione peroxidase, respectively.[220]

Several modification were made to improve sensitivity and specificity of the triiodide assay including the use different catalyst such as Cu(II),[221] Al(III),[222] or Mo(VI).[223] The use of nonionic and cationic detergents in 18% methanol made the addition of hydroperoxides with varying polarity in either water or ethanol, under room lighting, and with exposure to air possible. Longer incubation time and elevated temperature allow for the analysis of less reactive hydroperoxides with the same robustness as those for highly reactive hydroperoxides.[224] Total elimination of water from biological samples through azeoptropic distillation using ethanol/water mixture allows for the analysis of peroxides from liposomes as a function of time in as little sample volume of 5–50 μL.[225] Commercially available kit used for the estimation of cholesterol employs modified triiodide method which is composed of KI, NaN_3, polyethylene glycol, alkylbenzyldimethylammonium chloride, and ammonium molybdate, that is insensitive to malondialdehyde interference, oxygen, glucose, ascorbic acid, uric acid, GSH, Hb, and bilirubin. This kit was employed for the measurement of H_2O_2, *t-butyl* hydroperoxide and cumeme hydroperoxide for up to 1 nmol/mL concentration, and also for the measurement of LDL LOOH but at an extended incubation time of 60 min.[223] Application of this cholesterol kit for peroxide measurement in blood plasma gave lower values than using the standard anaerobic iodometric method suggesting that this kit could only be employed for the analysis of H_2O_2 or at prolonged incubation time for oxidized LDL.[226,227]

Automated-continuous flow iodometric assay was employed for the analysis of hydroperoxides in biological samples with detection limit of <100 pmol of peroxide.[228] Iodometric assay showed similar sensitivity to FOX assay for the measurement of H_2O_2 and linoleic hydroperoxide.[229] Modified iodometric method was shown to be the most specific for lipid hydroperoxides assay in plasma compared to conjugated dienes, methylene blue reaction in the presence of Hb and simple iodometric assay.[230] Comparison of the five assays for the determination of peroxide value in ghee showed that FOX and ferric thiocyanate assays gave the best correlation with time of storage compared to the iodometric techniques but not significantly.[217]

6.8 CHEMILUMINESCENCE

6.8.1 Lucigenin (10,10′-dimethyl-9,9′-biacridinium dinitrate, DNB^{2+})

Early observation of the emission of light by lucigenin in the presence of oxygen during hypoxanthine oxidation by xanthine oxidase was made by Tooter et al.[231] Lucigenin is widely used for the chemiluminescence detection of $O_2^{\bullet-}$ not only from HX/XO but also from NADPH oxidoreductases,[232] mitochondria,[233,234] cell membrane,[235] neutrophils,[236] endothelial cells,[237] fibroblasts,[238,239] smooth muscle cells,[240,241] myocytes,[242] bacteria,[243] I–R injury perfusate,[244] cardiac tissue,[245] and vascular tissue.[246,247] Lucigenin undergoes an initial one-electron reduction by $O_2^{\bullet-}$ or through enzymatic means where the formed radical cation then reacts further with $O_2^{\bullet-}$ to yield the dioxetane (Fig. 6.20). Through internal rearrangement, the dioxetane yields both the excited and ground state *N*-methylacridinium where the excited state further relaxes to the ground state emitting light that can be followed at 470 nm using a luminometer. Electrochemically generated $O_2^{\bullet-}$ shows that chemiluminescence is only observed at the potential range where O_2 and lucigenin are reduced simultaneously on the electrode, and that SOD prevents chemiluminescence but not catalase, indicating that that chemiluminescence originated from the reaction of $O_2^{\bullet-}$ and $DBT^{\bullet+}$, and not from H_2O_2 and $DBT^{\bullet+}$.[248]

Light emission was found to last 6 times longer in air than in closed system and was shown to be enhanced by cyanide, bovine serum albumin (BSA), amino acids,

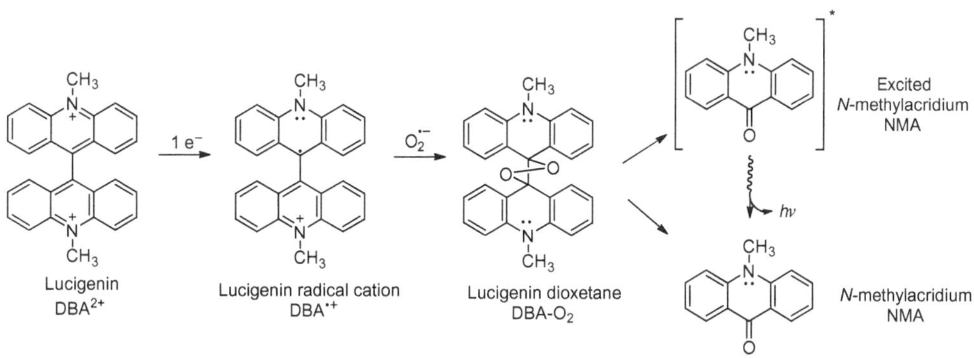

Figure 6.20 Mechanism of lucigenin chemiluminescence resulting from reaction of $O_2^{\bullet-}$ with the lucigenin radical cation to ultimately form the *N*-methylacridium excited state with light emission during relaxation.

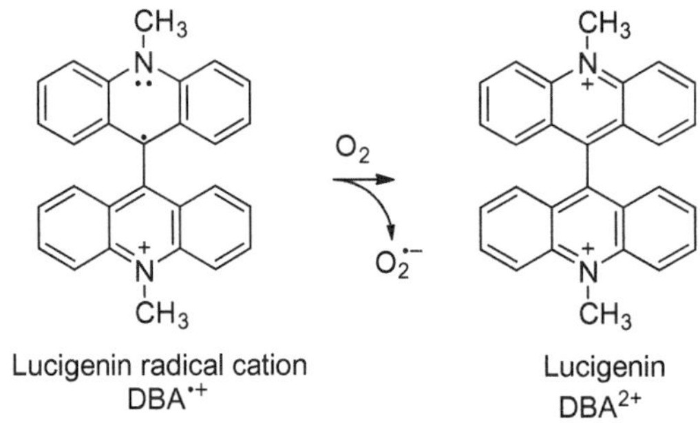

Figure 6.21 Electron transfer reaction between $DBT^{\bullet +}$ and O_2 to produce $O_2^{\bullet -}$ and lucigenin.

EDTA, and other chelating agents, and could be induced by other iron-flavoproteins enzymes. Also, under anaerobic conditions, lucigenin can be directly reduced by xanthine oxidase in the presence of xanthine oxidation.[249] In X/XO, NADH-XO, neutrophils, monocytes, and white blood cells, strong correlation between lucigenin-amplified chemiluminescence and cyt *c* reduction was observed.[250] In fact, increased rate of NADPH oxidation and cyt *c* reduction by the microsomes were observed in the presence of lucigenin and this was suggested to arise from several factors such as microsomal NAD(P)H-oxidoreductases, redox cycling of lucigenin radicals, dioxetane formation by (di)oxygenases, and its catalytic decomposition by cyt p450 heme.[251] Moreover, lucigenin was shown to mediate in $O_2^{\bullet -}$ production in glucose/glucose oxidase where $O_2^{\bullet -}$ production does not typically occur. This was proposed to be due to the direct reduction of lucigenin by the enzyme to $DBT^{\bullet +}$ which in turn reduces O_2 to $O_2^{\bullet -}$ via electron transfer mechanism according to Fig. 6.21.[252]

The contention over the years had focused on whether $DBT^{\bullet +}$ has the ability to reduced O_2 to $O_2^{\bullet -}$ to regenerate lucigenin, DBT^{2+}, also referred to as the redox cycling by lucigenin. Lucigenin in the presence of NADH reductases and NADH shows production of $HO^{\bullet}$ which is derived from $O_2^{\bullet -}$ via nonspecific coupling of $O_2^{\bullet -}$ with NADPH reductases that catalyzes O_2 reduction.[253] This was further demonstrated in endothelial cell homogenates showing that NADH leads to ~15-fold increase in the production of $O_2^{\bullet -}$ in the presence of lucigenin compared to in the absence of lucigenin, therefore, measurement of $O_2^{\bullet -}$ from NAD(P)H-dependent oxidase systems using lucigenin was suggested to be not reliable.[254] In eNOS/NADH system, $DBT^{\bullet +}$ can reduce O_2 to $O_2^{\bullet -}$ and warned that this technique may not be used for $O_2^{\bullet -}$ detection.[255] Lucigenin added to seminal plasma in the presence of ascorbate reduces lucigenin to $DBT^{\bullet +}$ which in turn can reduce O_2 to form $O_2^{\bullet -}$ giving false positive results.[256] Nonsuperoxide source that is not inhibitable by SOD was also reported to cause lucigenin chemiluminescence from phospholipids and human saphenous veins, although oxygen-dependent, the mechanism remained not fully understood.[257]

Confirmation of the redox cycling property of lucigenin has proven to be difficult due to the very fast conversion of $DBT^{\bullet+}$ to DBT and that several cyclic voltammetric measurements were not successful in separating $DBT^{2+}/DBT^{\bullet+}$ from the irreversible two-electron couple DBT^{2+}/DBT. The reported two-electron reduction potential for lucigenin ranges from -0.14 to -0.51 V depending on the type of solvent and electrode used.[258–262] Using dichloromethane at $-50°C$, the two-electron transfer process was resolved showing two reversible one-electron couples (i.e., $E_{1/2} = -0.35$ and -0.27 V vs Ag/Ag^{+}) corresponding to the formation of stable radical cation ($DBT^{\bullet+}$) and a neutral molecule (DBT), respectively. The equilibrium for the Eq. (6.6) was determined to be shifted to the right with an estimated equilibrium constant of $K = 2.17$.[263]

$$DBT^{2+} + DBT \overset{K=2.17}{\rightleftharpoons} 2\,DBT^{\bullet+} \tag{6.6}$$

The peak wave potential at -0.90 V for the reduction of O_2 to $O_2^{\bullet-}$ in DMSO in the presence of lucigenin[264] was also assumed to be the reduction potential of DBT^{2+} to $DBT^{\bullet+}$.[265] By correlating the one-electron redox potential of organic compounds in aprotic with that in aqueous medium, the E_o for the $DBT^{2+}/DBT^{\bullet+}$couple was approximated to be $+0.19$ V.[265] Therefore, using the redox potential for the $O_2/O_2^{\bullet-}$ couple of -0.16 V (1.0 M O_2)[266–268] gives $K = 10^6$ indicating that the equilibrium shown in Eq. (6.7) is mainly shifted to the right.

$$DBT^{2+} + O_2^{\bullet-} \overset{K=10^6}{\rightleftharpoons} DBT^{\bullet+} + O_2 \tag{6.7}$$

Electrostatic attraction between the positively charged, DBT^{2+} (or $DBT^{\bullet+}$) with $O_2^{\bullet-}$ is expected to be more favorable than $DBT^{\bullet+}$ and the neutral, O_2. It was suggested that chemiluminescence only reflects the lucigenin-independent $O_2^{\bullet-}$ production at "safe" lucigenin concentrations that could only be determined by carefully measuring O_2 consumption by polarography and $O_2^{\bullet-}$ production by EPR spin trapping upon stimulation, and this may vary using different experimental conditions and $O_2^{\bullet-}$-generating system. Nevertheless, it was suggested that lucigenin is still an appropriate probe for $O_2^{\bullet-}$ detection provided that the right lucigenin concentration is used.[269] Moreover, in the presence of xanthine oxidase, optimum pO_2 is required for the reduction of the two competing electron acceptors (i.e., O_2 and lucigenin), and determines the behavior of lucigenin-induced chemiluminescence.[270] In vascular tissue, lucigenin concentration could also affect chemiluminescence interpretation such that the NADH-driven vascular oxidase activity could be overestimated due to lucigenin redox cycling at 5 μM lucigenin concentration but is more pronounced at 250 μM and is significantly increased in the presence NADH substrate. At this high lucigenin concentration, chemiluminescence could be overestimated, not from $O_2^{\bullet-}$ production, but from oxidase-dependent two-electron transfer process to lucigenin.[271]

Regardless, whether redox cycling of lucigenin is significant or not in most of these assays, one thing is certain, i.e., $O_2^{\bullet-}$ is necessary for light emission from lucigenin and, therefore, is still the preferred method for assessing low levels of $O_2^{\bullet-}$ especially in vascular systems.[272–275]

Other limitations of lucigenin include its sensitivity to light at pH 10.4, while luciferin alone in organic solvents (e.g., alcohol and pyridine) spontaneously oxidizes with light emission.[231] In another study, UV irradiation in general can significantly increase luminescence in borate buffer but not in distilled H_2O alone or DMSO, and that temperature above 25°C can induce luminescence.[276] The pH also affects lucigenin luminescence where optimal signal is achieved at both pH 9.5 and 10.0 using KO_2 and HX/XO, respectively, while two maxima at pH 7.4 and 9.6 were observed for HRP/H_2O_2 system.[277]

Therefore, lucigenin with its shortcomings must not be used exclusively to assess $O_2^{\bullet-}$ production and should be employed in conjunction with other techniques. Comparison of lucigenin performance with other chemiluminescence probes is worth noting. For example, using PMN cells, lucigenin only detects ROS extracellularly and this correlates that of cyt *c* assay, while luminol is more appropriate for intracellular ROS detection and correlates with dichlorodihydrofluorescein diacetate (DCFH_2-DA) flow cytometry.[278] In monitoring $O_2^{\bullet-}$ production from perfused rat heart during I–R, the time course of lucigenin chemiluminescence is more consistent with the decrease in ROS production during hypoxic condition and increase ROS during reperfusion than the DHE fluorescence data.[244] Comparison of the various $O_2^{\bullet-}$ detection techniques such as cyt *c* reduction, ESR spin trapping, and DHE fluorescence indicate that these methods are satisfactory for the detection of extracellular endothelium-derived $O_2^{\bullet-}$ production while lucigenin gave artifactual results even at low concentrations.[33]

6.8.2 Cyprodina luciferin analogue and methoxy-CLA

Early observations of the light emitting properties of Cypridina luciferin were reported from xanthine–xanthine oxidase system[279] and activated human granulocytes[280] suggesting $O_2^{\bullet-}$-mediated luminescence. Hence, luciferin analogues were later employed to detect singlet oxygen ($^1O_2$) from peroxidase–H_2O_2–halide system[281] and H_2O_2-dependent generation of $^1O_2$ from human saliva[282] as well as in the detection of $O_2^{\bullet-}$ in various organs such as rat gastric mucosa during I–R,[283] intact vascular tissue,[284] perfused liver, and rat lung from necrotized pancreatitis.[285] Superoxide or singlet oxygen reacts with luciferin analogues such as 2-methyl-6-phenyl-3,7-dihydroimidazo[1,2-a]pyrazin-3-one (CLA) and 2-methyl-6-(4-methoxyphenyl)-3,7-dihydroimidazo[1,2-a]pyrazin-3-one hydrochloride (MCLA) with rate constants of 10.8 and 2.56×10^8 M^{-1} s^{-1},[286,287] respectively. This reaction produces a

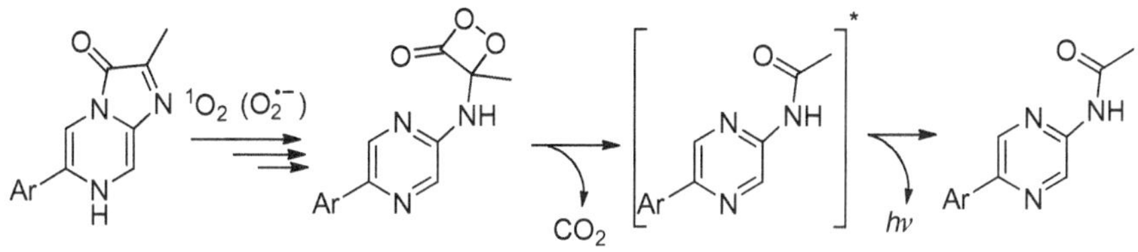

Figure 6.22 Light emission from CLA and MCLA probes via decarboxylation reaction after reaction with superoxide or singlet oxygen.

dioxetanone intermediate which undergoes decarboxylation generating the singlet amidopyrazine excited state which relaxes to the ground state emitting light in the process (Fig. 6.22). Light emission for CLA and MCLA is typically observed at 410 and 430 nm, respectively.[288–290] Kinetic studies show that O_2 reaction with CLA is slow with rate constant of 0.4 M^{-1} s^{-1} at 20°C in which this autoxidation forms the CLA–$O_2^{\bullet-}$ pair but the exact mechanism by which this occurs is not well understood.[291] Comparison of the selectivity of chemiluminescence probes to various radicals shows that only some acridinium derivatives to be highly selective to $O_2^{\bullet-}$ and $HO^{\bullet}$ compared to other probes that were investigated. This is followed by CLA and MCLA which gave moderate selectivity to $O_2^{\bullet-}$ and high selectivity to NO. In general, none of the probes were found to be selective to H_2O_2, ClO^-, or $^1O_2$.[292] In the presence of nitroxide radical, the chemiluminescence of MCLA and CLA were greatly enhanced owing to the catalytic property of nitroxide. For example, MCLA is converted to MCLA$^{\bullet}$ by the nitroxide allowing for a more facile reaction between the MCLA$^{\bullet}$ and $O_2^{\bullet-}$.[293]

The enzyme inhibitor, sodium azide, was shown to be a specific quencher of $^1O_2$ from three H_2O_2 decomposition systems (i.e., H_2O_2-HRP; H_2O_2-$CuSO_4$; and H_2O_2-NaOCl) using luminol and CLA analogues.[294] Other known chemiluminescence quenchers using MCLA independent of competition reaction with $O_2^{\bullet-}$ were sulfur-containing compounds such as thiocarbonyl, thiocarbamoyl cyclic molecules, and other thiol and disulfides compounds.[295]

6.8.3 Luminol

Luminol is employed for ROS detection under basic condition. Luminol-amplified luminescence measurement of ROS from the respiratory burst of activated phagocytes was shown to parallel results obtained from cyt *c* oxidation and *p*-hydroxyphenyl acetate oxidation.[296] Luminol-dependent luminescence was also found to be in agreement with fluorometric assays using stimulated human neutrophils.[297] Compared to other chemiluminescence probes such as CLA, MCLA, and acridinium, luminol is

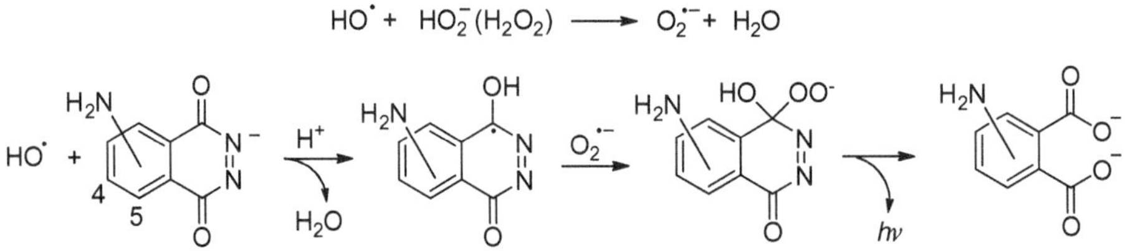

luminol: X = 5-amino
isoluminol : X = 4-amino

Figure 6.23 Light emission from the reaction of ROS with luminol via initial formation of hydrazyl radical and its subsequent reaction with superoxide to ultimately the diacid.

not as sensitive to most ROS species.[292] However, using activated rat blood phagocytes, the order of decrease luminescence signal was found to be: luminol > isoluminol > lucigenin.[298] Electrochemical technique was also shown to be as sensitive as isoluminol for monitoring ROS production from neutrophil-like cells and human neutrophils at as low as 500 cells.[299]

Luminol and its analogues are phthaloyl hydrazide-based probes. The mechanism of luminol chemiluminescence upon reaction with $O_2^{\bullet -}$ involves initial fast reaction of $HO^{\bullet}$ (usually generated from Fenton chemistry) with the cyclic hydrazide to form the hydrazyl radical that upon reaction with $O_2^{\bullet -}$ forms the peroxide adduct. The reaction of $O_2^{\bullet -}$ with hydrazyl radical is very fast and pH dependent with $k = 1.4 \times 10^9\ M^{-1}\ s^{-1}$ pH 7.7 and $\sim 2 \times 10^8\ M^{-1}\ s^{-1}$ at pH 11. The peroxide adduct then decomposes to the diacid with concomitant emission of light that is observable at 450 nm at pH 7.7 (Fig. 6.23).[300] Optimum chemiluminescence from luminol could be achieved at pH 9.0 and 9.4 in KO_2 and HX/XO, respectively, but luminol is less sensitive compared to MCLA and lucigenin. In the presence of H_2O_2 alone, no luminescence between pH of 4 and 9.5 was observed but optimal luminescence can be achieved at basic pH in HRP/H_2O_2 and MPO/H_2O_2 system, and $^1O_2$ generation from HOCl/H_2O_2 and sodium 1,4 napthylidine-3,3′ dipropionate endoperoxide.[277] It is also worth noting that luminol and isoluminol only gave weak luminescence in acetonitrile alone, or in water with tetra-*n*-propylammonium hydroxide or H_2O_2.[301] This low response to H_2O_2 was further supported by previous studies where the specificity of luminol-enhanced luminescence due to H_2O_2 from polymorphonuclear leukocytes raised some questions. The use of stimulants such as concanavalin A which only produces H_2O_2 as well as formylated peptide (formylmethionyl-leucyl-phenylalanine, fMLP) which induces both H_2O_2 and $O_2^{\bullet -}$ production revealed that for concanavalin A, no luminescence was observed while luminescence was suppressed by SOD for fMLP. This indicates that luminol is not specific to H_2O_2 in HRP-azide-luminol system.[302] The presence of Fe^{2+} can also induce luminescence using luminol, but not with lucigenin. Ferrous-induced luminescence can be suppressed by antioxidants such as urate, ascorbate, and methionine, and by $HO^{\bullet}$ scavengers such as DMSO and mannitol, as well as by catalase and SOD indicating H_2O_2 or $O_2^{\bullet-}$ -mediated $HO^{\bullet}$ formation.[303] However, the specificity of isoluminol is not only limited to ROS but was also used to detect serum Hb at < 5 mg/L with lesser interference by nonheme Fe compared to the colorimetric method typically used for Hb analysis using tetramethylbenzine.[304] Also, antioxidant activity in biological fluids such as rat serum, human saliva, and green tea were assessed using isoluminol in the presence of microperoxidase and cumene hydroperoxide.[305]

Detection of ROS using luminol requires the use of HRP especially in the study of oxidative burst from neutrophils in order to remove H_2O_2 thus preventing its accumulation and minimizing further H_2O_2 side reaction, and at the same time allows for the detection of H_2O_2. Azide is also added to inhibit cell-derived MPO and catalase that may compete for the ROS but not for the purpose of inhibiting extracellularly added SOD and HRP. Hence, addition of HRP/azide improves signal-to-noise ratio and faster rate of reaching the maximum signal.[297] Luminol is more lipophilic than isoluminol and, therefore, the former is a membrane permeable probe while isoluminol does not permeate into the cell. Using isoluminol as an extracellular probe along with luminol, one would be able to measure total H_2O_2 and extracellular H_2O_2, respectively, and is useful for assessing extracellular vs intracellular H_2O_2 generation.[306–308] This is further supported by other studies using PMN showing that luminol-dependent luminescence gave good correlation with intracellular ROS detection with $DCFH_2$-DA, while lucigenin correlates with extracellular ROS as detected using cyt *c* reduction assay.[278] Moreover, it was shown that luminol being lipophilic can exert inhibitory effect on NADPH activity of neutrophils at the surface of aggregated IgG upon incubation but not isoluminol. There was no inhibitory effect in cell-free NADPH-oxidase system alone indicating that the mechanism of luminol inhibition is mediated through signal transduction pathway leading to the assembly of oxidases.[309]

Isoluminol-enhanced luminescence was found to detect $O_2^{\bullet-}$ from as few as 250 neutrophils[307] but no consistent trend in achieving high luminescence yield from using different cell lines and activators was found and, therefore, optimization of experimental conditions was recommended.[237] However, the time it takes for luminescence to form could also be influenced by the type of stimulant used, for example, stimulation takes longer and that the onset times are concentration-dependent when using PMA compared to fMLP where luminescence is detectable within 2 s of stimulation.[297] In rat blood leukocytes, highest luminescence response was observed for luminol and isoluminol vs lucigenin with opsonized zymozan compared to using PMA, calcium ionophore, and fMLP activators.[298] Other compounds were employed to enhance luminescence such as the use of Co(II) and β-CD in human hemoglobin and human blood samples[310]; through photosensitization reaction that uses quinones to generate 3,6-dihydroxyphthalic acid[311]; diimine ligands in the presence of CH_3CN, H_2O_2, tetra-*n*-propylammonium hydroxide (TPA), and Fe (III) ion[301]; L-carnosine for chemiluminescence assay of blood factors and transglutamase[312]; 4-iodophenylboronic acid and 4-phenylboronic acid,[313] or hydroxybenzothiozoles[314] in HRP-catalyzed luminol or isolumnol oxidation. Conversely, certain peptides such as the annexin AI reduces luminescence of isoluminol but not luminol in stimulated cells or H_2O_2/HRP.[315] Interference from hydroquinone ubiquinol-10 was reported but can be prevented by using reducing agents such as sodium borohydride or triphenylphosphine.[316]

Other instrumental methods were employed for ROS detection using luminol probes. HPLC technique was used for peroxides detection with isoluminol exhibiting ultrasensitivity at picomole levels.[317,318] The HPLC technique was also employed for the analysis of peroxides from the methanol and hexane extracts of human blood plasma where H_2O_2 was suggested to have not come from the plasma itself due to its facile degradation by endogenous catalase but was formed during the sample preparation suggesting that only the lipid hydroperoxides were detectable by the HPLC method but not H_2O_2. Immunoassay technique using polyclonal antibodies was applied using amino-derivatized isoluminol which was used to label steroids, or carboxylated-luminol derivatives were used to couple to the amino residues of proteins and were employed for the detection of progesterone, estradiol, estriol, and prolactin from biological fluids with sensitivity ranging from 0.54 to 0.04 nmol/L.[319–322]

REFERENCES

1. Knowles PF, Gibson JF, Pick FM, Bray RC. Electron-spin-resonance evidence for enzymic reduction of oxygen to a free radical, the superoxide ion. *Biochem J* 1969;**111**(1):53–8.
2. McCord JM, Fridovich I. Superoxide dismutase. Enzymic function for erythrocuprein (hemocuprein). *J Biol Chem* 1969;**244**(22):6049–55.
3. Nauseef WM. Detection of superoxide anion and hydrogen peroxide production by cellular NADPH oxidases. *Biochim Biophys Acta, Gen Subj* 2014;**1840**(2):757–67.
4. Lippitt B, Fridovich I. Tetrazolium oxidase and superoxide dismutase. Evidence for identity. *Arch Biochem Biophys* 1973;**159**(2):738–41.
5. Patriarca P, Dri P, Rossi F. Superoxide dismutase in leukocytes. *FEBS (Fed Eur Biochem Soc) Lett* 1974;**43**(3):247–51.
6. Henry LEA, Halliwell B, Hall DO. The superoxide dismutase activity of various photosynthetic organisms measured by a new and rapid assay technique. *FEBS Letters* 1976;**66**(2):303–6.
7. Bielski BHJ, Shiue GG, Bajuk S. Reduction of nitro blue tetrazolium by CO2- and O2- radicals. *J Phys Chem* 1980;**84**(8):830–3.
8. Rajagopalan KV, Handler P. Hepatic aldehyde oxidase. II. Differential inhibition of electron transfer to various electron acceptors. *J Biol Chem* 1964;**239**(6):2022–6.
9. Abugo OO, Rifkind JM. Oxidation of hemoglobin and the enhancement produced by nitroblue tetrazolium. *J Biol Chem* 1994;**269**(40):24845–53.
10. Pistelli A, Mollace V, Nistico G, Salvemini D, Vane J. Nitroblue tetrazolium inhibits oxidation of glyceryl trinitrate to nitric oxide in bovine aortic smooth muscle cells. *Biochem Pharmacol* 1994;**47**(10):1737–42.

11. Wang C, Cai Z-X, You Z-L, Guo H-S, Shang D-J, Wang X-L, et al. Free Radical Scavenging Activity and Neuroprotective Potentials of D138, One Cu(II)/Zn(II) Schiff-Base Complex Derived from N,N'-bis(2-Hydroxynaphthylmethylidene)-1,3-propanediamine. *Neurochem Res* 2014;**39**(9): 1834–44.
12. Samhan-Arias AK, Gutierrez-Merino C. Purified NADH-cytochrome b5 reductase is a novel superoxide anion source inhibited by apocynin: sensitivity to nitric oxide and peroxynitrite. *Free Radical Biol Med* 2014;**73**:174–89.
13. Serrander L, Cartier L, Bedard K, Banfi B, Lardy B, Plastre O, et al. NOX4 activity is determined by mRNA levels and reveals a unique pattern of ROS generation. *Biochem J* 2007;**406**(1):105–14.
14. Halliwell B. Superoxide dismutase activity of iron complexes. *FEBS Lett* 1975;**56**(1):34–8.
15. Younes M, Weser U. Inhibition of nitroblue tetrazolium reduction by cuprein (superoxide dismutase), Cu(Tyr)2 and Cu(Lys)2. *FEBS Lett* 1976;**61**(2):209–12.
16. Basarali KM, Yazar H, Buyukbas S, Kilinc C, Kayrak M. Acute myocardial infarction ischemia of the erythtrocyte oxidative stress in the role of nitric oxide. *J Chem Pharm Res* 2011;**3**(6):271–6.
17. Weydert C, Cullen JJ. Measurement of superoxide dismutase, catalase and glutathione peroxidase in cultured cells and tissue. *Nat Protoc* 2010;**5**(1):51–66.
18. Spitz DR, Oberley LW. An assay for superoxide dismutase activity in mammalian tissue homogenates. *Anal Biochem* 1989;**179**(1):8–18.
19. Peskin AV, Winterbourn CC. A microtiter plate assay for superoxide dismutase using a water-soluble tetrazolium salt (WST-1). *Clin Chim Acta* 2000;**293**(1-2):157–66.
20. Valinoti JM, Zinkl JG, Jain NC. Combined Phagocytosis-Nitroblue Tetrazolium Reduction Test in Bovine Neutrophils. *Veterinary Clinical Pathology* 1988;**17**(4):99–104.
21. Bielski BHJ, Richter HW. A study of the superoxide radical chemistry by stopped-flow radiolysis and radiation induced oxygen consumption. *J Am Chem Soc* 1977;**99**(9):3019–23.
22. Butler J, Koppenol WH, Margoliash E. Kinetics and mechanism of the reduction of ferricytochrome c by the superoxide anion. *J Biol Chem* 1982;**257**(18):10747–50.
23. Butler J, Jayson GG, Swallow AJ. Reaction between the superoxide anion radical and cytochrome c. *Biochim Biophys Acta, Bioenerg* 1975;**408**(3):215–22.
24. Bielski BHJ, Allen AO. Mechanism of the disproportionation of superoxide radicals. *J Phys Chem* 1977;**81**(11):1048–50.
25. Heijman MGJ, Nauta H, Levine YK. The influence of a detergent on the reactivity of cytochrome c towards the superoxide radical as measured by pulse radiolysis. *Biochimica et Biophysica Acta (BBA) - Protein Structure and Molecular Enzymology* 1982;**704**(3):560–3.
26. Turrens JF, McCord JM. How relevant is the reoxidation of ferrocytochrome c by hydrogen peroxide when determining superoxide anion production? *FEBS Lett* 1988;**227**(1):43–6.
27. Misra HP, Fridovich I. Univalent reduction of oxygen by reduced flavins and quinones. *J Biol Chem* 1971;**247**(1):188–92.
28. Boveris A, Cadenas E. Mitochondrial production of superoxide anions and its relation to the antimycin insensitive respiration. *FEBS Lett* 1975;**54**(3):311–14.
29. Hoffmeyer MR, Jones SP, Ross CR, Sharp B, Grisham MB, Laroux FS, et al. Myocardial ischemia/reperfusion injury in NADPH oxidase-deficient mice. *Circ Res* 2000;**87**(9):812–17.
30. Liu D, Sybert TE, Qian H, Liu J. Superoxide production after spinal injury detected by microperfusion of cytochrome c. *Free Radic Biol Med* 1998;**25**(3):298–304.
31. Roubaud V, Sankarapandi S, Kuppusamy P, Tordo P, Zweier JL. Quantitative measurement of superoxide generation using the spin trap 5-(diethoxyphosphoryl)-5-methyl-1-pyrroline-N-oxide. *Anal Biochem* 1997;**247**(2):404–11.
32. Dikalov SI, Li W, Mehranpour P, Wang SS, Zafari AM. Production of extracellular superoxide by human lymphoblast cell lines: comparison of electron spin resonance techniques and cytochrome c reduction assay. *Biochem Pharmacol* 2007;**73**(7):972–80.
33. Barbacanne MA, Souchard JP, Darblade B, Iliou JP, Nepveu F, Pipy B, et al. Detection of superoxide anion released extracellularly by endothelial cells using cytochrome c reduction, ESR, fluorescence and lucigenin-enhanced chemiluminescence techniques. *Free Radic Biol Med* 2000;**29**(5):388–96.
34. Schejter A, Aviram I. Reaction of cytochrome c with imidazole. *Biochemistry* 1969;**8**(1):149–53.

35. Kunitomo R, Miyauchi Y, Inoue M. Synthesis of a cytochrome c derivative with prolonged in vivo half-life and determination of ascorbyl radicals in the circulation of the rat. *J Biol Chem* 1992; **267**(13):8732–8.
36. Tomoda A, Ida M, Yoneyama Y, Kitajima S, Minakami S. Reduction of ferricytochrome c by human red cells. *Experientia* 1980;**36**(12):1345–7.
37. Rodkey FL, Ball EG. Oxidation-Reduction Potentials of the Cytochrome c System. *Journal of Biological Chemistry* 1950;**182**(1):17–28.
38. Creutz C, Sutin N. Reduction of ferricytochrome c by dithionite ion: electron transfer by parallel adjacent and remote pathways. *Proc Natl Acad Sci U S A* 1973;**70**(6):1701–3.
39. Kertesz D. The reduction of ferricytochrome c during the enzymic oxidation of 3,4-dihydroxyphenylalanine. *Biochim Biophys Acta* 1968;**167**(2):250–6.
40. Polacow I, Cilento G. The reduction of ferricytochrome c by pyridine coenzymes mediated by thyroid hormones and related compounds. *Arch Biochem Biophys* 1970;**139**(2):401–5.
41. Guillemette JG, Barker PD, Eltis LD, Lo TP, Smith M, Brayer GD, et al. Analysis of the bimolecular reduction of ferricytochrome c by ferrocytochrome b5 through mutagenesis and molecular modelling. *Biochimie* 1994;**76**(7):592–604.
42. Cheung E, English AM. Temperature dependence of the unimolecular reduction of ferricytochrome c by ferrocytochrome c peroxidase at low and high ionic strengths. *Inorganic Chemistry* 1988;**27**(6):1078–81.
43. Simic MG, Taub IA, Tocci J, Hurwitz PA. Free radical reduction of ferricytochrome c. *Biochem Biophys Res Commun* 1975;**62**(2):161–7.
44. Shafferman A, Stein G. Reduction of ferricytochrome c by some free radical agents. *Science* 1974;**183**(4123):428–30.
45. Van Leeuwen JW, Raap A, Koppenol WH, Nauta H. A tunnelling model to explain the reduction of ferricytochrome c by H and OH radicals. *Biochimica et Biophysica Acta (BBA) - Bioenergetics* 1978;**503**(1):1–9.
46. Dikalov S, Khramtsov V, Zimmer G. Determination of rate constants of the reactions of thiols with superoxide radical by electron paramagnetic resonance: critical remarks on spectrophotometric approaches. *Arch Biochem Biophys* 1996;**326**(2):207–18.
47. Winterbourn CC. Superoxide dismutase-inhibitible reduction of cytochrome c by the alloxan radical. Implications for alloxan cytotoxicity. *Biochem J* 1982;**207**(3):609–12.
48. Vandewalle PL, Petersen NO. Oxidation of reduced cytochrome c by hydrogen peroxide: Implications for Superoxide assays. *FEBS Letters* 1987;**210**(2):195–8.
49. Dikalov S, Griendling KK, Harrison DG. Measurement of reactive oxygen species in cardiovascular studies. *Hypertension* 2007;**49**(4):717–27.
50. Misra HP, Fridovich I. The role of superoxide anion in the autoxidation of epinephrine and a simple assay for superoxide dismutase. *J Biol Chem* 1972;**247**(10):3170–5.
51. Alhasan R, Njus D. The epinephrine assay for superoxide: why dopamine does not work. *Anal Biochem* 2008;**381**(1):142–7.
52. Misra HP, Fridovich I. Role of superoxide anion in the autoxidation of epinephrine and a simple assay for superoxide dismutase. *J Biol Chem* 1972;**247**(10):3170–5.
53. Prough RA, Masters BSS. NADPH oxidase reaction of NADPH-cytochrome c reductase. I. Role of superoxide anion. *Ann N Y Acad Sci* 1973;**212**:89–93.
54. Salimath BP, Savitha G. Mechanism of inhibition by cyclic AMP of protein kinase C-triggered respiratory burst in Ehrlich ascites tumor cells. *Cell Signalling* 1992;**4**(6):651–63.
55. Siraki AG, Smythies J, O'Brien PJ. Superoxide radical scavenging and attenuation of hypoxia-reoxygenation injury by neurotransmitter ferric complexes in isolated rat hepatocytes. *Neurosci Lett* 2000;**296**(1):37–40.
56. Sun M, Zigman S. An improved spectrophotometric assay for superoxide dismutase based on epinephrine autoxidation. *Anal Biochem* 1978;**90**(1):81–9.
57. Dikalov S, Khramtsov V, Zimmer G. Determination of rate constants of the reactions of thiols with superoxide radical by electron paramagnetic resonance: critical remarks on spectrophotometric approaches. *Arch Biochem Biophys* 1996;**326**(2):207–18.

58. Mehta JL, Li D. Epinephrine upregulates superoxide dismutase in human coronary artery endothelial cells. *Free Radical Biol Med* 2001;**30**(2):148–53.
59. Rizzi C, Samouilov A, Kutala VK, Parinandi NL, Zweier JL, Kuppusamy P. Application of a trityl-based radical probe for measuring superoxide. *Free Radic Biol Med* 2003;**35**(12):1608–18.
60. Kutala VK, Parinandi NL, Zweier JL, Kuppusamy P. Reaction of superoxide with trityl radical: implications for the determination of superoxide by spectrophotometry. *Arch Biochem Biophys* 2004;**424**(1):81–8.
61. Kutala VK, Villamena FA, Ilangovan G, Maspoch D, Roques N, Veciana J, et al. Reactivity of superoxide anion radical with a perchlorotriphenylmethyl (trityl) radical. *J Phys Chem B* 2008;**112**(1): 158–67.
62. Liu Y, Song Y, De Pascali F, Liu X, Villamena FA, Zweier JL. Tetrathiatriarylmethyl radical with a single aromatic hydrogen as a highly sensitive and specific superoxide probe. *Free Radic Biol Med* 2012;**53**(11):2081–91.
63. Warwar N, Mor A, Fluhr R, Pandian RP, Kuppusamy P, Blank A. Detection and Imaging of Superoxide in Roots by an Electron Spin Resonance Spin-Probe Method. *Biophysical Journal* 2011;**101**(6):1529–38.
64. Ghiselli A. Aromatic hydroxylation: salicylic acid as a probe for measuring hydroxyl radical production. *Methods Mol Biol (Totowa, NJ)* 1998;**108**:89–100 (Free Radical and Antioxidant Protocols)
65. Anbar M, Neta P. A compilation of specific bimolecular rate constants for the reactions of hydrated electrons, hydrogen atoms, and hydroxyl radicals with inorganic and organic compounds in aqueous solution. *Int J Appl Radiat Isot* 1967;**18**(7):493–523.
66. Halliwell B. Superoxide-dependent formation of hydroxyl radicals in the presence of iron chelates: is it a mechanism for hydroxyl radical production in biochemical systems? *FEBS Lett* 1978;**92**(2): 321–6.
67. Nappi AJ, Vass E. Hydroxyl radical production by ascorbate and hydrogen peroxide. *Neurotoxic Res* 2000;**2**(4):343–55.
68. Kaur H, Whiteman M, Halliwell B. Peroxynitrite-dependent aromatic hydroxylation and nitration of salicylate and phenylalanine. Is hydroxyl radical involved? *Free Radic Res* 1997;**26**(1):71–82.
69. Bojic M, Sedgeman CA, Nagy LD, Guengerich FP. Aromatic hydroxylation of salicylic acid and aspirin by human cytochromes P450. *Eur J Pharm Sci* 2015;**73**:49–56.
70. Maskos Z, Rush JD, Koppenol WH. The hydroxylation of the salicylate anion by a fenton reaction and Γ-radiolysis: A consideration of the respective mechanisms. *Free Radical Biology and Medicine* 1990;**8**(2):153–62.
71. Halliwell B, Kaur H, Ingelman-Sundberg M. Hydroxylation of salicylate as an assay for hydroxyl radicals: a cautionary note. *Free Radic Biol Med* 1991;**10**(6):439–41.
72. Panchompoo J, Aldous L, Kabeshov M, Pilgrim BS, Donohoe TJ, Compton RG. A green approach to Fenton chemistry: mono-hydroxylation of salicylic acid in aqueous medium by the electrogeneration of Fenton's reagent. *New J Chem* 2012;**36**(5):1265–72.
73. Yoshimura Y, Otsuka K, Uchiyama K, Tanaka H, Tamura K, Ohsawa K, et al. Detection of hydroxyl radicals with salicylic acid. *Anal Sci* 1989;**5**(2):161–4.
74. Valgimigli M, Merli E, Malagutti P, Soukhomovskaia O, Cicchitelli G, Antelli A, et al. Hydroxyl radical generation, levels of tumor necrosis factor-alpha, and progression to heart failure after acute myocardial infarction. *Journal of the American College of Cardiology* 2004;**43**(11):2000–8.
75. Grienberger HJ, Pillai DR, Schlachetzki F, Gruber M, Dittmar MS. Detection of free radicals by isolated perfusion of the rat brain following hemorrhagic stroke: a novel approach to cerebrovascular biomarker research. *Exp Brain Res* 2010;**206**(3):311–17.
76. Powell SR, Hall D. Use of salicylate as a probe for hydroxyl radical formation in isolated ischemic rat hearts. *Free Radical Biol Med* 1990;**9**(2):133–41.
77. Onodera T, Ashraf M. Detection of hydroxyl radicals in the postischemic reperfused heart, using salicylate as a trapping agent. *J Mol Cell Cardiol* 1991;**23**(3):365–70.
78. Piantadosi CA, Zhang J, Demchenko IT. Production of hydroxyl radical in the hippocampus after CO hypoxia or hypoxic hypoxia in the rat. *Free Radical Biol Med* 1997;**22**(4):725–32.

79. Floyd RA, Henderson R, Watson JJ, Wong PK. Use of salicylate with high pressure liquid chromatography and electrochemical detection (LCED) as a sensitive measure of hydroxyl free radicals in adriamycin treated rats. *J Free Radicals Biol Med* 1986;**2**(1):13–18.
80. Solar S. Reaction of OH with phenylalanine in neutral aqueous solution. *Radiation Physics and Chemistry (1977)* 1985;**26**(1):103–8.
81. Kaur H, Fagerheim I, Grootveld M, Puppo A, Halliwell B. Aromatic hydroxylation of phenylalanine as an assay for hydroxyl radicals: application to activated human neutrophils and to the heme protein leghemoglobin. *Anal Biochem* 1988;**172**(2):360–7.
82. Fujimoto S, Kawakami N, Ohara A. Hydroxylation of phenylalanine and salicylate by stimulated polymorphonuclear leukocytes and the accelerating effect of glutathione on their hydroxylation. *Biol Pharm Bull* 1994;**17**(6):767–72.
83. van der Vliet A, O'Neill CA, Halliwell B, Cross CE, Kaur H. Aromatic hydroxylation and nitration of phenylalanine and tyrosine by peroxynitrite. Evidence for hydroxyl radical production from peroxynitrite. *FEBS Lett* 1994;**339**(1-2):89–92.
84. Biondi R, Xia Y, Rossi R, Paolocci N, Ambrosio G, Zweier JL. Detection of hydroxyl radicals by D-phenylalanine hydroxylation: a specific assay for hydroxyl radical generation in biological systems. *Anal Biochem* 2001;**290**(1):138–45.
85. Biondi R, Ambrosio G, Liebgott T, Cardounel AJ, Bettini M, Tritto I, et al. Hydroxylation of D-phenylalanine as a novel approach to detect hydroxyl radicals: Application to cardiac pathophysiology. *Cardiovasc Res* 2006;**71**(2):322–30.
86. Themann C, Teismann P, Kuschinsky K, Ferger B. Comparison of two independent aromatic hydroxylation assays in combination with intracerebral microdialysis to determine hydroxyl free radicals. *J Neurosci Methods* 2001;**108**(1):57–64.
87. Lubec B, Hermon M, Hoeger H, Lubec G. Aromatic hydroxylation in animal models of diabetes mellitus. *FASEB J* 1998;**12**(14):1581–7.
88. Sun J-Z, Kaur H, Halliwell B, Li XY, Bolli R. Use of aromatic hydroxylation of phenylalanine to measure production of hydroxyl radicals after myocardial ischemia in vivo. Direct evidence for a pathogenic role of the hydroxyl radical in myocardial stunning. *Circ Res* 1993;**73**(3):534–49.
89. Fujimoto S, Ishimitsu S, Hirayama S, Kawakami N, Ohara A. Hydroxylation of phenylalanine by myeloperoxidase-hydrogen peroxide system. *Chem Pharm Bull* 1991;**39**(6):1598–600.
90. Ishimitsu S, Fujimoto S, Ohara A. The photochemical decomposition and hydroxylation of phenylalanine in the presence of riboflavin. *Chem Pharm Bull* 1985;**33**(4):1552–6.
91. Ishimitsu S, Fujimoto S, Ohara A. Hydroxylation of phenylalanine by the hypoxanthine-xanthine oxidase system. *Chem Pharm Bull* 1984;**32**(11):4645–9.
92. Ishimitsu S, Fujimoto S, Ohara A. Studies on the hydroxylation of phenylalanine by 6,7-dimethyl-5,6,7,8-tetrahydropteridine. *Chem Pharm Bull* 1984;**32**(2):752–6.
93. Ishimitsu S, Fujimoto S, Ohara A. Studies on the hydroxylation of phenylalanine by the ascorbic acid-hydrogen peroxide system. *Chem Pharm Bull* 1981;**29**(7):2089–92.
94. Prior RL, Wu X, Schaich K. Standardized Methods for the Determination of Antioxidant Capacity and Phenolics in Foods and Dietary Supplements. *J Agric Food Chem* 2005;**53**(10):4290–302.
95. Liang X-L, Wang X-L, Li Z, Hao Q-H, Wang S-Y. Improved in Vitro Assays of Superoxide Anion and 1,1-Diphenyl- 2-picrylhydrazyl (DPPH) Radical-Scavenging Activity of Isoflavones and Isoflavone Metabolites. *J Agric Food Chem* 2010;**58**(22):11548–52.
96. Blois MS. Antioxidant determinations by the use of a stable free radical. *Nature (London, U K)* 1958;**181**:1199–200.
97. Park J-M, Han Y-M, Lee J-S, Ko KH, Hong S-P, Kim E-H, et al. Nrf2-mediated mucoprotective and anti-inflammatory actions of Artemisia extracts led to attenuate stress related mucosal damages. *J Clin Biochem Nutr* 2015;**56**(2):132–42.
98. Mueller L, Catalano A, Simone R, Cittadini A, Froehlich K, Boehm V, et al. Antioxidant Capacity of Tomato Seed Oil in Solution and Its Redox Properties in Cultured Macrophages. *J Agric Food Chem* 2013;**61**(2):346–54.

99. McDermott GP, Noonan LK, Mnatsakanyan M, Shalliker RA, Conlan XA, Barnett NW, et al. High-performance liquid chromatography with post-column 2,2'-diphenyl-1-picrylhydrazyl radical scavenging assay: Methodological considerations and application to complex samples. *Anal Chim Acta* 2010;**675**(1):76–82.
100. Bandoniene D, Murkovic M. The detection of radical scavenging compounds in crude extract of borage (Borago officinalis L.) by using an on-line HPLC-DPPH method. *J Biochem Biophys Methods* 2002;**53**(1-3):45–9.
101. Chrzczanowicz J, Gawron A, Zwolinska A, de Graft-Johnson J, Krajewski W, Krol M, et al. Simple method for determining human serum 2,2-diphenyl-1-picryl-hydrazyl (DPPH) radical scavenging activity - possible application in clinical studies on dietary antioxidants. *Clin Chem Lab Med* 2008;**46**(3):342–9.
102. Magalhaes LM, Segundo MA, Reis S, Lima JLFC. Automatic method for determination of total antioxidant capacity using 2,2-diphenyl-1-picrylhydrazyl assay. *Anal Chim Acta* 2006; **558**(1-2):310–18.
103. Pyrzynska K, Pekal A. Application of free radical diphenylpicrylhydrazyl (DPPH) to estimate the antioxidant capacity of food samples. *Anal Methods* 2013;**5**(17):4288–95.
104. Xie J, Schaich KM. Re-evaluation of the 2,2-diphenyl-1-picrylhydrazyl free radical (DPPH) assay for antioxidant activity. *J Agric Food Chem* 2014;**62**(19):4251–60.
105. Miller NJ, Rice-Evans CA, Davies MJ. A new method for measuring antioxidant activity. *Biochem Soc Trans* 1993;**21**(2):95S.
106. Childs RE, Bardsley WG. The steady-state kinetics of peroxidase with 2,2'-azino-di-(3-ethyl-benzthiazoline-6-sulphonic acid) as chromogen. *Biochemical Journal* 1975;**145**(1):93–103.
107. Re R, Pellegrini N, Proteggente A, Pannala A, Yang M, Rice-Evans C. Antioxidant activity applying an improved ABTS radical cation decolorization assay. *Free Radic Biol Med* 1999;**26**(9-10): 1231–7.
108. Labrinea EP, Georgiou CA. Stopped-flow method for assessment of pH and timing effect on the ABTS total antioxidant capacity assay. *Anal Chim Acta* 2004;**526**(1):63–8.
109. Perez-Jimenez J, Saura-Calixto F. Anti-oxidant capacity of dietary polyphenols determined by ABTS assay: a kinetic expression of the results. *Int J Food Sci Technol* 2008;**43**(1):185–91.
110. Zheng L, Lin L, Su G, Zhao Q, Zhao M. Pitfalls of using 1,1-diphenyl-2-picrylhydrazyl (DPPH) assay to assess the radical scavenging activity of peptides: Its susceptibility to interference and low reactivity towards peptides. *Food Res Int* 2015;**76**(Part_3):359–65.
111. Kalili KM, De Smet S, van Hoeylandt T, Lynen F, de Villiers A. Comprehensive two-dimensional liquid chromatography coupled to the ABTS radical scavenging assay: a powerful method for the analysis of phenolic antioxidants. *Anal Bioanal Chem* 2014;**406**(17):4233–42.
112. Cao G, Alessio HM, Cutler RG. Oxygen-radical absorbance capacity assay for antioxidants. *Free Radical Biol Med* 1993;**14**(3):303–11.
113. Nakajima A, Matsuda E, Masuda Y, Sameshima H, Ikenoue T. Characteristics of the spin-trapping reaction of a free radical derived from AAPH: further development of the ORAC-ESR assay. *Anal Bioanal Chem* 2012;**403**(7):1961–70.
114. Dorta E, Fuentes-Lemus E, Aspee A, Atala E, Speisky H, Bridi R, et al. The ORAC (oxygen radical absorbance capacity) index does not reflect the capacity of antioxidants to trap peroxyl radicals. *RSC Adv* 2015;**5**(50):39899–902.
115. Sueishi Y, Yoshioka D, Oowada S, Endoh N, Kohri S, Fujii H, et al. Is the oxygen radical absorbance capacity (ORAC) method a peroxyl-radical scavenging assay? *Z Phys Chem (Muenchen, Ger)* 2010;**224**(6):921–8.
116. Nakajima A, Masuda Y, Matsuda E, Tajima K, Sameshima H, Ikenoue T. Abilities of Antioxidants to Eliminate the Peroxyl Radical Derived from 2,2'-Azobis(2-Amidinopropane) Dihydrochloride (AAPH). *Appl Magn Reson* 2013;**44**(9):997–1005.
117. Zhang D, Liu Y, Chu L, Wei Y, Wang D, Cai S, et al. Relationship Between the Structures of Flavonoids and Oxygen Radical Absorbance Capacity Values: A Quantum Chemical Analysis. *J Phys Chem A* 2013;**117**(8):1784–94.

118. Martin I, Aspee A, Torres P, Lissi E, Lopez-Alarcon C. Influence of the Target Molecule on the Oxygen Radical Absorbance Capacity Index: A Comparison Between Alizarin Red- and Fluorescein-Based Methodologies. *J Med Food* 2009;**12**(6):1386–92.
119. Lopez-Alarcon C, Lissi E. A novel and simple ORAC methodology based on the interaction of Pyrogallol Red with peroxyl radicals. *Free Radical Res* 2006;**40**(9):979–85.
120. Guclu K, Kibrislioglu G, Ozyurek M, Apak R. Development of a fluorescent probe for measurement of peroxyl radical scavenging activity in biological samples. *J Agric Food Chem* 2014;**62**(8):1839–45.
121. Atala E, Aspee A, Speisky H, Lissi E, Lopez-Alarcon C. Antioxidant capacity of phenolic compounds in acidic medium: A pyrogallol red-based ORAC (oxygen radical absorbance capacity) assay. *J Food Compos Anal* 2013;**32**(2):116–25.
122. Barriga-Gonzalez G, Aguilera-Venegas B, Folch-Cano C, Perez-Cruz F, Olea-Azar C. Electron Spin Resonance as a Powerful Tool for Studying Antioxidants and Radicals. *Curr Med Chem* 2013;**20**(37):4731–43.
123. Cao G, Prior RL. Comparison of different analytical methods for assessing total antioxidant capacity of human serum. *Clin Chem (Washington, D C)* 1998;**44**(6):1309–15.
124. Nkhili E, Brat P. Reexamination of the ORAC assay: effect of metal ions. *Anal Bioanal Chem* 2011;**400**(5):1451–8.
125. Cooper CE. Nitric oxide and iron proteins. *Biochim Biophys Acta* 1999;**1411**(2-3):290–309.
126. Noack E, Kubitzek D, Kojda G. Spectrophotometric determination of nitric oxide using hemoglobin. *NeuroProtocols* 1992;**1**(2):133–9.
127. Kelm M, Dahmann R, Wink D, Feelisch M. The nitric oxide/superoxide assay. Insights into the biological chemistry of the NO/O-.2 interaction. *J Biol Chem* 1997;**272**(15):9922–32.
128. Goode HF, Webster NR, Howdle PD, Walker BE. Nitric oxide production by human peripheral blood polymorphonuclear leukocytes. *Clin Sci* 1994;**86**(4):411–15.
129. Zhang Y, Samson FE, Nelson SR, Pazdernik TL. Nitric oxide detection with intracerebral microdialysis: Important considerations in the application of the hemoglobin-trapping technique. *J Neurosci Methods* 1996;**68**(2):165–73.
130. Samouilov A, Zweier JL. Development of chemiluminescence-based methods for specific quantitation of nitrosylated thiols. *Anal Biochem* 1998;**258**(2):322–30.
131. Bates JN. Nitric oxide measurement by chemiluminescence detection. *NeuroProtocols* 1992;**1**(2):141–9.
132. Hausladen A, Rafikov R, Angelo M, Singel DJ, Nudler E, Stamler JS. Assessment of nitric oxide signals by triiodide chemiluminescence. *Proc Natl Acad Sci USA* 2007;**104**(7):2157–62.
133. Rogers SC, Khalatbari A, Gapper PW, Frenneaux MP, James PE. Detection of Human Red Blood Cell-bound Nitric Oxide. *J Biol Chem* 2005;**280**(29):26720–8.
134. Basu S, Wang X, Gladwin MT, Kim-Shapiro DB. Chemilumescent detection of S-nitrosated proteins: Comparison of tri-iodide, copper/CO/cysteine, and modified copper/cysteine methods. *Methods Enzymol* 2008;**440**:137–56 (Nitric Oxide, Part F)
135. Gladwin MT, Ognibene FP, Pannell LK, Nichols JS, Pease-Fye ME, Shelhamer JH, et al. Relative role of heme nitrosylation and beta-cysteine 93 nitrosation in the transport and metabolism of nitric oxide by hemoglobin in the human circulation. *Proc Natl Acad Sci USA* 2000;**97**(18):9943–8.
136. Gladwin MT, Shelhamer JH, Schechter AN, Pease-Fye ME, Waclawiw MA, Panza JA, et al. Role of circulating nitrite and S-nitrosohemoglobin in the regulation of regional blood flow in humans. *Proc Natl Acad Sci USA* 2000;**97**(21):11482–7.
137. Channon KM, Blazing MA, Shetty GA, Potts KE, George SE. Adenoviral gene transfer of nitric oxide synthase: High level expression in human vascular cells. *Cardiovasc Res* 1996;**32**(5):962–72.
138. Duling LC, Cherng TW, Griego JR, Perrine MF, Kanagy NL. Loss of α2B-adrenoceptors increases magnitude of hypertension following nitric oxide synthase inhibition. *Am J Physiol* 2006;**291**(5, Pt. 2):H2403–8.
139. Liu H-B, Zhang J, Sun Y-Y, Li X-Y, Jiang S, Liu M-Y, et al. Dietary salt regulates epithelial sodium channels in rat endothelial cells: adaptation of vasculature to salt. *Br J Pharmacol* 2015. Ahead of Print.

140. Palumbo A, Napolitano A, Carraturo A, Russo GL, d'Ischia M. Oxidative Conversion of 6-Nitrocatecholamines to Nitrosating Products: A Possible Contributory Factor in Nitric Oxide and Catecholamine Neurotoxicity Associated with Oxidative Stress and Acidosis. *Chem Res Toxicol* 2001;**14**(9):1296–305.
141. Chang CF, Chau YP, Kung HN, Lu KS. The lipopolysaccharide-induced pro-inflammatory response in RAW264.7 cells is attenuated by an unsaturated fatty acid-bovine serum albumin complex and enhanced by a saturated fatty acid-bovine serum albumin complex. *Inflammation Res* 2012;**61**(2):151–60.
142. Oh YS, Lee Y-J, Park EY, Jun H-S. Interleukin-6 treatment induces beta-cell apoptosis via STAT-3-mediated nitric oxide production. *Diabetes/Metab Res Rev* 2011;**27**(8):813–19.
143. Richeson CE, Mulder P, Bowry VW, Ingold KU. The Complex Chemistry of Peroxynitrite Decomposition: New Insights1. *Journal of the American Chemical Society* 1998;**120**(29):7211–19.
144. Shiva S. Nitrite: A physiological store of nitric oxide and modulator of mitochondrial function. *Redox Biology* 2013;**1**(1):40–4.
145. Lundberg JO, Weitzberg E, Gladwin MT. The nitrate-nitrite-nitric oxide pathway in physiology and therapeutics. *Nat Rev Drug Discov* 2008;**7**(2):156–67.
146. Griess P. Diazo-compounds. *Ber* 1883;**15**:2183–201.
147. Griess P. Diazo-derivatives. *Ber* 1884;**17**:603–8.
148. Locigno EJ, Zweier JL, Villamena FA. Nitric oxide release from the unimolecular decomposition of the superoxide radical anion adduct of cyclic nitrones in aqueous medium. *Org Biomol Chem* 2005;**3**(17):3220–7.
149. Miranda KM, Espey MG, Wink DA. A Rapid, Simple Spectrophotometric Method for Simultaneous Detection of Nitrate and Nitrite. *Nitric Oxide* 2001;**5**(1):62–71.
150. Casey TE, Hilderman RH. Modification of the Cadmium Reduction Assay for Detection of Nitrite Production Using Fluorescence Indicator 2,3-Diaminonaphthalene. *Nitric Oxide* 2000;**4**(1):67–74.
151. Sastry KV, Moudgal RP, Mohan J, Tyagi JS, Rao GS. Spectrophotometric determination of serum nitrite and nitrate by copper-cadmium alloy. *Anal Biochem* 2002;**306**(1):79–82.
152. Gilliam MB, Sherman MB, Griscavage JM, Ignarro LJ. A spectrophotometric assay for nitrate using NADPH oxidation by Aspergillus nitrate reductase. *Anal Biochem* 1993;**212**(2):359–65.
153. Indika PN, Bayachou M. Eliminating absorbing interference using the H-point standard addition method: case of Griess assay in the presence of interferent heme enzymes such as NOS. *Anal Bioanal Chem* 2004;**379**(7-8):1055–61.
154. Hunter RA, Storm WL, Coneski PN, Schoenfisch MH. Inaccuracies of Nitric Oxide Measurement Methods in Biological Media. *Anal Chem (Washington, DC, US)* 2013;**85**(3):1957–63.
155. Giustarini D, Rossi R, Milzani A, Dalle-Donne I. Nitrite and nitrate measurement by Griess reagent in human plasma: evaluation of interferences and standardization. *Methods Enzymol* 2008;**440**:361–80 (Nitric Oxide, Part F)
156. Greenberg SS, Xie J, Sptizer JJ, Wang J-f, Lancaster J, Grisham MB, et al. Nitro containing L-arginine analogs interfere with assays for nitrate and nitrite. *Life Sci* 1995;**57**(21):1949–61.
157. Tsikas D, Fuchs I, Gutzki F-M, Frolich JC. Measurement of nitrite and nitrate in plasma, serum and urine of humans by high-performance liquid chromatography, the Griess assay, chemiluminescence and gas chromatography-mass spectrometry: Interferences by biogenic amines and NG-nitro-l-arginine analogs. *J Chromatogr B: Biomed Sci Appl* 1998;**715**(2):441–4.
158. Nithipatikom K, Pratt PF, Campbell WB. Nitro-L-arginine inteferes with the cadmium reduction of nitrate/griess reaction method of measuring nitric oxide production. *Eur J Clin Chem Clin Biochem* 1996;**34**(2):133–7.
159. Tsikas D. Analysis of nitrite and nitrate in biological fluids by assays based on the Griess reaction: Appraisal of the Griess reaction in the L-arginine/nitric oxide area of research. *J Chromatogr B: Anal Technol Biomed Life Sci* 2007;**851**(1-2):51–70.
160. Romitelli F, Santini SA, Chierici E, Pitocco D, Tavazzi B, Amorini AM, et al. Comparison of nitrite/nitrate concentration in human plasma and serum samples measured by the enzymatic batch Griess assay, ion-pairing HPLC and ion-trap GC-MS: The importance of a correct removal of proteins in the Griess assay. *J Chromatogr B: Anal Technol Biomed Life Sci* 2007;**851**(1-2):257–67.

161. Ellman GL. Tissue sulfhydryl groups. *Arch Biochem Biophys* 1959;**82**:70–7.
162. Whitesides GM, Houk J, Patterson MAK. Activation parameters for thiolate-disulfide interchange reactions in aqueous solution. *J Org Chem* 1983;**48**(1):112–15.
163. Riddles PW, Blakeley RL, Zerner B. Ellman's reagent: 5,5'-dithiobis(2-nitrobenzoic acid) - a reexamination. *Anal Biochem* 1979;**94**(1):75–81.
164. Whitesides GM, Lilburn JE, Szajewski RP. Rates of thiol-disulfide interchange reactions between mono- and dithiols and Ellman's reagent. *J Org Chem* 1977;**42**(2):332–8.
165. Shaked Ze, Szajewski RP, Whitesides GM. Rates of thiol-disulfide interchange reactions involving proteins and kinetic measurements of thiol pKa values. *Biochemistry* 1980;**19**(18):4156–66.
166. Legler G. 4,4'-Dinitrodiphenyldisulfides of different charge type as probes for the electrostatic environment of sulfhydryl groups. *Biochim Biophys Acta, Protein Struct* 1975;**405**(1):136–43.
167. Cline DJ, Thorpe C, Schneider JP. General method for facile intramolecular disulfide formation in synthetic peptides. *Anal Biochem* 2004;**335**(1):168–70.
168. Egwim IOC, Gruber HJ. Spectrophotometric Measurement of Mercaptans with 4,4'-Dithiodipyridine. *Anal Biochem* 2001;**288**(2):188–94.
169. Riener C, Kada G, Gruber H. Quick measurement of protein sulfhydryls with Ellman's reagent and with 4,4′-dithiodipyridine. *Analytical and Bioanalytical Chemistry* 2002;**373**(4-5):266–76.
170. Luthra NP, Dunlap RB, Odom JD. Characterization of new sulfhydryl group reagent: 6,6'-diselenobis(3-nitrobenzoic acid), a selenium analog of Ellman's reagent. *Anal Biochem* 1981;**117**(1): 94–102.
171. Singh R, Blattler WA, Collinson AR. An amplified assay for thiols based on reactivation of papain. *Anal Biochem* 1993;**213**(1):49–56.
172. Singh R. A sensitive assay for maleimide groups. *Bioconjug Chem* 1994;**5**(4):348–51.
173. Deponte M. Glutathione catalysis and the reaction mechanisms of glutathione-dependent enzymes. *Biochimica et Biophysica Acta (BBA) - General Subjects* 2013;**1830**(5):3217–66.
174. Hansen RE, Ostergaard H, Norgaard P, Winther JR. Quantification of protein thiols and dithiols in the picomolar range using sodium borohydride and 4,4'-dithiodipyridine. *Anal Biochem* 2007;**363**(1):77–82.
175. Chen W, Zhao Y, Seefeldt T, Guan X. Determination of thiols and disulfides via HPLC quantification of 5-thio-2-nitrobenzoic acid. *J Pharm Biomed Anal* 2008;**48**(5):1375–80.
176. Walmsley TA, Abernethy MH, Fitzgerald HP. Effect of daylight on the reaction of thiols with Ellman's reagent, 5,5'-dithiobis(2-nitrobenzoic acid). *Clin Chem (Winston-Salem, N C)* 1987;**33** (10):1928–31.
177. Beales D, Finch R, McLean AEM, Smith M, Wilson ID. Determination of penicillamine and other thiols by combined high-performance liquid chromatography and post-column reaction with Ellman's reagent: application to human urine. *J Chromatogr, Biomed Appl* 1981;**226**(2):498–503.
178. Moss RA, Swarup S. Transvesicular reactions of thiols with Ellman's reagent. *J Org Chem* 1988;**53** (25):5860–6.
179. Laragione T, Gianazza E, Tonelli R, Bigini P, Mennini T, Casoni F, et al. Regulation of redox-sensitive exofacial protein thiols in CHO cells. *Biol Chem* 2006;**387**(10/11):1371–6.
180. Jovanovic VB, Penezic-Romanjuk AZ, Pavicevic ID, Acimovic JM, Mandic LM. Improving the reliability of human serum albumin-thiol group determination. *Anal Biochem* 2013;**439**(1):17–22.
181. Guan X, Hoffman B, Dwivedi C, Matthees DP. A simultaneous liquid chromatography/mass spectrometric assay of glutathione, cysteine, homocysteine and their disulfides in biological samples. *J Pharm Biomed Anal* 2003;**31**(2):251–61.
182. Janero DR. Malondialdehyde and thiobarbituric acid-reactivity as diagnostic indexes of lipid peroxidation and peroxidative tissue injury. *Free Radical Biol Med* 1990;**9**(6):515–40.
183. Neifakh EA. Mechanism of unsaturated fatty acid oxidation by malignant and normal tissues. *Acta Unio Int Cancrum* 1964;**20**(4-5):972–5.
184. Marks F, Hecker E. Dependence of lipid peroxidation in rat liver microsomes on ferrous ions. *Z Physiol Chem* 1967;**348**(6):727–9.
185. Anderson PJ. Sensitive reagent for detecting 2-deoxysugars and 3-deoxypolyols. *J Chromatogr* 1966;**21**(1):163–4.

186. Green J, Bunyan J, Cawthorne MA, Diplock AT. Vitamin E and hepatotoxic agents. I. Carbon tetrachloride and lipid peroxidation in the rat. *Brit J Nutr* 1969;**23**(2):297–307.
187. Jain SK, McVie R, Duett J, Herbst JJ. Erythrocyte membrane lipid peroxidation and glycosylated hemoglobin in diabetes. *Diabetes* 1989;**38**(12):1539–43.
188. Slatter DA, Bolton CH, Bailey AJ. The importance of lipid-derived malondialdehyde in diabetes mellitus. *Diabetologia* 2000;**43**(5):550–7.
189. Duner P, To F, Berg K, Alm R, Bjoerkbacka H, Engelbertsen D, et al. Immune responses against aldehyde-modified laminin accelerate atherosclerosis in Apoe -/- mice. *Atherosclerosis (Amsterdam, Neth)* 2010;**212**(2):457–65.
190. Kaya Y, Ari E, Demir H, Soylemez N, Cebi A, Alp H, et al. Accelerated atherosclerosis in hemodialysis patients; correlation of endothelial function with oxidative DNA damage. *Nephrol, Dial, Transplant* 2012;**27**(3):1164–9.
191. Lee R, Antoniades C. Thiobarbituric acid reactive substances as a biomarker for coronary heart disease. *J Atheroscler Thromb* 2011;**18**(12):1127–8.
192. Carlstroem M, Persson AEG, Larsson E, Hezel M, Scheffer PG, Teerlink T, et al. Dietary nitrate attenuates oxidative stress, prevents cardiac and renal injuries, and reduces blood pressure in salt-induced hypertension. *Cardiovasc Res* 2011;**89**(3):574–85.
193. Marnett LJ. Oxy radicals, lipid peroxidation and DNA damage. *Toxicology* 2002;**181-182**:219–22.
194. Chen C-M, Liu J-L, Wu Y-R, Chen Y-C, Cheng H-S, Cheng M-L, et al. Increased oxidative damage in peripheral blood correlates with severity of Parkinson's disease. *Neurobiol Dis* 2009;**33**(3): 429–35.
195. Aoki M, Oshita T, Sakaguchi M. The comparison of thiobarbituric acid reactive substances (TBARS) concentrations in plasma and serum from dairy cattle. *J Vet Med Sci* 2008;**70**(1):107–10.
196. Nicolescu AC, Zavorin SI, Turro NJ, Reynolds JN, Thatcher GRJ. Inhibition of Lipid Peroxidation in Synaptosomes and Liposomes by Nitrates and Nitrites. *Chem Res Toxicol* 2002; **15**(7):985–98.
197. Moselhy HF, Reid RG, Yousef S, Boyle SP. A specific, accurate, and sensitive measure of total plasma malondialdehyde by HPLC. *J Lipid Res* 2013;**54**(3):852–8.
198. Seljeskog E, Hervig T, Mansoor MA. A novel HPLC method for the measurement of thiobarbituric acid reactive substances (TBARS). A comparison with a commercially available kit. *Clin Biochem* 2006;**39**(9):947–54.
199. Wasowicz W, Neve J, Peretz A. Optimized steps in fluorometric determination of thiobarbituric acid-reactive substances in serum: importance of extraction pH and influence of sample preservation and storage. *Clin Chem (Washington, D C)* 1993;**39**(12):2522–6.
200. Grintzalis K, Zisimopoulos D, Grune T, Weber D, Georgiou CD. Method for the simultaneous determination of free/protein malondialdehyde and lipid/protein hydroperoxides. *Free Radical Biology and Medicine* 2013;**59**:27–35.
201. Gay C, Gebicki JM. A critical evaluation of the effect of sorbitol on the ferric-xylenol orange hydroperoxide assay. *Anal Biochem* 2000;**284**(2):217–20.
202. Wolff SP. *Ferrous ion oxidation in presence of ferric ion indicator xylenol orange for measurement of hydroperoxides. Methods in Enzymology,* Volume 233. Academic Press; 1994p. 182–9.
203. Nourooz-Zadeh J, Tajaddini-Sarmadi J, Wolff SP. Measurement of plasma hydroperoxide concentrations by the ferrous oxidation-xylenol orange assay in conjunction with triphenylphosphine. *Anal Biochem* 1994;**220**(2):403–9.
204. Deiana L, Carru C, Pes G, Tadolini B. Spectrophotometric measurement of hydroperoxides at increased sensitivity by oxidation of Fe^{2+} in the presence of xylenol orange. *Free Radical Res* 1999;**31**(3):237–44.
205. Gay CA, Gebicki JM. Perchloric acid enhances sensitivity and reproducibility of the ferric-xylenol orange peroxide assay. *Anal Biochem* 2002;**304**(1):42–6.
206. Gay C, Collins J, Gebicki JM. Hydroperoxide Assay with the Ferric-Xylenol Orange Complex. *Anal Biochem* 1999;**273**(2):149–55.
207. Yin H, Porter NA. Specificity of the ferrous oxidation of xylenol orange assay: analysis of autoxidation products of cholesteryl arachidonate. *Anal Biochem* 2003;**313**(2):319–26.

208. Winterbourn CC, Parsons-Mair HN, Gebicki S, Gebicki JM, Davies MJ. Requirements for superoxide-dependent tyrosine hydroperoxide formation in peptides. *Biochem J* 2004;**381**(1):241–8.
209. Field SM, Villamena FA. Theoretical and Experimental Studies of Tyrosyl Hydroperoxide Formation in the Presence of H-Bond Donors. *Chem Res Toxicol* 2008;**21**(10):1923–32.
210. Hirsch I, Prell E, Weiwad M. Assessment of cell death studies by monitoring hydrogen peroxide in cell culture. *Anal Biochem* 2014;**456**:22–4.
211. Yu Z, Wang J, Zhou N, Zhao C, Qiu J. A highly sensitive method for quantitative determination of l-amino acid oxidase activity based on the visualization of ferric-xylenol orange formation. *PLoS One* 2013;**8**(12). e82483/1-e82483/7, 7 pp
212. Winterbourn CC, Pichorner H, Kettle AJ. Myeloperoxidase-dependent generation of a tyrosine peroxide by neutrophils. *Arch Biochem Biophys* 1997;**338**(1):15–21.
213. Hermes-Lima M, Willmore WG, Storey KB. Quantification of lipid peroxidation in tissue extracts based on Fe(III)xylenol orange complex formation. *Free Radical Biol Med* 1995;**19**(3):271–80.
214. Gay CA, Gebicki JM. Measurement of protein and lipid hydroperoxides in biological systems by the ferric-xylenol orange method. *Anal Biochem* 2003;**315**(1):29–35.
215. Bleau G, Giasson C, Brunette I. Measurement of hydrogen peroxide in biological samples containing high levels of ascorbic acid. *Anal Biochem* 1998;**263**(1):13–17.
216. Banerjee D, Jacob J, Kunjamma G, Madhusoodanan UK, Ghosh S. Measurement of urinary hydrogen peroxide by FOX-1 method in conjunction with catalase in diabetes mellitus - sensitive and specific approach. *Clin Chim Acta* 2004;**350**(1-2):233–6.
217. Long LH, Evans PJ, Halliwell B. Hydrogen Peroxide in Human Urine: Implications for Antioxidant Defense and Redox Regulation. *Biochem Biophys Res Commun* 1999;**262**(3):605–9.
218. DeLong JM, Prange RK, Hodges DM, Forney CF, Bishop MC, Quilliam M. Using a Modified Ferrous Oxidation-Xylenol Orange (FOX) Assay for Detection of Lipid Hydroperoxides in Plant Tissue. *J Agric Food Chem* 2002;**50**(2):248–54.
219. Jiang ZY, Woollard ACS, Wolff SP. Lipid hydroperoxide measurement by oxidation of iron in the presence of xylenol orange. Comparison with the TBA assay and an iodometric method. *Lipids* 1991;**26**(10):853–6.
220. Mehta BM, Darji VB, Aparnathi KD. Comparison of five analytical methods for the determination of peroxide value in oxidized ghee. *Food Chem* 2015;**185**:449–53.
221. Conlon CA, Hill MHE, Powers HJ. An improved assay for peroxides in lipid emulsion, adapted for a plate reader. *Anal Biochem* 2002;**311**(2):191–2.
222. Sugino K. Simultaneous determination of different classes of lipid hydroperoxides by high-performance liquid chromatography with post column detection by a ferrous/xylenol orange reagent. *Biosci, Biotechnol, Biochem* 1999;**63**(4):773–5.
223. Cramer GL, Miller Jr JF, Pendleton RB, Lands WEM. Iodometric measurement of lipid hydroperoxides in human plasma. *Anal Biochem* 1991;**193**(2):204–11.
224. Wallin B, Camejo G. Lipoprotein oxidation and measurement of hydroperoxide formation in a single microtiter plate. *Scand J Clin Lab Invest* 1994;**54**(4):341–6.
225. Asakawa T, Matsushita S. A colorimetric microdetermination of peroxide values utilizing aluminum chloride as the catalyst. *Lipids* 1980;**15**(11):965–7.
226. el-Saadani M, Esterbauer H, el-Sayed M, Goher M, Nassar AY, Jurgens G. A spectrophotometric assay for lipid peroxides in serum lipoproteins using a commercially available reagent. *J Lipid Res* 1989;**30**(4):627–30.
227. Darrow RA, Organisciak DT. An improved spectrophotometric triiodide assay for lipid hydroperoxides. *Lipids* 1994;**29**(8):591–4.
228. Cimato A, Facorro G, Aguirre F, Hager A, De Paoli T, Ihlo J, et al. A spectrophotometric method for the determination of hydroperoxides in liposomes. *Spectrochim Acta, Part A* 1998;**54A**(12): 2001–8.
229. Gebicki JM, Collins J, Baoutina A, Phair P. Failure of an aerobic iodometric peroxide assay. *Spec Publ - R Soc Chem* 1996;**181**:87–92 (Natural Antioxidants and Food Quality in Atherosclerosis and Cancer Prevention)

230. Gebicki JM, Collins J, Baoutina A, Phair P. The limitations of an iodometric aerobic assay for peroxides. *Anal Biochem* 1996;**240**(2):235–41.
231. Thomas SM, Jessup W, Gebicki JM, Dean RT. A continuous-flow automated assay for iodometric estimation of hydroperoxides. *Anal Biochem* 1989;**176**(2):353–9.
232. Jiang ZY, Hunt JV, Wolff SP. Ferrous ion oxidation in the presence of xylenol orange for detection of lipid hydroperoxide in low density lipoprotein. *Anal Biochem* 1992;**202**(2):384–9.
233. Chajes V, Sattler W, Stultschnig M, Kostner GM. Photometric evaluation of lipid peroxidation products in human plasma and copper oxidized low density lipoproteins: correlation of different oxidation parameters. *Atherosclerosis (Shannon, Irel)* 1996;**121**(2):193–203.
234. Totter JR, Medina VJ, Scoseria JL. Luminescence during the Oxidation of Hypoxanthine by Xanthine Oxidase in the Presence of Dimethylbiacridylium Nitrate. *Journal of Biological Chemistry* 1960;**235**(1):238–41.
235. Nikiforova AB, Fadeev RS, Kruglov AG. Rapid fluorescent visualization of multiple NAD(P)H oxidoreductases in homogenate, permeabilized cells, and tissue slices. *Anal Biochem* 2013;**440**(2): 189–96.
236. Li Y, Stansbury KH, Zhu H, Trush MA. Biochemical Characterization of Lucigenin (Bis-N-methylacridinium) as a Chemiluminescent Probe for Detecting Intramitochondrial Superoxide Anion Radical Production. *Biochem Biophys Res Commun* 1999;**262**(1):80–7.
237. Li Y, Zhu H, Trush MA. Detection of mitochondria-derived reactive oxygen species production by the chemilumigenic probes lucigenin and luminol. *Biochim Biophys Acta, Gen Subj* 1999; **1428**(1):1–12.
238. Kervinen M, Patsi J, Finel M, Hassinen IE. Lucigenin and coelenterazine as superoxide probes in mitochondrial and bacterial membranes. *Anal Biochem* 2004;**324**(1):45–51.
239. Myhre O, Andersen JM, Aarnes H, Fonnum F. Evaluation of the probes 2',7'-dichlorofluorescin diacetate, luminol, and lucigenin as indicators of reactive species formation. *Biochem Pharmacol* 2003;**65**(10):1575–82.
240. Kopprasch S, Pietzsch J, Graessler J. Validation of different chemilumigenic substrates for detecting extracellular generation of reactive oxygen species by phagocytes and endothelial cells. *Luminescence* 2003;**18**(5):268–73.
241. An SJ, Boyd R, Zhu M, Chapman A, Pimentel DR, Wang HD. NADPH oxidase mediates angiotensin II-induced endothelin-1 expression in vascular adventitial fibroblasts. *Cardiovasc Res* 2007; **75**(4):702–9.
242. Pagano PJ, Chanock SJ, Siwik DA, Colucci WS, Clark JK. Angiotensin II induces p67phox mRNA expression and NADPH oxidase superoxide generation in rabbit aortic adventitial fibroblasts. *Hypertension* 1998;**32**(2):331–7.
243. Mistry Y, Poolman T, Williams B, Herbert KE. A role for mitochondrial oxidants in stress-induced premature senescence of human vascular smooth muscle cells. *Redox Biol* 2013;**1**(1):411–17.
244. Pang X, Liu J, Zhao J, Mao J, Zhang X, Feng L, et al. Homocysteine induces the expression of C-reactive protein via NMDAr-ROS-MAPK-NF-κB signal pathway in rat vascular smooth muscle cells. *Atherosclerosis (Amsterdam, Neth)* 2014;**236**(1):73–81.
245. Traynham CJ, Roof SR, Wang H, Prosak RA, Tang L, Viatchenko-Karpinski S, et al. Diesterified nitrone rescues nitroso-redox levels and increases myocyte contraction via increased SR Ca^{2+} handling. *PLoS One* 2012;**7**(12):e52005.
246. Liochev SI, Fridovich I. Lucigenin luminescence as a measure of intracellular superoxide dismutase activity in Escherichia coli. *Proc Natl Acad Sci USA* 1997;**94**(7):2891–6.
247. Napankangas JP, Liimatta EV, Joensuu P, Bergmann U, Ylitalo K, Hassinen IE. Superoxide production during ischemia-reperfusion in the perfused rat heart: A comparison of two methods of measurement. *J Mol Cell Cardiol* 2012;**53**(6):906–15.
248. Kinugawa S, Huang H, Wang Z, Kaminski PM, Wolin MS, Hintze TH. A Defect of Neuronal Nitric Oxide Synthase Increases Xanthine Oxidase-Derived Superoxide Anion and Attenuates the Control of Myocardial Oxygen Consumption by Nitric Oxide Derived From Endothelial Nitric Oxide Synthase. *Circ Res* 2005;**96**(3):355–62.

249. Skatchkov MP, Sperling D, Hink U, Mulsch A, Harrison DG, Sindermann I, et al. Validation of lucigenin as a chemiluminescent probe to monitor vascular superoxide as well as basal vascular nitric oxide production. *Biochem Biophys Res Commun* 1999;**254**(2):319–24.
250. Xie H, Ray PE, Short BL. NF-κB activation plays a role in superoxide-mediated cerebral endothelial dysfunction after hypoxia/reoxygenation. *Stroke* 2005;**36**(5):1047–52.
251. Okajima T, Ohsaka T. Chemiluminescence of lucigenin by electrogenerated superoxide ions in aqueous solutions. *Luminescence* 2003;**18**(1):49–57.
252. Greenlee L, Fridovich I, Handler P. Chemiluminescence Induced by Operation of Iron-Flavoproteins*. *Biochemistry* 1962;**1**(5):779–83.
253. Afanas'ev IB, Ostrakhovitch EA, Mikhal'chik EV, Korkina LG. Direct enzymatic reduction of lucigenin decreases lucigenin-amplified chemiluminescence produced by superoxide ion. *Luminescence* 2001;**16**(5):305–7.
254. Schepetkin IA. Lucigenin as a substrate of microsomal NAD(P)H-oxidoreductases. *Biochemistry (Moscow)* 1999;**64**(1):25–32.
255. Liochev SI, Fridovich I. Lucigenin (bis-N-methylacridinium) as a mediator of superoxide anion production. *Arch Biochem Biophys* 1997;**337**(1):115–20.
256. Heiser I, Muhr A, Elstner EF. Production of OH-radical-type oxidant by lucigenin. *Z Naturforsch, C: Biosci* 1998;**53**(1/2):9–14.
257. Sohn H-Y, Keller M, Gloe T, Crause P, Pohl U. Pitfalls of using lucigenin in endothelial cells: implications for NAD(P)H dependent superoxide formation. *Free Radical Res* 2000;**32**(3):265–72.
258. Vasquez-Vivar J, Hogg N, Pritchard Jr KA, Martasek P, Kalyanaraman B. Superoxide anion formation from lucigenin: an electron spin resonance spin-trapping study. *FEBS Lett* 1997;**403**(2):127–30.
259. Ochsendorf FR, Goy C, Fuchs J, Morke W, Beschmann HA, Bromme HJ. Lucigenin chemiluminescence in human seminal plasma. *Free Radical Res* 2001;**34**(2):153–65.
260. Barber DA, Do NH, Tackett RL, Capomacchia AC. Nonsuperoxide lucigenin-enhanced chemiluminescence from phospholipids and human saphenous veins. *Free Radical Biol Med* 1995;**18**(3):565–9.
261. Legg KD, Shive DW, Hercules DM. Electrochemistry of lucigenin. *Anal Chem* 1972;**44**(9):1650–5.
262. Wada S, Maeda K, Nakada K. Voltammetric study of N,N'-dimethyl-9,9'-biacridinium dinitrate (lucigenin) in dimethyl sulfoxide and aqueous solutions. *Nippon Kagaku Kaishi* 1977;(5):639–45.
263. Murphy RJ, Svehla G. Voltammetric study of the electrochemical reduction of lucigenin in aqueous medium. *Anal Chim Acta* 1981;**125**:73–83.
264. Rodriquez-Amaro R, Sanchez M, Munoz E, Ruiz JJ, Camacho L. Electrochemical reduction of lucigenin on mercury in aqueous media. *J Electrochem Soc* 1996;**143**(7):2132–6.
265. Spasojevic I, Liochev SI, Fridovich I. Lucigenin: Redox Potential in Aqueous Media and Redox Cycling with O-2 Production. *Arch Biochem Biophys* 2000;**373**(2):447–50.
266. Ahlberg E, Hammerich O, Parker VD. Electron-transfer reactions accompanied by large structural changes. 1. Lucigenin-10,10'-dimethyl-9,9'-biacridylidene redox system. *J Am Chem Soc* 1981;**103**(4):844–9.
267. Legg KD, Hercules DM. Electrochemically generated chemiluminescence of lucigenin. *Journal of the American Chemical Society* 1969;**91**(8):1902–7.
268. Afanas'ev IB, Ostrachovitch EA, Korkina LG. Lucigenin Is a Mediator of Cytochrome C Reduction but Not of Superoxide Production. *Arch Biochem Biophys* 1999;**366**(2):267–74.
269. Wood PM. The potential diagram for oxygen at pH 7. *Biochemical Journal* 1988;**253**(1):287–9.
270. Buettner GR. The pecking order of free radicals and antioxidants: Lipid peroxidation, α-tocopherol, and ascorbate. *Archives of Biochemistry and Biophysics* 1993;**300**(2):535–43.
271. Meisel D, Czapski G. One-electron transfer equilibriums and redox potentials of radicals studied by pulse radiolysis. *The Journal of Physical Chemistry* 1975;**79**(15):1503–9.
272. Li Y, Zhu H, Kuppusamy P, Roubaud V, Zweier JL, Trush MA. Validation of lucigenin (bis-N-methylacridinium) as a chemilumigenic probe for detecting superoxide anion radical production by enzymatic and cellular systems. *J Biol Chem* 1998;**273**(4):2015–23.

273. Faulkner K, Fridovich I. Luminol and lucigenin as detectors for superoxide radical. *Free Radical Biol Med* 1993;**15**(4):447–51.
274. Janiszewski M, Souza HP, Liu X, Pedro MA, Zweier JL, Laurindo FRM. Overestimation of NADH-driven vascular oxidase activity due to lucigenin artifacts. *Free Radical Biol Med* 2002;**32**(5):446–53.
275. Moreira JD, Pernomian L, Gomes MS, Pernomian L, Moreira RP, do Prado AF, et al. Acute restraint stress increases carotid reactivity in type-I diabetic rats by enhancing Nox4/NADPH oxidase functionality. *Eur J Pharmacol* 2015;**765**:503–16.
276. Sanchez A, Martinez P, Munoz M, Benedito S, Garcia-Sacristan A, Hernandez M, et al. Endothelin-1 contributes to endothelial dysfunction and enhanced vasoconstriction through augmented superoxide production in penile arteries from insulin-resistant obese rats: role of ETA and ETB receptors. *Br J Pharmacol* 2014;**171**(24):5682–95.
277. Tanaka LY, Bechara LRG, dos Santos AM, Jordao CP, de Sousa LGO, Bartholomeu T, et al. Exercise improves endothelial function: A local analysis of production of nitric oxide and reactive oxygen species. *Nitric Oxide* 2015;**45**:7–14.
278. Wong PS, Randall MD, Roberts RE. Sex differences in the role of NADPH oxidases in endothelium-dependent vasorelaxation in porcine isolated coronary arteries. *Vasc Pharmacol* 2015;**72**:83–92.
279. Hyrsl P, Lojek A, Ciz M, Kubala L. Chemiluminescence of lucigenin is dependent on experimental conditions. *Luminescence* 2004;**19**(2):61–3.
280. Oosthuizen MMJ, Engelbrecht ME, Lambrechts H, Greyling D, Levy RD. The effect of pH on chemiluminescence of different probes exposed to superoxide and singlet oxygen generators. *J Biolumin Chemilumin* 1997;**12**(6):277–84.
281. Caldefie-Chezet F, Walrand S, Moinard C, Tridon A, Chassagne J, Vasson M-P. Is the neutrophil reactive oxygen species production measured by luminol and lucigenin chemiluminescence intra or extracellular? Comparison with DCFH-DA flow cytometry and cytochrome c reduction. *Clin Chim Acta* 2002;**319**(1):9–17.
282. Goto T, Takagi T. Cypridina bioluminescence. X. Chemiluminescence of a Cypridina luciferin analog, 2-methyl-6-phenyl-3,7-dihydroimidazo[1,2-a]pyrazin-3-one, in the presence of the xanthine-xanthine oxidase system. *Bull Chem Soc Jpn* 1980;**53**(3):833–4.
283. Nakano M, Sugioka K, Ushijima Y, Goto T. Chemiluminescence probe with Cypridina luciferin analog, 2-methyl-6-phenyl-3,7-dihydroimidazo[1,2-a]pyrazin-3-one, for estimating the ability of human granulocytes to generate superoxide anion. *Anal Biochem* 1986;**159**(2):363–9.
284. Nakano M. Determination of superoxide radical and singlet oxygen based on chemiluminescence of luciferin analogs. *Methods Enzymol* 1990;**186**:585–91 (Oxygen Radicals Biol. Syst., Pt. B)
285. Sun S, Li X, Zhang G, Ma H, Zhang D, Bao Z. Determination of H2O2-dependent generation of singlet oxygen from human saliva with a novel chemiluminescence probe. *Biochim Biophys Acta, Gen Subj* 2006;**1760**(3):440–4.
286. Sakurai K, Ito T, Nakano M, Takahashi A. Application of CLA- and MCLA-dependent chemiluminescence method for the detection of superoxide generation in the rat gastric mucosa during ischemia/reperfusion. *Photomed Photobiol* 1995;**17**:83–6.
287. Skatchkov MP, Sperling D, Hink U, Anggard E, Munzel T. Quantification of superoxide radical formation in intact vascular tissue using a Cypridina luciferin analog as an alternative to lucigenin. *Biochem Biophys Res Commun* 1998;**248**(2):382–6.
288. Nakano M, Kikuyama M, Hasegawa T, Ito T, Sakurai K, Hiraishi K, et al. The first observation of O2- generation at real time in vivo from non-Kupffer sinusoidal cells in perfused rat liver during acute ethanol intoxication. *FEBS Lett* 1995;**372**(2-3):140–3.
289. Akutsu K, Nakajima H, Katoh T, Kino S, Fujimori K. Chemiluminescence of Cipridina luciferin analogs. Part 2. Kinetic studies on the reaction of 2-methyl-6-phenylimidazo[1,2-a]pyrazin-3(7H)-one (CLA) with superoxide: hydroperoxyl radical is an actual active species used to initiate the reaction. *J Chem Soc, Perkin Trans* 1995;**2**(8):1699–706.
290. Suzuki N, Suetsuna K, Mashiko S, Yoda B, Nomoto T, Toya Y, et al. Reaction rates for the chemiluminescence of Cypridina luciferin analogs with superoxide: a quenching experiment with superoxide dismutase. *Agric Biol Chem* 1991;**55**(1):157–60.

291. Hirano T, Takahashi Y, Kondo H, Maki S, Kojima S, Ikeda H, et al. The reaction mechanism for the high quantum yield of Cypridina (Vargula) bioluminescence supported by the chemiluminescence of 6-aryl-2-methylimidazo[1,2-a]pyrazin-3(7H)-ones (Cypridina luciferin analogues). *Photochem Photobiol Sci* 2008;**7**(2):197–207.
292. Kondo H, Igarashi T, Maki S, Niwa H, Ikeda H, Hirano T. Substituent effects on the kinetics for the chemiluminescence reaction of 6-arylimidazo[1,2-a]pyrazin-3(7H)-ones (Cypridina luciferin analogues): support for the single electron transfer (SET)-oxygenation mechanism with triplet molecular oxygen. *Tetrahedron Lett* 2005;**46**(45):7701–4.
293. Teranishi K, Hisamatsu M, Yamada T. Chemiluminescence of 2-methyl-6-arylimidazo[1,2-a]pyrazin-3(7H)-one in protic solvents: electron-donating substituent effect on the formation of the neutral singlet excited-state molecule. *Luminescence* 1999;**14**(6):297–302.
294. Fujimori K, Nakajima H, Akutsu K, Mitani M, Sawada H, Nakayama M. Chemiluminescence of Cypridina luciferin analogs. Part 1. Effect of pH on rates of spontaneous autoxidation of CLA in aqueous buffer solutions. *J Chem Soc, Perkin Trans* 1993;**2**(12):2405–9.
295. Yamaguchi S, Kishikawa N, Ohyama K, Ohba Y, Kohno M, Masuda T, et al. Evaluation of chemiluminescence reagents for selective detection of reactive oxygen species. *Anal Chim Acta* 2010; **665**(1):74–8.
296. Takeshita K, Okazaki S, Itoda A. Nitroxyl radicals remarkably enhanced the superoxide anion radical-induced chemiluminescence of Cypridina luciferin analogues. *Anal Chem* 2013;**85**(14):6833–9.
297. Bancirova M. Sodium azide as a specific quencher of singlet oxygen during chemiluminescent detection by luminol and Cypridina luciferin analogues. *Luminescence* 2011;**26**(6):685–8.
298. Kruglov AG, Nikiforova AB, Shatalin YV, Shubina VV, Fisyuk AS, Akatov VS. Sulfur-containing compounds quench 3,7-dihydro-2-methyl-6-(4-methoxyphenyl)imidazol[1,2-a]pyrazine-3-one chemiluminescence: Discrimination between true antioxidants and quenchers using xanthine oxidase. *Anal Biochem* 2010;**406**(2):230–2.
299. Bylund J, Bjoernsdottir H, Sundqvist M, Karlsson A, Dahlgren C. Measurement of Respiratory Burst Products, Released or Retained, During Activation of Professional Phagocytes. *Methods Mol Biol (NY, US)* 2014;**1124**:321–38 (Neutrophil Methods and Protocols)
300. Wymann MP, von Tscharner V, Deranleau DA, Baggiolini M. Chemiluminescence detection of H2O2 produced by human neutrophils during the respiratory burst. *Analytical Biochemistry* 1987;**165**(2):371–8.
301. Pavelkova M, Kubala L. Luminol-, isoluminol- and lucigenin-enhanced chemiluminescence of rat blood phagocytes stimulated with different activators. *Luminescence* 2004;**19**(1):37–42.
302. Ashkenazi A, Abu-Rabeah K, Marks RS. Electrochemistry and chemiluminescence techniques compared in the detection of NADPH oxidase activity in phagocyte cells. *Talanta* 2009;**77**(4): 1460–5.
303. Merenyi G, Lind JS. Role of a peroxide intermediate in the chemiluminescence of luminol. A mechanistic study. *Journal of the American Chemical Society* 1980;**102**(18):5830–5.
304. Smanmoo C, Yamasuji M, Sagawa T, Shibata T, Kabashima T, Yuan DQ, et al. Diimine ligand as a novel chemiluminescence enhancer of luminol-containing compounds. *Talanta* 2009;**77**(5):1761–6.
305. Lock R, Johansson A, Orselius K, Dahlgren C. Analysis of horseradish peroxidase-amplified chemiluminescence produced by human neutrophils reveals a role for the superoxide anion in the light emitting reaction. *Analytical Biochemistry* 1988;**173**(2):450–5.
306. Yildiz G, Demiryurek AT. Ferrous iron-induced luminol chemiluminescence: a method for hydroxyl radical study. *J Pharmacol Toxicol Methods* 1998;**39**(3):179–84.
307. Olsson T, Bergstrom K, Thore A. A sensitive method for determination of serum hemoglobin based on iso-luminol chemiluminescence. *Clin Chim Acta* 1982;**122**(2):125–33.
308. Hirayama O, Takagi M, Hukumoto K, Katoh S. Evaluation of antioxidant activity by chemiluminescence. *Anal Biochem* 1997;**247**(2):237–41.
309. Dahlgren C, Karlsson A, Bylund J. Measurement of respiratory burst products generated by professional phagocytes. *Methods Mol Biol (Totowa, NJ, US)* 2007;**412**:349–63 (Neutrophil Methods and Protocols)

310. Lundqvist H, Dahlgren C. Isoluminol-enhanced chemiluminescence: a sensitive method to study the release of superoxide anion from human neutrophils. *Free Radical Biol Med* 1996;**20**(6):785–92.
311. Stenfeldt A-L, Dahlgren C. An attempt to design an isoluminol-hydrogen peroxidase-amplified CL that measures intracellularly produced H2O2 in phagocytes: sensitivity for H2O2 is not high enough to allow detection. *Luminescence* 2007;**22**(5):507–10.
312. Faldt J, Ridell M, Karlsson A, Dahlgren C. The phagocyte chemiluminescence paradox: luminol can act as an inhibitor of neutrophil NADPH-oxidase activity. *Luminescence* 1999;**14**(3):153–60.
313. Maeztu R, Tardajos G, Gonzalez-Gaitano G. Natural Cyclodextrins as Efficient Boosters of the Chemiluminescence of Luminol and Isoluminol: Exploration of Potential Applications. *J Phys Chem B* 2010;**114**(8):2798–806.
314. Ahmed S, Kishikawa N, Ohyama K, Maki T, Kurosaki H, Nakashima K, et al. An ultrasensitive and highly selective determination method for quinones by high-performance liquid chromatography with photochemically initiated luminol chemiluminescence. *J Chromatogr A* 2009;**1216**(18): 3977–84.
315. Achyuthan KE. Enhancement of aminophthalhydrazides chemiluminescence by N-beta-alanyl-L-histidine (L-carnosine). *Luminescence* 1999;**14**(2):75–81.
316. Kricka LJ, Ji X, Thorpe GH, Edwards B, Voyta J, Bronstein I. Comparison of 5-hydroxy-2, 3-dihydrophthalazine-1, 4-dione and luminol as co-substrates for detection of horseradish peroxidase in enhanced chemiluminescent reactions. *J Immunoassay* 1996;**17**(1):67–83.
317. Thorpe GH, Kricka LJ, Gillespie E, Moseley S, Amess R, Baggett N, et al. Enhancement of the horseradish peroxidase-catalyzed chemiluminescent oxidation of cyclic diacyl hydrazides by 6-hydroxybenzothiazoles. *Anal Biochem* 1985;**145**(1):96–100.
318. Dahlberg M, Dahlgren C, Hellstrand K, Movitz C. A new chemiluminescence paradox: selective inhibition of isoluminol-amplified activity in phagocytes by peptides from annexin AI. *Luminescence* 2008;**23**(3):139–43.
319. Frei B, Yamamoto Y, Niclas D, Ames BN. Evaluation of an isoluminol chemiluminescence assay for the detection of hydroperoxides in human blood plasma. *Anal Biochem* 1988;**175**(1):120–30.
320. Yamamoto Y, Ames BN. Detection of lipid hydroperoxides and hydrogen peroxide at picomole levels by an HPLC and isoluminol chemiluminescence assay. *Free Radic Biol Med* 1987;**3**(5): 359–61.
321. Yamamoto Y, Brodsky MH, Baker JC, Ames BN. Detection and characterization of lipid hydroperoxides at picomole levels by high-performance liquid chromatography. *Anal Biochem* 1987; **160**(1):7–13.
322. De Boever J, Kohen F, Leyseele D, Vandekerckhove D. Isoluminol as a marker in direct chemiluminescence immunoassays for steroid hormones. *J Biolumin Chemilumin* 1990;**5**(1):5–10.
323. De Boever JG, Kohen F, Bosmans E. Antibody binding efficiency of differently labeled steroid hormones. *Anal Chim Acta* 1993;**275**(1-2):81–7.
324. De Boever JG, Kohen F, Bosmans E. Binding of homologous and heterologous isoluminol- and enzyme-labeled progesterone conjugates to monoclonal antibodies. *Anal Chim Acta* 1994; **290**(1-2):239–45.
325. Messeri G, Orlandini A, Pazzagli M. Luminescent immunoassay using isoluminol derivatives. *J Biolumin Chemilumin* 1989;**4**(1):154–8.

CHAPTER 7

Electrochemical, Mass Spectroscopic, Immunochemical, and Nuclear Magnetic Resonance Techniques

7.1 INTRODUCTION

Other modes of reactive species (RS) detection have become increasingly necessary due to the limitations of conventional spectrophotometric and fluorometric techniques in their ability to unambiguously identify RS. The use of electron paramagnetic resonance (EPR) technique has been limited due the poor reactivity of superoxide to spin traps and the relatively short adduct half-lives. Several techniques have been developed with improved sensitivity to detect, identify, or quantify RS such as in the case of immunochemical and electrochemical methods. For the identification of the site of oxidative lesion in DNA or protein systems, the use of immunochemical coupled with liquid chromatography—mass spectrometry (LC—MS) methods has proven to be the most accurate. Confocal microscopy using immunochemical staining and magnetic resonance imaging (MRI) using contrast agents that report reduction—oxidation (redox) modification allow for the spatial visualization of the RS modification from cells to whole animals. While these other techniques may have limitations compared to conventional spectroscopic techniques, they offer advantages such as high sensitivity up to subnanomolar concentrations of analytes in the case of immunochemical and electrochemical techniques, the ability to spatiospectrally visualize the site of RS generation in the cell (MRI), measure the RS generation in real-time (for electrochemical technique) and high-throughput assay (enzyme-linked immunosorbent assay, ELISA). Although the decision to use any of these techniques will greatly depend on the type of sample (i.e., enzymatic, organelle, cell, tissue, organ, or whole animal), the desired goal (i.e., quantification versus identification or both), cost and speed, and the availability of expertise such as in the case of MS analysis and MRI operation, they are complementary techniques that can elucidate and help understand the mechanism of oxidative stress at the molecular level.

7.2 ELECTROCHEMICAL TECHNIQUES

Electrochemical detection exploits the redox reaction at the interface of the electrode and the electrolyte. The electrolyte, which is mainly composed of an electrochemically

Reactive Species Detection in Biology
DOI: http://dx.doi.org/10.1016/B978-0-12-420017-3.00008-6

inert solvent and its supporting salts, contains the redox-active analytes—in this case, a RS. RS possess electrons, and their interaction with the electrode rich in electrons (cathode) or deficient in electrons (anode) results their reduction or oxidation, respectively. In a two-electrode electrochemical cell, the working electrode, reference electrode, and electrolyte are housed in a miniaturized sensor probe for ease of application. Therefore, most electrochemical detection applications could only measure RS production extracellularly since the analyte must be in contact with the electrode surface. The intracellular source of RS could be deduced by using enzyme inhibitors (e.g., azide for mitochondria, allopurinol for xanthine oxidase, or diphenyleneiodonium chloride for nicotinamide adenine dinucleotide phosphate (NADPH) oxidase).

The working electrode is where the redox reaction takes place and could be anodic or cathodic. Electrodes are typically made of carbon fiber (CF), gold, silver, or platinum, while Ag–AgCl is used as a reference electrode and, as the name implies, has a known constant electrode potential for the half-cell reduction reaction of AgCl to Ag^{o} and Cl^{-}, thus allowing for the determination of the potential of an unknown half-reaction. The potential for the Ag–AgCl reference electrode vs the standard hydrogen electrode of 0.0 volt (V) ranges from 0.197 to 0.25 V at saturated to 0.6 M KCl.[1] The current from the reference electrode (i.e., 0 V) is held constant and allows for the investigation of the reduction (or oxidation) of the RS at the working electrode surface. At the working electrode, voltage is raised in a stepwise manner until the desired voltage is reached to cause the redox reaction; an instantaneous rise in current is then observed. Since the reaction occurs on the electrode surface, the current will subsequently fall as the analyte is chemically transformed and, therefore, has to be replenished through diffusion of a new analyte onto the electrode surface. However, since the half-life of these known and targeted RS analyte are very short (μs to ms) and their diffusion distance from the site of their production is very short, single-potential or direct-current amperometry detection is typically employed. Here a known potential (V) is applied between the electrodes, and the change in current between the electrodes is measured during reduction or oxidation of the analyte in the cathode or anode, respectively. This applied potential corresponds to the standard potential of the analyte's redox couple and could be adjusted for optimal redox reaction with the electrode. However, although the RS could be measured directly at a particular potential with high sensitivity, the specificity can be compromised because, in the presence of biological milieu, other analytes could also be oxidized or reduced at the same potential and therefore could interfere with the measurement. This problem was remedied with the advent of biosensors and use of gas-permeable membranes allowed for a more specific detection of RS. Several approaches to the chemical modification of electrode surfaces for increased specificity are discussed.

7.2.1 Superoxide detection

Direct measurement of $O_2^{\bullet-}$ can be achieved at a standard redox potential based on the half-reactions: $-0.33V$ vs NHE $(O_2|O_2^{\bullet-})$[2] and $+0.89$ V vs NHE $(O_2^{\bullet-}|H_2O_2)$[3] at pH 7. Current electrochemical sensors are designed based on the immobilization of $O_2^{\bullet-}$-specific enzymes such as superoxide dismutase (SOD) and cytochrome c onto the electrode surface. Several motifs for SOD-based biosensors can be classified as first-, second-, or third-generation biosenserors as shown in Fig. 7.1.[4] In the first generation, H_2O_2 was electrochemically oxidized as a by-product of the $O_2^{\bullet-}$ dismutation by SOD, which is entrapped in a gel matrix or alginate matrix[5] between the electrode and semipermeable membrane but which suffers from selectivity due to the high potential used for H_2O_2 oxidation to O_2 (Fig. 7.1A). The second generation employs indirect detection of $O_2^{\bullet-}$ via an electron relay–type approach in which $O_2^{\bullet-}$ undergoes selective oxidation to O_2 by SOD (or reduction to H_2O_2) where the oxidation–reduction by SOD, for example, can be mediated by immobilized ferrocene–carboxaldehyde at the electrode surface through its oxidation at the anode

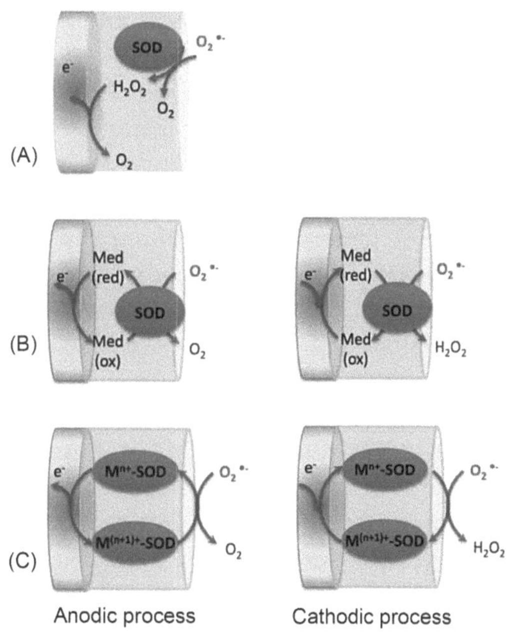

Figure 7.1 Sensor modalities for a SOD-based biosensor for $O_2^{\bullet-}$ showing (A) oxidation of the H_2O_2 product from $O_2^{\bullet-}$ dismutation from the matrix; (B) oxidation–reduction of $O_2^{\bullet-}$ by SOD as mediated by a redox-active molecule (e.g., ferrocene); and (C) oxidation–reduction of $O_2^{\bullet-}$ by immobilized SOD on the electrode surface. *Adapted from Alas-Blanchard et al.* Electroanalysis *2014;**26**:1277–86.*

(Fig. 7.1B).[6] In the third-generation probes, direct modification of the electrode surface with SOD allows for direct electron transfer from the electrode to the SOD, resulting in either anodic oxidation or cathodic reduction of $O_2^{\bullet -}$ to O_2 or H_2O_2, respectively. For example, SOD can be immobilized directly onto the gold electrode using thiol groups introduced as cysteine mutants in human Cu,Zn−SOD. The electron from $O_2^{\bullet -}$ is passed onto SOD to form the SOD^- and then reoxidation of the SOD^- at the anode (Fig. 7.1C).[7]

The early cytochrome c−based sensor employed its immobilization in carbon paste and used protoporphyrin membrane (immobilized in cellulose triacetate)[8] or real-time detection of $O_2^{\bullet -}$ from activated human glioblastoma cells using a cytochrome c−modified gold electrode.[9]Cytochrome c-based biosensor for $O_2^{\bullet -}$ was achieved through self-assembled monolayers using a variety of alkanethiols,[10] thioctic acid,[11] or thioctic amine[11] or dithiobis(succinimidyl)propionate[10] where the thiol end is attached to the gold electrode surface and the other end can be derivatized and tethered to cytochrome c.[12] However, pinholes or spaces between the assembled alkenethiols results in the detection of $O_2^{\bullet -}$ via direct oxidation by the gold electrode and therefore could affect the sensor's selectivity to $O_2^{\bullet -}$ because of its oxidation independently of cytochrome c.[12] Iron porphyrin complexes that mimic cytochrome c could be electrodeposited on a carbon micrelectrode surface as a polymeric complex to give high catalytic activity for the oxidation of $O_2^{\bullet -}$ by Fe(III).[13]

Improved electrode material composition that is not enzyme based allows for better electrode stability. Bimetallic nanoparticles have been developed with enhanced electrocatalytic properties with high sensitivities and short response times, improved electrochemical stabilities from interference and fouling as well as excellent selectivity. Examples are well-dispersed PtPd[14] or PtAu[15] nanoparticles; in the case of PtPd nanoparticles, these could be supported on a reduced graphene oxide nanocomposite modified with multifunctional polydopamine layers[14] or combined with carbon nanotubes (CNTs).[16] Also, a nanocomposite sheet has been designed for biomimetic sensing of $O_2^{\bullet -}$ at a nanoscale that is composed of manganous phosphate ($Mn_3(PO_4)_2$) nanosheets that are template synthesized with DNA and further assembled on CNTs to form unique DNA−$Mn_3(PO_4)_2$−CNT nanocomposite sheets.[17] This composite was shown to catalyze dismutation of $O_2^{\bullet -}$ and exhibited fast electron transfer, high sensitivity, and electrode stability. It was employed in the detection of $O_2^{\bullet -}$ from stimulated mouse cancel cells and normal skin cells.

Aside from SOD and cytochrome c, other biocatalytic agents included multiwalled CNTs, $Mn_2P_2O_7$−formylstyrylpyridine biomimetic SOD,[18] *Pseudomonas aeruginosa* azurin,[19] and a naringin−copper complex.[20] Also, other platforms for enzyme immobilization were designed such as mesoporous SiO_2−(l)-lysine hybrid nanodisks,[21] a composite electrospun nanofiber electrode made of dispersed gold nanoparticles onto a poly (methyl-methacrylate)−polyaniline core shell,[22] a ZnO nanodisk[23] for SOD, and nanostructured mesoporous titania film bound to pyridine dicarboxylic acid for cytochrome c[24] that showed improved durability and excellent biocompatibility. An opsonized electrode with a graphite surface was modified with human immunoglobulin G (IgG) and was employed for the direct detection of $O_2^{\bullet -}$ from stimulated neutrophils.[25] Superoxide release from stimulated breast cancer cells can be reliably and robustly monitored using highly innovative integrated sensor chips that can be mounted at the bottom of a standard 25-cm^2 cell culture flask and fabricated using 100-mm diameter borosilicate glass wafer with gold deposition on platinum as the working electrode and an Ag−AgCl reference electrode, all of which were integrated by an electrodeposition process. The electrode was modified by a linear polyethylenimine layer as a semipermeable membrane for improved specificity.[26] Using scanning electrochemical microscopy, the spatial topography of ROS release from live macrophages was made possible without the influence of cell morphology and the addition of external redox mediators, thus allowing for the detection of ROS production from a single cell.[27]

7.2.2 Hydrogen peroxide detection

Direct H_2O_2 detection in rat plasma can be carried out using a platinum electrode where the addition of catalase provided a baseline current due to the elimination of all the H_2O_2. The addition of azide inhibits the catalase and myeloperoxidase (MPO) that can contribute to the degradation of H_2O_2.[28] Unmodified carbon as well as horseradish peroxidase (HRP)−modified graphite electrode[29] were employed to detect H_2O_2 from stimulated neutrophils and were combined with $O_2^{\bullet -}$ detection using luminol-dependent chemiluminescence assay.[30] Hydrogen peroxide detection from isolated mitochondria was demonstrated using a platinized CF electrode with simultaneous detection of O_2 using a Clark-type electrode. An oxidation potential of 0.6 V and Ag−AgCl as reference electrode was used, and various substrates and drugs (e.g., glutamate−malate, adenosine diphosphate (ADP), carbonyl cyanide-4-(trifluoromethoxy)phenylhydrazone, antimycin, and rotenone) were employed to stimulate or inhibit mitochondrial respiration. Adding these agents causes a spike in current intensity, while the addition of H_2O_2 gave a nice linear response as a function of H_2O_2 concentration with detection limits of 0.5−5 μM (Fig. 7.2).[31] Concurrent measurement of H_2O_2 levels with O_2 showed consumption and formation of H_2O_2 when treated with ADP (Fig. 7.3) or glutamate−malate and oligomycin. This simple setup allowed for the investigation of aerobic metabolism in isolated mitochondria showing adenosine triphosphate (ATP) activation and led to increased H_2O_2 production.[31] Possible interference is the presence of nitric oxide (NO) since $HO^{\bullet}$ was known to be produced from the reaction of H_2O_2 with NO independently of transition metal ions.[32] This is supported by a previous observation that H_2O_2 was shown to interfere with the electrochemical detection of NO.[33]

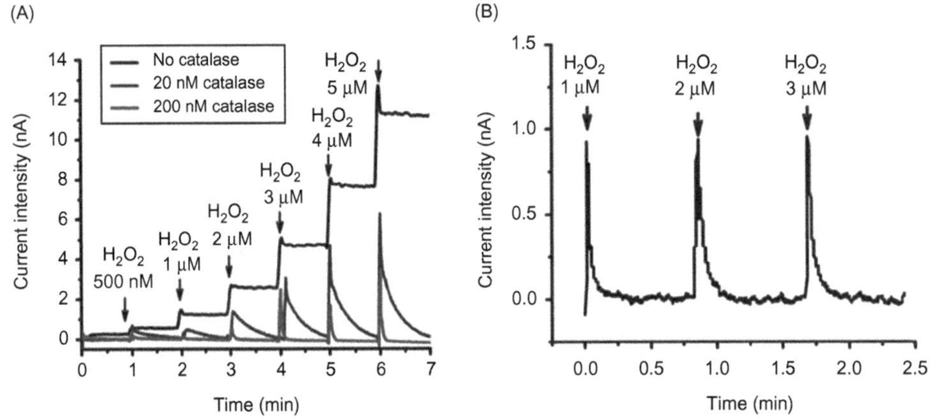

Figure 7.2 (A) Linear response of exogenously added H_2O_2 using platinized CF electrode in the absence and presence of a 20-nM catalase electrode; (B) addition of various concentrations of H_2O_2 in the presence of nonenergized mitochondria, which indicates the presence of endogenous-active catalase. *Adapted from Rapino et al.* Methods Enzymol *2013;**526**:123–34.*

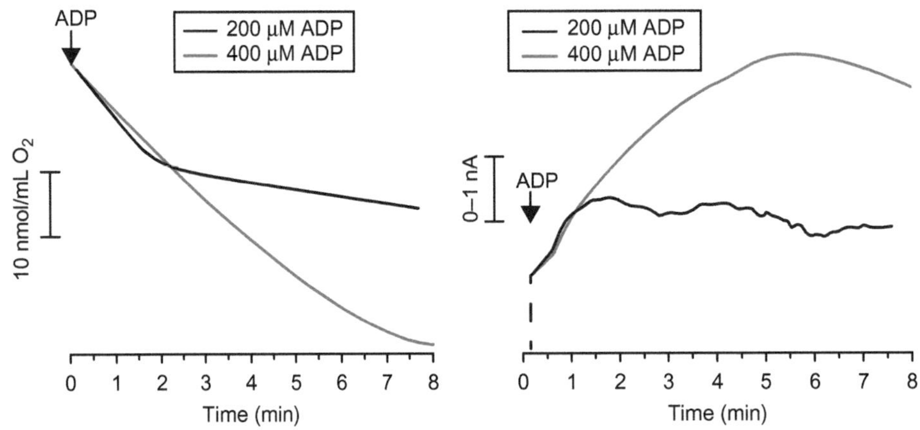

Figure 7.3 Simultaneous measurement of O_2 consumption (top) and H_2O_2 production (bottom) by isolated mouse liver mitochondria on addition of ADP. *Adapted from Rapino et al.* Methods Enzymol *2013;**526**:123–34.*

Nanotechnology has been gaining ground as a means of nonenzymatic sensing of H_2O_2 because of its high sensitivity, reproducibility, stability, and specificity. Nonenzymatic detection of H_2O_2 using nanostructured Au–MnO_2–graphene-modified CF was employed for the real-time monitoring of H_2O_2 production from cervical cancer cells.[34] Other nonenzymatic materials used for direct H_2O_2 sensing were an indium tin oxide nanocomposite with silver nanoparticles and reduced grapheme oxide,[35] Pt nanoparticles supported on porous graphene,[36] microporous Co_3O_4 hollow nanospheres containing encapsulated Pd nanoparticles,[37] CeO_2−hexamethylene−tertaamine or fructose,[38] gold nanoparticles on nitrogen-doped graphene quantum dots,[39] on mesoporous silica-covered reduced graphene oxide,[40] or monodispersed surfactant-free MoS_2 nanoparticles.[41] Organic-based molecules can also be used to increase electrocatalytic reduction of H_2O_2; for example, Prussian blue can be deposited layer by layer on multiwalled CNTs with poly(diallyldimethylammonium chloride).[42]

Protein-based electrochemical biosensors that involve direct electron transfer from protein to H_2O_2 were also developed. Several strategies have been employed to immobilize proteins on electrode surfaces such as physical adsorption, chemical binding through protein chemical coupling on the functionalized electrode surface, and site-directed immobilization for more uniform enzyme orientation.[43] Similar to the design of $O_2^{\bullet -}$ biosensors, several sensor modalities were employed similar to those shown in Fig. 7.1 such as electron-transfer from cytochrome c on nanostructured ZnO surface[44]; myoglobin immobilized on gold nano structures[45]; hemoglobin encapsulation in phosphonate−chitosan−CNTs[46]; horseradish peroxidase (HRP) hydroxyapatite nanocomposite[47]; or ferridoxin cross-linked with 11-mercaptoundecanoic acid on Au electrode[48]; and HRP entrapped in nanodiamond-based sponges.[49]

Another major design involves the miniaturization of biosensors in the form of an on-chip that uses photolithography and chemical etching techniques where electrodes are deposited onto glass slides. This is followed by encapsulation of biomolecules using a prepolymer hydrogel composed of HRP, polyethylene glycol, and glutaraldehyde. Using a three-electrode design (i.e., miniaturized gold working electrode, Pt counter electrode, and Ag−AgCl reference electrode), amperometric measurement of H_2O_2 from cells can be performed.[50,51] Hydrogen peroxide was also measured directly from tissue such as the rat hippocampus, where a sensor array was fabricated and placed at multiple positions in the hippocampal slice, which allowed for the detection of H_2O_2 directly from neuronal brain tissue via kainite stimulation and endogenous catalase inhibition.[52]

7.2.3 Nitric oxide detection

The design of NO-selective electrochemical sensors is based on three major strategies: (1) use of Clark-type electrodes; (2) use of carbon or metal electrodes whose surface could be further modified; and (3) use of solid electrodes with surfaces layered by electrocatalysts.[53] The NO-oxidation potentials employed in all of these electrode designs range from 0.4 to 0.95 V, depending on the reference electrode used.[54] While these designs have evolved over the years, one important feature is common: the use of gas-permeable or permselective membranes to isolate the working electrode from other potential interferences, specifically that of nitrite endogenously produced from NO or as an end product of NO electrooxidation. Both hydrophilic and hydrophobic membranes are employed, but the most commonly used is the hydrophilic membrane, Nafion, which is a sulfonated tetrafluoroethylene-based fluoropolymer–coplymer. The sulfonate side chain, unlike in the gas-permeable membranes, is highly conductive to cations and neutral species in solution when applied in thin layers (~10 nm).[55] A combination of Nafion precoat layers and electropolymerized polymer film of *o*-phenylendiamene (*o*-PD) Pt electrode surface was shown to exhibit better performance, with the latter preventing larger interfering molecules from accessing the Pt surface.[56] Hydrophobic membranes such as chlorophene are those that are gas permeable and similar to those used in Clark-type O_2 electrodes.

The Clark-type NO electrode was initially designed by Shibuki[57–60] using Pt working electrode and Ag–AgCl as the reference electrode. Compared to the negative potential for the reduction of O_2, NO detection was accomplished by applying a positive potential for the oxidation of NO to NO_3^- via Eq. (7.1):

$$NO + HO^- + H_2O \rightarrow NO_3^- + 3H^+ + 3e^- \tag{7.1}$$

However, Shibuki electrodes exhibit major drawbacks such as slow response time and strong background signal fluctuation.[54] Commercially available NO microsensors[61] are descendants of the Shibuki-style NO sensor, which uses CF wire and Ag–AgCl as working and reference electrodes, respectively, and utilizes a proprietary novel surface membrane that amplifies NO signal with fast response times and eliminates interference from nitrite, nitrate, ascorbate, H_2O_2, and catecholamines.[62] A nano-sized electrode version with up to 7-μm tip diameter is also available to measure NO in arteries and blood vessels.[63]

Electrode surfaces such as graphite, iridium, palladium, platinum, and ruthenium can also be electrochemically activated for optimal NO sensitivity. Electrodes that underwent potential cycling gave increased response only for graphite and ruthenium. However, unlike graphite, for Ru the cycling potential does not form soluble species metal salts that could indicate corrosion of the electrode surface.[64] Other modes of electrode surface modification were demonstrated such as Pt–Fe(III) nanoparticle electrodeposition on glassy carbon electrode (GCE),[65] platinized CF,[66] Pt wire coated with multiwalled carbon nantotubes and subsequently with Ru via chemical vapor deposition,[67] Pt–Ir alloy,[68] Au nanoparticles deposited on electrochemically reduced graphene oxide,[69] graphene oxide nanocomposing with ceria (CeO_2),[70] reduced graphene oxide–cobalt oxide nanocube–platinum composite,[71] amine–thiol functionalized boron-doped diamond,[72] Ni–hexacyanoferrate layer on Pt,[73] and bimetallic Au–Ag nanocluster (NC)-modified electrode stabilized by poly(acrylamide-co-diallyldimethylammonium chloride) matrix.[74] A microfluidic device was also developed using Au–hexacyanoferrate layer on a porous Au–indium–tin oxide electrode, which allowed detection of NO from macrophages within the microchannels on stimulation.[75] Bedioui et al. suggested that electrode surface chemistry should not be the only factor to consider when optimizing electrode efficiency; the most appropriate amperometric technique should be considered, including the use of differential normal pulse amperometry or pulsed chronoamperometry or both.[76]

Surface modification of NO sensors with metal complexes could also increase electrode selectivity and sensitivity via enhanced catalytic oxidation of NO. Commonly used electrocatalysts are metalloporphyrins and metallocyanines, which can be coated on CFs via oxidative polymerization and finally coated with polymer membranes for better resistance to degradation. Other electrocatalysts include myoglobin[77] and hemoglobin.[78,79] Electroactive metal centers of phthalocyanine (Pc) complexes show better catalytic activity toward NO than the electroinactive metal centers where the order of increasing catalytic oxidative activity for NO are as follow: $H_2Pc < CuPc < ZnPc < MnPc < CoPc \sim NiPc < FePc$.[80] Also, the method of depositing metallophthalocyanine film on the electrode surface could affect the magnitude of oxidation current and potential of NO.[81] Nickel (II) tetrasulfonated phthalocyanine complex (NiTSPc) was electropolymerized on CF and was employed for NO detection in human blood platelets and umbilical vein endothelial cells,[82] or nonadherent cells such as U937 promonocytic cells.[83] Nickel phtalocyanine-modified Pt–Ir electrode was employed for the in vivo detection of NO in tumor-bearing mice,[84] while Ni(II) porphyrinic-based microsensor was employed for beat-to-beat NO production from isolated intact cardiomyocytes[85] or endothelial cells.[86] Pyrrole-functionalized Mn(II) trimethoxycarboxyphenylporphyrin on Pt-disk electrode was applied for NO detection on endothelial cells.[87] GCE can be modified with 5,10,15,20-tetrakis(4-methoxyphenyl)porphyrin and was employed for the detection of NO from cancer cells.[88] Nickel tetrasulfonated phthalocyanine (NiTSPc) film compared with Ni porphyrin–based film showed better efficiency, and better sensitivity was achieved using potential scans rather than controlled potential electrolysis as a method for thin film deposition.[89] A three-electrode design was also employed to detect NO from human umbilical vein endothelial cells (HUVECs) using three different NO-sensing chemistries that are integrated in an array-type sensor—i.e., NiTSPc, pyrrole-substituted manganese porphyrin, and Ni tetramethylpyridyl porphyrin (NiTmPyP)—and deposited on platinized gold electrode, where NiTSPc provided the highest sensitivity with NiTmPyP having the lowest sensitivity.[90]

Several challenges are encountered in the application of electrochemical NO detection in biological system such as the short half-life of NO in solution, especially in the presence of O_2 or $O_2^{\bullet -}$, and their reactivity with endogenous heme proteins, low bioavailability that ranges from pmolar to micromolar, and interference from NO by-products such as nitrite.[91] Especially with the use of a Pt-Ir–based electrode, nitrite was shown to induce baseline current variation compared to a nickel–porphyrin–Nafion–based CF electrode where the latter showed better specificity and selectivity against nitrite oxidation.[92] Hydrogen peroxide was also shown to interfere with NO detection, showing an increase in current as a function of H_2O_2 concentration in cultured cells using a porphyrinic-based microelectrode.[33] This is also further supported by the observed reaction of H_2O_2, with NO producing $HO^{\bullet}$.[32] Tyrosine was detected at the same potential alongside NO using a porphyrinic microsensor in a dog brain and thus is an important consideration in data interpretation.[93] Other interfering species include ascorbic acid, uric acid, acetaminophen, serotonin, 5-hydroxyindole-3-acid, and some catocholamines such as dopamine and norepinephrine[62,94,95] as well as carbon monoxide.[96]

Micrometric sensors in various designs such as array, planar, or needle-type had been fabricated that allowed detection of NO in cell cultures or tissues and for the simultaneous detection of other analytes such as ONOO, CO, O_2, glutamate, and $O_2^{\bullet -}$. In lipopolysaccharide (LPS)-activated human glioblastoma cells, both $O_2^{\bullet -}$ and NO were directly measured in real time, with $O_2^{\bullet -}$ showing a lag time of ~1 min due to the distance of the electrode from the cell surface and the response time after stimulation.[9] A planar-type microsensor made up of a Pt microdisk with poly(tetrafluoroethylene) was designed and could detect both CO and NO; the Pt electrode was electrochemically deposited with or without Sn exhibiting better sensitivity to both NO and CO with Pt–Sn electrode but better selectivity to NO with Pt alone.[96] Dopaminergic transmission in vivo in rat striatum suggests NO involvement,[97] and recent microelectrode designs made of CF were employed for dual NO–dopamine detection.[98] Simultaneous detection of NO–$ONOO^-$ using an ultramicroelectrodes array[99] or microchannels[100] had been demonstrated; NO–O_2 and NO–glutamate were also described.[91] Indirect detection of NO as peroxynitrite can be selectively determined at −0.1 V vs standard calomel electrode (SCE) via direct reduction of peroxynitrous acid on a bare Au electrode at pH 7.1.[101] Nitric oxide in the form of S-nitrothiols such as nitrosocysteine or nitrosoglutathione were detected using a modified NO gas sensor with immobilized organoselenium catalyst in thin hydrogel,[102,103] or as nitrite by electrophoresis using electrochemical ultramicroelectrodes device made of polydimethylsiloxane–CF microchip capillary.[104]

In vivo NO measurement through the use of microsensors in healthy human patients was made possible by inserting the sensor into the hand vein and then providing stimulation with acetylcholine or bradykanin showing dose-dependent signal formation and attenuation of signal by L-*N*-monomethyl arginine.[105] A subnanomolar detection limit of <0.3 nM was achieved using microchips based on a multielectrode array modified with layers of cationic exchangers and NO-selective membranes on activated C, which was deposited on a Si chip.[63] Microsensors paved the way for the real-time detection of NO in vascular walls[106] but could be limited by interference from other substances, physical forces, temperature, and pH. In conscious and unrestrained rats, real-time NO detection was also performed using an NO sensor positioned in the right hippocampus; this also allowed the exogenous administration of substrates such as L-arginine or 7-nitroindazole through intravenous infusion or intraperitoneal cannulal administration, respectively.[107]

7.2.4 Hydrogen sulfide detection

Electrochemical methods and spectroscopic techniques are perhaps the two most widely employed methods for analyzing hydrogen sulfide (H_2S). Hydrogen sulfide detection employs two common strategies to detect the electroactive thiol adduct and direct oxidation of the sulfide on the modified ion-selective electrode surface. In the first strategy, detection of H_2S is electrochemically initiated based on a prior two-electron oxidation of a solution of methylene blue precursor *N,N*-dimethylphenylene-1,4-diamine (DMPD) to $DMPD^+$ and subsequent nucleophilic addition of H_2S to yield $DMPD^+{-}SH$, as shown in Eq. (7.2). In an aqueous system, amperometric sensing of H_2S uses a GCE that shows a linear response from a 3- to 150-μM concentration within a pH range of 3–6 at an oxidation potential of +0.4 V.[108]

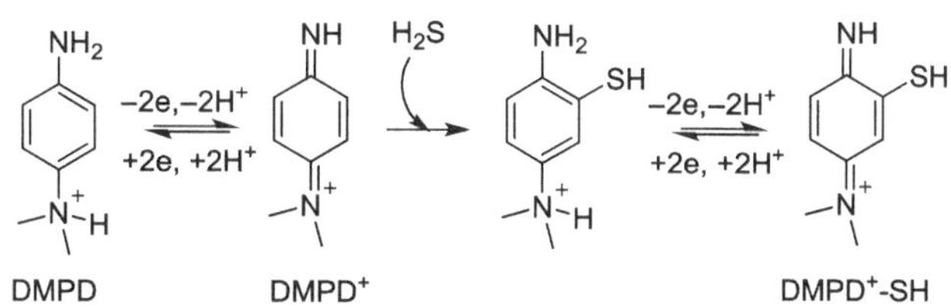

(7.2)

H_2S analysis in organic solvent such as dimethylformamide (DMF) was also demonstrated using the ethyl analog *N,N*-diethylphenylene-1,4-diamine.[109] Other than the nucleophile addition of H_2S to $DMPD^+$, electrophilic compounds such as hydroquinone, 4-aminophenol, catechol, and dopamine were also exploited for H_2S monitoring at oxidation potentials between 0.3 and 0.39 V. Similar to DMPD, these electrophilic compounds are also capable of forming a quinonoid intermediate on oxidation and could undergo a nucleophilic addition reaction with sulfide. However, with the exception of DMPD, all of these compounds were susceptible to oligomer deposition on the electrode surface, which results in fouling that requires frequent polishing.[110] Potential interference could also arise from the low-molecular-weight thiols but has been shown to exhibit more positive potential for reoxidation compared to the thiol adduct formed with DMPD.[111]

Second strategy is that employed in commercially available H_2S sensors.[112] In particular, an ion-selective $Ag{-}Ag_2S$ working microelectrode response to HS^- concentration is based on the equilibrium shown in Eq. (7.3),

$$2Ag + S^{2-} \rightleftharpoons Ag_2S + 2e^- \qquad (7.3)$$

where the concentration of S^{2-} is governed by the equilibrium shown in Eq. (7.4):[113]

$$HS^- \rightleftharpoons H^+ + S^{2-} \tag{7.4}$$

However, although a Ag–Ag_2S electrode has been shown to have no interferences with limits of detection in the 100 nM, it suffers several disadvantages. In aqueous solution, H_2S acts as a weak acid that forms the electroactive anion HS^- with known pK_a of 6.6–7.0. However, to favor formation of S^{2-}, such measurements are performed in alkaline conditions because the pK_{a2} of HS^- based on the equilibrium shown in Eq. (7.4) is ~13.8.[112] A polarographic H_2S sensor (PHSS) allows for the real-time measurement of H_2S in blood and tissues. Based on a modified Clark-type oxygen sensor, a commercially available PHSS uses platinum wires as the anode and cathode; H_2S permeability allows selective passage of H_2S through a hole and an internal electrolyte solution of $K_3Fe(CN)_6$. According to Eqs. (7.5)–(7.7), the HS^- is oxidized to elemental sulfur by ferrocyanide to form the ferricyanide. The ferricyanide is then reduced back to ferrocyanide on the surface of the platinum electrode at 0.1–0.2 V vs SCE producing a current that is proportional to the H_2S concentration.[112]

$$H_2S \rightleftharpoons HS^- + H^+ \tag{7.5}$$

$$HS^- + 2\,Fe(CN)_6^{3-} \rightarrow S^0 + 2\,Fe(CN)_6^{4-} \tag{7.6}$$

$$2\,Fe(CN)_6^{4-} + 2\,e^- \rightarrow 2\,Fe(CN)_6^{3-} \tag{7.7}$$

Several biological applications that employ micro and macro PHSS electrodes have been compiled using samples ranging from blood plasma to cells to tissues.[112]

A modified electrode surface also includes electrodeposited films of metal macrocylic complexes such as cobalt phthalocyanine derivatives but is prone to interferences from varying O_2 levels or the presence of dithionite, sulfite, iodine, and cyanide.[111] Other examples of electrocatalytic materials such as dicyano-ferriprotoporphyrin deposited on to the Pt electrode surface exhibit higher selectivity for H_2S compared to other gases (e.g., NO and CO).[114] Electrochemiluminescence (ECL) detection of H_2S in rat brain microdialysates uses an H_2S-selective ruthenium bipyridine complex—$[Ru(bpy)_2(bpyDPA)Cu]^{4+}$—immobilized on a GCE and coated with Nafion and exhibiting enhanced ECL on reaction with H_2S that is free of interference from higher-molecular-weight nonvolatile thiols and bicarbonate with a detection limit of 0.25 μM.[115] If nanocrystalline CdS (CdSNC) film is on a GCE, H_2S can be detected in serum via ECL quenching. ECL enhancement via CdSNC–GCE can be activated with H_2O_2 and stabilized by citrate prior to H_2S quenching.[116]

Other innovative electrochemical modalities were employed for H_2S sensing. For example, a dual-band electrode device made of boron-doped diamond-flow electrode was designed that could monitor H_2S across a wide pH range. The system allows for the controlled generation of HO^- from an upstream electrode, and the subsequent formation of the electroactive HS^- allows for its electrochemical detection at a downstream detector electrode. Although still yet to be applied in biological systems, this electrochemical sensor design offers enhanced capability for monitoring H_2S in an online electrochemical detection system.[117] Dual-analyte sensing was also demonstrated for the analysis of acetone along with H_2S and thus has potential for the analysis of these substances from exhaled breath for the diagnosis of halitosis and diabetes, respectively, using gas sensor based on sensitized electrospun semiconductor metal oxides such as SnO_2 nanofibers with reduced graphene oxide nanosheets (RGO-NS). At low RGO-NS composition, SnO_2 nanofibers provided higher selectivity to H_2S compared to other gases such as acetone, ethanol, toluene, CO, NH_3, and pentane.[118] Vanadium pentoxide film electrodeposited on Pt-Ir wire was employed for the monitoring of H_2S levels in solution and exhibited higher sensitivity compared to bare Pt-Ir surfaces without interference from CO and NO gases.[119] Polyaniline nanofibers decorated with Au nanoparticles exhibited good detection properties for H_2S and CH_3SH, allowing analysis of human breath after ingestion of raw garlic.[120]

7.2.5 Oxygen detection

Conventional amperometric detection of O_2 employs a Clark-type electrode composed of a cathodic Pt working electrode and anodic Ag–AgCl reference electrode based on the following half-reactions (Eqs. (7.8)–(7.9)):

$$O_2 + 2H_2O + 4e \rightarrow 4HO^- \tag{7.8}$$

$$4Ag + 4CI^- \rightarrow 4AgCI + 4e \tag{7.9}$$

where a cathode potential of −0.7 V is applied to give a linear response to O_2 concetration.[121] Since a bare Pt electrode is susceptible to fouling and interference from other substrates, electrodes are typically isolated from the biological milieu by a gas--permeable membrane, therefore only allowing the diffusion of O_2 to the electrode. Poly(hexafluoropropylene-co-tetrafluoroethylene) or polytetrafluoroethylene membranes are typically used for fast response, which is appropriate for the investigation of enzymatic or cellular respiration, specifically from neutrophils or isolated mitochondria,[122,123] as well as in other cellular systems such as human pluripotent stem cells,[124] rat hepatocytes,[125,126] HUVECs,[127] $CD4^+$ T lymphocytes immune cells,[128] K562 tumor cells,[129] or plant respiration.[130] In in vivo systems, O_2 tension in tissues have been monitored such as in rat hippocampus,[131] corneas,[132] intestines,[133] human adipose tissue,[134] kidney and hind limb skeletal muscle,[135] pig jejunal mucosal tissue,[136] and rabbit renal cortical tissue.[137]

Ionic liquid electrolyte was incorporated in a Clark-type online electrode to improve electrode conductivity and was also employed for amperometric O_2 sensing because it imparted an improved limit of detection for O_2 of 0.05 volume percent with good selectivity, reversibility, stability, and performance compared to an aqueous electrolyte.[138] Comparison of a Clark electrode with fluorescence optodes for the measurement of renal cortical tissue oxygen tension showed that the latter were less variable but gave readings that are proportionally less than a Clark electrode provides. However, it was found that the true interstitial oxygen tension was somewhere between the values obtained from fluorescence and electrochemical techniques.[137] Others designed Clark electrodes that measure low concentrations of NO and N_2O in addition to O_2 and with a high sensitivity of 5–50 nM.[139]

Comparison of a carbon paste and a Pt microelectrode for real-time detection of O_2 in brain tissue showed higher sensitivity for the latter with minimal change in sensitivity when polyphenylenediamine polymer coated the Pt surface to increase electrode biocompatibility with negligible interferences from biological milieu, pH, and ions.[140] There have been several approaches in O_2 sensor design. For example, single-cell[141] and single-embryo[142] oxygen consumption measurements were accomplished using modulated amperometric sensors in a self-referencing mode or microfluidic chips with built-in amperometric detector arrays, respectively. In single photosynthetic cells, O_2 can be measure by immobilizing the cell on a microfluidic system and inserting a nanoprobe with a submicron-sized ultramicroelectrode into the cytosolic space, allowing for the in vivo amperometric measurement of O_2 under light irradiation.[143] In freely moving animals such as rats, oxygen dynamics in the brain striatum can be detected in real time using an implantable device consisting of a carbon microsensor coupled to a current-to-voltage converter with the oxygen current digitally transmitted.[144]

Other analytes can be simultaneously measured along with O_2. For example, dopamine and O_2 can be simultaneously detected using pyrolyzed photoresist carbon microarray structures; this enables multiple fast-scan cyclic voltammetries combined with amperometry for dopamine detection.[145] Oxygen and glucose can also be simultaneously monitored using a nanocomposite of dispersed CNTs in ionic liquid on a modified GCE surface. This mediator-free sensor showed electrocatalytic activity toward O_2 and NADH while glucose dehydrogenase trapped in chitosan on the nanocomposite allowed for the detection of glucose, thus showing better voltammetric and amperometric performance in real blood samples.[146] A two-dimensional (2D) imaging of O_2, H_2O_2, and glucose distribution using a monolithic array of 400 individually addressable microelectrodes was fabricated made up of Pt electrodes arranged in a square matrix for the amperomtric detection of oxygen at -8 V and H_2O_2 at $+0.8$V vs Ag—AgCl while glucose was detected using entrapped glucose oxidase in conducting polymer polypyrrole integrated that is into all the array. However, compared to a scanning electron microscope, the time-resolved imaging of the analytes only gave a resolution of 500 μm but still offers great potential for future silicon thin film technology designs with resolutions of as much as 1 μm.[147]

Innovative materials or biomaterials were employed for amperometric detection of O_2, including microelectrodes of boron-doped diamond film on a SiO_2 substrate.[148] Using the same boron-doped diamond electrode modified with Au nanoparticles allowed for sensing of O_2 demand from immobilized *Rhodotorula mucilaginosa* UICC Y-181 in a sodium alginate matrix showing improved current stability and sensitivity compared to free yeast.[149] Extracellular release of O_2 from human erythrocytes can be monitored using a system comprising laccase (Lac) and 2,2-azino-bis(3-ethylbenzothiazoline-6-sulfonic acid) (ABTS) that are assembled on a graphene surface; the ABTS acts as a redox mediator on bioelectrocatalytic four-electron reduction of O_2 to water by the Lac.[150] With electropolymerized eugenol as a protective film for Au electrodes, in vitro detection of O_2 showed less fouling in biological milieu and impermeability to ascorbic acid.[151] Hemin adsorbed onto porous carbon felt was also employed for O_2 detection by exploiting the redox wave due to Fe(III)—Fe(II).[152]

7.3 MASS SPECTROSCOPY

The reactions of reactive oxygen species to biomolecules such as proteins, DNA, lipids, and even carbohydrates could lead to cascades of events that result in cell signaling, cellular injury, and even cell death. As analyzed by mass spectroscopy, these events most often leave a trail of evidence as to the fate of these biomolecules during oxidative insult. Mass spectroscopy is an important analytical tool that exploits the inherent masses of particles and molecules to elucidate their molecular compositions or structures. Analysis begins with the ionization of oxidatively modified biomolecules or formed oxidation by-products and the generation of charged molecules or charged fragments that are separated via deflection under electrical or magnetic field according to their mass-to-charge (m/z) ratios. Ions pass through an electron multiplier where they are detected and displayed as spectra of relative abundance vs m/z ratio. Molecular compositions or structures are then deduced based on the expected masses or from the fragmentation pattern. Two major MS approaches are commonly employed to identify post-translational modifications (PTMs): the bottom-up approach and the top-down approach.

The bottom-up or shotgun approach employs either a 2D gel or a nongel approach. In the 2D gel bottom-up approach, proteins are denatured and thiols are reduced and digested. PTMs are stabilized using chemical-labeling tag-specific binding systems, thus reducing the complexity of samples and facilitating detection of the oxidative PTM of interest. Gels are transferred onto a 2D Western blot and specific antibodies are used to visualize oxidized spots. Gel-based methods are advantageous because they target a specific subset of the proteome, and numerous gels can be ran and aligned with sophisticated software tools. Proteomes are then excised from the gel, trypsinized, and run through a matrix-assisted laser desorption and ionization (MALDI) system or electrospray ionization–mass spectrometry (ESI–MS). The peptide mass fragmentation pattern is fingerprinted and compared against human and other species' genomic protein databases using available search tools.[153] Nongel redox proteomics have been widely employed in the identification of protein carbonylation due to its abundance in blood plasma. Avidin affinity chromatography is employed to enrich biotin hydrazide–derivatized carbonyl proteins allowing minimal interference from unmodified proteins during LC or MS.[153]

The top-down approach involves analysis of intact proteins or peptides (or both) generated from enzymatic digestion. MALDI–MS or ESI–MS are typically employed to analyze intact and modified proteins. Instruments providing high mass accuracy and resolution are therefore employed using Fourier transform ion cyclotron resonance, Orbitrap, quadrupole-time-of-flight (Q-TOF), or TOF–TOF instruments, which allow detection of modified protein isoforms and their relative abundance.[154] Regardless of the approach used, the overall aim is to quantify the level and identify the site of oxidation in post-translationally modified proteins.

7.3.1 Proteins system

PTM is widely used as biomarker of oxidative stress. A variety of PTMs such as phosphorylation, thionylation, acetylation, and glycosylation have been investigated and were found to be relevant in the pathogenesis of various ROS-mediated diseases and therefore can act as indirect biomarkers of ROS formation. Furthermore, oxidation of protein such as the formation of carbonyls, nitration, and thiol oxidation have also be been used as direct biomarkers of oxidative stress. From using complex sample matrices such as cell culture, body fluids, and tissues, various proteomics techniques have been employed to analyze PTMs for the purpose of early diagnosis. As the name implies, redox proteomics involves identifying markers of oxidative or nitrosative stress. These oxidatively modified proteins are either reversible (e.g., S-nitrosylation and S-glutathionylation (SSG)) or irreversible (e.g., carbonylation, nitration, or thiol oxidation), where protein aggregation occurs for the latter.[155]

7.3.1.1 Reversible thiol oxidation, SSG, SNO, and S-OH

Reversible thiol PTMs such as S-sulfenylation (S-OH), S-nitrosylation (SNO), and S-glutathionylation were also found to be responsible for modulating such physiological functions as muscle contraction and relaxation, gene expression, metabolism, and apoptosis. Modification of cysteine by NO to form S-NO serves as one key intermediate in the formation of disulfide bonds and has been found to be important for signal transduction. Also, reversible thiol oxidation has been implicated but is not exclusive in regulating pump activity of Na^{+}−K^{+}-ATPase and ion channels such as RyR2 and SERCA2, which are vital for normal cellular physiological function. Therefore, identifying these PTMs can provide mechanistic insights into their role in the pathogenesis of chronic diseases. A biotin switch technique (BST) was employed to analyze SNO−Cys for S-nitrosylation of soluble guanylyl cyclase. In BST, SNO−Cys is indirectly detected through the initial reduction of SNO−Cys with ascorbic acid to yield the free thiol and subsequent biotinylation using biotin−hexylpyridyldithiopropionamide followed by avidin affinity enrichment and sample analysis using LC and tandem mass spectrometic (MS−MS) detection. Both collision-induced dissociation (CID) and high-energy collisional dissociation were effective activation methods for protein fragmentation in MS−MS analysis. By using protein database searches and bioinformatics, identifying SNO−Cys in cells showed higher sensitivity compared to other techniques.[156]

Conventionally employed biotin−avidin−based capture technique suffers from interference from natural biotinylation. To address the interference problem, a tandem mass tag (TMT) switch-labeling strategy is employed in which the reversible thiol modifications are tagged using various cysteine-reactive iodo-TMT labels and have been shown to be more specific than the BST. TMT tags have varying number of isotopic substitutions. For example, TMT sixplex tag reagents (iodo TMT^{6}) shown in Fig. 7.4 contain two heavy isotopes of ^{13}C and ^{15}N at five different positions. The tags comprise a sulfhydryl-reactive iodoacetyl group that binds to the cysteine moiety, a mass normalizing spacer arm, and a mass reporter that has a unique m/z of 126−131. Fig. 7.5 shows a typical labeling procedure, and Fig. 7.6 shows a common workflow employed for a TMT switch-labeling strategy. Protein lysates are treated with an alkylating agent using iodoacetamide to block the free thiols and prevent further modification of free thiol groups. Proteins are precipitated and washed to remove excess reagents. To differentiate S−NO from S−SG and S−SP, a sequential iodo-TMT switch is required.[157] The −SNO is first freed of the NO by using ascorbate,[158] and the free thiol is tagged, for example, with a specific iodo-TMT1. After excess reagents are removed, the remaining reversible thiols such as the disulfides SSG and SSP can be freed of SG or SP using tris(2-carboxyethyl)phosphine (TCEP) and the free thiols are then tagged, for example, with iodo-TMT2. The labeled samples are then pooled, precipitated, and trypsinized. The peptides are then fractionated by LC and analyzed by mass spectroscopy. Using the fragmentation data, labeled peptides are identified against the database, thus allowing the identification of the corresponding proteins. The attached tags also fragment to generate the low-molecular-mass reporter ion ($m/z = 126-131$), thus allowing quantification of the peptides relative to their proteins of origin.[157] Identification and quantification of more than 100 cysteine S-nitrosylated and S-sulfenylated peptides were demonstrated using TMT strategy in

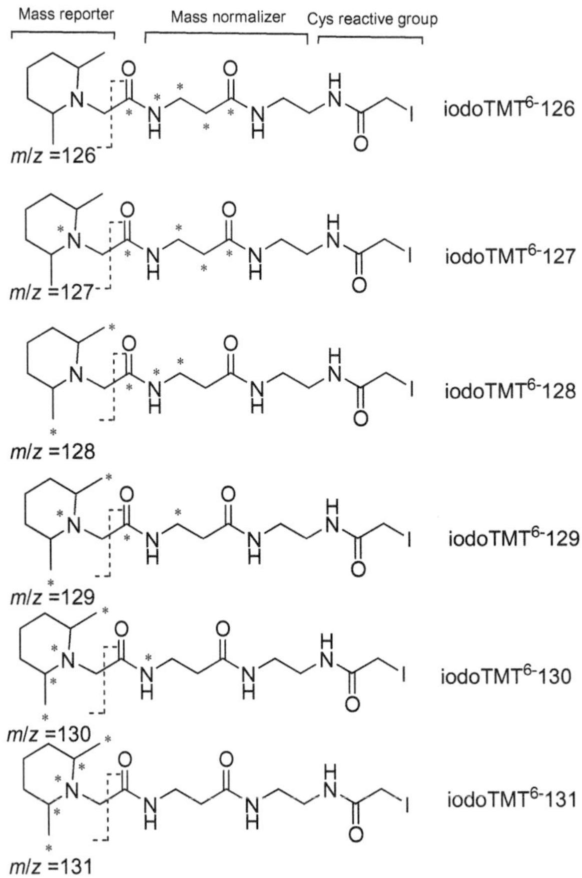

Figure 7.4 The TMT switch-labeling strategy uses TMT tags with varying number of isotopic substitution. Shown here are commercially available TMT 6-plex (iodo-TMTsixplex) tag reagents.

Escherichia coli under mild oxidative stress by subjecting the sample to anaerobic conditions in the presence of fumarate or nitrate or by short-term treatment with H_2O_2.[159] Commercially available thiol-affinity resin (Thiopropyl Sepharose 6B) could be used to enrich peptides for the analyses of S-nitrosylation and the level of total reversible cysteine modifications from RAW 264.7 cells.[160,161]

Other applications of TMT strategy coupled with resin enrichment include nanoparticle-induced oxidative stress on macrophages where reversible protein S-glutathionylation was quantitatively profiled to reveal ~2500 unique SSG-modified cysteine sites from more than 1200 proteins using LC–MS–MS.[162] In tissue, global reversible oxidized cysteine was identified using LC–MS–MS in proteins isolated from the left ventricle of mice with metabolic syndrome–associated diastolic dysfunction and fed a high-fat and high-sucrose diet. A more than 1.5-fold

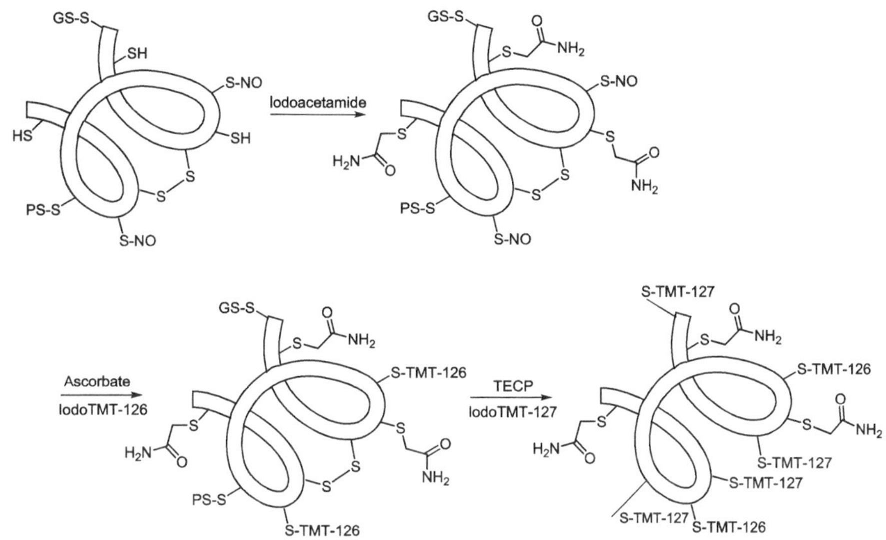

Figure 7.5 Isotopic labeling of reversible thiol modification using TMT tags involving prior amide protection of unmodified thiols followed by labeling of S-nitroso and disulfides using ascorbate and *tris*(2-carboxyethyl)phosphine (TCEP), respectively.

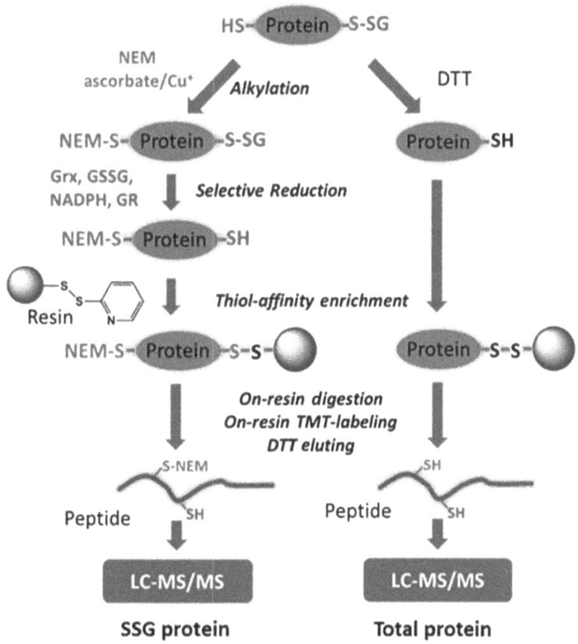

Figure 7.6 The overall workflow scheme for enrichment and site-specific identification of SSG-modified Cys peptides. *Reprinted with permission from Duan, J. Kodali, V.K. Gaffrey, M.J. Guo, J. Chu, R.K. Camp, D.G., et al. Quantitative Profiling of Protein S-Glutathionylation Reveals Redox-Dependent Regulation of Macrophage Function during Nanoparticle-Induced Oxidative Stress ACS Nano 2016;* ***10****: 524. Copyright © 2017 American Chemical Society.*

increase was seen in cysteine modification in 173 proteins,[163] and cardiac-specific overexpression of catalase showed a decreased thiol occupancy by more than 1.3-fold in 82 proteins.[164]

7.3.1.2 Protein carbonylation

One of the hallmarks of oxidative damage in protein systems is carbonylation, which is an irreversible oxidative PTM resulting from the direct oxidation of side chains of lysine, arginine, proline, threonine, and tryptophan (Fig. 7.7). Protein carbonylation also includes oxidative cleavage of peptide bonds via α-amidation and Michael adduction of thiols to lipid peroxidation by-products such as α,β-unsaturated aldehydes (e.g., 4-hydroxy-trans-2-nonenal and acrolein), while the ketoaldehydes (e.g., 4-oxo-trans-2-nonenal, ONE) can undergo 1,2-addition with lysine to form irreversible ketoamide adducts.[165] The lipid peroxidation product malondialdehyde and the glycooxidation product (advanced glycation product) glyoxal undergo Schiff base

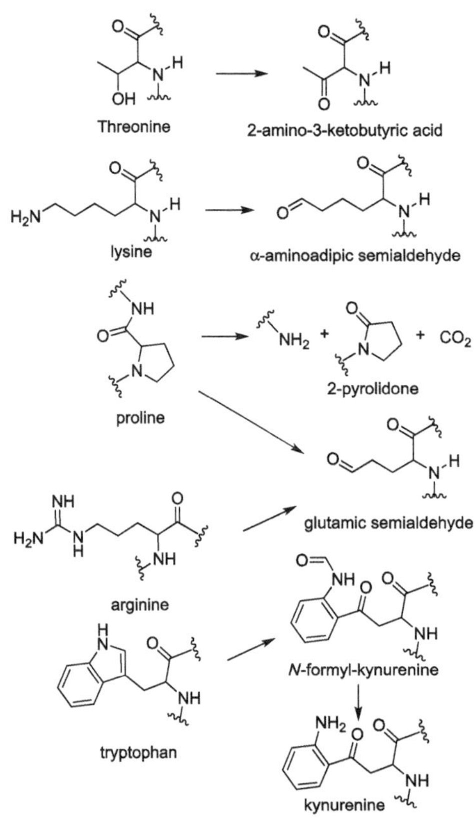

Figure 7.7 Post-translational protein carbonylation showing the formation of various oxidized amino acid side chains with carbonyl moiety.

condensation with amino acids with $-NH_2$ side chains and can be reversible. Carbonylation of proteins could also occur via nonlipid peroxidation or advanced glycation end product (AGE) product derived modification such as through metal-catalyzed oxidation by $HO^{\bullet}$ of amino acid residues via Fenton reaction. For example, proline and arginine oxidation lead to the formation of glutamic semialdehyde, threonine to 2-amino-3-ketobutyric acid, and lysine to aminoadipic semialdehyde, while direct tryptophan oxidation leads to the formation of kynurenine and *N*-formyl kynurenine.[166] Therefore, protein carbonylation could originate from direct oxidation of amino acid side chains, oxidative cleavage of the protein backbone, or adduction to Cys, His, or Lys residue of lipid peroxidation aldehydic decomposition products and AGEs.

Also demonstrated was identification of carbonylated protein levels as a function of oxidative stress in the plasma of diabetic rats using reversed-phase LC−MS−MS. Three major types of carbonylation were identified that showed direct oxidation as the predominant mode of carbonylation. In particular, hemoglobin and muringlobulin-1 homologues were heavily oxidized proteins, which indicates the role of proximate metal ions in facilitating such direct oxidation processes.[167] While conventional detection techniques for carbonyls involve the use of probes such as 2,4-dinitrophenylhydrazine, tritiated sodium borohydride, biotinylated and fluorescent probes using spectrophotometric and fluorescence techniques, and immunoblot or immunoassay methods, they were not able to identify specific carbonylated proteins and the nature of such oxidations, only bulk carbonyl content. The use of 2D electrophoresis followed by immunoblot assay using specific antibodies allowed for the visualization of separated proteins where they are excised and identified using MALDI−MS or ESI−MS. The nongel-based bottom-up proteomic approach uses prior separation with reversed-phase LC−MS−MS and was shown to be indispensable for identifying carbonylated proteins from biological milieu.[168]

In most of the proteomic approaches, carbonyl groups are derivatized through treatment with hydrazine to form the hydrazone adducts prior to MS identification of the modified sites (Fig. 7.8). Hydrazine conjugates such as 2,4-dintiorophenylhydrazine (DNPH) had been extensively used for the derivatization of carbonyl groups but showed nonspecificity due to their reactivity to oxidized thiols[153] and the nonspecificity of secondary antibodies to DNP−anti-DNP complexes that also detect IgG contaminants in tissue; they suffer from nonspecific background noise and contaminating endogenous immunoglobulins.[169] Biotin hydrazides were used as an alternative for DNPH, which can be visualized by chemiluminescence using a streptavidin-dependent detection method[169] or by using an avidin affinity column where the biotinylated oxidized protein can be purified, digested, fractionated, and identified by MALDI−TOF−TOF, MALDI−MS−MS, ESI−MS−MS, ESI− Q-TOF, or ESI−selected reaction monitoring (SRM).[167] An extensive survey of various hydrazine-based carbonyl derivatization agents as well as isotopically labeled hydrazine tags is shown elsewhere.[154]

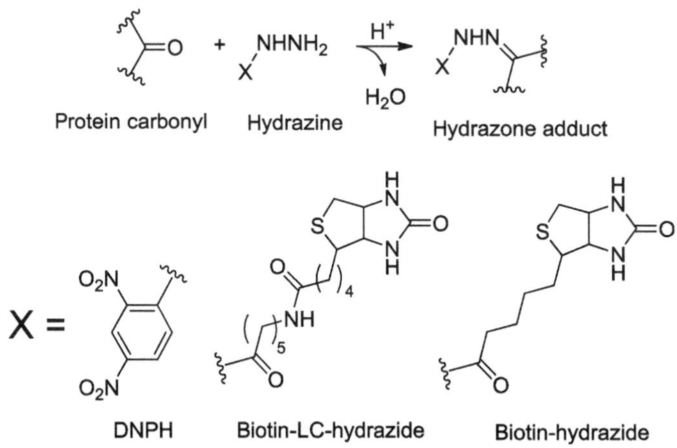

Figure 7.8 Derivatization of protein carbonyl by hydrazine conjugates to form hydrazone adduct.

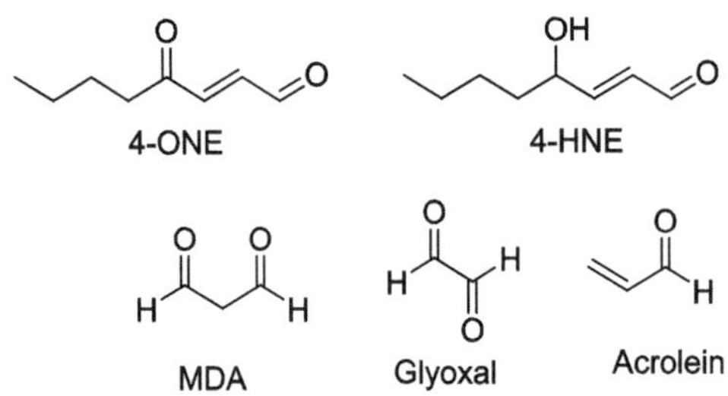

Figure 7.9 Reactive PUFA- or sugar-derived aldehyde end products for the formation of protein carbonyl adducts.

Both the 2D gel and gel-free bottom-up approaches have been used to identify direct protein carbonylation by reactive oxidized polyunsaturated fatty acid (PUFA) or sugar end products in vivo. Among the highly reactive aldehydes produced in biological systems and shown in Fig. 7.9 are 4-hydroxynonenal (HNE), 4-ONE, acrolein, glyoxal, and malonaldialdehyde (MDA). Protein carbonylation can occur via a Schiff base condensation, a Michael addition, or by amidation reactions through the amino acid side chain moieties such as seen with lysine in Fig. 7.10. Proteomic methods employing the use of DNPH assay for 2D gel or stable isotope labeling of peptides from enriched-carbonylated proteins allowed for the identification and quantification of protein carbonylation using LC–MS–MS analysis and sequence database searching.[170]

7.3.1.3 Protein nitration

Protein nitration is mediated mostly by the nitrogen dioxide radical, NO_2. When it is added to aromatic moieties of the amino acid side chain of tyrosine it forms the 3-nitrotyrosine (3-NO_2–Tyr). Nitrogen dioxide is formed via decomposition of

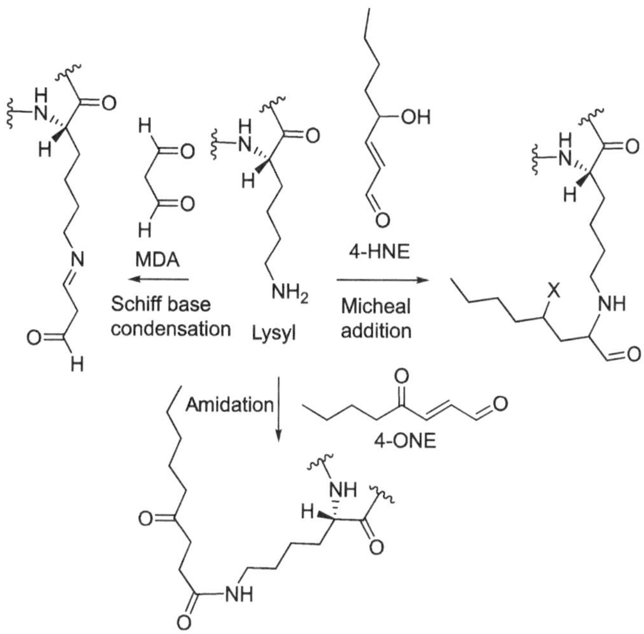

Figure 7.10 Various modes of lysine modification by reactive aldehydes showing Schiff base formation via condensation, Michael addition, and amidation reactions.

peroxynitrite, nitrosoperoxycarbonate, or dinitrogen trioxide or via MPO-mediated oxidation of nitrite. While studies show that under a low steady-state concentration of peroxynitrite, dityrosine formation is more favored than 3-NO_2−Tyr; formation of the latter is more favored at a higher flux of NO−$O_2^{\bullet -}$.[171] Using MnSOD as a protein substrate, a 20-fold increase in the efficiency of tyrosine nitration by peroxynitrite was observed in the presence of CO_2 which suggests that the carbonate radical anion, $CO_3^{\bullet -}$, plays a crucial role in the nitration process and subsequent SOD deactivation.[172,173] This validates previous studies on the role of CO_2 in enhancing peroxynitrite-mediated protein nitration of glutamine synthetase, bovine serum albumin (BSA),[174] or human plasma proteins.[175] Protein nitration in BSA or tissue homogenates was also observed from MPO-mediated and HRP−mediated oxidation of nitrite in the presence of H_2O_2 or HOCl,[176] which demonstrates the relevance of this process in inflammatory disorders involving leukocytes.[177] Although the low levels of protein-bound 3-NO_2−Tyr have been observed in vivo, protein nitration has been used as an indicator of nitrosative stress along with cysteine oxidation in disease conditions. The accessibility of tyrosine within the protein structure is one of the major limitations in protein nitration identification. Several precautionary measures are applied to prevent misidentification of tyrosine nitration such as collecting multiple MS−MS data and confirming the presence of random noise or isobaric co-eluting peptides, as well as manual analysis of the MS—MS data as compared to the use of an automated database search algorithm.[178] Mass spectroscopic techniques such as MALDI—MS have been employed, however, under these conditions: protein-bound 3-NO_2—Tyr is labile and its identification has been problematic, while high-performance liquid chromatograpny and tandem mass spectroscopy (HPLC—MS—MS) and other tandem MS—MS techniques such as CID, electron-capture dissociation, infrared multiphoton dissociation, and metastable atom-activated dissociation have been employed with improved sensitivity.[178]

Free 3-NO_2—Tyr is detected in body fluids such as plasma, bronchoalveolar lining fluid, and urine and represents the turnover of nitrated proteins, which are elevated in most diseases. Commonly employed methods for 3-NO_2—Tyr detection are gas chromatography or LC—MS—MS. Due to the low abundance of 3-NO_2—Tyr, preconcentration of the sample is essential before a complete MS analysis. In urine samples, 3-NO_2—Tyr and its metabolite, 3-nitro-4-hydroxyphenylacetic acid, were preconcentrated using manual solid-phase extraction and subsequent LC—MS—MS showing limits of quantification of <3 pg/mL for both analytes.[179] Using ultraperformance LC—MS—MS, protein-bound oxidation (*o*-Tyr—Phe and *m*-Tyr—Phe), nitration (3-NO_2—Tyr—*p*-Tyr), and chlorination (3Cl—Tyr—*p*-Tyr) biomarkers in plasma and tissue samples were determined from hypoxic newborn piglets. Results are sample specific where 3-NO_2—Tyr—*p*-Tyr ratio was significantly increased compared to normoxic condition but showed no difference in hypoxic compared to normoxic tissue samples using a series of protein extraction and protein hydrolysis prior to MS analysis. The *m*-Tyr—Phe and 3Cl—Tyr—*p*-Tyr ratios were significantly increased in hypoxic liver but not in plasma and brain tissues under the same hypoxic condition.[180] From human astrocytomas, nitroproteins were analyzed using a 2D gel—LC—MS—MS method. Nitroproteins were enriched using 2D gel electrophoresis followed by nitrotyrosine immunoaffinity-based Western blotting, where blots are excised and subjected to LC—ESI—quadrupole TOF—MS—MS analysis to obtain the first nitroproteome map of a human astrocytoma that revealed varying degrees of nitration with 18 nitroproteins and 20 nitrotyrosine sites.[181]

Several factors could lead to the formation of artifactual 3-NO_2—Tyr and protein-bound 3-NO_2—Tyr in biological samples such as during sample acidification, which is typically employed in protein precipitation or protein hydrolysis; proteins are preferably hydrolyzed enzymatically under nonacidic conditions. Also, during protein denaturation, release of 3-NO_2—Tyr from protein could be minimized by gentle blood draws, immediate plasma separation, filtration, and analysis while limiting plasma storage to a single instance. GC—MS exhibits 3-NO_2—Tyr interferences and overestimates 3-NO_2—Tyr in plasma compared to the GC—MS—MS method. It is also worth noting that basal 3-NO_2—Tyr is not detectable using LC—MS—MS but is detectable with GC—MS—MS. It was concluded that GC—MS—MS is the most appropriate technique for the assessment of free and protein-bound 3-NO_2−Tyr, while LC−MS−MS and LC−MS are better suited for the investigation of selective tyrosine nitration in proteins.[182]

7.3.2 Lipids system

Oxidized PUFAs have been one of the most established indicators of oxidative stress. The multiple *cis* double bonds in particular present in PUFA are the most susceptible to propagative peroxidation process that could result in damage to cell membrane. Moreover, the aldehydic by-products generated from lipid peroxidation such as MDA and 4-HNE exhibit reactivity to DNA-forming adducts imparting mutagenicity and carcinogenicity. Analysis of lipid peroxidation and its by-products has been well established by spectrophotometric and fluorometric techniques. However, the use of MS allows for oxidized lipid detection from complex matrix such as body fluids and tissues, which provide more accurate diagnosis of oxidative damage. The level of lipid peroxidation is highly dependent on the extent of radical production, thus resulting in lack of specificity. Moreover, fatty acid distribution depends on the type of tissue and fatty acid structure; the levels of biomarkers of lipid oxidation may not reflect the progression of a particular disease. Hence, it was suggested that multiple markers of lipid peroxidation must be analyzed to get a better representation of their tissue distribution and analyzed for variations in their levels relative to other tissues.[183] Commonly employed MS techniques include prior separation either by GC or LC and detection using soft-ionization methods that allow for minimal degradation of the molecules where ESI−MS or electron capture−atmospheric pressure chemical ionization−mass spectrometry (EC−APCI−MS) are the most commonly employed.[184]

7.3.2.1 Lipid hydroperoxides

Hydroperoxides are unstable intermediates of lipid peroxidation by-products (e.g., lipid hydroxides and isoprostanes) formed from oxidation of PUFAs or cholesterol through either nonradical or radical mechanisms. A nonradical mechanism involves direct addition of singlet oxygen to the unsaturated lipid to form the peroxyl radical adduct.[185] In a biological system, a formed superoxide was shown to produce singlet oxygen[186] and is thermodynamically favorable.[187] In a radical-mediated mechanism, the initial H-atom abstraction from the unsaturated lipid leads to the formation of a pentadienyl radical that on addition of molecular oxygen gives the diene peroxyl radical; further H abstraction yields hydroperoxyl, which leads to propagation of the peroxidation process.[188] Initiation of lipid peroxidation by H-atom abstraction can occur via physical or chemical means such as in the presence of xenobiotics or metal ions or under ultraviolet (UV) light or ionizing radiation. Furthermore, enzymes such as NOXs, XO, NOS, cyt P450, and metC can initiate lipid peroxidation via ROS formation. Other known initiators are lipoxygenase, MPO, and a tyrosyl radical from myoglobin or hemoglobin as well as normal turnover of prostaglandin H2 synthase (PGHS).[184]

In spite of the instability of hydroperoxides, identification of phospholipid hydroperoxides is possible through soft-ionization MS such as ESI via direct infusion of the lipids into the quadrupole linear ion-trap mass spectrometer after prior separation by 2D high-performance thin-layer chromatography using silica gel plates.[189] Coordination ion spray–mass spectrometry (CIS–MS) eliminates the chemical and thermal instability problem of peroxides associated with the use of ESI–MS. This technique uses an Ag ion and was shown to be a powerful technique for the analysis of intact lipid peroxide as Ag complexes giving m/z ratios of $[M + Ag^{107}]^+$ and $[M + Ag^{109}]^+$ as doublets in the mass spectrum. Complexes exhibit a fragmentation pattern that is characteristic of unsaturated lipid hydroperoxides and cyclic peroxides.[190] The CIS–MS could also be coupled with LC for prior separation of a complex sample matrix. Among the lipids detected using this technique were cyclic peroxides and hydroperoxides of cholesterol linoleate and cholesterol arachidonate,[191] phospholipid hydroperoxides,[192] monocyclic peroxides, bicyclic endoperoxides, serial cyclic peroxides, and a novel class of endoperoxides (dioxolane–isoprostane peroxides) from the oxidation of arachidonate.[193]

7.3.2.2 Lipid hydroxides

Shown in Fig. 7.11, the lipid hydroxides such as hydroxyoctadecadienoic acids (HODEs), hydroxyeicosatetraenoic acids (HETEs), and hydroxycholesterols (COHs) were recommended to be reliable biomarkers of lipid peroxidation[183] since their hydroperoxide precursors are unstable. However, lipid hydroxides have also been shown to be metabolic products of prostaglandin H synthases, lipoxygenases, and cytochrome P450 isoforms.[194] Total (E,E)–HODE (E,Z)–HODE and 7-COH were determined by GC–MS from plasma, erythrocytes, and urine from human and animal samples. Tissue homogenization followed by reduction with sodium borohydride and saponification by NaOH converts free and ester forms of hydroperoxide-, ketone- and hydroxide-linoleate and cholesterol to HODE and 7-COH, respectively.[195] However, the use of borohydride or any hydride as a reducing agent was discouraged to avoid ex vivo reduction of metal ions that could induce lipid peroxidation through redox cycling. Therefore, the use of mild reductants such as triphenylphosphine, trimethylphosphite, or thioethers were instead recommended for HODE analysis from tissue and plasma using HPLC–MS–MS.[196] Total plasma HETEs (a mixture of 5(S)-, 12(S)-, 15(S)-, and 20-HETE) was analyzed using GC–MS in Parkinson's disease patients and showed elevated levels compared to controls.[197]

7.3.2.3 Isoprostanes and neuroprostanes

Isoprostanes are stable products from the peroxidation of arachidonic acid via a free radical mechanism.[198] Isoprostanes form from the abstraction of bis-allylic H-atoms to form arachidonoyl radicals. Oxygen addition results in the formation of peroxyl

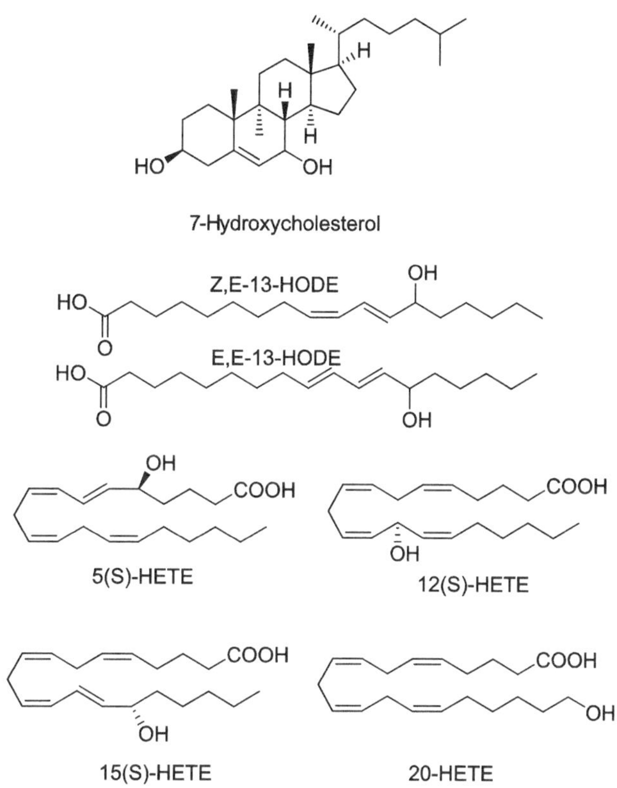

Figure 7.11 Hydroperoxyl-derived lipid hydroxides as reliable markers of lipid peroxidation.

adducts that undergo endocyclization. Through further oxygen addition, this leads to the formation of a prostaglandin G2–like bicyclic endoperoxide hydroperoxide intermediates whose reduction to hydroxyl results in the formation of prostaglandin F2–like compounds, the F-type isoprostanes (F_2-IsoPs). F-type isoprotanes comprise four major regioisomers (5, 8, 12, and 15 series) (Fig. 7.12), each having multiple stereoisomeric forms of 16 diasteriomers giving a total of 64 regioisomers and stereoisomers of F_2-isoPs. The F_2-IsoPs are gold-standard biomarkers of risk factors for coronary heart disease and neurological, renal, liver, lung, and other diseases such as osteoporosis, Down syndrome, Crohns' disease, and scleroderma.[199,200] In contrast to arachidonic acid, docosahexaenoic acid (DHA) is the most abundant PUFA in the brain. Iso–P-like compounds formed from DHA known as neuroprostanes (NPs) were shown to provide unique markers of oxidative injury to brain tissue. NPs were identified in both in vitro and in vivo systems and have eight possible regioisomeric groups (4, 7, 10, 11, 13, 14, 17, and 20 series) with the 4 and 20 series (F_4-NPs and F_{20}-NPs) (Fig. 7.13) being present in higher levels compared to other series.[201,202]

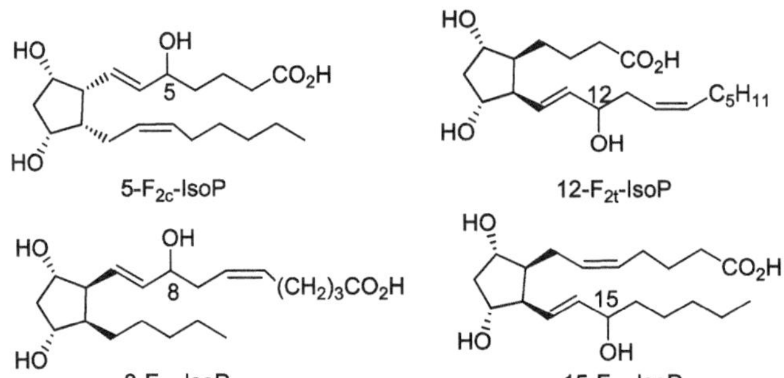

Figure 7.12 Four major regioisomers of F-type isoprostanes.

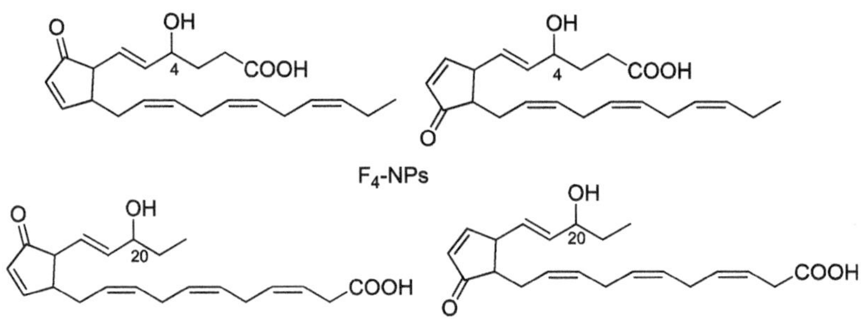

Figure 7.13 Representative NPs of the 4 and 20 series.

Current protocol involves basic extraction from a tissue sample and purification through solid-phase extraction. The isolated F_2-IsoPs were then esterified via the carboxylic acid with pentafluorobenzyl (PFB) moiety for increased MS sensitivity, while the hydroxyl group is protected by trimethylsilyl (TMS). By using an internal standard, molecular ions can be monitored at $m/z = 569$ (as M-181 from loss of PFB) (Fig. 7.14) using a GC–negative-ion chemical ionization (NICI)–MS technique.[203] NPs, along with F_2-IsoPs, were analyzed using a GC–NICI–MS technique[204] from either cerebrospinal fluid (CSF) or brain tissue and showed elevated levels in Alzheimer's disease (AD),[205,206] Rett syndrome,[207] traumatic brain injury,[208] the mouse model of Smith–Lemli–Opitz syndrome,[209] asphyxia and reoxygenation of the brain,[210] aneurysmal subarachnoid hemorrhages,[211] ethanol withdrawal,[212] but showed lower F_4-NP levels in vitamin E–deficient mouse models[213].

While IsoPs and F_4-NPs have been shown to provide accurate assessment of AD progression in brain and CSF, results in peripheral fluids such as blood and urine have been conflicting and reflect no correlation with AD patients, which indicates that body fluids outside the central nervous system are not appropriate samples for investigating AD status.[214] An improper analytical procedure could produce

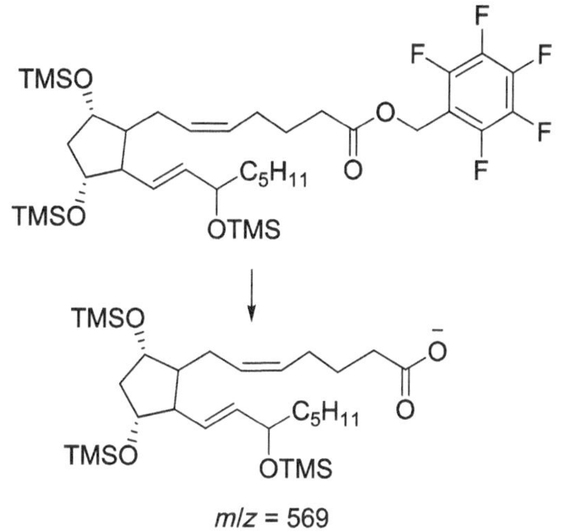

Figure 7.14 Pentafluorobenzyl-esterified and TMS-protected F_2-isoP yield a distinct MS molecular ion with $m/z = 569$ atomic mass units.

interference for the simultaneous analysis of F_2-IsoPs and F_4-NPs from CSF using GC–NICI–MS.[215] By using a sample-preparation procedure for F_4-NPs could cause poor chromatographic separation from CSF samples giving erroneously high F_2-IsoPs values. Possible interference of unknown substances during F_4-NPs analysis was also reported from CSF and from plasma during F_2-IsoPs analysis. Recommendations related to sample-processing procedures, instrument components, and settings were proposed for the analysis of isoprotanes and NPs using GC–NICI–MS.[216]

LC–MS was also employed for the analysis of F-isoPs and offers advantages over GC–MS such that no derivatization is necessary, hence simplifying sample preparation. However, LC–MS suffers from decreased sensitivity compared to GC–MS,[217] but some studies showed greater precision with lower coefficient of variations when LC–MS–MS is used for the analysis of prostaglandin F (PGFs) in plasma.[218] Analysis of F_2-isoPs from urine is made complicated by the possible F_2-isoP production from the kidney, which suggests that 2,3-dinor-5,6-dihydro-15–F_{2t}-IsoP is the primary urinary metabolite of 15-F_{2t}-IsoP[203,219] and was therefore suggested as a more appropriate marker of oxidative stress in urine samples. NPs isolated from rat brains or rat livers were analyzed using EC–APCI–MS with prior derivatizations as in Iso-Ps showing 4 and 20 series NP regioisomers as predominant products.[201]

7.3.2.4 Electrophilic aldehydes

PUFA-derived reactive aldehydes form advanced lipoxygenation end products (ALEs) and are analogous to AGEs, which are sugar-derived late-stage oxidation products. ALEs are collectively referred to as *adducts*. Crosslinks formed from the reaction of

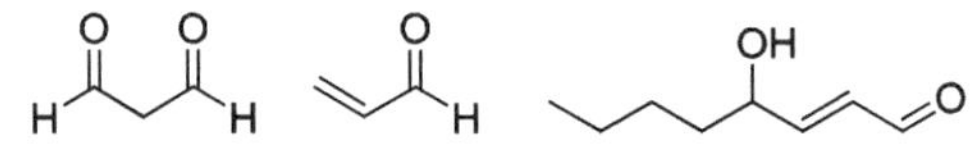

Figure 7.15 Reactive PUFA-derived dialdehyde (MDA) and α,β-unsaturated aldehyde (acrolein and 4-HNE).

proteins and aldehydes are the most widely investigated biomarkers. Among the most abundant reactive aldehydes are 4-HNE, malondialdehyde, and acrolein (Fig. 7.15). Thiobarbituric acid is commonly employed as a chromogenic agent for the spectrophotometric and fluorometric detection of protein-unbound MDA. However, MDA is not a reliable direct index of lipid peroxidation due to the fact that MDA is not an exclusive product of lipid peroxidation; it is also produced from DNA's 2-deoxyribose oxidation via stand breakage and only a few specific lipid peroxidation reactions lead to the formation of MDA. Therefore, analysis of MDA requires caution in interpreting its levels in biological systems. Spectroscopic analysis of thiobarbituric acid reactive substance lacks specificity even with prior HPLC separation. Using MDA-d_2 as an internal standard, GC–MS or GC–MS–MS methods produced more accurate and precise analysis of MDA with a high sensitivity that is free of interference from human plasma with limit of detection (LOD) of 2 amol (2×10^{-18} mol).[220] Analysis of protein-unbound HNE has always been a challenge due to its high reactivity and instability. HNE can be derived from pentafluorobenzylhydroxylamine-HCl followed by trimethylsilylation to TMS ethers. HNE-d_{11} as an internal standard showed high calibration accuracy for HNE when a a GC–MS technique was used. The methodology employed was further applied to HNE detection in ethylenediaminetetraacetate plasma from migraine patients showing elevated HNE levels compared to a control.[221] Using GC–MS, a comparison of 4-HNE, protein carbonyl, and MDA plasma levels in CCl_4-challenged rats showed that 4-HNE was the most sensitive and viable toxicological biomarker.[222]

7.3.3 Nucleotides

Due to the poor accessibility of DNA to oxidants, low amounts of oxidized DNA are formed, so the percent oxidation of DNA compared to protein oxidation is significantly lower, which makes detection of DNA lesions challenging. One of the most extensively studied phenomena of DNA oxidation is the ability of resulting radical cations to efficiently and reversibly migrate across the duplex DNA until they oxidize a nucleobase with structural, kinetic, and thermodynamic preference for oxidation.[223] DNA lesions typically occur on the guanine residues due to their high oxidation potential to form the most commonly studied product, 8-hydroxy-2′-deoxyguanosine (8-OHdG), or its tautomer, 8-oxo-7,8-dihydro-2′-deoxyguanosine (8-oxodG) (Fig. 7.16), but oxidation has also been observed on the cytosine, thymine, and adenine bases.

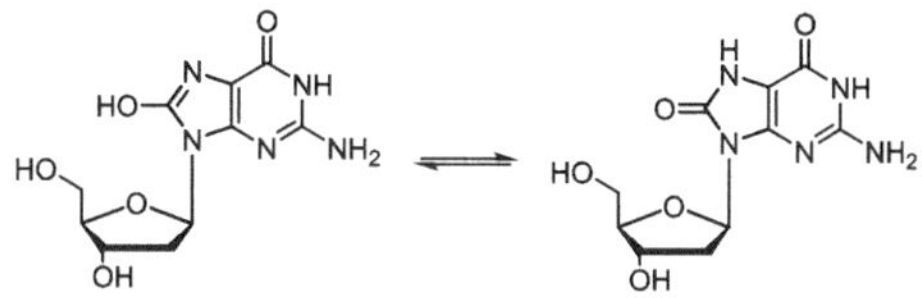

Figure 7.16 Gold-standard biomarker of DNA oxidation (8-OHdG) and its tautomeric form, 8-oxodG.

The degree of guanine oxidation was also found to be dependent on the DNA sequence—i.e., (8-oxodG = O): COA ≈ AOG > GOG ≥ COT > TOC > AOC as determined using the Q-TOF mass spectrometric method.[224] Moreover, detection of DNA lesions is complicated by the presence of complex patterns in DNA lesions as well as the presence of artifactual products. For example, the most commonly measured 8-oxo-dG requires derivatization prior to GC−MS analysis for increased volatility and thermal stability, although this methodology results in the formation of artifactual signals and could be problematic.[225] While prepurification of the 8-oxodG using HPLC can be carried out prior to GC−MS analysis for reduced artifactual signal formation,[226] the use of the LC−MS−MS method offers a milder condition for lesion analysis since it avoids derivatization and prevents potential hydrolysis as well as thermal decomposition during GC analysis.[227,228] Artifactual generation of 8-oxo-dG from cellular DNA can be minimized using an isolation method that employs the commercially available DNAzol instead of the chaotropic- and phenol-based extraction method along with the use of chelex-treated water and desferrioxamine to prevent Fenton-mediated 8-oxo-dG formation.[229] However, the addition of an iron-chelating agent such as desferrioxamine does not completely protect analytes from further oxidation during workups involving DNA hydrolysis and purification. The use of an online solid-phase extraction method removes excess dG and directly introduces the samples for ESI−MS analysis, thus allowing for the simultaneous determination of 8-oxodG and dG, which demonstrates an efficient way of preventing artifactual 8-oxodG formation.[230]

Analysis of DNA base lesions such as levels of 8-oxo-dG and other oxidized bases from tissue or cells reflects steady-state damage arising from the dynamic equilibrium between the rates of damage and repair, while 8-oxo-dG analysis from urine samples assesses the rate of whole-body DNA damage. Its interpretation should be taken with caution, however, and a thorough study that includes both steady state and total DNA damage has been suggested.[225,227,231,232] The steady-state levels of 8-OHdG and the 8-oxo-Gua could also arise from dietary influence or cell death.[233] For example, 8-OHdG and the 8-oxo-Gua were analyzed from urine samples using an automated online in-tube solid-phase microextraction to preconcentrate the samples; this was followed by stable isotope dilution LC to assess the effect of smoking, green tea, and alcohol consumption on 8-OHdG levels as a measure of total oxidized DNA level.[234] A screening method using ultraperformance LC–MS–MS was developed to validate levels of oxidative stress from urine samples of extremely low-birth-weight infants using 8-OH-dG in relation to levels of oxidized protein and lipid as biomarkers.[235] Saliva is a convenient sample source for the measurement of exocyclic DNA adducts and lipid peroxidation–mediated etheno-adducts as biomarkers for cancer. Using a highly sensitive and specific assay based on isotope dilution nanoflow LC–nanospray ionization tandem mass spectrometry (nanoLC–NSI–MS–MS), simultaneous detection and quantification of various DNA lesions in human salivary DNA was demonstrated.[236] Also, in both CSF and urine of mixed Alzheimer's and dementia patients, levels of 8-OHdG were identified using GC–MS and with prior HPLC prepurification.[226]

In a typical experiment, a DNA sample is isolated from cells or tissues and then purified and hydrolyzed using nucleases to generate the free nucleotides. This is followed by dephosphorylation using alkaline phosphatase to yield the nucleoside. Internal standardization of analyte samples employs isotope dilution using isotopically labeled surrogate standards of the analytes, which thus allow quantification of the oxidized bases. In urine samples, base lesion levels are expressed as ng/mg creatine, so the measurement of creatine levels is preformed[237,238] or is the ratio of the number of 8-oxodG to the number of dG, hence simultaneous analysis of 8-oxodG and dG is necessary in this case.[230,239] Various base lesions have been analyzed in various cell types (fibroblast cells, hepatocytes), biological fluids (plasma, urine, saliva, CSF), and disease states (cognitive disorders, cancer, environmental exposures) as products of the oxidation of thymine, cytosine, guanine, adenine, and 2-deoxyribose from various one-electron oxidants.[240,241] Shown in Fig. 7.17 are the various DNA lesion identified in vivo and resulting from ROS–RNS-mediated oxidation, nitrosative deamination, lipid oxidation end product formation, and DNA fragmentation. It has been shown that lipid peroxidation dominates the chemistry of DNA adduct formation in the mouse model of inflammation.[242] Using nanoflow LC–nanoESI–MS–MS coupled with isotope dilution method, lipid peroxidation derivative products such as the etheno-DNA adducts, 1,*N*(2)-etheno-2′-deoxyguanosine (εdG), 1,*N*(6)-etheno-2′-deoxyadenosine (εdA), and 3,*N*(4)-etheno-2′-deoxycytidine (εdC) have been identified.[243,244] Nitrosative deaminated bases such as 2′-deoxyuridine (dU), 2′-deoxyxanthosine (dX), and 2′-deoxyinosine (dI) have also been detected in mouse models of inflammation.[244,245] Trans-4-HNE-derived cyclic DNA adduct as HNE-dG was identified from human brain tissue by employing a prior separation using capillary reversed-phase chromatography; it was quantified using a high-capacity ion-trap mass spectrometer in the MS–MS mode via a nonospray interface.[246] The DNA lesion, 5-chloro-deoxycytosine (5-Cl-dC), is a HOCl-mediated biomarker of chronic inflammation during neutrophil activation and has mutagenic properties whose identification from purified oligonucleotide was made possible using HPLC–ESI–Q-TOF

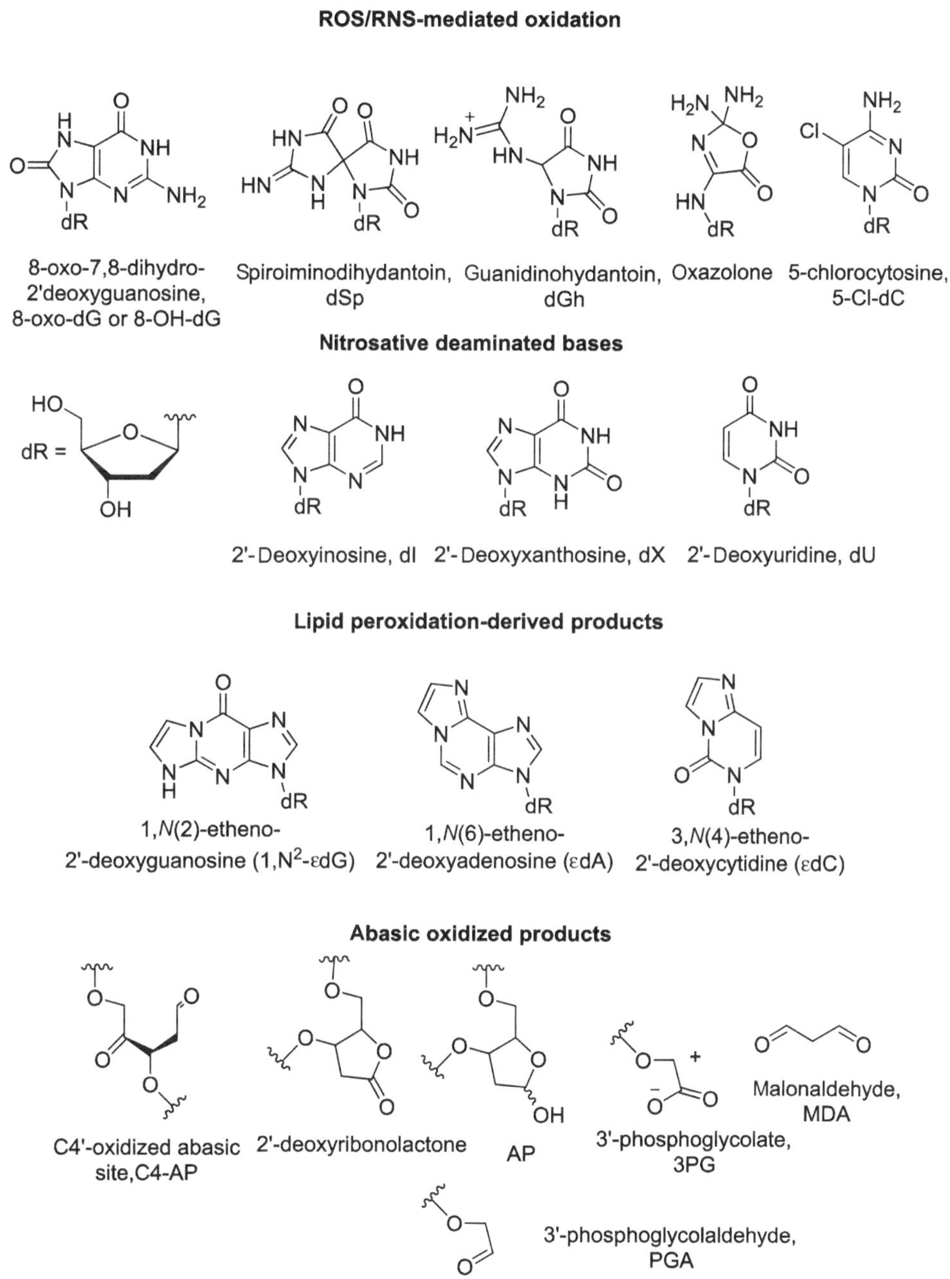

Figure 7.17 Markers of DNA damage from various oxidative mechanisms.

mass spectrometry.[247] Four-electron oxidation of 2′-deoxyguanine (dG) yields 5-guanidinohydantoin (dGh) as the major product and spiroiminodihydantoin (dSP) as the minor one via formation of 8-oxo-dG at netural pH.[248] This hyperoxidized dG was identified from oligodeoxynucleotides through its reaction with Ir^{4+} using HPLC purification followed by MALDI–TOF mass analysis; it is a putative potential mutagen when transfected in an in vitro system.[249] Also, both dGH and dSp along with oxazolone were identified from peroxynitrite-challenged calf thymus DNA as analyzed by isotope-labeling HPLC–ESI–MS–MS, which revealed artifactual Gh and Sp of <3.7% of the measured 8-oxodG with no artifactual oxazolone observed.[250] The formation of Sp lesion was also identified using an LC–ESI–MS method and was demonstrated to destabilize DNA helix in its vicinity and affect nucleosome stability and positioning.[251]

Oxidative damage to DNA sugars was also measured as a biomarker of tumor formation from bleomycin and ionizing radiation. In particular, the C4′-oxidized abasic site (C4-AP) (Fig. 7.17) was shown to be formed from Fe-bleomycin–induced DNA damage and was detected at the cytosine and thymine sites using direct MS analysis. Using an LC–ESI–MS-based approach, C4-AP can be detected along with other C4′-oxidation end products such as 3′-phosphoglycolate, thus allowing to map C4-AP in the presence of other oxidized apurinic sites in intact DNA fragments that do not require hydrolysis of DNA to nucleosides.[252] GC–MS was employed to analyze 2-deoxyribose oxidation products such as C4-AP, 3'-phosphoglycolate (3PG), MDA, free base, and base propenal from γ-irradiation and bleomycin-induced DNA damage. Abasic sites were converted to the 3′-phosphoro-3-pyridazinylmethylate derivative by treatment of the damaged DNA with hydrazine and are released as 3-hydroxymethylpyridazine (HMP) by enzymatic hydrolysis. Using HPLC prepurification, 3PG and HMP were silylated and analyzed by GC–MS.[253] Quantification of 3′phophoglycolaldehyde was also achieved through its derivatization as a stable oxime with pentafluorobenzylhydroxylamine followed by solvent extraction and GC–NCI–MS analysis.[254]

Chemical derivatization was found to increase sensitivity for the detection of base lesions. Although derivatization has been discouraged because of guanine's increased susceptibility to further oxidation, this technique has a great advantage for identifying the oxidized forms of the important epigenetic marker 5-methylcytosine or 5-mC—namely, 5-hydroxymethylcytosine (5-hmC), 5-formylcytosine (5-foC), and 5-carboxylcytosine (5-caC)—due to their very low concentrations, which makes their simultaneous determination a major limitation (Fig. 7.18). Using 2-bromo-1-(4-dimethylamino-phenyl)-ethanone coupled with the LC–ESI–MS–MS technique, simultaneous determination of these cytosine modifications in genomic DNA is possible with limits of detection ranging from 0.06 to 0.2 fmol, thus allowing for their quantification in human colorectal carcinoma and normal tissue.[255] In urine

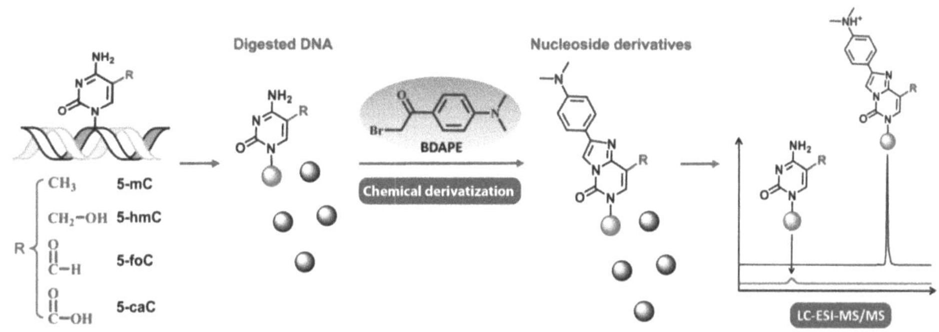

Figure 7.18 Determination of 5-mC and its oxidation products from DNA by chemical derivatization coupled with LC–ESI–MS–MS analysis. Source: *Reprinted under a Creative Commons Attribution License from Tang, J. Zhu, X. Niu, X. Liu, T. Zhao, H. Lan, M. An amperometric superoxide anion radical biosensor based on SOD/PtPd-PDARGO modified electrode Talanta 2015;* ***137****: 18. Copyright © 2015 American Chemical Society.*

samples using an online solid phase extraction (SPE) LC–MS–MS method, 5-mC and 5-deoxymC were identified at the nanolevels and showed positive correlation with other DNA lesions using the same method, which suggests a link between genetic DNA lesions and epigenetic DNA methylation.[237]

7.4 IMMUNOCHEMICAL TECHNIQUE

As its name implies, the immunochemical technique is based on a fundamental mechanism underlying the immune system—i.e., recognition of the epitope by an antibody. In this case, the oxidatively modified part of an antigen is the epitope, and several antibodies had been developed to recognize such oxidative lesions in DNA, lipids, and proteins as visualized or detected using Western blot, immunohistochemical, or ELISA detection methods, depending on the type of sample used. For example, ELISA, which is based on enzymatic color reaction resulting from the formation of an antibody–antigen complex, is best suited for all types of samples such as cells, urine, serum, and tissue. The Western blot is typically employed to detect modified proteins in cell or tissue samples, and the immunohistochemical technique is mainly used to visualize oxidative lesions on tissue samples. Commercially available antibodies are either monoclonal or polyclonal; each has its own advantages and disadvantages in terms of cost, specificity, and variability. Monoclonal antibodies have higher specificity and sensitivity compared to the polyclonal analogs and have minimal cross-reactivity to other analogous lesions. Although monoclonal antibodies are more expensive than polyclonal ones, they give high reproducibility, which only allows detection of specific epitopes on the antigen and results in less background stain, making them ideal as primary antibodies and in affinity purification. Three major strategies have been used to detect oxidation lesions in biomolecular systems: (1) direct recognition of the late-stage oxidative lesion by the primary antibody, (2) recognition of the chemically derivatized late-stage oxidative lesion, and (3) recognition of the early-stage protein or DNA radical formation as adducts of 5,5-dimethyl-pyrroline-*N*-oxide (DMPO) using anti-DMPO primary antibody. The use of antibodies can be used to purify antigens using affinity column chromatography or immune precipitation for their later identification using LC–MS–MS analysis as well as to visualize using confocal microscopy, MRI imaging, immunohistochemistry, autoradiography, or immunoblot and to quantify using direct colorimetric, chemiluminescent, or fluorescent detection with ELISA, flow cytometry, or spectrophotometric and fluorometric assays.

7.4.1 DNA damage antibodies

Detection of 8-oxo-7,8-dihydro-29-deoxyguanosine (8-oxodG or 8-OHdG) (Fig. 7.16) is the most studied oxidized nucleoside due to its abundance, and it has become one of the main biomarkers of disease development. Using ELISA and immunohistochemical methods, 8-oxodG has been detected in various cell lines in human and animal tissues.[256] A commercially available anti-8-OHdG monoclonal antibody (mAb) was shown to have no cross-reactivity against 8-OHdG analogs such as guanosine(G),7-methyl-G, 6-SH-G, 8-bromo-G, dA, dC, dT, dI, dU, dG, O6-methyl-dG,8-OHdA, guanine(Gua),O6-methyl-Gua, 8-OH-Gua, uric acid, urea, creatine, creatinine, 8-sulfhydryl-G, and 8-OH-G. The principal advantages of the immunochemical technique over the other techniques such as HPLC–EC, GC–MS, and ^{32}P-postlabeling methods are, of course, its high sensitivity, which allows for the measurement of 8-OHdG lesion levels in the ng/mL range, its ability to visualize DNA lesion in intact tissue, and its ability to prevent the formation of artifactual 8-OHdG that may arise during sample pretreatment.[257] The application of anti-OHdG is not only limited to cells and tissues but also can be applied to body fluids. For example, the effect of breast-feeding vs formula feeding on the urinary 8-OHdG excretion was investigated using a competitive ELISA kit and levels expressed as 8-OHdG–creatinine ratio.[258]

The epitope for the nitrosative DNA lesion is in the form of 8-nitroguanine (8-nitrodG) (Fig. 7.19) and was shown to be a potential biomarker for the evaluation of risk for inflammation-mediated carcinogenesis. A highly specific and sensitive anti-8-nitroG antibody was developed and applied to detect 8-nitroG in cervical biopsies, and the 8-oxodG was covisualized using the anti-8-oxodG Ab.[259] Cross-reactivity of the anti-8-nitroG polyclonal antibody with normal nucleotide bases is also negligible for bromo, chloro, and hydroxyl bases and only recognizes 8-nitroguanine and 8-nitroguanosine. In hamsters infected with *Opisthorchis viverrini*, 8-nitroG formation was implicated in the pathogenesis of cholangiocarcinoma and was found to be mostly localized in the cytoplasm and slightly in the nucleus of inflammatory cells. Using dual immunofluorescence staining with anti-8-oxodG Ab showed the presence of

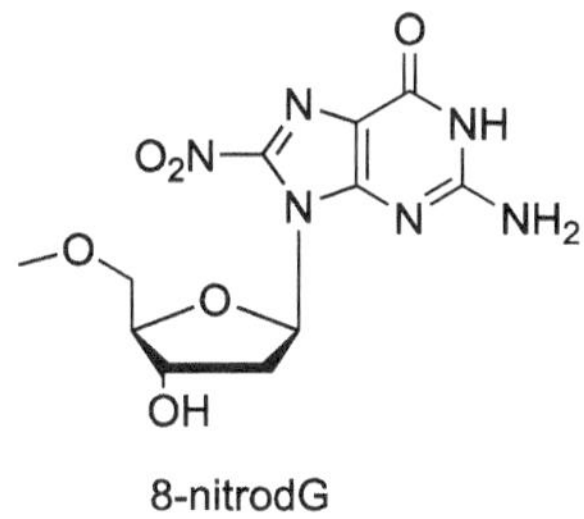

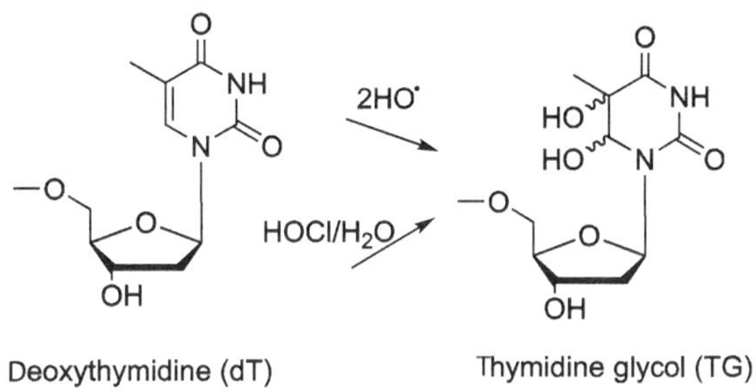

Figure 7.19 Other DNA oxidation lesion epitopes, 8-nitrodG, and TG.

both 8-nitroG and 8-oxodG in epithelium of bile ducts.[260,261] In urine samples of cigarette smokers, 8-nitroG was purified and quantitated using an HPLC–EC technique.[262] The most common thymidine oxidation lesion is the thymidineglycol as shown in Fig. 7.19, which is formed via direct oxidation of deoxythymidine by a hydroxyl radical or through the thymidine intermediate formation via HOCl–MPO–mediated oxidation. Anti–thymidine glycol (anti-TG) mAb was developed and, through ELISA, found application in the detection of TG in the fmol levels from DNA lesions generated after DNA irradiation or OsO_4-treatment.[263]

7.4.2 Protein oxidation antibodies

Carbonyls are the most widely studied protein oxidation products. Unlike the other oxidized protein lesions, however, the mode of carbonyl lesion analysis involves its derivatization to 2,4 dinitrophenylhydrazone (DNP-hydrazone) through its reaction with DNPH. A primary anti-DNP antibody can be detected using an immunoblotting technique by incubating with HRP-conjugated secondary antibody, which allows for subsequent quantification using chemiluminescence through luminol and H_2O_2. Also, using the secondary peroxidase-conjugated rat antimouse immunoglobulin E antibody, *o*-phenylenediamine, and H_2O_2, spectrophotometric analysis using ELISA could also be carried out. Individually oxidized proteins can be separated and identified using sodium dodecyl sulfate–polyacrylamide gel electrophoresis (SDS–PAGE) followed by Western blotting. Several commercially available anti-DNP assay kits were tested for their robustness, and it was recommended that analysis of protein concentrations should be performed with caution by including BSA as an internal control for every analysis. For Western blot and MS analysis, the risk of losing oxidized proteins in highly oxidized samples was observed and an increase in protein oxidation was evident using peptidases prior to MS analysis. Therefore, ELISA was found to be the best method for protein carbonyl quantification, while oxyblotting, Western blot, and MS could provide insights into the molecular mass of the oxidized protein and sites of carbonylation (for MS) but are less quantitative compared to ELISA.[264]

Shown in Fig. 7.20 are the various oxidized tyrosine and tryptophan epitopes whose recognition by antibodies had been demonstrated. Antibodies for nitrative and oxidative PTM of tyrosine side chains were developed and are available commercially, including anti-3-nitrotyrosine (anti-3NT), anti-dibromo tyrosine (anti-DiBrY), and anti-dityrosine (anti-DiY). The formation of 3-NT from intact human HepG2 and SW1353 cells or cell lysates exposed to peroxynitrite was investigated using a variety of techniques, including Western blotting, ELISA, flow cytometry, and confocal microscopy.[265] Slot-blot analysis of 3-NT-modified brain proteins was described,[266] and immunohistochemical staining of 2-NT in ischemic cerebrovasculature showed modest 3-NT staining compared to sham.[267] Using an iNOS antibody, iNOS expression can be visualized along with 3-NT expression and was demonstrated in a mouse brain under subarachnoidal haemorrage[268] and in the heart under severe septic shock.[269] Another nitration product, 6-nitrotryptophan, is formed from a peroxynitrite reaction with tryptophan protein residue and was found to be formed in human Cu,Zn SOD. Anti-6-nitrotryptophan antiserum showed no reactivity to peroxynitrite-untreated Cu,Zn-SOD and MnSOD, 3-NT residue containing Mn-SOD or 6-aminotryptophan containing Cu,Zn-SOD. Using Western blot for 2D gel analysis and subsequent LC–ESI–MS–MS analysis of the immunostained spots, various 6-nitrotryptophan-modified proteins were identified in PC12 cells treated with peroxynitrite [270–272] or the skin of atopic dermatitis model mice.[273]

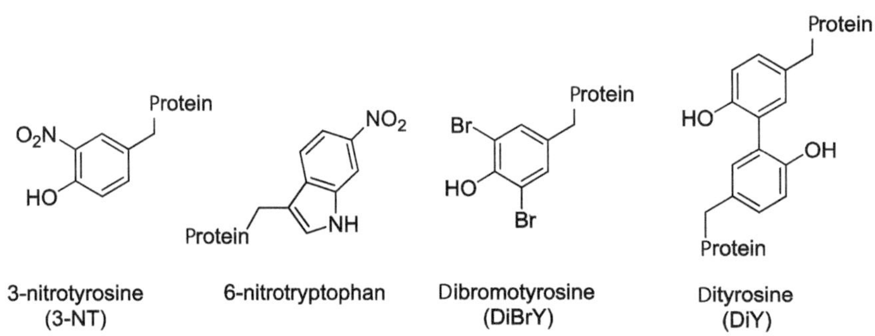

Figure 7.20 Common tyrosine and tryptophan oxidation epitopes.

Another biomarker of protein oxidation is the ditryosine (DiY) formed from a tyrosyl radical intermediate. Using the immunohistochemical technique, positive staining for DiY in atherosclerotic lesions of apolipoprotein E (Apo E)−deficient mice was observed[274] as well as lens proteins oxidized by a H_2O_2−Cu system, which was suppressed by the addition of reducing agents such as glutathione (GSH) and ascorbic acid.[275] Dihalogenated tyrosine is formed from eisonophils using peroxidases such as MPO and eosinophil peroxidase that catalyze the formation of halogenating species and hypohalous acids (e.g., hypochlorous acid, HOCl, or hypobromous acid, HOBr) from H_2O_2 and Br^-−Cl^-. Reaction of these hypohalous acids on proteins bearing tyrosine forms 3,5-dichlorotyrosine or 3,5-dibromotyrosine (DiBrY) or both. Monoclonal antibodies were raised from mice immunized with DiBrY−keyhole limpet hemocyanin (KLH). Recognition by the antibody showed specificity to DiBrY−BSA using ELISA compared to N-bromosuccinimide−BSA, BrY−BSA, HOCl−BSA, nitrated BSA, or DiY−BSA.[276] MPO activity has been used as index of neutrophil infiltration and in indomethacin-mediated intestinal mucosa injuries. Dihalogenated proteins were separated using 2D−PAGE and subsequently identified by MALDI−TOF to show increased levels of DiBrY in the presence of indomethacin with enolase-1 and albumin as the DiBrY-modified proteins.[277] Eosinophil activity has been implicated in airway inflammation, so the formation of halogenated protein can serve as biomarkers for asthma. Through the use of ELISA, the BTK-94C monoclonal antibody was utilized and designated as a general halotyrosine antibody with selectivity toward free and protein-bound mono- and DiBrY and, to a lesser degree, chlorotyrosine.[278]

7.4.3 Lipid peroxidation antibodies

Commercially available antibodies for the investigation of lipid peroxidation are mostly designed to recognize early and late-stage lipid peroxidation products such as 4-HNE, acrolein, malonaldehyde, 4-hydroxy-2-hexenal, methylglyoxal, crotonaldehyde, and hexanoic acid as well as the oxidation product of cholesterol, 7-ketocholestreol. In all of these, the epitopes are detected predominantly as protein adducts formed via Michael or nucleophilic addition to the lysine amino group, cysteine thiol, or histidine imidazole-N. Shown in Figs. 7.21−7.25 are the epitopes formed from lysine. The adducts formed from late-stage peroxidation products are traditionally detected using monoclonal antibodies specific to carbonyl-derived protein modifications. These antibodies could detect both AGE-derived and advanced lipoxidation end product (ALE)−derived carbonyl adducts, but several antibodies specific to lipid peroxidation products are able to differentiate between sugar-derived and lipid-derived carbonyl products.[279] In fact, the anti-4-HNE-histidine antibody was found to be more effective for in vitro detection of 4-HNE−modified apomyoglobin and

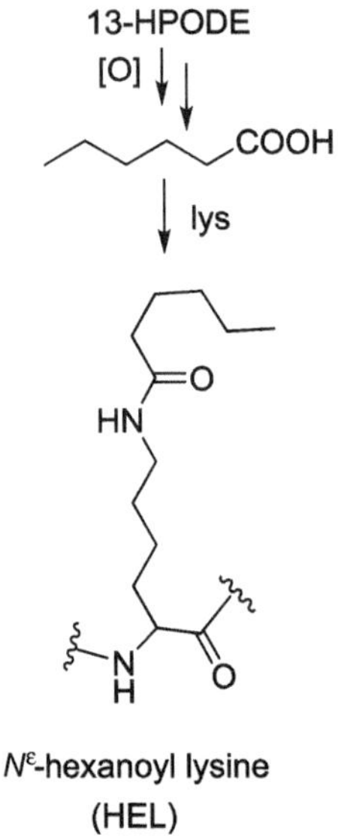

Figure 7.21 N^{ε}-HEL adduct as a novel biomarker of an early-stage lipid peroxidation end product.

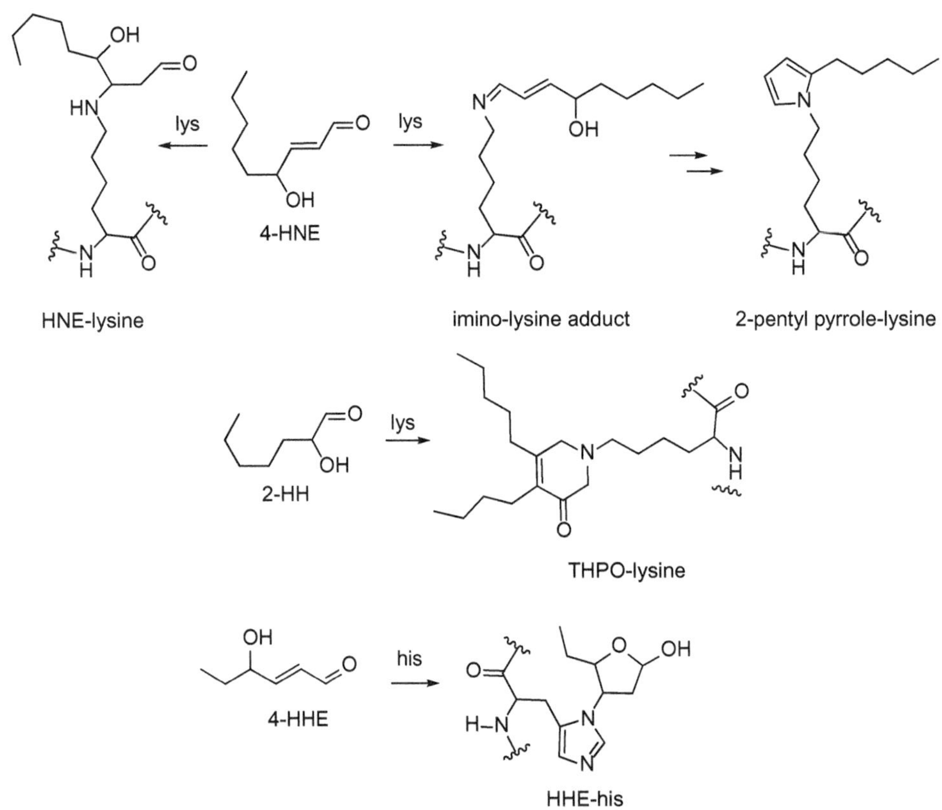

Figure 7.22 Common epitopes for late-stage advanced lipid peroxidation end products from 4-HNE, 2-HH, and 4-HHE.

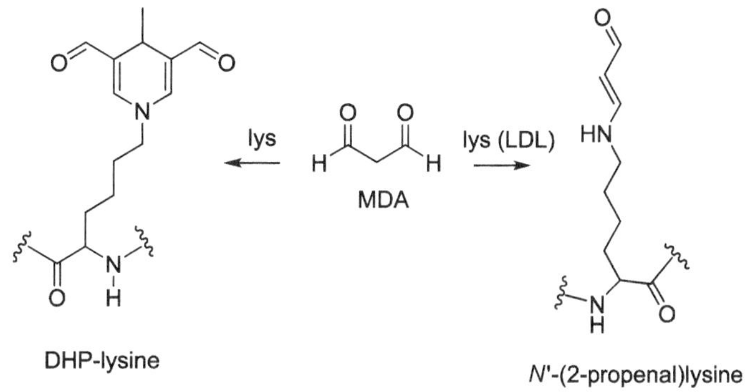

Figure 7.23 MDA-derived lysine adducts forming the N^{ε}-(2-propenal)lysine and DHP–lysine epitopes.

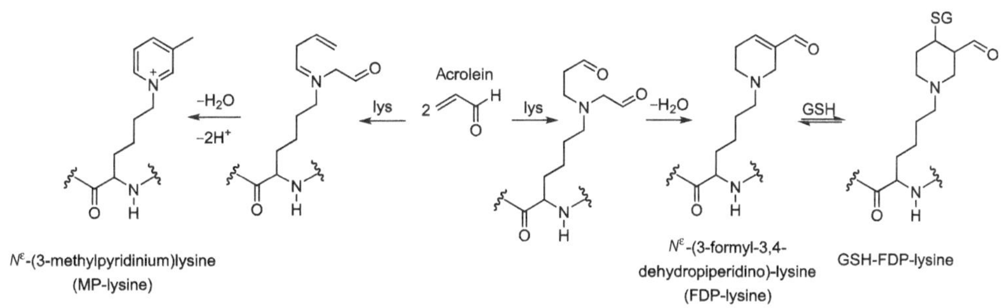

Figure 7.24 Acrolein-derived lysine adducts forming the MP–lysine and FDP–lysine epitopes and glutathiolation of the latter GSH–FDP–lysine.

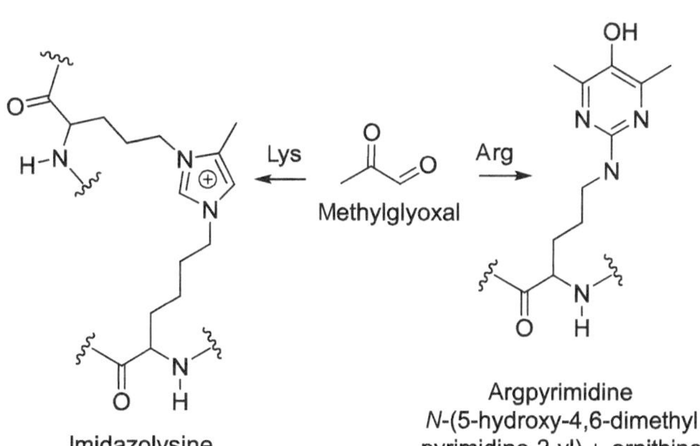

Figure 7.25 MG-derived lysine and arginine adducts forming the imidazolysine and argpyrimidine epitopes, respectively.

cytosolic proteins than direct carbonylation using biotin-coupled hydrazide or dinitrophenylhydrazine derivatization of carbonyls.[280]

The N^{ε}-hexanoyl–lysine (HEL) adduct is a novel biomarker of an early-stage lipid peroxidation product of arachidonic acid or linoleic acid (Fig. 7.21). Anti-HEL was developed and proved useful for the estimation of amide-type adducts in biological samples.[281] Using competitive ELISA, HEL can be assayed from 2 to 700 nmol/L in human urine, serum, and isolated low-density lipoprotein (LDL), and has been shown to be a good biomarker for the assessment of omega-6 PUFA oxidation, exhibiting high reproducibility and linearity.[282] Levels of HEL in bone and urine were examined after steroid administration in the rabbit model of osteonecrosis using immunohistochemical and ELISA techniques, respectively.[283] In neoplastic transformed gastric mucosal cells, HEL-modified proteins were identified using SDS–PAGE or 2D gel electrophoresis and Western blotting followed by MS.[284]

Late-stage peroxidation products such as 4-HNE, 2-hydroxyheptanal (4-HH), 4-hydroxyhexenal (4-HHE), MDA, acrolein, and methyl glyoxal (also from glycolysis) form detectable epitopes with amino acids such as Lys, Arg or His (Figs. 7.22–7.25). Protocols were described on the use of anti-4-HNE–lysine for Western blot analysis, ELISA, and immunohistochemical methods from a variety of biological samples (Fig. 7.22).[285,286] The antisera against HNE–lysine-rich protein haptens such as 4-HNE–LDL and 4-HNE–KLH, showed specificity toward aldehyde–protein adducts formed via *a* Michael adduction reaction.[287] A Schiff base condensation reaction of 4-HNE with lysine forms the imino products and subsequent cyclization form 2-pentyl pyrrole–lysine adduct (HNE–pyrrole) (Fig. 7.22), which is considered as an ALE product marker whose immunogen was prepared from 2-pentylpyrrole and KLH protein hapten conjugate.[288] It has been shown that radical-mediated oxidation of LDL proteins mainly produces the 2-pentyl pyrrole epitopes via 4-HNE formation as detected using ELISA and HNE–pyrrole–KLH immunogen. HNE–pyrrole epitopes were detected in human plasma and found to be elevated in renal failure and atherosclerotic patients compared to healthy subjects. Although HNE–pyrrole epitopes were detected in atherosclerotic plaque at high levels in extracellular connective tissues, they are not associated with ApoB–LDL proteins in plasma.[289] Other 4-HNE–related lipid peroxidation products are 2-hydroxyheptanal (2-HH) and 4-hydroxyhexenal (4-HHE). Antibody 3C8 (mAb3C8) for lipid peroxidation products such as 2-HH was developed from reaction of 2-HH with KLH and the epitope identified as a trihydropyridinone (THPO) structure (Fig. 7.22), 1-alkyl-4-butyl-5-pentyl-1,2,6-trihydropyridin-3-one, which forms from the reaction of 2-HH with the lysine residue. This Maillard reaction–like lysine adduct (3-HP) was detected in BSA treated with 2-HH, and it showed no cross-reactivity when BSA was reacted with other related aldehydes. The antibody mAb3C8 was successfully used to detect atherosclerotic lesions in the human aorta.[290] The protein-bound 4-HHE was detected using the MAb

HHE53 antibody in Cu-oxidized LDL,[291] in gray matter of sporadic amyotrophic lateral sclerosis (ALS) spinal cords,[291] and in insulin[292]; this Michael addition–type HHE-his adduct is the major epitope (Fig. 7.22).

MDA is also known to form adducts with lysine, histidine, and cysteine side chain residues. In the case of lysine, for example, at least two possible epitopes were reported: N^{ε}-(2-propenal)lysine and dihydropyridine (DHP)–lysine (Fig. 7.23). The N^{ε}-(2-propenal)lysine is considered to be the predominant epitope formed from MDA-modified LDL, and the antibody is specific to the length of the chain and to the enaminal form of the epitope but not to the imine, amino, and amino-aldehyde analogs.[293] Using LC–MS–MS, characterization of MDA-modified insulin B led to the identification of another MDA–lysine epitope other than the *N*-propenal observed in MDA–LDL—i.e., the DHP-type adduct. ELISA analysis of the HPLC peptide fractions from atherosclerotic lesions of human aorta originating from macrophage-derived foam cells showed a significantly higher immunoreactivity of monoclonal antibody (mAb1F83) directed to the lipofuscin-like DHP–lysine adduct compared to the *N*-propenal type.[294] The MDA and DHP epitopes were found to increase by 20-fold in human skin under acute exposure to solar UV radiation and were also shown to be more prevalent in nonmelanoma skin cancer and in squamous cell carcinoma tissue compared to adjacent normal tissue.[295] Moreover, DHP–lysine showed to be a potent endogenous UVA photosensitizer of human skin cells exhibiting ROS formation.[296]

Acrolein rapidly reacts with the lysine side chain of a protein. The antibody mAb (mAb5F6), raised against acrolein-modified KLH, was employed to detect acrolein-modified proteins in vivo. The epitope formed was identified as N^{ε}-(3-formyl-3,4-dehydropiperidino)lysine (FDP–lysine) (see Fig. 7.24) and was detected in atherosclerotic lesions using immunohistochemistry, which suggests that FDP–lysine is a potential marker of long-term oxidative stress in aging, atherosclerosis, and diabetes.[297] Further analysis using LC–ESI–MS of the acrolein–lysine epitope from an acrolein-modified insulin B chain identified a major epitope aside from the FDP–lysine, a pyridinium-type lysine adduct, N^{ε}-(3-methylpyridinium)lysine (MP–lysine) (Fig. 7.24), which indicates a condensation reaction to form a Schiff base product as the major mechanism of acrolein reaction with lysine.[298] Using the same antibody, FDP–lysine was detected in neurofibrillary tangles, which is a major diagnostic lesion in AD brains.[299] High levels of acrolein-modified proteins were observed in photo-damaged skin,[300] diabetic glomerular lesions,[301] the saliva of patients with primary Sjögren's syndrome,[302] and the ApoE of tobacco smokers,[303] which also impairs ApoE function.[304] In patients with alcoholic liver disease, antibodies raised from BSA modified by MDA, 4-HNE, 2-hexenal, acrolein, methylglyoxal, and oxidized arachidonic and linoleic acids were only able to recognize high levels of epitopes derived from MDA, HNE, and oxidized fatty acids in the human sera of cirrhotic patients compared to nonalcoholic patients, and with no significant difference in the levels of epitopes derived from acrolein, 2-hexenal, and methylglyoxal.[305] Glutathiolation of FDP—lysine to form GSH—FDP—lysine (see Fig. 7.24) has been observed, resulting in decreased immunoreactivity to the anti-FDP—lysine monoclonal antibody; this could account for the potentiation of oxidative stress through depletion of GSH.[306] Antibodies raised from GS—FDP—lysine showed recognition of acrolein—lysine adduct, FDP—lysine, and S-glutathiolated proteins, which could find application in the detection of irreversibly bound GSH and other RS to the oxidized LDL in atherogenic models.[307]

Methylgloxal (MG) is produced mostly from glycolysis but is also a product of lipid peroxidation. The MG antibody is useful for investigating AGEs-related modification in biological systems. Two epitopes were identified from the Maillard reaction of MG with proteins: the lysine—lysine protein cross-linking structure (imidazolysine) and a fluorescent arginine modification (argpyrimidine) (Fig. 7.25). Antibodies raised from methylglyoxal-modified proteins such RNase A via the Maillard reaction of sugars showed no reactivity to N^{ε}-carboxymethyllysine but exhibited higher antibody reactivity to the serum protein of diabetic patients compared to normal subjects, which indicates that N^{ε}-carboxymethyllysine is not the major MG—lysine adduct formed in these diabetic subjects.[308] Also, higher levels of MG-modified protein were observed in the corneal collagen of older subjects compared to younger ones, indicating that the Maillard reaction occurred in vivo.[308] Under high D-glucose concentration (30—100 mM), MG modification of proteins was observed in yeast,[309] endothelial cells,[310] human malignant melanoma cells,[311] gastrointestinal cancer epithelial cells,[312] and the atherosclerotic aortic walls of hemodialysis patients.[313] Note that the cell culture condition affects the extent of modification of arginine by MG such that this could lead to the formation of new epitopes such as dihydroxyimidazolidine and hydroimidazolone.[308]

The oxysterol 7-ketocholestrol is one of the most widely studied oxidized forms of cholesterols and has been shown to be involved in the pathophysiology of atherosclerosis (Fig. 7.26). Monoclonal IgG3 antibodies AC1 and AC8 raised for 7-ketochelostrol also showed cross-reactivity to cholesterol and ergosterol.[314] IgG antibodies were also used to measure 7-ketoCh from oxLDL from the sera of patients after

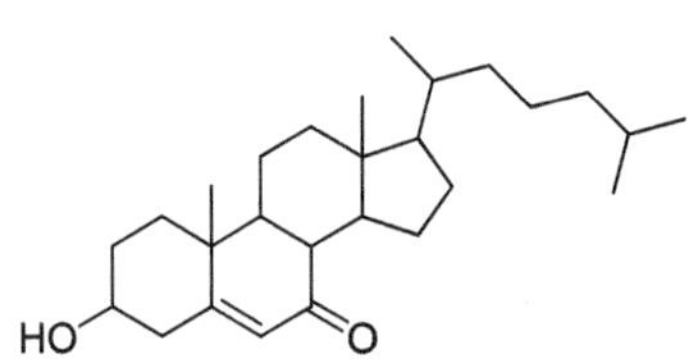

7-Ketocholesterol

Figure 7.26 Epitope of oxysterol 7-ketocholesterol.

orthotopic heart transplantation and showed higher anti-7-ketoCh levels compared to a control group, which indicates the role of 7-ketoCh in the development of atherosclerosis.[315]

7.4.4 Anti-DMPO antibody (immuno–spin trapping)

Immuno–spin trapping detects the adduct formed from the protein[316,317] or DNA[318,319] radical and the spin trap DMPO. Similar to spin trapping, the reaction of DMPO to biomolecule radicals such as DNA or proteins forms a paramagnetic DMPO–biomolecule radical adduct that could be detected by EPR. The EPR technique is highly dependent on the persistence of the paramagnetic DMPO adduct, which are most often short-lived and thus limited in their application for the detection of biomolecular radicals. Immuno–spin trapping offers advantages over fluorescence or EPR techniques such that it is a million times more sensitive than ESR and is not subject to the formation of artefactual signals.[320] It has also been employed in the study of redox biochemistry of a variety of biological systems such as cells, tissues, and whole animals.[321] Shown in Fig. 7.27 is a comparison of the nature of the various species detected, their lifetimes, and methods of detection showing the advantage of employing immuno–spin trapping for biomolecular radical detection using an array of detection

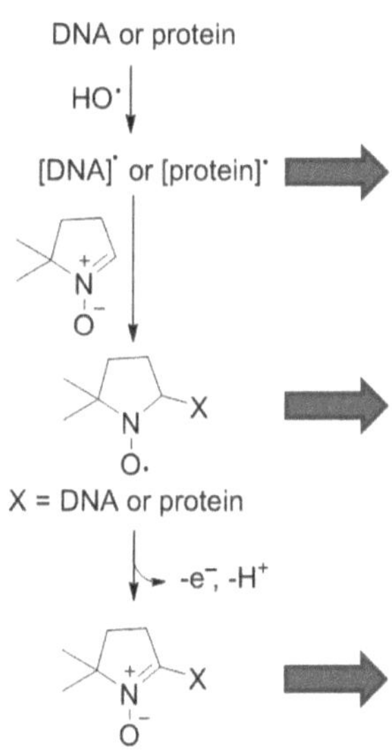

Species	Species lifetime	Method of detection
DNA/protein radical	Persistent for up to μs-s	EPR-rapid freezing technique
DMPO-DNA/protein radical adduct	Persistent for up to min-h	EPR-spin trapping
DMPO-DNA/protein nitrone adduct	Stable for up to an indefinite period	Immuno-spin trapping coupled with: ELISA, western blot, immunoprecipitation, immunocyto/hystochemistry, mass spectroscopy, magnetic resonance imaging

Figure 7.27 Comparison of the lifetimes and modes of detection for a biomolecular radical and its DMPO–nitroxide and –nitrone adducts. *Adapted from* Ramirez, D. et al. *Nature Protocols 2* 2007; **526**, 512-22.

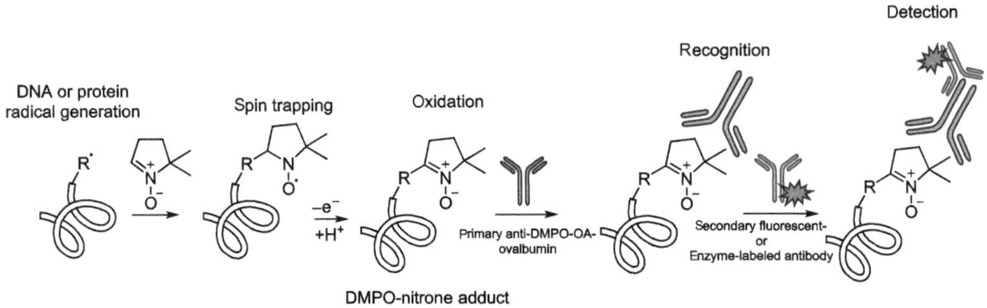

Figure 7.28 Schematic representation of immuno–spin trapping involving biomolecular radical trapping by DMPO, its oxidation to the DMPO–nitrone adduct, recognition by anti-DMPO antibodies, and detection using secondary antibodies.

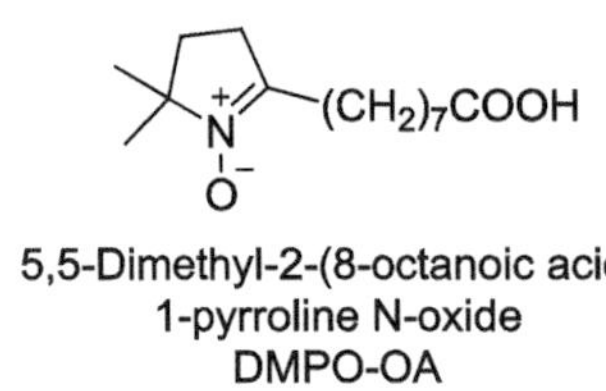

Figure 7.29 Structure of the DMPO epitope used for the development of anti-DMPO.

techniques. In this technique, macromolecular radicals, either DNA or protein, are generated, and DMPO is added to form the paramagnetic spin adduct. In an in vivo environment, the paramagnetic radical adduct is oxidized to the diamagnetic DMPO–nitrone adduct via a mechanism that still yet to be identified but has been suggested as probably due to oxidation by the peroxidase heme centers (Fig. 7.28).[316] The anti-DMPO polyclonal antibody was raised from the immunization of rabbits with the antigen ovalbumin-conjugated 5,5-dimethyl-2-(8-octanoic acid)-1-pyrroline N-oxide (DMPO–OA–ovalbumin) (Fig. 7.29).[322] This anti-DMPO antibody then binds to the DMPO–nitrone adducts with high affinity. Subsequent detection of the anti-DMPO antibody can be carried out using a secondary antibody that is conjugated to a fluorescent tag or enzyme labeled for chemiluminescence detection.

DNA radical detection from calf thymus DNA of hepatocytes after exposure of the cells to Cu^{2+}, H_2O_2, and DMPO involves DNA extraction and then detection of DNA–DMPO radical adduct by immune-slot(dot) blot, Western blot, ELISA, or fluorescence microscopy.[318] DNA radicals were also detected from macrophages in the presence of ferric or cupric citrate and *t*-BuOOH,[319] from WT and cytosolic peroxiredoxin–null cells in the presence of Cu^{2+} and H_2O_2,[323] in rat liver and human prostate epithelial tumorgenic cells and sodium arsenite,[324] MPO activity in PMA- or LPS-activated macrophages in the presence of H_2O_2, glucose–glucose oxidase-treated human leukemia cells or PMA–neutrophil-activated MPO-loaded

A549 epithelial cells,[325] and follicular dendritic cells from septic mice or human tonsil-derived HK cells.[326] Cellular localization of gold-nanoparticle–labeled DMPO nitrone adducts (both protein and DNA) was found to be distributed in the mitochondria, cytoplasm, and nucleus of hepatocytes isolated from rat liver and subjected to hepatic ischemia and reperfusion (I–R), and elevated DMPO-adduct levels were observed in I–R livers compared to controls.[327]

Using the MS–MS technique, the DNA–DMPO lesion was identified as the aminyl radical arising from dehydration of the hydroxyl radical adduct at C4 of the purine moiety of the 2′-deoxyadenosine (Fig. 7.30). This corroborates previous findings that adenosine radical intermediates are indeed formed under one-electron oxidation of DNA and can form other DNA lesions via the hole-transfer process.[328,329]

In protein systems, protein radical formation was detected using anti-DMPO and was shown to be mediated by H_2O_2,[316] nitrite and red blood cells,[330] copper- and $CO_3^{\bullet -}$-mediated oxidations,[331] hypochlorite,[332] LPS treatment,[333] bisulfite oxidation by eosinophil peroxidase–H_2O_2,[334,335] UVA irradiation of nano-TiO_2 particles,[336] and riboflavin sensitization.[337] Protein systems in which DMPO adducts were identified include hemoglobin and myoglobin,[330] SOD1,[331] mitochondrial complex I,[338] thyroid peroxidase,[339] catalase,[332] neuroglobin,[340] hemoprotein *Mycobacterium tuberculosis* catalase–peroxidase,[341] carboxypeptidase B1,[333] Ras GTPase,[342] α-lactalbumin,[337] MPO,[343] α-synuclein,[344] and LDL apo B-100.[345] Immuno–spin trapping of protein found applications in the investigation of chronic inflammatory conditions such as obesity[346] and Parkinson's disease[344] and in acute conditions such as liver I–R injury.[327]

The sites of protein radical formation were identified using MALDI–MS from heart myglobin in the presence of H_2O_2. Evidence showed that oxidation occurs at the Tyr-103 to form the Tyr-O radical, the site of DMPO adduct formation. Cross-linkage of Tyr-151 to form the myoglobin dimer was inhibited by the addition of DMPO; although no Try-151-DMPO adduct was observed, this study suggests that formation of DMPO adduct at Tyr-103 prevents dityrosine cross-linkage formation at Tyr-151.[347] Evidence for Try–DMPO adduct formation was also observed in hemoglobin–H_2O_2 system.[348]

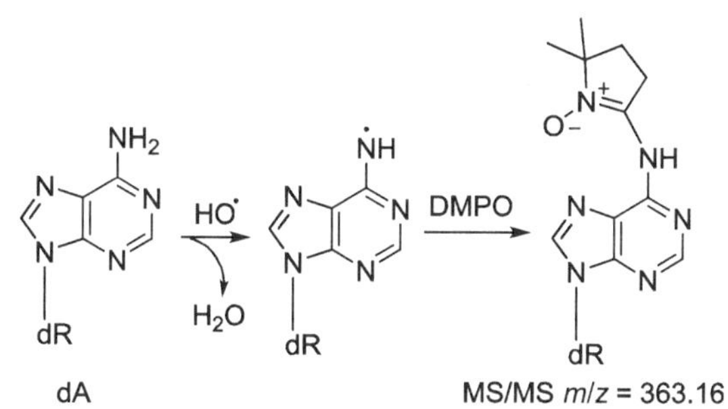

Figure 7.30 Identified DNA–DMPO lesion as 2′-deoxyadenosine-DMPO.

The formation of myoglobin—DMPO adduct could also be inhibited by NO and NO_2 through formation of nitrotyrosine, which indicates that tyrosine nitration is a more competitive oxidative process than formation of the Try-O radical. Tryptic digestion of the hemoglobin—H_2O_2—DMPO system followed by LC—MS—MS analyses, peptide mapping, and MS sequencing revealed formation of DMPO adducts at the Cys-93 of the β-chain and Tyr-42, Tyr-24, and His-20 of the α-chain.[349] Similar Tyr—DMPO and His—DMPO adducts were observed in other globin—H_2O_2 systems such as horse heart myoglobin and sperm whale myoglobin except that these myoglobins have no Cys—DMPO adducts.[350] Not only could immuno—spin trapping provide evidence of biomolecular radical formation, but also this technique could be exploited to investigate intra- or intermolecular electron-transfer processes by varying the added DMPO concentration. For example, using oxyHb—peroxynitrite or Mb—H_2O_2 systems in the presence of high DMPO concentrations, the Try—DMPO adduct was predominantly formed along with the secondary Cys—DMPO adduct. In the presence of low DMPO concentration, the secondary Cys—DMPO adduct was the only product, which indicates that the initial site of radical formation is the Tyr followed by its oxidation of cysteine to form the thiyl radical.[351]

7.5 NUCLEAR MAGNETIC RESONANCE SPECTROSCOPY AND IMAGING

Nuclear magnetic resonance (NMR) spectroscopy exploits the magnetic moment of the nuclei as opposed to EPR, which exploits the magnetic moment of the electron. NMR measurements are typically performed on diamagnetic compounds, but measurements of open-shell systems or molecules with unpaired electrons could only be observed as a paramagnetic shift of the magnetic nuclei (e.g., ^{1}H, ^{13}C, ^{14}N, ^{19}F, ^{17}O, and ^{31}P), which is also known as paramagnetic NMR spectroscopy. Due to the strong electron—nuclei coupling, paramagnetic NMR spectrum is characterized by a wide range of chemical shifts (or contact shifts) as well as broadened signals, therefore diminishing spectral resolution. Direct detection of biological radicals or paramagnetic radical adducts using NMR has thus not been forthcoming due to diminished spectral resolution and the short half-lives of these paramagnetic species in biological milieu. The use of paramagnetic contrast agents in MRI applications has been widely exploited since contrast agents can shorten the spin-lattice (or longitudinal) relaxation times (T1) of the proton nuclei of water. Since water is distributed in different tissues with unique microenvironment characteristics, water protons could then exhibit different T1 values, hence giving rise to image contrasts seen in MRI. Contrasts agents can serve as redox reporters where signal enhancement can be correlated with the presence of redox-active centers. Although MRIs of endogenously overexpressed paramagnetic ferritin molecules had been obtained, they do not provide substantial information into the nature and site of radical production in vivo.[352–354]

7.5.1 Gadolinium-based contrast agent

Gadolinium (III) complexes represent a class of T1 agents that enhance magnetic resonance (MR) water signals in tissues. Molecular MRI coupled with the immuno–spin trapping technique was applied for the imaging of protein–lipid radicals in a variety of disease conditions. The anti-DMPO probe was designed and raised as a multifunctional MRI contrast agent that is composed of the anti-DMPO antibody for the DMPO–adduct recognition, biotin moiety for non–MRI imaging, and Gd–DTPA as the contrast agent for MRI imaging as shown in Fig. 7.31. Analogous IgG isotype probes with contrast agents as negative controls were also designed and raised for nonspecific detection of radicals. Using the anti-DMPO probe, Streptozotocin-exposed mice showed significant increases in MRI signal intensity from the left-ventricular

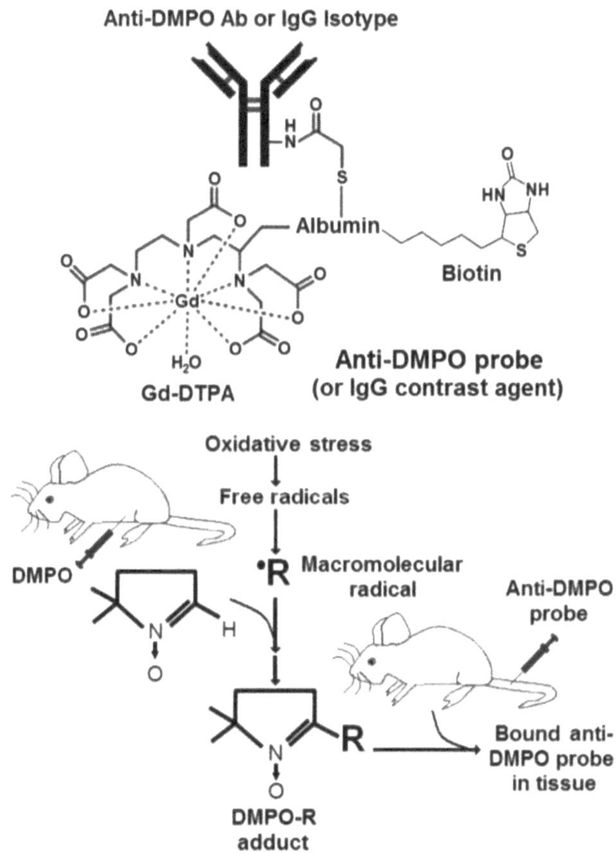

Figure 7.31 Combined in vivo molecular MRI and immuno–spin trapping using anti-DMPO probe (anti-DMPO antibody–albumin–Gd–DTPA–biotin). The biomolecular free radical (•R) reacts with DMPO to form the DMPO–R adducts. Injected anti-DMPO mMRI probe target DMPO–R adducts, which can be visualized by mMRI. *Adapted from Towner et al.* Free Radic Biol Med *2013;**63**:351–60.*

cardiac tissue of diabetic murine animals, indicating increased levels of radical formation in diabetic tissue compared to normal. The biotin moiety was then targeted with fluorescently labeled streptavidin–Cy3 to locate the anti-DMPO probe in excised cardiac tissues and, as in MRI data, showed elevated red fluorescence in the cardiac muscles from diabetic mice compared to nondiabetic mice.[355,356] In vivo visualization of free radical formation using the same anti-DMPO probe previously used for diabetic mice models was also demonstrated in the spinal cord of SOD1^{G93A}-transgenic (Tg) ALS mice model,[357] in the brain of mice-bearing orthotopic GL261 gliomas,[358] and those with septic encephalopathy.[359]

7.5.2 Iron–MGD-based contrast agent

Aside from gadolinium-based contrast agents, MRIs of RS such as ROS, O_2, and NO in biological system exploits the use of persistent or stable radical probes such as trityl radical or nitroxide, or the NO spin trap, Fe(II) complex with *N*-methyl-D-glucamine dithiocarbamate (MGD), $Fe(MGD)_2$. The interaction of these paramagnetic probes with the surrounding water protons in tissues results in enhanced MR images due to shortened T1, or spin–spin or transverse (T2) relaxation times of the MR-excited proton spins. The T1 relaxivity of the NO-complexed $Fe(MGD)_2$–NO is higher than the $Fe(MGD)_2$, and the difference in image contrast between these two complexes can be exploited to assess NO generation in vivo. MRI imaging of NO using $Fe(MGD)_2$ in rats under septic shock allowed for the visualization and mapping of the site of NO production in vivo.[360]

7.5.3 Trityl- and nitroxide-based contrast agents

One sophisticated imaging modality is the use of low-field MRI scanner along with EPR imaging to provide an image that combines high temporal and spatial anatomical resolution with spatial distribution of the stable radical probe with high sensitivity for radical detection. Commonly known as *Overhauser-enhanced magnetic resonance imaging* (OMRI),[361,362] this technique requires coregistration of EPR and MR images using specialized instrumentation composed of an MRI scanner operating at very low field and an EPR resonator. OMRI is based on the Overhauser effect whereby saturation by irradiation of the EPR transition of the stable radical probe transfers polarization from the unpaired electron to the coupled proton spins. This electron–nuclear interaction translates into the enhancement of the NMR signal intensities of water or of the surrounding protons by a significant factor via the Overhauser effect. This electron–nucleus cross-relaxation serves as the basis for dynamic nuclear polarization phenomenon and has been employed in a variety of NMR applications. Stable radical probes such as trityl radicals or nitroxides are typically employed as OMRI polarizing (contrast) agents; since the spectral property of these stable radical probes is affected by oxygen concentration leading to line-width broadening as well as presence of RS that leads to spectral quenching, they could also serve as reporter molecules for tissue oxygenation or tissue redox status, respectively. The trityl Oxo63 (Fig. 7.32) is one of the most commonly used contrast imaging agents for in vivo OMRI imaging of tissue oxygenation due to its high solubility in water, low toxicity, simple spectrum, oxygen-dependent line width, and biological stability.[361,363] EPR line width of the contrast agent can be estimated from the change in MRI intensity at two EPR power levels. Since spin–spin exchange interaction between the contrast agent and oxygen causes line-width broadening by influencing the $T_2{}^*$ relaxation time of the probe, the EPR line width can be correlated with pO_2 after appropriate calibration.[364] OMRI has been employed for the study of tumor oximetry allowing coregistration of tumor anatomy along with tumor oxygenation with spatial and temporal resolutions of ~1 mm and 2 min, respectively.[364] Tumor oxygenation and microvascular permeability were simultaneously imaged by OMRI using the Ox63 probe and showing as much as a 7000% increase in MRI signal at Ox63 concentrations of 1.5 mM compared to a control.[365] Pharmacokinetic studies of Oxo63 in mice revealed a half-life of 17–21 min and clearance excretion of 0.033–0.040 min^{-1}, which therefore could provide reliable pO_2 estimation in vivo.[363]

Other paramagnetic probes such as the proxyl- and TEMPO-derivatized nitroxides (Fig. 7.32) were employed for both MRI and OMRI applications.[366,367] Nitroxides are converted to the diamagnetic species on reaction with ROS–RNS. This loss of paramagnetism translates to a decrease of image intensity and can be exploited to assess the redox state of a tissue. For example, nitroxyl contrast agents such as the proxyl and TEMPO derivatives were employed for the imaging of whole-body animals[368] or the

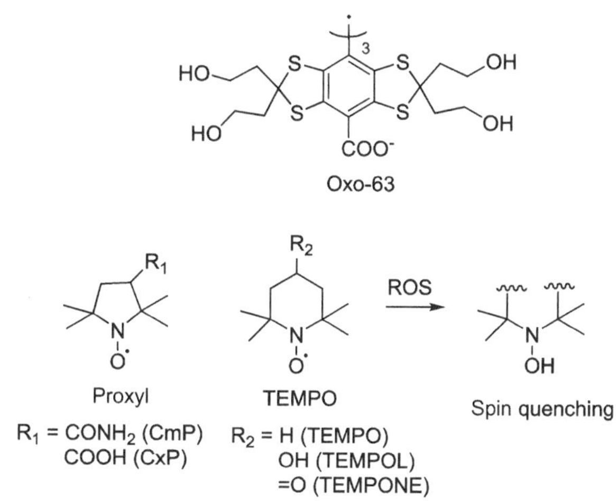

Figure 7.32 Trityl- and nitroxide-based OMRI contrast agents.

brain redox state.[369] At the same concentration, ^{15}N-labeled proxyl exhibited higher proton signal enhancement compared with ^{14}N-labeled analog due to the reduced spectral multiplicity of the nucleus (doublet vs triplet) that translates to higher signal intensity for the former.[370] However, stability of nitroxides in biological milieu is diminished due to their bioreduction and that their broader EPR line width than trityl radicals make them less sensitive for the measurement of oxygen.

7.5.4 Solution NMR spectroscopy

Solution ^{31}P MR and ^{19}F NMR spectroscopy were employed to identify small molecule radical intermediates using phosphorous-containing nitrone, DEPMPO, and DIPPMO, as well as the fluorine-containing spin trap 4-hydroxy-5,5-dimethyl-2-trifluoromethylpyrroline-1-oxide, respectively.[371,372] Similar to immuno–spin trapping, the nitrone spin trap initially forms a spin adduct, but in this case from radicals such as $^{\bullet}OH$, $^{\bullet}CH_3$, and $^{\bullet}CH_2OH$. Although $^{\bullet}SO_3^-$ was identified by NMR spin trapping, the formation of nitrone–SO_3 adduct could also occur via nonradical mechanism, so proper precautions should be observed when interpreting the origin of $^{\bullet}SO_3^-$.[373] The paramagnetic spin adducts then undergo disproportionation reaction to the diamagnetic NMR-detectable hydroxylamine and nitrone adduct forms. However, the NMR solution spectroscopic technique is limited by its poor sensitivity and need for high concentrations of radicals that are not physiologically relevant. However, the generation of oxygen- and carbon-centered radicals from the oxidation of linoleic acid by soybean lipoxygenases-1 was demonstrated using ^{31}P NMR spin trapping.[374] Although this technique was employed using purified enzyme, its application in the study of free radical generation in cellular system has yet to be realized.

REFERENCES

1. Sawyer DT, Sobkowiak A, Roberts JL. *Electrochemistry for Chemists*. 2nd ed. J. Wiley and Sons Inc.; 1995
2. Sawyer DT, Gibian MJ. The chemistry of superoxide ion. *Tetrahedron* 1979;**35**:1471.
3. Sawyer DT, Valentine JS. How super is superoxide? *Acc Chem Res* 1981;**14**:393.
4. Calas-Blanchard C, Catanante G, Noguer T. Electrochemical Sensor and Biosensor Strategies for ROS/RNS Detection in Biological Systems. *Electroanalysis* 2014;**26**:1277.
5. Moschopoulou, G.; Kintzios, S. Application of "membrane-engineering" to bioelectric recognition cell sensors for the ultra-sensitive detection of superoxide radical: A novel biosensor principle Anal Chim Acta **2006**, 573 + 574, 90.
6. Endo K, Miyasaka T, Mochizuki S, Aoyagi S, Himi N, Asahara H, et al. Development of a superoxide sensor by immobilization of superoxide dismutase. *Sens Actuators, B* 2002;**83**:30.
7. Kapp A, Beissenhirtz MK, Geyer F, Scheller F, Viezzoli MS, Lisdat F. Electrochemical and sensorial behavior of SOD mutants immobilized on gold electrodes in aqueous/organic solvent mixtures. *Electroanalysis* 2006;**18**:1909.
8. Campanella L, Favero G, Tomassetti M. A modified amperometric electrode for the determination of free radicals. *Sens Actuators, B* 1997;**44**:559.
9. Manning P, McNeil CJ, Cooper JM, Hillhouse EW. Direct, real-time sensing of free radical production by activated human glioblastoma cells. *Free Radic Biol Med* 1998;**24**:1304.

10. Chen XJ, West AC, Cropek DM, Banta S. Detection of the Superoxide Radical Anion Using Various Alkanethiol Monolayers and Immobilized Cytochrome c. *Anal Chem* 2008;**80**:9622.
11. Ji X, Ren J, Jin J, Nakamura T. A sensor for superoxide in aqueous and organic/aqueous media based on immobilized cytochrome c on binary self-assembled monolayers. *Biosens Bioelectron* 2007;**23**:241.
12. Gaspar S, Marty JL, Gheorghiu E. Cytochrome c-Based Amperometric Sensors for Superoxide Detection: Where Their Signal Comes From? *Electroanalysis* 2013;**25**:448.
13. Yuasa M, Oyaizu K, Yamaguchi A, Ishikawa M, Eguchi K, Kobayashi T, et al. Electrochemical sensor for superoxide anion radical using polymeric iron porphyrin complexes containing axial 1-methylimidazole ligand as cytochrome c mimics. *Polym Adv Technol* 2005;**16**:287.
14. Tang J, Zhu X, Niu X, Liu T, Zhao H, Lan M. An amperometric superoxide anion radical biosensor based on SOD/PtPd-PDARGO modified electrode. *Talanta* 2015;**137**:18.
15. Lu D, Zhang Y, Lin S, Wang L, Wang C. Synthesis of PtAu bimetallic nanoparticles on graphene-carbon nanotube hybrid nanomaterials for nonenzymatic hydrogen peroxide sensor. *Talanta* 2013;**112**:111.
16. Zhu X, Niu X, Zhao H, Tang J, Lan M. Immobilization of superoxide dismutase on Pt—Pd/MWCNTs hybrid modified electrode surface for superoxide anion detection. *Biosens Bioelectron* 2015;**67**:79.
17. Ma X, Hu W, Guo C, Yu L, Gao L, Xie J, et al. DNA-Templated Biomimetic Enzyme Sheets on Carbon Nanotubes to Sensitively In situ Detect Superoxide Anions Released from Cells. *Adv Funct Mater* 2014;**24**:5897.
18. Yuan L, Liu S, Tu W, Zhang Z, Bao J, Dai Z. Biomimetic Superoxide Dismutase Stabilized by Photopolymerization for Superoxide Anions Biosensing and Cell Monitoring. *Anal Chem* 2014;**86**:4783.
19. Shleev S, Wetteroe J, Magnusson K-E, Ruzgas T. Electrochemical characterization and application of azurin-modified gold electrodes for detection of superoxide. *Biosens Bioelectron* 2006;**22**:213.
20. Madhurantakam S, Selvaraj S, Nesakumar N, Sethuraman S, Balaguru Rayappan JB, Maheswari Krishnan U. Electrochemical enzymeless detection of superoxide employing naringin-copper decorated electrodes. *Biosens Bioelectron* 2014;**59**:134.
21. Han M, Guo P, Wang X, Tu W, Bao J, Dai Z. Mesoporous SiO2-(l)-lysine hybrid nanodisks: direct electron transfer of superoxide dismutase, sensitive detection of superoxide anions and its application in living cell monitoring. *RSC Adv* 2013;**3**:20456.
22. Santhosh P, Manesh KM, Lee S-H, Uthayakumar S, Gopalan AI, Lee K-P. Sensitive electrochemical detection of superoxide anion using gold nanoparticles distributed poly(methyl methacrylate)-polyaniline core-shell electrospun composite electrode. *Analyst* 2011;**136**:1557.
23. Deng Z, Rui Q, Yin X, Liu H, Tian Y. In Vivo Detection of Superoxide Anion in Bean Sprout Based on ZnO Nanodisks with Facilitated Activity for Direct Electron Transfer of Superoxide Dismutase. *Anal Chem* 2008;**80**:5839.
24. Kim H, Park SS, Seo J, Ha C-S, Moon C, Kim Y. Stable Protein Device Platform Based on Pyridine Dicarboxylic Acid-Bound Cubic-Nanostructured Mesoporous Titania Films. *ACS Appl Mater Interfaces* 2013;**5**:6873.
25. Green MJ, Hill HAO, Tew DG, Walton NJ. An opsonised electrode. The direct electrochemical detection of superoxide generated by human neutrophils. *FEBS Lett* 1984;**170**:69.
26. Flamm H, Kieninger J, Weltin A, Urban GA. Superoxide microsensor integrated into a Sensing Cell Culture Flask microsystem using direct oxidation for cell culture application. *Biosens Bioelectron* 2015;**65**:354.
27. Zhao X, Diakowski PM, Ding Z. Deconvoluting Topography and Spatial Physiological Activity of Live Macrophage Cells by Scanning Electrochemical Microscopy in Constant-Distance Mode. *Anal Chem* 2010;**82**:8371.
28. Swei A, Lacy F, Delano FA, Parks DA, Schmid-Schonbein GW. A mechanism of oxygen free radical production in the Dahl hypertensive rat. *Microcirculation* 1999;**6**:179.
29. Shleev S, Wetteroe J, Magnusson K-E, Ruzgas T. Simultaneous use of electrochemistry and chemiluminescence to detect reactive oxygen species produced by human neutrophils. *Cell Biol Int* 2008;**32**:1486.

30. Ashkenazi A, Abu-Rabeah K, Marks RS. Electrochemistry and chemiluminescence techniques compared in the detection of NADPH oxidase activity in phagocyte cells. *Talanta* 2009;**77**:1460.
31. Rapino S, Marcu R, Paolucci F, Giorgio M. Electrochemical detection of H2O2 formation in isolated mitochondria. *Methods Enzymol* 2013;**526**:123.
32. Nappi AJ, Vass E. Hydroxyl radical formation resulting from the interaction of nitric oxide and hydrogen peroxide. *Biochim Biophys Acta* 1998;**1380**:55.
33. Nagase S, Ohkoshi N, Ueda A, Aoyagi K, Koyama A. Hydrogen peroxide interferes with detection of nitric oxide by an electrochemical method. *Clin Chem* 1997;**43**:1246.
34. Abdurhman AAM, Zhang Y, Zhang G, Wang S. Hierarchical nanostructured noble metal/metal oxide/graphene-coated carbon fiber: in situ electrochemical synthesis and use as microelectrode for real-time molecular detection of cancer cells. *Anal Bioanal Chem* 2015;**407**:8129.
35. Jamil A, Lim HN, Yusof NA, Tajudin AA, Huang NM, Pandikumar A, et al. Preparation and characterization of silver nanoparticles-reduced graphene oxide on ITO for immunosensing platform. *Sens Actuators, B* 2015;**221**:1423.
36. Liu J, Bo X, Zhao Z, Guo L. Highly exposed Pt nanoparticles supported on porous graphene for electrochemical detection of hydrogen peroxide in living cells. *Biosens Bioelectron* 2015;**74**:71.
37. Xi J, Zhang Y, Wang N, Wang L, Zhang Z, Xiao F, et al. Ultrafine Pd Nanoparticles Encapsulated in Microporous Co3O4 Hollow Nanospheres for In Situ Molecular Detection of Living Cells. *ACS Appl Mater Interfaces* 2015;**7**:5583.
38. Ujjain SK, Das A, Srivastava G, Ahuja P, Roy M, Arya A, et al. Nanoceria based electrochemical sensor for hydrogen peroxide detection. *Biointerphases* 2014;**9**:031011/1.
39. Ju J, Chen W. In Situ Growth of Surfactant-Free Gold Nanoparticles on Nitrogen-Doped Graphene Quantum Dots for Electrochemical Detection of Hydrogen Peroxide in Biological Environments. *Anal Chem* 2015;**87**:1903.
40. Maji SK, Sreejith S, Mandal AK, Ma X, Zhao Y. Immobilizing Gold Nanoparticles in Mesoporous Silica Covered Reduced Graphene Oxide: A Hybrid Material for Cancer Cell Detection through Hydrogen Peroxide Sensing. *ACS Appl Mater Interfaces* 2014;**6**:13648.
41. Wang T, Zhu H, Zhuo J, Zhu Z, Papakonstantinou P, Lubarsky G, et al. Biosensor Based on Ultrasmall MoS2 Nanoparticles for Electrochemical Detection of H2O2 Released by Cells at the Nanomolar Level. *Anal Chem* 2013;**85**:10289.
42. Li J, Yao Y, Shiu K-K. Layer-by-layer assembly of Prussian blue and carbon nanotube composites with poly(diallyldimethylammonium chloride) for the sensitive detection of hydrogen peroxide. *Anal Sci* 2010;**26**:431.
43. Yagati AK, Choi J-W. Protein Based Electrochemical Biosensors for H2O2 Detection Towards Clinical Diagnostics. *Electroanalysis* 2014;**26**:1259.
44. Rui Q, Komori K, Tian Y, Liu H, Luo Y, Sakai Y. Electrochemical biosensor for the detection of H2O2 from living cancer cells based on ZnO nanosheets. *Anal Chim Acta* 2010;**670**:57.
45. Dengale SM, Yagati AK, Chung Y-H, Min J, Choi J-W. An electrochemical H2O2 detection method based on direct electrochemistry of myoglobin immobilized on gold deposited ITO electrode. *J Nanosci Nanotechnol* 2013;**13**:6424.
46. Liu H, Cui Y, Li P, Zhou Y, Chen Y, Tang Y, et al. Polyphosphonate induced coacervation of chitosan: Encapsulation of proteins/enzymes and their biosensing. *Anal Chim Acta* 2013;**776**:24.
47. Li C, Zhang H, Wu P, Gong Z, Xu G, Cai C. Electrochemical detection of extracellular hydrogen peroxide released from RAW 264.7 murine macrophage cells based on horseradish peroxidase-hydroxyapatite nanohybrids. *Analyst* 2011;**136**:1116.
48. Yagati AK, Lee T, Min J, Choi J-W. Amperometric sensor for hydrogen peroxide based on direct electron transfer of spinach ferredoxin on Au electrode. *Bioelectrochemistry* 2011;**80**:169.
49. Gopalan AI, Komathi S, Sai Anand G, Lee K-P. Nanodiamond based sponges with entrapped enzyme: A novel electrochemical probe for hydrogen peroxide. *Biosens Bioelectron* 2013;**46**:136.
50. Enomoto J, Matharu Z, Revzin A. Electrochemical biosensors for on-chip detection of oxidative stress from cells. *Methods Enzymol* 2013;**526**:107.
51. Yan J, Pedrosa VA, Enomoto J, Simonian AL, Revzin A. Electrochemical biosensors for on-chip detection of oxidative stress from immune cells. *Biomicrofluidics* 2011;**5**:032008.

52. Kasai N, Han C, Torimitsu K. Multichannel detection of hydrogen peroxide released from a rat hippocampal slice. *Chem Sens* 2004;**20**:764.
53. Ciszewski A, Milczarek G. Electrochemical detection of nitric oxide using polymer modified electrodes. *Talanta* 2003;**61**:11.
54. Bedioui F, Villeneuve N. Electrochemical nitric oxide sensors for biological samples - principle, selected examples and applications. *Electroanalysis* 2002;**15**:5.
55. Allen BW, Coury Jr LA, Piantadosi CA. Electrochemical detection of physiological nitric oxide: materials and methods. *Methods Enzymol* 2002;**359**:125.
56. Wynne AM, Reid CH, Finnerty NJ. In vitro characterization of ortho phenylenediamine and Nafion-modified Pt electrodes for measuring brain nitric oxide. *J Electroanal Chem* 2014;**732**:110.
57. Shibuki K. A nitric oxide microprobe. *Seitai no Kagaku* 1991;**42**:57.
58. Shibuki K, Okada D. Endogenous nitric oxide release required for long-term synaptic depression in the cerebellum. *Nature* 1991;**349**:326.
59. Shibuki K. An electrochemical microprobe for detecting nitric oxide release in brain tissue. *Neurosci Res* 1990;**9**:69.
60. Shibuki K. Detection of nitric oxide by an electrochemical microprobe. *NeuroProtocols* 1992;**1**:151.
61. Hurst RD, Clark JB. The utility of the nitric oxide electrochemical sensor in biomedical research. *Sensors* 2003;**3**:321.
62. Serpe MJ, Zhang X. The Principles, Development and Application of Microelectrodes for the In Vivo Determination of Nitric Oxide. In: Michael AC, Borland LM, editors. *Electrochemical Methods for Neuroscience*. Boca Raton, FL: CRC Press/Taylor & Francis; 2007
63. Zhang X, Kislyak Y, Lin J, Dickson A, Cardosa L, Broderick M, et al. Nanometer size electrode for nitric oxide and S-nitrosothiols measurement. *Electrochem Commun* 2002;**4**:11.
64. Allen BW, Piantadosi CA. Electrochemical activation of electrodes for amperometric detection of nitric oxide. *Nitric Oxide* 2003;**8**:243.
65. Wang S, Lin X. Electrodeposition of Pt-Fe(III) nanoparticle on glassy carbon electrode for electrochemical nitric oxide sensor. *Electrochim Acta* 2005;**50**:2887.
66. Amatore C, Arbault S, Bouret Y, Cauli B, Guille M, Rancillac A, et al. Nitric oxide release during evoked neuronal activity in cerebellum slices: Detection with platinized carbon-fiber microelectrodes. *Chem Phys Chem* 2006;**7**:181.
67. Allen BW, Liu J, Piantadosi CA. Electrochemical detection of nitric oxide in biological fluids. *Methods Enzymol* 2005;**396**:68.
68. Dalbasti T, Kilinc E. Microelectrode for in vivo real-time detection of NO. *Methods Enzymol* 2005;**396**:584.
69. Ting SL, Guo CX, Leong KC, Kim D-H, Li CM, Chen P. Gold nanoparticles decorated reduced graphene oxide for detecting the presence and cellular release of nitric oxide. *Electrochim Acta* 2013;**111**:441.
70. Hu FX, Xie JL, Bao SJ, Yu L, Li CM. Shape-controlled ceria-reduced graphene oxide nanocomposites toward high-sensitive in situ detection of nitric oxide. *Biosens Bioelectron* 2015;**70**:310.
71. Shahid MM, Rameshkumar P, Pandikumar A, Lim HN, Ng YH, Huang NM. An electrochemical sensing platform based on a reduced graphene oxide-cobalt oxide nanocube@platinum nanocomposite for nitric oxide detection. *J Mater Chem A* 2015;**3**:14458.
72. Sund JB, Wolter SD, Parker CB, Natishan PM, Glass JT. Diamond for Biosensing: Electrochemical Detection of NOx Species with Thiol-Amine Functionalized Diamond. *J Electrochem Soc* 2015;**162**:B225.
73. Krylov AV, Lisdat F. Nickel hexacyanoferrate-based sensor electrode for the detection of nitric oxide at low potentials. *Electroanalysis* 2007;**19**:23.
74. Viswanathan P, Manivannan S, Ramaraj R. Polyelectrolyte stabilized bi-metallic Au/Ag nanoclusters modified electrode for nitric oxide detection. *RSC Adv* 2015;**5**:54735.
75. Cha W, Tung Y-C, Meyerhoff ME, Takayama S. Patterned Electrode-Based Amperometric Gas Sensor for Direct Nitric Oxide Detection within Microfluidic Devices. *Anal Chem* 2010;**82**:3300.
76. Bedioui F, Trevin S. Practical aspects and methodological approaches to achieve electrochemical detection of submicromolar NO in biological systems. *Biosens Bioelectron* 1998;**13**:227.

77. Kroening S, Scheller FW, Wollenberger U, Lisdat F. Myoglobin-clay electrode for nitric oxide (NO) detection in solution. *Electroanalysis* 2004;**16**:253.
78. Guo Z, Chen J, Liu H, Cha C. Direct electrochemistry of hemoglobin and myoglobin at didodecyldimethylammonium bromide-modified powder microelectrode and application for electrochemical detection of nitric oxide. *Anal Chim Acta* 2008;**607**:30.
79. Wu Y, Shen Q, Hu S. Direct electrochemistry and electrocatalysis of heme-proteins in regenerated silk fibroin film. *Anal Chim Acta* 2006;**558**:179.
80. Nyokong T, Vilakazi S. Phthalocyanines and related complexes as electrocatalysts for the detection of nitric oxide. *Talanta* 2003;**61**:27.
81. Oni J, Diab N, Reiter S, Schuhmann W. Metallophthalocyanine-modified glassy carbon electrodes: effects of film formation conditions on electrocatalytic activity towards the oxidation of nitric oxide. *Sens Actuators, B* 2005;**105**:208.
82. Bedioui F, Trevin S, Devynck J, Lantoine F, Brunet A, Devynck M-A. Elaboration and use of nickel planar macrocyclic complex-based sensors for the direct electrochemical measurement of nitric oxide in biological media. *Biosens Bioelectron* 1997;**12**:205.
83. Rodrigues NP, Bedioui F, Deutsch A, Zurgil N, Afrimzon E, Shafran Y, et al. Construction and use of an integrated electrochemical device for the detection of biologically relevant compounds released from non-adherent cells: Application for the electrochemical determination of nitric oxide produced by human U937 cells. *Electrochem Commun* 2006;**8**:341.
84. Griveau S, Dumezy C, Seguin J, Chabot GG, Scherman D, Bedioui F. In Vivo Electrochemical Detection of Nitric Oxide in Tumor-Bearing Mice. *Anal Chem* 2007;**79**:1030.
85. Kanai AJ, Pearce LL, Clemens PR, Birder LA, VanBibber MM, Choi S-Y, et al. Identification of a neuronal nitric oxide synthase in isolated cardiac mitochondria using electrochemical detection. *Proc Natl Acad Sci U S A* 2001;**98**:14126.
86. Isik S, Etienne M, Oni J, Bloechl A, Reiter S, Schuhmann W. Dual microelectrodes for distance control and detection of nitric oxide from endothelial cells by means of scanning electrochemical microscope. *Anal Chem* 2004;**76**:6389.
87. Oni J, Diab N, Radtke I, Schuhmann W. Detection of NO release from endothelial cells using Pt micro electrodes modified with a pyrrole-functionalized Mn(II) porphyrin. *Electrochim Acta* 2003;**48**:3349.
88. Chandra S, Mende C, Bahadur D, Hildebrandt A, Lang H. Fabrication of a porphyrin-based electrochemical biosensor for detection of nitric oxide released by cancer cells. *J Solid State Electrochem* 2015;**19**:169.
89. Pontie M, Lecture H, Bedioui F. Improvement in the performance of a nickel complex-based electrochemical sensor for the detection of nitric oxide in solution. *Sens Actuators, B* 1999;**56**:1.
90. Oni J, Pailleret A, Isik S, Diab N, Radtke I, Bloechl A, et al. Functionalised electrode array for the detection of nitric oxide released by endothelial cells using different NO-sensing chemistries. *Anal Bioanal Chem* 2004;**378**:1594.
91. Griveau S, Bedioui F. Overview of significant examples of electrochemical sensor arrays designed for detection of nitric oxide and relevant species in a biological environment. *Anal Bioanal Chem* 2013;**405**:3475.
92. Villeneuve N, Bedioui F, Voituriez K, Avaro S, Vilaine JP. Electrochemical detection of nitric oxide production in perfused pig coronary artery: comparison of the performances of two electrochemical sensors. *J Pharmacol Toxicol Methods* 1998;**40**:95.
93. Stingele R, Wilson DA, Traystman RJ, Hanley DF. Tyrosine confounds oxidative electrochemical detection of nitric oxide. *Am J Physiol* 1998;**274**:H1698.
94. Privett BJ, Shin JH, Schoenfisch MH. Tutorial Review: Electrochemical Nitric Oxide Sensors for Physiological Measurements. *Chem Soc Rev* 2010;**39**:1925.
95. Bedioui F, Quinton D, Griveau S, Nyokong T. Designing molecular materials and strategies for the electrochemical detection of nitric oxide, superoxide and peroxynitrite in biological systems. *Phys Chem Chem Phys* 2010;**12**:9976.
96. Lee Y, Kim J. Simultaneous Electrochemical Detection of Nitric Oxide and Carbon Monoxide Generated from Mouse Kidney Organ Tissues. *Anal Chem* 2007;**79**:7669.

97. Lin AMY, Kao LS, Chai CY. Involvement of nitric oxide in dopaminergic transmission in rat striatum: an in vivo electrochemical study. *J Neurochem* 1995;**65**:2043.
98. Huffman ML, Venton BJ. Carbon-fiber microelectrodes for in vivo applications. *Analyst* 2009;**134**:18.
99. Thi Kim LTO, Escriou V, Griveau S, Girard A, Griscom L, Razan F, et al. Array of ultramicroelectrodes for the simultaneous detection of nitric oxide and peroxynitrite in biological systems. *Electrochim Acta* 2014;**140**:33.
100. Li Y, Sella C, Lemaitre F, Guille-Collignon M, Thouin L, Amatore C. Electrochemical Detection of Nitric Oxide and Peroxynitrite Anion in Microchannels at Highly Sensitive Platinum-Black Coated Electrodes. Application to ROS and RNS Mixtures prior to Biological Investigations. *Electrochim Acta* 2014;**144**:111.
101. Quinton D, Griveau S, Bedioui F. Electrochemical approach to detect the presence of peroxynitrite in aerobic neutral solution. *Electrochem Commun* 2010;**12**:1446.
102. Cha W, Meyerhoff ME. S-nitrosothiol detection via amperometric nitric oxide sensor with surface modified hydrogel layer containing immobilized organoselenium catalyst. *Langmuir* 2006;**22**:10830.
103. Wu Y, Zhang F, Wang Y, Krishnamoorthy M-K, Roy-Chaudhury P, Bleske BE, et al. Photoinstability of S-nitrosothiols during sampling of whole blood: a likely source of error and variability in S-nitrosothiol measurements. *Clin Chem* 2008;**54**:916.
104. Kikura-Hanajiri R, Martin RS, Lunte SM. Indirect Measurement of Nitric Oxide Production by Monitoring Nitrate and Nitrite Using Microchip Electrophoresis with Electrochemical Detection. *Anal Chem* 2002;**74**:6370.
105. Vallance P, Patton S, Bhagat K, MacAllister R, Radomski M, Moncada S, et al. Direct measurement of nitric oxide in human beings. *Lancet* 1995;**346**:153.
106. Rodriguez-Rodriguez R, Simonsen U. Measurement of nitric oxide and reactive oxygen species in the vascular wall. *Curr Anal Chem* 2012;**8**:485.
107. Heinzen EL, Pollack GM. Use of an electrochemical nitric oxide sensor to detect neuronal nitric oxide production in conscious, unrestrained rats. *J Pharmacol Toxicol Methods* 2003;**48**:139.
108. Lawrence NS, Davis J, Jiang L, Jones TGJ, Davies SN, Compton RG. The electrochemical analog of the methylene blue reaction: a novel amperometric approach to the detection of hydrogen sulfide. *Electroanalysis* 2000;**12**:1453.
109. Giovanelli D, Lawrence NS, Klymenko OV, Jiang L, Jones TGJ, Compton RG. The electrochemically initiated reaction of sulfide with N,N-diethyl-p-phenylenediamine in dimethylformamide. Part II: Implications for sensing strategies. *Electroanalysis* 2003;**15**:961.
110. Lawrence NS, Davis J, Jiang L, Jones TGJ, Davies SN, Compton RG. Electrochemically initiated 1,4-nucleophilic substitutions: a general strategy for the analytical detection of hydrogen sulfide. *Electroanalysis* 2001;**13**:432.
111. Lawrence NS, Davis J, Compton RG. Analytical strategies for the detection of sulfide: a review. *Talanta* 2000;**52**:771.
112. Xu T, Scafa N, Xu LP, Zhou S, Abdullah Al-Ghanem K, Mahboob S, et al. Electrochemical hydrogen sulfide biosensors. *Analyst* 2016;**141**:1185.
113. Li G, Polk BJ, Meazell LA, Hatchett DW. ISE Analysis of Hydrogen Sulfide in Cigarette Smoke. *J Chem Ed* 2000;**77**:1049.
114. Bennett JA, Wheeler CD, Sterling KL, Chiodo AM. Exploring dicyano-ferriprotoporphyrin as a novel electrocatalytic material for selective H2S gasotransmitter detection. *Electrochim Acta* 2013;**88**:86.
115. Yue X, Zhu Z, Zhang M, Ye Z. Reaction-Based Turn-on Electrochemiluminescent Sensor with a Ruthenium(II) Complex for Selective Detection of Extracellular Hydrogen Sulfide in Rat Brain. *Anal Chem* 2015;**87**:1839.
116. Zhang Y-Y, Zhou H, Wu P, Zhang H-R, Xu J-J, Chen H-Y. In Situ Activation of CdS Electrochemiluminescence Film and Its Application in H2S Detection. *Anal Chem* 2014;**86**:8657.
117. Bitziou E, Joseph MB, Read TL, Palmer N, Mollart T, Newton ME, et al. In Situ Optimization of pH for Parts-Per-Billion Electrochemical Detection of Dissolved Hydrogen Sulfide Using Boron Doped Diamond Flow Electrodes. *Anal Chem* 2014;**86**:10834.

118. Choi S-J, Jang B-H, Lee S-J, Min BK, Rothschild A, Kim I-D. Selective Detection of Acetone and Hydrogen Sulfide for the Diagnosis of Diabetes and Halitosis Using SnO2 Nanofibers Functionalized with Reduced Graphene Oxide Nanosheets. *ACS Appl Mater Interfaces* 2014;**6**:2588.
119. Bennett JA, Pander III JE, Neiswonger MA. Investigating the viability of electrodeposited vanadium pentoxide as a suitable electrode material for in vivo amperometric hydrogen sulfide detection. *J Electroanal Chem* 2011;**654**:1.
120. Liu C, Hayashi K, Toko K. Au nanoparticles decorated polyaniline nanofiber sensor for detecting volatile sulfur compounds in expired breath. *Sens Actuators, B* 2012;**161**:504.
121. Clark Jr LC, Wolf R, Granger D, Taylor Z. Continuous recording of blood oxygen tensions by polarography. *J Appl Physiol* 1953;**6**:189.
122. Silva A, Oliveira P. Evaluation of Respiration with Clark Type Electrode in Isolated Mitochondria and Permeabilized Animal Cells. In: Palmeira CM, Moreno AJ, editors. *Mitochondrial Bioenergetics*, Vol. 810. Humana Press; 2012. p. 7.
123. Li Z, Graham BH. Measurement of mitochondrial oxygen consumption using a Clark electrode. *Methods Mol Biol* 2012;**837**:63.
124. Zhang J, Nuebel E, Wisidagama DRR, Setoguchi K, Hong JS, Van Horn CM, et al. Measuring energy metabolism in cultured cells, including human pluripotent stem cells and differentiated cells. *Nat Protoc* 2012;**7**:1068.
125. Marchandeau J-P, Labbe G. Evaluation of mitochondrial respiration in cultured rat hepatocytes. *Methods Mol Biol* 2011;**691**:243.
126. Shi Q, Greenhaw J, Salminen WF. Inhibition of cytochrome P450s enhances (+)-usnic acid cytotoxicity in primary cultured rat hepatocytes. *J Appl Toxicol* 2014;**34**:835.
127. Ferrero E, Fulgenzi A, Belloni D, Foglieni C, Ferrero ME. Cell food improves respiratory metqabolism of endothelial cells and inhibits hypoxia-induced reactive oxygen species (ROS) generation. *J Physiol Pharmacol* 2011;**62**:287.
128. Dziurla R, Gaber T, Fangradt M, Hahne M, Tripmacher R, Kolar P, et al. Effects of hypoxia and/or lack of glucose on cellular energy metabolism and cytokine production in stimulated human CD4+ T lymphocytes. *Immunol Lett* 2010;**131**:97.
129. Diepart C, Verrax J, Calderon PB, Feron O, Jordan BF, Gallez B. Comparison of methods for measuring oxygen consumption in tumor cells in vitro. *Anal Biochem* 2010;**396**:250.
130. Renger G, Hanssum B. Oxygen detection in biological systems. *Photosynth Res* 2009;**102**:487.
131. Gegentonglaga, Yoshizato H, Higuchi Y, Toyota Y, Hanai Y, Ando Y, et al. Variable alteration of regional tissue oxygen pressure in rat hippocampus by acute swimming exercise. *Life Sci* 2013;**93**:773.
132. Takatori SC, de la Jara PL, Holden B, Ehrmann K, Ho A, Radke CJ. In vivo corneal oxygen uptake during soft-contact-lens wear. *Invest Ophthalmol Visual Sci* 2013;**54**:3472.
133. Hiltebrand LB, Koepfli E, Kimberger O, Sigurdsson GH, Brandt S. Hypotension during Fluid-restricted Abdominal Surgery: Effects of Norepinephrine Treatment on Regional and Microcirculatory Blood Flow in the Intestinal Tract. *Anesthesiology* 2011;**114**:557.
134. Pasarica M, Rood J, Ravussin E, Schwarz J-M, Smith SR, Redman LM. Reduced oxygenation in human obese adipose tissue is associated with impaired insulin suppression of lipolysis. *J Clin Endocrinol Metab* 2010;**95**:4052.
135. Nematbakhsh M, Eppel GA, Goddard D, O'Connor PM, Evans RG. Local maximum oxygen disappearance rate has limited utility as a measure of local renal tissue oxygen consumption. *J Pharmacol Toxicol Methods* 2010;**61**:297.
136. Maier S, Hasibeder W, Pajk W, Hengl C, Ulmer H, Hausdorfer H, et al. Arginine-vasopressin attenuates beneficial norepinephrine effect on jejunal mucosal tissue oxygenation during endotoxinaemia. *Br J Anaesth* 2009;**103**:691.
137. Leong C-L, O'Connor PM, Eppel GA, Anderson WP, Evans RG. Measurement of Renal Tissue Oxygen Tension: Systematic Differences between Fluorescence Optode and Microelectrode Recordings in Anesthetized Rabbits. *Nephron* 2008;**108**:p11.

138. Wang Z, Lin P, Baker GA, Stetter J, Zeng X. Ionic Liquids as Electrolytes for the Development of a Robust Amperometric Oxygen Sensor. *Anal Chem* 2011;**83**:7066.
139. Pouvreau LAM, Strampraad MJF, Van Berloo S, Kattenberg JH, de Vries S. NO, N2O, and O2 reaction kinetics: scope and limitations of the Clark electrode. *Methods Enzymol* 2008;**436**:97.
140. Bolger FB, Bennett R, Lowry JP. An in vitro characterisation comparing carbon paste and Pt microelectrodes for real-time detection of brain tissue oxygen. *Analyst* 2011;**136**:4028.
141. Osbourn DM, Sanger RH, Smith PJS. Determination of Single-Cell Oxygen Consumption with Impedance Feedback for Control of Sample-Probe Separation. *Anal Chem* 2005;**77**:6999.
142. Wu C-C, Saito T, Yasukawa T, Shiku H, Abe H, Hoshi H, et al. Microfluidic chip integrated with amperometric detector array for in situ estimating oxygen consumption characteristics of single bovine embryos. *Sens Actuators, B* 2007;**125**:680.
143. Bai S-J, Ryu WH, Fasching RJ, Grossman AR, Prinz FB. In vivo O2 measurement inside single photosynthetic cells. *Biotechnol Lett* 2011;**33**:1675.
144. Bazzu G, Puggioni GGM, Dedola S, Calia G, Rocchitta G, Migheli R, et al. Real-Time Monitoring of Brain Tissue Oxygen Using a Miniaturized Biotelemetric Device Implanted in Freely Moving Rats. *Anal Chem* 2009;**81**:2235.
145. Zachek MK, Takmakov P, Moody B, Wightman RM, McCarty GS. Simultaneous Decoupled Detection of Dopamine and Oxygen Using Pyrolyzed Carbon Microarrays and Fast-Scan Cyclic Voltammetry. *Anal Chem* 2009;**81**:6258.
146. Bai L, Wen D, Yin J, Deng L, Zhu C, Dong S. Carbon nanotubes-ionic liquid nanocomposites sensing platform for NADH oxidation and oxygen, glucose detection in blood. *Talanta* 2012;**91**:110.
147. Meyer H, Drewer H, Gruendig B, Cammann K, Kakerow R, Manoli Y, et al. Two-Dimensional Imaging of O2, H2O2, and Glucose Distributions by an Array of 400 Individually Addressable Microelectrodes. *Anal Chem* 1995;**67**:1164.
148. Cheirsirikul S, Mongkolkiartichai T. Diamond microelectrodes for amperometric oxygen sensor. Adv Mater Res 2014;**911**:276.
149. Ivandini TA, Harmesa, Saepudin E, Einaga Y. Yeast-based biochemical oxygen demand sensors using gold-modified boron-doped diamond electrodes. *Anal Sci* 2015;**31**:643.
150. Wu X, Hu Y, Jin J, Zhou N, Wu P, Zhang H, et al. Electrochemical Approach for Detection of Extracellular Oxygen Released from Erythrocytes Based on Graphene Film Integrated with Laccase and 2,2-Azino-bis(3-ethylbenzothiazoline-6-sulfonic acid). *Anal Chem* 2010;**82**:3588.
151. Paul DW, Prajapati I, Reed ML. Electropolymerized eugenol: Evaluation as a protective film for oxygen sensing. *Sens Actuators, B* 2013;**183**:129.
152. Wang Y, Hosono T, Hasebe Y. Hemin-adsorbed carbon felt for sensitive and rapid flow-amperometric detection of dissolved oxygen. *Microchim Acta* 2013;**180**:1295.
153. Butterfield DA, Gu L, Di Domenico F, Robinson RA. Mass spectrometry and redox proteomics: applications in disease. *Mass Spectrom Rev* 2014;**33**:277.
154. Colzani M, Aldini G, Carini M. Mass spectrometric approaches for the identification and quantification of reactive carbonyl species protein adducts. *J Proteomics* 2013;**92**:28.
155. Fratelli M, Demol H, Puype M, Casagrande S, Villa P, Eberini I, et al. Identification of proteins undergoing glutathionylation in oxidatively stressed hepatocytes and hepatoma cells. *Proteomics* 2003;**3**:1154.
156. Beuve A, Wu C, Cui C, Liu T, Jain MR, Huang C, et al. Identification of novel S-nitrosation sites in soluble guanylyl cyclase, the nitric oxide receptor. *J Proteomics* 2016;**138**:40.
157. Pan KT, Chen YY, Pu TH, Chao YS, Yang CY, Bomgarden RD, et al. Mass spectrometry-based quantitative proteomics for dissecting multiplexed redox cysteine modifications in nitric oxide-protected cardiomyocyte under hypoxia. *Antioxid Redox Signal* 2014;**20**:1365.
158. Kirsch M, Buscher A-M, Aker S, Schulz R, de Groot H. New insights into the S-nitrosothiol-ascorbate reaction. The formation of nitroxyl. *Org Biomol Chem* 2009;**7**:1954.
159. Wojdyla K, Williamson J, Roepstorff P, Rogowska-Wrzesinska A. The SNO/SOH TMT strategy for combinatorial analysis of reversible cysteine oxidations. *J Proteomics* 2015;**113**:415.

160. Guo J, Gaffrey MJ, Su D, Liu T, Camp DG, Smith RD, et al. Resin-assisted enrichment of thiols as a general strategy for proteomic profiling of cysteine-based reversible modifications. *Nat Protoc* 2014;**9**:64.
161. Su D, Gaffrey MJ, Guo J, Hatchell KE, Chu RK, Clauss TR, et al. Proteomic identification and quantification of S-glutathionylation in mouse macrophages using resin-assisted enrichment and isobaric labeling. *Free Radic Biol Med* 2014;**67**:460.
162. Duan J, Kodali VK, Gaffrey MJ, Guo J, Chu RK, Camp DG, et al. Quantitative Profiling of Protein S-Glutathionylation Reveals Redox-Dependent Regulation of Macrophage Function during Nanoparticle-Induced Oxidative Stress. *ACS Nano* 2016;**10**:524.
163. Behring JB, Kumar V, Whelan SA, Chauhan P, Siwik DA, Costello CE, et al. Does reversible cysteine oxidation link the Western diet to cardiac dysfunction? *FASEB J* 2014;**28**:1975.
164. Yao C, Behring JB, Shao D, Sverdlov AL, Whelan SA, Elezaby A, et al. Overexpression of Catalase Diminishes Oxidative Cysteine Modifications of Cardiac Proteins. *Plos One* 2015;**10**:e0144025.
165. Galligan JJ, Rose KL, Beavers WN, Hill S, Tallman KA, Tansey WP, et al. Stable histone adduction by 4-oxo-2-nonenal: a potential link between oxidative stress and epigenetics. *J Am Chem Soc* 2014;**136**:11864.
166. Fedorova M, Bollineni RC, Hoffmann R. Protein carbonylation as a major hallmark of oxidative damage: update of analytical strategies. *Mass Spectrom Rev* 2014;**33**:79.
167. Madian AG, Myracle AD, Diaz-Maldonado N, Rochelle NS, Janle EM, Regnier FE. Differential carbonylation of proteins as a function of in vivo oxidative stress. *J Proteome Res* 2011;**10**:3959.
168. Baraibar MA, Ladouce R, Friguet B. Proteomic quantification and identification of carbonylated proteins upon oxidative stress and during cellular aging. *J Proteomics* 2013;**92**:63.
169. Hensley K. Detection of Protein Carbonyls by Means of Biotin Hydrazide-Streptavidin Affinity Methods. In: Kurien BT, Scofield RH, editors. *Protein Blotting and Detection*, Vol. 536. New York, NY: Humana Press; 2009. p. 457.
170. Grimsrud PA, Xie H, Griffin TJ, Bernlohr DA. Oxidative stress and covalent modification of protein with bioactive aldehydes. *J Biol Chem* 2008;**283**:21837.
171. Pfeiffer S, Schmidt K, Mayer B. Dityrosine formation outcompetes tyrosine nitration at low steady-state concentrations of peroxynitrite. Implications for tyrosine modification by nitric oxide/superoxide in vivo. *J Biol Chem* 2000;**275**:6346.
172. Surmeli NB, Litterman NK, Miller AF, Groves JT. Peroxynitrite mediates active site tyrosine nitration in manganese superoxide dismutase. Evidence of a role for the carbonate radical anion. *J Am Chem Soc* 2010;**132**:17174.
173. Lymar SV, Jiang Q, Hurst JK. Mechanism of carbon dioxide-catalyzed oxidation of tyrosine by peroxynitrite. *Biochemistry* 1996;**35**:7855.
174. Tien M, Berlett BS, Levine RL, Chock PB, Stadtman ER. Peroxynitrite-mediated modification of proteins at physiological carbon dioxide concentration: pH dependence of carbonyl formation, tyrosine nitration, and methionine oxidation. *Proc Natl Acad Sci U S A* 1999;**96**:7809.
175. Gow A, Duran D, Thom SR, Ischiropoulos H. Carbon dioxide enhancement of peroxynitrite-mediated protein tyrosine nitration. *Arch Biochem Biophys* 1996;**333**:42.
176. Sampson JB, Ye Y, Rosen H, Beckman JS. Myeloperoxidase and horseradish peroxidase catalyze tyrosine nitration in proteins from nitrite and hydrogen peroxide. *Arch Biochem Biophys* 1998;**356**:207.
177. Eiserich JP, Hristova M, Cross CE, Jones AD, Freeman BA, Halliwell B, et al. Formation of nitric oxide-derived inflammatory oxidants by myeloperoxidase in neutrophils. *Nature* 1998;**391**:393.
178. Yeo WS, Kim YJ, Kabir MH, Kang JW, Ahsan-Ul-Bari M, Kim KP. Mass spectrometric analysis of protein tyrosine nitration in aging and neurodegenerative diseases. *Mass Spectrom Rev* 2015;**34**:166.
179. Chao MR, Hsu YW, Liu HH, Lin JH, Hu CW. Simultaneous Detection of 3-Nitrotyrosine and 3-Nitro-4-hydroxyphenylacetic Acid in Human Urine by Online SPE LC-MS/MS and Their Association with Oxidative and Methylated DNA Lesions. *Chem Res Toxicol* 2015;**28**:997.
180. Torres-Cuevas I, Kuligowski J, Carcel M, Chafer-Pericas C, Asensi M, Solberg R, et al. Protein-bound tyrosine oxidation, nitration and chlorination by-products assessed by ultraperformance liquid chromatography coupled to tandem mass spectrometry. *Anal Chim Acta* 2016;**913**:104.

181. Peng F, Li J, Guo T, Yang H, Li M, Sang S, et al. Nitroproteins in Human Astrocytomas Discovered by Gel Electrophoresis and Tandem Mass Spectrometry. *J Am Soc Mass Spectrom* 2015;**26**:2062.
182. Tsikas D, Schwedhelm E, Frölich JC. Methodological Considerations on the Detection of 3-Nitrotyrosine in the Cardiovascular System. *Circ Res* 2002;**90**:e70.
183. Niki E. Biomarkers of lipid peroxidation in clinical material. *Biochim Biophys Acta* 2014;**1840**:809.
184. Yin H, Xu L, Porter NA. Free radical lipid peroxidation: mechanisms and analysis. *Chem Rev* 2011;**111**:5944.
185. Schafer FQ, Wang HP, Kelley EE, Cueno KL, Martin SM, Buettner GR. Comparing beta-carotene, vitamin E and nitric oxide as membrane antioxidants. *Biol Chem* 2002;**383**:671.
186. Kellogg III EW, Fridovich I. Superoxide, hydrogen peroxide, and singlet oxygen in lipid peroxidation by a xanthine oxidase system. *J Biol Chem* 1975;**250**:8812.
187. Koppenol WH. Reactions involving singlet oxygen and the superoxide anion. *Nature* 1976;**262**:420.
188. Stratton SP, Liebler DC. Determination of singlet oxygen-specific versus radical-mediated lipid peroxidation in photosensitized oxidation of lipid bilayers: effect of beta-carotene and alpha-tocopherol. *Biochemistry* 1997;**36**:12911.
189. Tyurin VA, Tyurina YY, Ritov VB, Lysytsya A, Amoscato AA, Kochanek PM, et al. Oxidative Lipidomics of Apoptosis: Quantitative Assessment of Phospholipid Hydroperoxides in Cells and Tissues. In: Uppu MR, Murthy NS, Pryor AW, Parinandi LN, editors. *Free Radicals and Antioxidant Protocols*. Totowa, NJ: Humana Press; 2010. p. 353.
190. Yin H, Porter NA. Identification of intact lipid peroxides by Ag+ coordination ion-spray mass spectrometry (CIS-MS). *Methods Enzymol* 2007;**433**:193.
191. Havrilla CM, Hachey DL, Porter NA. Coordination (Ag+) Ion Spray – Mass Spectrometry of Peroxidation Products of Cholesterol Linoleate and Cholesterol Arachidonate: High-Performance Liquid Chromatography – Mass Spectrometry Analysis of Peroxide Products from Polyunsaturated Lipid Autoxidation. *J Am Chem Soc* 2000;**122**:8042.
192. Milne GL, Porter NA. Separation and identification of phospholipid peroxidation products. *Lipids* 2001;**36**:1265.
193. Yin H, Porter NA. New insights regarding the autoxidation of polyunsaturated fatty acids. *Antioxid Redox Signal* 2005;**7**:170.
194. O'Donnell VB. Free radicals and lipid signaling in endothelial cells. *Antioxid Redox Signal* 2003;**5**:195.
195. Yoshida Y, Niki E. Detection of lipid peroxidation in vivo: Total hydroxyoctadecadienoic acid and 7-hydroxycholesterol as oxidative stress marker. *Free Radical Res* 2004;**38**:787.
196. Liu W, Yin H, Akazawa YO, Yoshida Y, Niki E, Porter NA. Ex vivo oxidation in tissue and plasma assays of hydroxyoctadecadienoates: Z,E/E,E stereoisomer ratios. *Chem Res Toxicol* 2010;**23**:986.
197. Seet RCS, Lee C-YJ, Lim ECH, Tan JJH, Quek AML, Chong W-L, et al. Oxidative damage in Parkinson disease: Measurement using accurate biomarkers. *Free Radic Biol Med* 2010;**48**:560.
198. Roberts II LJ, Morrow JD. The generation and actions of isoprostanes. *Biochim Biophys Acta* 1997;**1345**:121.
199. Montuschi P, Barnes PJ, Roberts II LJ. Isoprostanes: markers and mediators of oxidative stress. *FASEB J* 2004;**18**:1791.
200. Davies SS, Roberts II LJ. F2-isoprostanes as an indicator and risk factor for coronary heart disease. *Free Radic Biol Med* 2011;**50**:559.
201. Yin H, Musiek ES, Gao L, Porter NA, Morrow JD. Regiochemistry of neuroprostanes generated from the peroxidation of docosahexaenoic acid in vitro and in vivo. *J Biol Chem* 2005;**280**:26600.
202. Reich EE, Zackert WE, Brame CJ, Chen Y, Roberts II LJ, Hachey DL, et al. Formation of novel D-ring and E-ring isoprostane-like compounds (D4/E4-neuroprostanes) in vivo from docosahexaenoic acid. *Biochemistry* 2000;**39**:2376.
203. Liu W, Morrow JD, Yin H. Quantification of F2-isoprostanes as a reliable index of oxidative stress in vivo using gas chromatography-mass spectrometry (GC-MS) method. *Free Radic Biol Med* 2009;**47**:1101.

204. Arneson KO, Roberts Ii,LJ. *Measurement of Products of Docosahexaenoic Acid Peroxidation, Neuroprostanes, and Neurofurans. Methods Enzymol*, Volume 433. Academic Press; 2007p. 127.
205. Markesbery WR, Kryscio RJ, Lovell MA, Morrow JD. Lipid peroxidation is an early event in the brain in amnestic mild cognitive impairment. *Ann Neurol* 2005;**58**:730.
206. Montine TJ, Neely MD, Quinn JF, Beal MF, Markesbery WR, Roberts LJ, et al. Lipid peroxidation in aging brain and Alzheimer's disease. *Free Radic Biol Med* 2002;**33**:620.
207. De Felice C, Della Ragione F, Signorini C, Leoncini S, Pecorelli A, Ciccoli L, et al. Oxidative brain damage in Mecp2-mutant murine models of Rett syndrome. *Neurobiol Dis* 2014;**68**:66.
208. Yen HC, Chen TW, Yang TC, Wei HJ, Hsu JC, Lin CL. Levels of F2-isoprostanes, F4-neuroprostanes, and total nitrate/nitrite in plasma and cerebrospinal fluid of patients with traumatic brain injury. *Free Radical Res* 2015;**49**:1419.
209. Korade Z, Xu L, Mirnics K, Porter NA. Lipid biomarkers of oxidative stress in a genetic mouse model of Smith-Lemli-Opitz syndrome. *J Inherit Metab Dis* 2013;**36**:113.
210. Solberg R, Longini M, Proietti F, Vezzosi P, Saugstad OD, Buonocore G. Resuscitation with supplementary oxygen induces oxidative injury in the cerebral cortex. *Free Radic Biol Med* 2012;**53**:1061.
211. Hsieh YP, Lin CL, Shiue AL, Yin H, Morrow JD, Hsu JC, et al. Correlation of F4-neuroprostanes levels in cerebrospinal fluid with outcome of aneurysmal subarachnoid hemorrhage in humans. *Free Radic Biol Med* 2009;**47**:814.
212. Milne GL, Morrow JD, Picklo Sr. MJ. Elevated oxidation of docosahexaenoic acid, 22:6 (n-3), in brain regions of rats undergoing ethanol withdrawal. *Neurosci Lett* 2006;**405**:172.
213. Cuddihy SL, Ali SS, Musiek ES, Lucero J, Kopp SJ, Morrow JD, et al. Prolonged alpha-tocopherol deficiency decreases oxidative stress and unmasks alpha-tocopherol-dependent regulation of mitochondrial function in the brain. *J Biol Chem* 2008;**283**:6915.
214. Montine TJ, Quinn JF, Milatovic D, Silbert LC, Dang T, Sanchez S, et al. Peripheral F2-isoprostanes and F4-neuroprostanes are not increased in Alzheimer's disease. *Ann Neurol* 2002;**52**:175.
215. Yen H-C, Wei H-J, Chen T-W. Analytical Variables Affecting Analysis of F2-Isoprostanes and F4-Neuroprostanes in Human Cerebrospinal Fluid by Gas Chromatography/Mass Spectrometry. *BioMed Res Int* 2013;**2013**:14.
216. Yen HC, Wei HJ, Lin CL. Unresolved issues in the analysis of F2-isoprostanes, F4-neuroprostanes, isofurans, neurofurans, and F2-dihomo-isoprostanes in body fluids and tissue using gas chromatography/negative-ion chemical-ionization mass spectrometry. *Free Radical Res* 2015;**49**:861.
217. Milne GL, Yin H, Brooks JD, Sanchez S, Jackson Roberts II L, Morrow JD. Quantification of F2-isoprostanes in biological fluids and tissues as a measure of oxidant stress. *Methods Enzymol* 2007;**433**:113.
218. Taylor AW, Bruno RS, Frei B, Traber MG. Benefits of prolonged gradient separation for high-performance liquid chromatography-tandem mass spectrometry quantitation of plasma total 15-series F-isoprostanes. *Anal Biochem* 2006;**350**:41.
219. Morales CR, Terry ES, Zackert WE, Montine TJ, Morrow JD. Improved assay for the quantification of the major urinary metabolite of the isoprostane 15-F(2t)-Isoprostane (8-iso-PGF(2alpha)) by a stable isotope dilution mass spectrometric assay. *Clin Chim Acta* 2001;**314**:93.
220. Tsikas D, Rothmann S, Schneider JY, Suchy MT, Trettin A, Modun D, et al. Development, validation and biomedical applications of stable-isotope dilution GC-MS and GC-MS/MS techniques for circulating malondialdehyde (MDA) after pentafluorobenzyl bromide derivatization: MDA as a biomarker of oxidative stress and its relation to 15(S)-8-iso-prostaglandin F2alpha and nitric oxide (NO). *J Chromatogr B* 2016;**1019**:95.
221. Zelzer S, Mangge H, Oberreither R, Bernecker C, Gruber HJ, Pruller F, et al. Oxidative stress: Determination of 4-hydroxy-2-nonenal by gas chromatography/mass spectrometry in human and rat plasma. *Free Radical Res* 2015;**49**:1233.
222. Kim DH, Kwack SJ, Yoon KS, Choi JS, Lee BM. 4-Hydroxynonenal: A Superior Oxidative Biomarker Compared to Malondialdehyde and Carbonyl Content Induced by Carbon Tetrachloride in Rats. *J Toxicol Environ Health A* 2015;**78**:1051.

223. Kanvah S, Joseph J, Schuster GB, Barnett RN, Cleveland CL, Landman U. Oxidation of DNA: damage to nucleobases. *Acc Chem Res* 2010;**43**:280.
224. Lim KS, Taghizadeh K, Wishnok JS, Babu IR, Shafirovich V, Geacintov NE, et al. Sequence-Dependent Variation in the Reactivity of 8-Oxo-7,8-dihydro-2'-deoxyguanosine toward Oxidation. *Chem Res Toxicol* 2012;**25**:366.
225. Cadet J, Poulsen H. Measurement of oxidatively generated base damage in cellular DNA and urine. *Free Radic Biol Med* 2010;**48**:1457.
226. Gackowski D, Rozalski R, Siomek A, Dziaman T, Nicpon K, Klimarczyk M, et al. Oxidative stress and oxidative DNA damage is characteristic for mixed Alzheimer disease/vascular dementia. *J Neurol Sci* 2008;**266**:57.
227. Halliwell B. Why and how should we measure oxidative DNA damage in nutritional studies? How far have we come? *Am J Clin Nutr* 2000;**72**:1082.
228. Ravanat J-L. Chromatographic methods for the analysis of oxidatively damaged DNA. *Free Radical Res* 2012;**46**:479.
229. Mangal D, Vudathala D, Park J-H, Lee SH, Penning TM, Blair IA. Analysis of 7,8-Dihydro-8-oxo-2'-deoxyguanosine in Cellular DNA during Oxidative Stress. *Chem Res Toxicol* 2009;**22**:788.
230. Chao M-R, Yen C-C, Hu C-W. Prevention of artifactual oxidation in determination of cellular 8-oxo-7,8-dihydro-2'-deoxyguanosine by isotope-dilution LC-MS/MS with automated solid-phase extraction. *Free Radic Biol Med* 2008;**44**:464.
231. Loft S, Danielsen P, Loehr M, Jantzen K, Hemmingsen JG, Roursgaard M, et al. Urinary excretion of 8-oxo-7,8-dihydroguanine as biomarker of oxidative damage to DNA. *Arch Biochem Biophys* 2012;**518**:142.
232. Mistry V, Teichert F, Sandhu JK, Singh R, Evans MD, Farmer PB, et al. Non-invasive assessment of oxidatively damaged DNA: liquid chromatography-tandem mass spectrometry analysis of urinary 8-oxo-7,8-dihydro-2'-deoxyguanosine. *Methods Mol Biol* 2011;**682**:279.
233. Olinski R, Rozalski R, Gackowski D, Foksinski M, Siomek A, Cooke MS. Urinary Measurement of 8-OxodG, 8-OxoGua, and 5HMUra: A Noninvasive Assessment of Oxidative Damage to DNA. *Antioxid Redox Signal* 2006;**8**:1011.
234. Kataoka H, Mizuno K, Oda E, Saito A. Determination of the oxidative stress biomarker urinary 8-hydroxy-2'-deoxyguanosine by automated on-line in-tube solid-phase microextraction coupled with liquid chromatography-tandem mass spectrometry. *J Chromatogr B* 2016;**1019**:140.
235. Kuligowski J, Torres-Cuevas I, Quintas G, Rook D, van Goudoever JB, Cubells E, et al. Assessment of oxidative damage to proteins and DNA in urine of newborn infants by a validated UPLC-MS/MS approach. *PLoS One* 2014;**9**:e93703/1.
236. Chen H-JC, Lin W-P. Quantitative Analysis of Multiple Exocyclic DNA Adducts in Human Salivary DNA by Stable Isotope Dilution Nanoflow Liquid Chromatography-Nanospray Ionization Tandem Mass Spectrometry. *Anal Chem* 2011;**83**:8543.
237. Hu C-W, Liu H-H, Li Y-J, Chao M-R. Direct Analysis of 5-Methylcytosine and 5-Methyl-2'-deoxycytidine in Human Urine by Isotope Dilution LC-MS/MS: Correlations with N-Methylated Purines and Oxidized DNA Lesions. *Chem Res Toxicol* 2012;**25**:462.
238. Jaruga P, Dizdaroglu M. Identification and quantification of (5'R)- and (5'S)-8,5'-cyclo-2'-deoxyadenosines in human urine as putative biomarkers of oxidatively induced damage to DNA. *Biochem Biophys Res Commun* 2010;**397**:48.
239. Li P, Ramm GA, Macdonald GA. Value of the 8-oxodG/dG ratio in chronic liver inflammation of patients with hepatocellular carcinoma. *Redox Biol* 2016;**8**:259.
240. Cadet J, Wagner JR. DNA base damage by reactive oxygen species, oxidizing agents, and UV radiation. *Cold Spring Harb Perspect Biol* 2013;**5**:a012559.
241. Cadet J, Wagner JR. Oxidatively generated base damage to cellular DNA by hydroxyl radical and one-electron oxidants: similarities and differences. *Arch Biochem Biophys* 2014;**557**:47.
242. Pang B, Zhou X, Yu H, Dong M, Taghizadeh K, Wishnok JS, et al. Lipid peroxidation dominates the chemistry of DNA adduct formation in a mouse model of inflammation. *Carcinogenesis* 2007;**28**:1807.

243. Yu Y, Guerrero CR, Liu S, Amato NJ, Sharma Y, Gupta S, et al. Comprehensive Assessment of Oxidatively Induced Modifications of DNA in a Rat Model of Human Wilson's Disease. *Mol Cell Proteomics* 2016;**15**:810.
244. Taghizadeh K, McFaline JL, Pang B, Sullivan M, Dong M, Plummer E, et al. Quantification of DNA damage products resulting from deamination, oxidation and reaction with products of lipid peroxidation by liquid chromatography isotope dilution tandem mass spectrometry. *Nat Protoc* 2008;**3**:1287.
245. Prestwich EG, Mangerich A, Pang B, McFaline JL, Lonkar P, Sullivan MR, et al. Increased levels of inosine in a mouse model of inflammation. *Chem Res Toxicol* 2013;**26**:538.
246. Liu X, Lovell MA, Lynn BC. Detection and Quantification of Endogenous Cyclic DNA Adducts Derived from trans-4-Hydroxy-2-nonenal in Human Brain Tissue by Isotope Dilution Capillary Liquid Chromatography Nanoelectrospray Tandem Mass Spectrometry. *Chem Res Toxicol* 2006;**19**:710.
247. Fedeles BI, Freudenthal BD, Yau E, Singh V, Chang SC, Li D, et al. Intrinsic mutagenic properties of 5-chlorocytosine: A mechanistic connection between chronic inflammation and cancer. *Proc Natl Acad Sci U S A* 2015;**112**:E4571.
248. Hah SS, Kim HM, Sumbad RA, Henderson PT. Hydantoin derivative formation from oxidation of 7,8-dihydro-8-oxo-2'-deoxyguanosine (8-oxodG) and incorporation of 14C-labeled 8-oxodG into the DNA of human breast cancer cells. *Bioorg Med Chem Lett* 2005;**15**:3627.
249. Henderson PT, Delaney JC, Muller JG, Neeley WL, Tannenbaum SR, Burrows CJ, et al. The hydantoin lesions formed from oxidation of 7,8-dihydro-8-oxoguanine are potent sources of replication errors in vivo. *Biochemistry* 2003;**42**:9257.
250. Yu H, Venkatarangan L, Wishnok JS, Tannenbaum SR. Quantitation of Four Guanine Oxidation Products from Reaction of DNA with Varying Doses of Peroxynitrite. *Chem Res Toxicol* 2005;**18**:1849.
251. Norabuena EM, Barnes Williams S, Klureza MA, Goehring LJ, Gruessner B, Radhakrishnan ML, et al. Effect of the Spiroiminodihydantoin Lesion on Nucleosome Stability and Positioning. *Biochemistry* 2016;**55**:2411.
252. Chowdhury G, Guengerich FP. Tandem Mass Spectrometry-Based Detection of C4'-Oxidized Abasic Sites at Specific Positions in DNA Fragments. *Chem Res Toxicol* 2009;**22**:1310.
253. Chen B, Zhou X, Taghizadeh K, Chen J, Stubbe J, Dedon PC. GC/MS Methods To Quantify the 2-Deoxypentos-4-ulose and 3'-Phosphoglycolate Pathways of 4' Oxidation of 2-Deoxyribose in DNA: Application to DNA Damage Produced by γ Radiation and Bleomycin. *Chem Res Toxicol* 2007;**20**:1701.
254. Collins C, Awada MM, Zhou X, Dedon PC. Analysis of 3'-phosphoglycolaldehyde residues in oxidized DNA by gas chromatography/negative chemical ionization/mass spectrometry. *Chem Res Toxicol* 2003;**16**:1560.
255. Tang Y, Zheng S-J, Qi C-B, Feng Y-Q, Yuan B-F. Sensitive and Simultaneous Determination of 5-Methylcytosine and Its Oxidation Products in Genomic DNA by Chemical Derivatization Coupled with Liquid Chromatography-Tandem Mass Spectrometry Analysis. *Anal Chem* 2015;**87**:3445.
256. Rossner Jr P, Sram RJ. Immunochemical detection of oxidatively damaged DNA. *Free Radical Res* 2012;**46**:492.
257. Toyokuni S, Tanaka T, Hattori Y, Nishiyama Y, Yoshida A, Uchida K, et al. Quantitative immunohistochemical determination of 8-hydroxy-2'-deoxyguanosine by a monoclonal antibody N45.1: its application to ferric nitrilotriacetate-induced renal carcinogenesis model. *Lab Invest* 1997;**76**:365.
258. Shoji H, Oguchi S, Shimizu T, Yamashiro Y. Effect of human breast milk on urinary 8-hydroxy-2'-deoxyguanosine excretion in infants. *Pediatr Res* 2003;**53**:850.
259. Hiraku Y, Kawanishi S. Immunohistochemical Analysis of 8-Nitroguanine, A Nitrative DNA Lesion, in Relation to Inflammation-Associated Carcinogenesis. In: Kozlov SV, editor. *Inflammation and Cancer: Methods and Protocols: Volume 2: Molecular Analysis and Pathways*. Totowa, NJ: Humana Press; 2009. p. 3.

260. Pinlaor S, Hiraku Y, Ma N, Yongvanit P, Semba R, Oikawa S, et al. Mechanism of NO-mediated oxidative and nitrative DNA damage in hamsters infected with Opisthorchis viverrini: a model of inflammation-mediated carcinogenesis. *Nitric Oxide* 2004;**11**:175.
261. Pinlaor S, Yongvanit P, Hiraku Y, Ma N, Semba R, Oikawa S, et al. 8-nitroguanine formation in the liver of hamsters infected with Opisthorchis viverrini. *Biochem Biophys Res Commun* 2003;**309**:567.
262. Sawa T, Tatemichi M, Akaike T, Barbin A, Ohshima H. Analysis of urinary 8-nitroguanine, a marker of nitrative nucleic acid damage, by high-performance liquid chromatography-electrochemical detection coupled with immunoaffinity purification: association with cigarette smoking. *Free Radic Biol Med* 2006;**40**:711.
263. Greferath R, Nehls P. Monoclonal antibodies to thymidine glycol generated by different immunization techniques. *Hybridoma* 1997;**16**:189.
264. Augustyniak E, Adam A, Wojdyla K, Rogowska-Wrzesinska A, Willetts R, Korkmaz A, et al. Validation of protein carbonyl measurement: A multi-centre study. *Redox Biol* 2015;**4**:149.
265. Whiteman M, Siau JL, Halliwell B. Lack of Tyrosine Nitration by Hypochlorous Acid in the Presence of Physiological Concentrations of Nitrite. *J Biol Chem* 2003;**278**:8380.
266. Sultana R, Butterfield DA. *Slot-Blot Analysis of 3-Nitrotyrosine-Modified Brain Proteins. Methods Enzymol*, Volume 440. Academic Press; 2008p. 309.
267. Maneen MJ, Hannah R, Vitullo L, DeLance N, Cipolla MJ. Peroxynitrite Diminishes Myogenic Activity and is Associated with Decreased Vascular Smooth Muscle F-Actin in Rat Posterior Cerebral Arteries. *Stroke* 2006;**37**:894.
268. Aladag MA, Turkoz Y, Parlakpinar H, Ozen H, Egri M, Unal SC. Melatonin Ameliorates Cerebral Vasospasm After Experimental Subarachnoidal Haemorrhage Correcting Imbalance of Nitric Oxide Levels in Rats. *Neurochem Res* 2009;**34**:1935.
269. Rossi MA, Celes MRN, Prado CM, Saggioro FP. Myocardial structural changes in long-term human severe sepsis/septic shock may be responsible for cardiac dysfunction. *Shock* 2007;**27**:10.
270. Ikeda K, Yukihiro HB, Iwai H, Matsumoto T, Mineki R, Taka H, et al. Detection of 6-nitrotryptophan in proteins by Western blot analysis and its application for peroxynitrite-treated PC12 cells. *Nitric Oxide* 2007;**16**:18.
271. Kawasaki H, Ikeda K, Shigenaga A, Baba T, Takamori K, Ogawa H, et al. Mass spectrometric identification of tryptophan nitration sites on proteins in peroxynitrite-treated lysates from PC12 cells. *Free Radic Biol Med* 2011;**50**:419.
272. Kawasaki H, Shigenaga A, Uda M, Baba T, Ogawa H, Takamori K, et al. Nitration of tryptophan in ribosomal proteins is a novel post-translational modification of differentiated and naive PC12 cells. *Nitric Oxide* 2011;**25**:176.
273. Kawasaki H, Tominaga M, Shigenaga A, Kamo A, Kamata Y, Iizumi K, et al. Importance of tryptophan nitration of carbonic anhydrase III for the morbidity of atopic dermatitis. *Free Radic Biol Med* 2014;**73**:75.
274. Kato Y, Wu X, Naito M, Nomura H, Kitamoto N, Osawa T. Immunochemical detection of protein dityrosine in atherosclerotic lesion of apo-E-deficient mice using a novel monoclonal antibody. *Biochem Biophys Res Commun* 2000;**275**:11.
275. Kato Y, Kitamoto N, Kawai Y, Osawa T. The hydrogen peroxide/copper ion system, but not other metal-catalyzed oxidation systems, produces protein-bound dityrosine. *Free Radic Biol Med* 2001;**31**:624.
276. Kato Y, Kawai Y, Morinaga H, Kondo H, Dozaki N, Kitamoto N, et al. Immunogenicity of a brominated protein and successive establishment of a monoclonal antibody to dihalogenated tyrosine. *Free Radic Biol Med* 2005;**38**:24.
277. Takagi T, Naito Y, Okada H, Okayama T, Mizushima K, Yamada S, et al. Indentification of dihalogenated proteins in rat intestinal mucosa injured by indomethacin. *J Clin Biochem Nutr* 2011;**48**:178.
278. Jin H, Hallstrand TS, Daly DS, Matzke MM, Nair P, Bigelow DJ, et al. A halotyrosine antibody that detects increased protein modifications in asthma patients. *J Immunol Methods* 2014;**403**:17.

279. Richter T, Munch G, Lueth H-J, Arendt T, Kientsch-Engel R, Stahl P, et al. Immunochemical crossreactivity of antibodies specific for "advanced glycation endproducts" with "advanced lipoxidation endproducts". *Neurobiol Aging* 2005;**26**:465.
280. Vasquez-Garzon VR, Rouimi P, Jouanin I, Waeg G, Zarkovic N, Villa-Trevino S, et al. Evaluation of three simple direct or indirect carbonyl detection methods for characterization of oxidative modifications of proteins. *Toxicol Mech Methods* 2012;**22**:296.
281. Kato Y, Osawa T. Detection of a lipid-lysine adduct family with an amide bond as the linkage: novel markers for lipid-derived protein modifications. *Methods Mol Biol* 2009;**580**:129.
282. Sakai K, Kino S, Masuda A, Takeuchi M, Ochi T, Osredkar J, et al. Determination of HEL (Hexanoyl-Lysine Adduct): A Novel Biomarker for Omega-6 PUFA Oxidation. In: Kato Y, editor. *Lipid Hydroperoxide-Derived Modification of Biomolecules*. Netherlands: Dordrecht: Springer; 2014. p. 61.
283. Ichiseki T, Kaneuji A, Ueda Y, Kaneko S, Ueda S, Matsumoto T. The initial phase of oxidative stress in a steroid-induced osteonecrosis rabbit model. *Adv Biosci Biotechnol* 2012;**3**:978.
284. Okada H, Naito Y, Takagi T, Takaoka M, Oya-Ito T, Fukumoto K, et al. Detection of Nε-(hexanoyl)lysine in the tropomyosin 1 protein in N-methyl-N'-nitro-N-nitrosoguanidine-induced rat gastric cancer cells. *J Clin Biochem Nutr* 2012;**50**:47.
285. Majima HJ, Nakanishi-Ueda T, Ozawa T. 4-Hydroxy-2-nonenal (4-HNE) staining by anti-HNE antibody. *Methods Mol Biol* 2002;**196**:31.
286. Szweda PA, Tsai L, Szweda LI. Immunochemical detection of a fluorophore derived from the lipid peroxidation product 4-hydroxy-2-nonenal and lysine. *Methods Mol Biol* 2002;**196**:277.
287. Hartley DP, Kroll DJ, Petersen DR. Prooxidant-Initiated Lipid Peroxidation in Isolated Rat Hepatocytes: Detection of 4-Hydroxynonenal- and Malondialdehyde-Protein Adducts. *Chem Res Toxicol* 1997;**10**:895.
288. Sayre LM, Sha W, Xu G, Kaur K, Nadkarni D, Subbanagounder G, et al. Immunochemical Evidence Supporting 2-Pentylpyrrole Formation on Proteins Exposed to 4-Hydroxy-2-nonenal. *Chem Res Toxicol* 1996;**9**:1194.
289. Salomon RG, Kaur K, Podrez E, Hoff HF, Krushinsky AV, Sayre LM. HNE-Derived 2-Pentylpyrroles Are Generated during Oxidation of LDL, Are More Prevalent in Blood Plasma from Patients with Renal Disease or Atherosclerosis, and Are Present in Atherosclerotic Plaques. *Chem Res Toxicol* 2000;**13**:557.
290. Itakura K, Furuhata A, Shibata N, Kobayashi M, Uchida K. Maillard reaction-like lysine modification by a lipid peroxidation product: immunochemical detection of protein-bound 2-hydroxyheptanal in vivo. *Biochem Biophys Res Commun* 2003;**308**:452.
291. Yamada S, Funada T, Shibata N, Kobayashi M, Kawai Y, Tatsuda E, et al. Protein-bound 4-hydroxy-2-hexenal as a marker of oxidized n-3 polyunsaturated fatty acids. *J Lipid Res* 2004;**45**:626.
292. Pillon NJ, Vella RE, Soulere L, Becchi M, Lagarde M, Soulage CO. Structural and Functional Changes in Human Insulin Induced by the Lipid Peroxidation Byproducts 4-Hydroxy-2-nonenal and 4-Hydroxy-2-hexenal. *Chem Res Toxicol* 2011;**24**:752.
293. Uchida K, Sakai K, Itakura K, Osawa T, Toyokuni S. Protein modification by lipid peroxidation products: formation of malondialdehyde-derived Nε-(2-propenal)lysine in proteins. *Arch Biochem Biophys* 1997;**346**:45.
294. Ishii T, Kumazawa S, Sakurai T, Nakayama T, Uchida K. Mass Spectroscopic Characterization of Protein Modification by Malondialdehyde. *Chem Res Toxicol* 2006;**19**:122.
295. Williams JD, Bermudez Y, Park SL, Stratton SP, Uchida K, Hurst CA, et al. Malondialdehyde-derived epitopes in human skin result from acute exposure to solar UV and occur in nonmelanoma skin cancer tissue. *J Photochem Photobiol, B* 2014;**132**:56.
296. Lamore SD, Azimian S, Horn D, Anglin BL, Uchida K, Cabello CM, et al. The malondialdehyde-derived fluorophore DHP-lysine is a potent sensitizer of UVA-induced photooxidative stress in human skin cells. *J Photochem Photobiol, B* 2010;**101**:251.
297. Uchida K, Kanematsu M, Sakai K, Matsuda T, Hattori N, Mizuno Y, et al. Protein-bound acrolein: potential markers for oxidative stress. *Proc Natl Acad Sci U S A* 1998;**95**:4882.

298. Furuhata A, Ishii T, Kumazawa S, Yamada T, Nakayama T, Uchida K. Nε-3-(Methylpyridinium) lysine, a Major Antigenic Adduct Generated in Acrolein-modified Protein. *J Biol Chem* 2003;**278**:48658.
299. Calingasan NY, Uchida K, Gibson GE. Protein-bound acrolein: a novel marker of oxidative stress in Alzheimer's disease. *J Neurochem* 1999;**72**:751.
300. Tanaka N, Tajima S, Ishibashi A, Uchida K, Shigematsu T. Immunohistochemical detection of lipid peroxidation products, protein-bound acrolein and 4-hydroxynonenal protein adducts, in actinic elastosis of photodamaged skin. *Arch Dermatol Res* 2001;**293**:363.
301. Suzuki D, Miyata T, Saotome N, Horie K, Inagi R, Yasuda Y, et al. Immunohistochemical evidence for an increased oxidative stress and carbonyl modification of proteins in diabetic glomerular lesions. *J Am Soc Nephrol* 1999;**10**:822.
302. Hirose T, Saiki R, Uemura T, Suzuki T, Dohmae N, Ito S, et al. Increase in acrolein-conjugated immunoglobulins in saliva from patients with primary Sjogren's syndrome. *Clin Chim Acta* 2015;**450**:184.
303. Tamamizu-Kato S, Wong JY, Jairam V, Uchida K, Raussens V, Kato H, et al. Modification by Acrolein, a Component of Tobacco Smoke and Age-Related Oxidative Stress, Mediates Functional Impairment of Human Apolipoprotein E. *Biochemistry* 2007;**46**:8392.
304. Tran TN, Kosaraju MG, Tamamizu-Kato S, Akintunde O, Zheng Y, Bielicki JK, et al. Acrolein Modification Impairs Key Functional Features of Rat Apolipoprotein E: Identification of Modified Sites by Mass Spectrometry. *Biochemistry* 2014;**53**:361.
305. Mottaran E, Stewart SF, Rolla R, Vay D, Cipriani V, Moretti M, et al. Lipid peroxidation contributes to immune reactions associated with alcoholic liver disease. *Free Radic Biol Med* 2001;**32**:38.
306. Furuhata A, Nakamura M, Osawa T, Uchida K. Thiolation of protein-bound carcinogenic aldehyde: an electrophilic acrolein-lysine adduct that covalently binds to thiols. *J Biol Chem* 2002;**277**:27919.
307. Furuhata A, Honda K, Shibata T, Chikazawa M, Kawai Y, Shibata N, et al. Monoclonal Antibody against Protein-Bound Glutathione: Use of Glutathione Conjugate of Acrolein-Modified Proteins as an Immunogen. *Chem Res Toxicol* 2012;**25**:1393.
308. Chumsae C, Gifford K, Lian W, Liu H, Radziejewski CH, Zhou ZS. Arginine Modifications by Methylglyoxal: Discovery in a Recombinant Monoclonal Antibody and Contribution to Acidic Species. *Anal Chem* 2013;**85**:11401.
309. Gomes RA, Sousa Silva M, Miranda HV, Ferreira AEN, Cordeiro CAA, Freire AP. Protein glycation in Saccharomyces cerevisiae. Argpyrimidine formation and methylglyoxal catabolism. *FEBS J* 2005;**272**:4521.
310. Schalkwijk CG, van Bezu J, van der Schors RC, Uchida K, Stehouwer CDA, van Hinsbergh VWM. Heat-shock protein 27 is a major methylglyoxal-modified protein in endothelial cells. *FEBS Lett* 2006;**580**:1565.
311. Bair III WB, Cabello CM, Uchida K, Bause AS, Wondrak GT. GLO1 overexpression in human malignant melanoma. *Melanoma Res* 2010;**20**:85.
312. Oya-Ito T, Naito Y, Takagi T, Handa O, Matsui H, Yamada M, et al. Heat-shock protein 27 (Hsp27) as a target of methylglyoxal in gastrointestinal cancer. *Biochim Biophys Acta* 2011;**1812**:769.
313. Takayama F, Aoyama I, Tsukushi S, Miyazaki T, Miyazaki S, Morita T, et al. Immunohistochemical detection of imidazolone and Nε-(carboxymethyl)lysine in aortas of hemodialysis patients. *Cell Mol Biol* 1998;**44**:1101.
314. Biro A, Cervenak L, Balogh A, Lorincz A, Uray K, Horvath A, et al. Novel anti-cholesterol monoclonal immunoglobulin G antibodies as probes and potential modulators of membrane raft-dependent immune functions. *J Lipid Res* 2007;**48**:19.
315. Wielkoszynski T, Szewczyk M, Zembala M, Szumska M. Concentration of Anti-7-ketocholesterol Antibodies in Patient Sera After Heart Transplantation. *Transplant Proc* 2009;**41**:3206.
316. Mason RP. Using anti-5,5-dimethyl-1-pyrroline N-oxide (anti-DMPO) to detect protein radicals in time and space with immuno-spin trapping. *Free Radic Biol Med* 2004;**36**:1214.
317. Ramirez DC, Mason RP. Immuno-spin trapping: detection of protein-centered radicals. *Curr Protoc Toxicol* 2005. *Chapter 17*, Unit 17.7

318. Ramirez DC, Mejiba SEG, Mason RP. Immuno-spin trapping of DNA radicals. *Nat Methods* 2006;**3**:123.
319. Ramirez DC, Gomez-Mejiba SE, Mason RP. Immuno-spin trapping analyses of DNA radicals. *Nat Protoc* 2007;**2**:512.
320. Mason RP. Imaging free radicals in organelles, cells, tissue, and in vivo with immuno-spin trapping. *Redox Biol* 2016;**8**:422.
321. Gomez-Mejiba SE, Zhai Z, Akram H, Deterding LJ, Hensley K, Smith N, et al. Immuno-spin trapping of protein and DNA radicals: "Tagging" free radicals to locate and understand the redox process. *Free Radic Biol Med* 2009;**46**:853.
322. Detweiler CD, Deterding LJ, Tomer KB, Chignell CF, Germolec D, Mason RP. Immunological identification of the heart myoglobin radical formed by hydrogen peroxide. *Free Radic Biol Med* 2002;**33**:364.
323. Ogusucu R, Rettori D, Netto LE, Augusto O. Superoxide dismutase 1-mediated production of ethanol- and DNA-derived radicals in yeasts challenged with hydrogen peroxide: molecular insights into the genome instability of peroxiredoxin-null strains. *J Biol Chem* 2009;**284**:5546.
324. Kojima C, Ramirez DC, Tokar EJ, Himeno S, Drobna Z, Styblo M, et al. Requirement of Arsenic Biomethylation for Oxidative DNA Damage. *J Natl Cancer Inst* 2009;**101**:1670.
325. Gomez-Mejiba SE, Zhai Z, Gimenez MS, Ashby MT, Chilakapati J, Kitchin K, et al. Myeloperoxidase-induced Genomic DNA-centered Radicals. *J Biol Chem* 2010;**285**:20062.
326. Chatterjee S, Lardinois O, Bhattacharjee S, Tucker J, Corbett J, Deterding L, et al. Oxidative stress induces protein and DNA radical formation in follicular dendritic cells of the germinal center and modulates its cell death patterns in late sepsis. *Free Radic Biol Med* 2011;**50**:988.
327. Dogan S, Ozlem Elpek G, Kirimlioglu Konuk E, Demir N, Aslan M. Measurement of intracellular biomolecular oxidation in liver ischemia-reperfusion injury via immuno-spin trapping. *Free Radic Biol Med* 2012;**53**:406.
328. Bhattacharjee S, Chatterjee S, Jiang JJ, Sinha BK, Mason RP. Detection and imaging of the free radical DNA in cells-Site-specific radical formation induced by Fenton chemistry and its repair in cellular DNA as seen by electron spin resonance, immuno-spin trapping and confocal microscopy. *Nucleic Acids Res* 2012;**40**:5477.
329. Bhattacharjee S, Deterding LJ, Chatterjee S, Jiang J, Ehrenshaft M, Lardinois O, et al. Site-specific radical formation in DNA induced by Cu(II)-H2O2 oxidizing system, using ESR, immuno-spin trapping, LC-MS, and MS/MS. *Free Radic Biol Med* 2011;**50**:1536.
330. Keszler A, Mason RP, Hogg N. Immuno-spin trapping of hemoglobin and myoglobin radicals derived from nitrite-mediated oxidation. *Free Radic Biol Med* 2006;**40**:507.
331. Ramirez DC, Gomez Mejiba SE, Mason RP. Mechanism of hydrogen peroxide-induced Cu,Zn-superoxide dismutase-centered radical formation as explored by immuno-spin trapping: the role of copper- and carbonate radical anion-mediated oxidations. *Free Radic Biol Med* 2005;**38**:201.
332. Bonini MG, Siraki AG, Atanassov BS, Mason RP. Immunolocalization of hypochlorite-induced, catalase-bound free radical formation in mouse hepatocytes. *Free Radic Biol Med* 2007;**42**:530.
333. Chatterjee S, Ehrenshaft M, Bhattacharjee S, Deterding LJ, Bonini MG, Corbett J, et al. Immuno-spin trapping of a post-translational carboxypeptidase B1 radical formed by a dual role of xanthine oxidase and endothelial nitric oxide synthase in acute septic mice. *Free Radic Biol Med* 2009;**46**:454.
334. Ranguelova K, Chatterjee S, Ehrenshaft M, Ramirez DC, Summers FA, Kadiiska MB, et al. Protein Radical Formation Resulting from Eosinophil Peroxidase-Catalyzed Oxidation of Sulfite. *J Biol Chem* 2010;**285**:24195.
335. Ranguelova K, Bonini MG, Mason RP. (Bi)sulfite oxidation by copper,zinc-superoxide dismutase: sulfite-derived, radical-initiated protein radical formation. *Environ Health Perspect* 2010;**118**:970.
336. Yin J-J, Liu J, Ehrenshaft M, Roberts JE, Fu PP, Mason RP, et al. Phototoxicity of nano titanium dioxides in HaCaT keratinocytes-Generation of reactive oxygen species and cell damage. *Toxicol Appl Pharmacol* 2012;**263**:81.
337. Dalsgaard TK, Triquigneaux M, Deterding L, Summers F, Ranguelova K, Mortensen G, et al. Site-Specific Detection of Radicals on α-Lactalbumin after a Riboflavin-Sensitized Reaction, Detected by Immuno-spin Trapping, ESR, and MS. *J Agric Food Chem* 2013;**61**:418.

338. Kang PT, Zhang L, Chen CL, Chen J, Green KB, Chen YR. Protein thiyl radical mediates S-glutathionylation of complex I. *Free Radic Biol Med* 2012;**53**:962.
339. Ehrenshaft M, Mason RP. Protein radical formation on thyroid peroxidase during turnover as detected by immuno-spin trapping. *Free Radic Biol Med* 2006;**41**:422.
340. Lardinois OM, Tomer KB, Mason RP, Deterding LJ. Identification of Protein Radicals Formed in the Human Neuroglobin-H2O2 Reaction Using Immuno-Spin Trapping and Mass Spectrometry. *Biochemistry* 2008;**47**:10440.
341. Ranguelova K, Suarez J, Magliozzo RS, Mason RP. Spin Trapping Investigation of Peroxide- and Isoniazid-Induced Radicals in Mycobacterium tuberculosis Catalase-Peroxidase. *Biochemistry* 2008;**47**:11377.
342. Davis MF, Zhou L, Ehrenshaft M, Ranguelova K, Gunawardena HP, Chen X, et al. Detection of Ras GTPase protein radicals through immuno-spin trapping. *Free Radic Biol Med* 2012;**53**:1339.
343. Ranguelova K, Rice AB, Lardinois OM, Triquigneaux M, Steinckwich N, Deterding LJ, et al. Sulfite-mediated oxidation of myeloperoxidase to a free radical: Immuno-spin trapping detection in human neutrophils. *Free Radic Biol Med* 2013;**60**:98.
344. Kumar A, Leinisch F, Kadiiska MB, Corbett J, Mason RP. Formation and Implications of Alpha-Synuclein Radical in Maneb- and Paraquat-Induced Models of Parkinson's Disease. *Mol Neurobiol* 2016;**53**:2983.
345. Ganini D, Mason RP. Absence of an effect of vitamin E on protein and lipid radical formation during lipoperoxidation of LDL by lipoxygenase. *Free Radic Biol Med* 2014;**76**:61.
346. Khoo NKH, Cantu-Medellin N, Devlin JE, St. Croix CM, Watkins SC, Fleming AM, et al. Obesity-induced tissue free radical generation: An in vivo immuno-spin trapping study. *Free Radic Biol Med* 2012;**52**:2312.
347. Detweiler CD, Lardinois OM, Deterding LJ, Ortiz de Montellano PR, Tomer KB, Mason RP. Identification of the myoglobin tyrosyl radical by immuno-spin trapping and its dimerization. *Free Radic Biol Med* 2005;**38**:969.
348. Ramirez DC, Chen Y-R, Mason RP. Immunochemical detection of hemoglobin-derived radicals formed by reaction with hydrogen peroxide: involvement of a protein-tyrosyl radical. *Free Radic Biol Med* 2003;**34**:830.
349. Deterding LJ, Ramirez DC, Dubin JR, Mason RP, Tomer KB. Identification of Free Radicals on Hemoglobin from its Self-peroxidation Using Mass Spectrometry and Immuno-spin Trapping: observation of a histidinyl radical. *J Biol Chem* 2004;**279**:11600.
350. Lardinois OM, Detweiler CD, Tomer KB, Mason RP, Deterding LJ. Identifying the site of spin trapping in proteins by a combination of liquid chromatography, ELISA, and off-line tandem mass spectrometry. *Free Radic Biol Med* 2008;**44**:893.
351. Bhattacharjee S, Deterding LJ, Jiang J, Bonini MG, Tomer KB, Ramirez DC, et al. Electron transfer between a tyrosyl radical and a cysteine residue in hemoproteins: spin trapping analysis. *J Am Chem Soc* 2007;**129**:13493.
352. Choi Y-S, Kim HS, Cho K-W, Lee K-M, Yi YJ, Eun S-J, et al. Noninvasive identification of viable cell populations in docetaxel-treated breast tumors using ferritin-based magnetic resonance imaging. *PLoS One* 2013;**8**:e52931.
353. Iordanova B, Hitchens TK, Robison CS, Ahrens ET. Engineered mitochondrial ferritin as a magnetic resonance imaging reporter in mouse olfactory epithelium. *PLoS One* 2013;**8**:e72720.
354. Kim HS, Woo J, Choi Y, Hwang EH, Choi SK, Cho K-W, et al. Noninvasive MRI and multilineage differentiation capability of ferritin-transduced human mesenchymal stem cells. *NMR Biomed* 2015;**28**:168.
355. Towner RA, Smith N, Saunders D, Henderson M, Downum K, Lupu F, et al. In vivo imaging of immuno-spin trapped radicals with molecular magnetic resonance imaging in a diabetic mouse model. *Diabetes* 2012;**61**:2405.
356. Towner RA, Smith N, Saunders D, Carrizales J, Lupu F, Silasi-Mansat R, et al. In vivo targeted molecular magnetic resonance imaging of free radicals in diabetic cardiomyopathy within mice. *Free Radical Res* 2015;**49**:1140.

357. Towner RA, Smith N, Saunders D, Lupu F, Silasi-Mansat R, West M, et al. In vivo detection of free radicals using molecular MRI and immuno-spin trapping in a mouse model for amyotrophic lateral sclerosis. *Free Radic Biol Med* 2013;**63**:351.
358. Towner RA, Smith N, Saunders D, De Souza PC, Henry L, Lupu F, et al. Combined molecular MRI and immuno-spin-trapping for in vivo detection of free radicals in orthotopic mouse GL261 gliomas. *Biochim Biophys Acta* 2013;**1832**:2153.
359. Towner RA, Garteiser P, Bozza F, Smith N, Saunders D, d'Avila JCP, et al. In vivo detection of free radicals in mouse septic encephalopathy using molecular MRI and immuno-spin trapping. *Free Radic Biol Med* 2013;**65**:828.
360. Fujii H, Wan X, Zhong J, Berliner LJ, Yoshikawa K. In vivo imaging of spin-trapped nitric oxide in rats with septic shock: MRI spin trapping. *Magn Reson Med* 1999;**42**:235.
361. Ardenkjaer-Larsen JH, Laursen I, Leunbach I, Ehnholm G, Wistrand LG, Petersson JS, et al. EPR and DNP Properties of certain novel single electron contrast agents intended for oximetric imaging. *J Magn Reson* 1998;**133**:1.
362. Golman K, Leunbach I, Ardenkjaer-Larsen JH, Ehnholm GJ, Wistrand LG, Petersson JS, et al. Overhauser-enhanced MR imaging (OMRI). *Acta Radiol* 1998;**39**:10.
363. Matsumoto K-i, English S, Yoo J, Yamada K-i, Devasahayam N, Cook JA, et al. Pharmacokinetics of a triarylmethyl-type paramagnetic spin probe used in EPR oximetry. *Magn Reson Med* 2004;**52**:885.
364. Krishna MC, English S, Yamada K, Yoo J, Murugesan R, Devasahayam N, et al. Overhauser enhanced magnetic resonance imaging for tumor oximetry: coregistration of tumor anatomy and tissue oxygen concentration. *Proc Natl Acad Sci U S A* 2002;**99**:2216.
365. Matsumoto S, Yasui H, Batra S, Kinoshita Y, Bernardo M, Munasinghe JP, et al. Simultaneous imaging of tumor oxygenation and microvascular permeability using overhauser enhanced MRI. *Proc Natl Acad Sci U S A* 2009;**106**:17898.
366. Utsumi H, Hyodo F. Free Radical Imaging Using In Vivo Dynamic Nuclear Polarization-MRI. *Methods Enzymol* 2015;**564**:553.
367. Grucker D. In vivo detection of injected free radicals by Overhauser effect imaging. *Magn Reson Med* 1990;**14**:140.
368. Kosem N, Naganuma T, Ichikawa K, Phumala Morales N, Yasukawa K, Hyodo F, et al. Whole-body kinetic image of a redox probe in mice using Overhauser-enhanced MRI. *Free Radic Biol Med* 2012;**53**:328.
369. Matsumoto K-i, Hyodo F, Anzai K, Utsumi H, Mitchell JB, Krishna MC. Brain redox imaging. *Methods Mol Biol* 2011;**711**:397.
370. Utsumi H, Yamada K-i, Ichikawa K, Sakai K, Kinoshita Y, Matsumoto S, et al. Simultaneous molecular imaging of redox reactions monitored by Overhauser-enhanced MRI with 14N- and 15N-labeled nitroxyl radicals. *Proc Natl Acad Sci U S A* 2006;**103**:1463.
371. Khramtsov VV, Clanton TL. NMR Spin Trapping: Insight into the Hidden Life of Free Radical Adducts. *Appl Magn Reson* 2011;**41**:305.
372. Khramtsov VV, Reznikov VA, Berliner LJ, Litkin AK, Grigor'ev IA, Clanton TL. NMR spin trapping: detection of free radical reactions with a new fluorinated DMPO analog. *Free Radic Biol Med* 2001;**30**:1099.
373. Potapenko DI, Clanton TL, Bagryanskaya EG, Gritsan NP, Reznikov VA, Khramtsov VV. Nonradical mechanism of (bi)sulfite reaction with DEPMPO: cautionary note for SO3*- radical spin trapping. *Free Radic Biol Med* 2003;**34**:196.
374. Zoia L, Perazzini R, Crestini C, Argyropoulos DS. Understanding the radical mechanism of lipoxygenases using 31P NMR spin trapping. *Bioorg Med Chem* 2011;**19**:3022.

INDEX

Note: Page numbers followed by "*f*" and "*t*" refer to figures and tables, respectively

P

Q

R

S

T

Kolibri 360 GmbH
Ferdinand-Jühlke-Straße 7
99095 Erfurt, Deutschland
produktsicherheit@kolibri360.de